Emerging Globalities and Civilizational Perspectives

This series documents the range of emerging globalities in the 21st century at the national, transnational and trans-civilizational levels of analysis. "Globality" refers to a global condition where people located at any point on Earth are aware of being part of the world as a whole—the world as a single interacting entity. Social interactions occur among actors belonging to different societies, different social strata and different cultural traditions so that the condition of "globality" is experienced in many different ways.

Examples of emerging globalities are social movements generated from the unfulfilled promises of neoliberalism and feelings of discrimination and marginalization of lower social strata; cultural otherization or the blaming of economic problems of certain geographical areas on a low level of cultural development; insecurities generated by technological risks, epidemics, and global terrorism; uncertainties generated by processes of transnational governance, outsourcing, unbalanced trade and massive migrations; biology-machine interfaces and impacts of non-human organisms and technologies on human consciousness and action; long-term threats of global warming, climate change and depletion of bio-diversity; increasing exploitation and marginalization of less industrialized regions.

We state that globalization entails encounters and often clashes among people and nations of different civilizational traditions. Hence, one of the exploratory questions of these volumes will be the extent to which negative or problematic globalities are reactions to failed promises and unrealized ideals of civilizational and national traditions and/or perhaps attempts to revive those traditions. Our notion of civilizational tradition takes inspiration from the classical works of Spengler and Toynbee, Benjamin Nelson, Vytautas Kavolis, Roland Robertson, Johann P. Arnason, Jeremy Smith, and others; a tradition which is in sharp contrast with the civilizationism recently promoted by authoritarian leaders with hegemonic ambitions. The volumes in this series aim to extend the inter-civilizational focus of classical civilizational thinkers from the analysis of the origins and development of civilizations to the fostering of contemporary inter-civilizational dialogues; the intent is to facilitate an international rapprochement in the contemporary atmosphere of global conflicts.

The volumes will reflect the diversity of theoretical perspectives and captures some of the novel thinking in social sciences, economics and humanities on intra- and inter-societal processes; the attention to novel thinking will extend to emerging policy formulations in dealing with threats, risks, insecurities and inequities and to strategic thinking for a sustainable global future. The historical perspective will also be an important component of analysis together with the avoidance of West-centric perspectives. The intended readership of this series is not just an academic audience but also policy decision-makers and the public at large; accessibility of language and clarity of discourse will be a key concern in the preparation of these volumes.

Thomas C. Burr

Societies Past and Present

A Social Science World History

Springer

Thomas C. Burr
Department of Sociology and Anthropology
Illinois State University
Normal, IL, USA

ISSN 2731-0620 ISSN 2731-0639 (electronic)
Emerging Globalities and Civilizational Perspectives
ISBN 978-3-031-91531-4 ISBN 978-3-031-91532-1 (eBook)
https://doi.org/10.1007/978-3-031-91532-1

This Springer imprint is published by the registered company Springer Nature Switzerland AG
The registered company address is: Gewerbestrasse 11, 6330 Cham, Switzerland

To Lisa

Preface

I have a long-standing interest in global issues. As an undergraduate double major in political science and Latin American studies at UC Berkeley in the 1980s, I wrote a research paper on world government, going into it with the optimism that I found world federalists had. By the time I finished it, I was more pessimistic. A few years later, in a graduate program in international relations at San Francisco State University, I turned to studying culture. As the Berlin Wall fell, I started thinking about global culture, as were many others at the time. My master's thesis on global political culture turned into a world history going back more than two millennia; it took three years to complete and was over five hundred pages long. I came to conclusions that, I discovered later, were similar to those of world polity scholars, though my thesis was more historically detailed and emphasized cultural variation.

On finishing that thesis, I went from international relations to history. After a year in a history PhD program at UC Davis, my professors advised me that I would fit better in a historical social science. They turned out to be correct, and in the year it took me to apply to and enter UC Davis's sociology PhD program, I took courses on world history and globalization and read Stephen Sanderson's *Social Transformations* (1995). Once in the program, I specialized in economic sociology and comparative-historical methods; I was twice a teaching assistant for a macrosociology/social change class; and eventually, I taught my own version of that class. However, I did not study world history for my dissertation; instead, I wanted to understand capitalism, so I decided to work on understanding markets and, by extension, market economies. I compared the U.S. and French national consumer markets for bicycles between 1875 and 1910 (later broadening it to the period from the 1860s to the 1920s). I defined "consumer markets" as ongoing relationships between producers and consumers (in this case, manufacturers and cyclists), relying mostly on primary historical sources. This definitely improves my understanding of capitalism.

Once on the tenure track at Illinois State University, I turned the dissertation into six peer-reviewed articles and planned further research to turn it all into a book. The central article (Burr 2013a) has unfortunately not attracted much notice, but two others have (Burr 2006, 2013b). Also on the tenure track, I regularly taught the macrosociology class using a social evolution framework. After a few iterations, using Sanderson's, Gerhard Lenski's, and Daniel Chirot's texts, plus many supplemental readings, I realized that social evolutionism in sociology, founded primarily on ethnographic inference, needed much more world history; after teaching it a few more years, I realized that it needed world prehistory even more. Chirot's book was the most historical, yet I knew something more was possible. On earning tenure in 2013, I started writing this book. Since it is a true world history, but one that relies primarily on social science concepts, it sits on the borders of history, sociology, political science, anthropology, archaeology, economics, psychology, and geography. My return to world history meant that the bicycle book slowly receded into the background. Ultimately, my career in bicycle history died a natural death, having accomplished its purpose (and given me a rather quirky research record).

After I completed a (messy and wordy) first draft of this book, its one informal reviewer asked me, "What is your research question?" That had never occurred to me; I had clearly thought that tracing the development of societies was self-evidently a good thing. Upon reflection, I realized that I had a collective version of the same question that David Byrne asks in "Once in a Lifetime": "Well, how did I get here?" (Talking Heads 1980) Yes—exactly: How *did* we get here? Long-term world history has always appealed to me because our ancestors started in such small, simply organized groups on the African savanna, but now we live in such complicated, diverse societies around the world, have a population that is a thousand times larger, can be in contact with each other at a moment's notice, and influence our very atmosphere. Really—how did that happen?

Telling a story of how we got from one situation to another seems to be inherently appealing to most humans. Telling this particular story requires discussing many factors, ones which influence us today. I have learned about the major changes that led those savannah-inhabiting hominins to become *us*, and I have needed to address many different factors. This has answered the main question: How did the descendants of people in a few small groups construct, over thousands of years, a global social system? Two other questions—one that I have been pondering for a while, another I did not know I had but which slowly became conscious—became relevant to this main story. The first has baffled people for centuries: Who are we? Why do we humans do what we do? What are we capable of? Ultimately, what is human nature—in fact, is there a human "nature"? A second question occurred to me later: we know, generally, how societies changed; what has remained the same?

Thus, I have not one but three research questions: Who are we? How did our original situation turn into our very different situation today? And what continuities were there throughout those changes? They are, really, three aspects of that one question: Well, how did we get here? The answer to this question, broken into three parts, constitutes the three arguments of the book: humans are groupish individualists, or "hypersocial"; three major social revolutions over human history, based on hypersocial human capacities, transformed our groups; and there are several very long continuities in that history, a few of which are reversing only now. These answers helped me understand the world today—the human world, which influences the entire physical world. In particular, the argument for continuities helps me, and I hope us, understand human societies in the flow of time, thus forcing us to relate that past to the future—to ponder how those past trends will influence that future.

I emphasize that I am not asking *why* the world is the way it is. Just as this book sits at the intersection of several scholarly disciplines, it also sits on the border of description and explanation—it is mostly descriptive, but it periodically offers reviews of others' explanations. If offering no theory yet disappoints some readers, so be it. Explaining why things turned out this way is too much for now; coming up with a set of concepts to tell the story, and then telling the story, has been plenty. Opposing the hostility to grand narratives of several decades ago, various authors have published a spate of world histories addressing this same enormous period in the last 15 years; others support their efforts by arguing explicitly for large-scale historical works that help contextualize more focused historical studies (Guldi and Armitage 2014). Still, nobody has pulled together the story in the way that I have here. It is based in a new(ish) paradigm about humans; it traces the development of multiple scales of human interaction over tens of thousands of years; and I strongly hope that it provides much-needed context for further historical studies.

The scale of such works tends to lead readers to suspect an author of hubris (Abu-Lughod 1989:ix); perhaps, it also reactivates a postmodernist skepticism of grand narratives (Lyotard [1979] 1984). Such reactions are understandable, but readers should know the other side. The act of producing a long-term world history creates the constantly nagging feeling of not being expert in anything, of being instead a poseur and a dilettante, because one has merely scratched the surface of so many literatures that are written by scholars with the kind of expertise that the author will never have. Furthermore, I simply could not read everything. Readers familiar with any of the literatures surveyed here will immediately see what is missing. (Most egregious was my neglect of the journal *Social Evolution and History.*) As Go (2020:91–92) argued eloquently, all research is partial, always leaving much out, and I am painfully aware of how partial this account is. Omissions of historical and geographic variation and depth, the constant collapsing of specific historical details into broad generalizations, and the awareness of how much more I could have read make this book seem appallingly oversimplified and extremely shallow to me.

Despite this permanent relationship with imposter syndrome, all world histories on this scale must be shallow, so I wrote it anyway, for what I think are very good reasons. It does what more focused and deeper studies cannot: it takes us far beyond the term of one human life—indeed, beyond many lifetimes—to see the larger patterns that have shaped us and in which we still live. Thus, being partial and shallow may be a strength; as this book is all context and background, I hope that it raises the kinds of questions that lead readers into those more focused studies mentioned above, and I hope this book equips them better for those studies. Furthermore, as I outline in the conclusion, I plan to theorize later, using this book as a foundation. Given the state of the human world, we certainly need better explanations of it than we have now. This book is a first step in that direction.

Normal, IL, USA Thomas C. Burr

References

Abu-Lughod, Janet. 1989. *Before European Hegemony: The World-System, A.D. 1250–1350*. New York: Oxford University Press.

Burr, Thomas C. 2006. Building Community, Legitimating Consumption: The Creation of the U.S. Bicycle Market, 1875-1884. *Socio-Economic Review* 4 (3): 417–446.

Burr, Thomas C. 2013a. Market Cycles: Bicycles, Riders, Industries, and Environments in France and the United Sates, 1865–1914. *American Journal of Economics and Sociology* 72 (2): 468–496.

Burr, Thomas C. 2013b. Market-Widening: Shaping Total Market Demand for French and American Bicycles *circa* 1890. *Marketing Theory* 15 (1): 19–34.

Go, Julian. 2020. Race, Empire, and Epistemic Exclusion: Or the Structures of Sociological Thought. *Sociological Theory* 38 (2): 79–100.

Guldi, Jo, and David Armitage. 2014. *The History Manifesto*. Cambridge, UK: Cambridge University Press.

Lyotard, Jean–François. (1979) 1984. *The Postmodern Condition: A Report on Knowledge*. Minneapolis: University of Minnesota Press.

Talking Heads. "Once in a Lifetime." Recorded July–August 1980. Track 1, side 2 on *Remain in Light*. Sire Records.

Acknowledgments

In writing this book, I learned just how true it is that nobody writes a book alone. Here I express my gratitude to the people who have helped me along the way, starting with those who aided me with its conceptual foundations and then mentioning those directly involved. I have many people to thank.

I will be forever grateful to the late L. Robert Kohls, Sanjoy Banerjee, and Raymond C. Miller for allowing me to research and write an extremely long MA thesis in international relations; that world history formed the foundation of my career. In addition, Bob Kohls persuaded this skeptical political science major of the importance of culture; Sanjoy Banerjee properly critiqued, and therefore greatly improved, my ability to argue; and my accidental path through several disciplines has helped me understand Raymond Miller's deep emphasis on interdisciplinarity much better than I understood it at the time.

While he probably does not remember me, I still appreciate the advice Jan de Vries gave me about graduate school when I was taking his senior seminar on capitalism through UC Berkeley Extension while finishing the thesis. Of course, I also learned much about capitalism.

At UC Davis, the late Cynthia Brantley influenced my thinking about gender more than I suspect she ever knew; I am sorry that I cannot thank her. In my year between history and sociology departments, I took three classes which deepened my understanding of world history. I appreciate William Hagen's advice and encouragement; his seminar on the twentieth-century world history; and his help in arranging my weekly attendance at UC Santa Cruz to study modern world history with Edmund "Terry" Burke III, who was stimulating and influential on my thinking. Christopher Lloyd, visiting UC Davis from the University of New England (Australia), graciously offered me his seminar on globalization as an independent study, deepening what I learned from Bill and Terry. In sociology, Jack A. Goldstone taught me comparative-historical methods, made sure I was his teaching assistant twice for his undergraduate Global Social Change class (the origin of this book), helped me attend the 1999 Rise of the West conference he organized at Davis, and guided me to a very useful dissertation research design. Jack's demographic/structural approach to political economy and Nicole W. Biggart's institutional approach to economic sociology influence this book throughout, and I am grateful to them both. I could not have arrived at my own definition of institutions without Nicole's questions, suggestions, mentoring, and support while I was her research assistant, and she helped me become a better scholar in general. Diane Felmlee's *bon mots* (in two footnotes below) changed my thinking about social psychology, as did her willingness to suggest readings to a graduate student not in her area of expertise. As the planned book on consumer markets will not happen, I gratefully acknowledge Fred L. Block's constant support, even metaphorical hand-holding, throughout my dissertation, and Catherine J. Kudlick's incisive and helpful interventions in both history and sociology.

My Illinois State University colleagues have had a direct influence on this book. Fred H. Smith, the late James M. Skibo, and Joan M. Brehm have been tremendously supportive department chairs. Fred's masterful summary of human origins, written with Matt Cartmill, influenced my thinking on what is human, and I am especially grateful for Joan's continual support for this project. When I asked beginner questions about social evolution, Maria O. Smith listened and shared ideas. I am grateful to Michael Dougherty not only for his valued friendship but also for suggesting that I seek a book contract for an unruly collection of drafted student readings; that suggestion changed the trajectory of my career. Peter Kivisto generously sat with me, a stranger, for over a half hour at the 2015 American Sociological Association book exhibit, giving me a great deal of advice on publishing, which helped me implement Mike's advice. I greatly appreciate Hemanth Kadambi's friendship, many discussions, citations, ideas, disagreements, and reassurances at our lunches; I also appreciate Chuck Springwood's support and encouragement. The late James Swindler's suggestion to learn social ontology reshaped my understanding of hypersociality. Ed Hines gave me other invaluable advice on dealing with publishers. Anna Muraco read part of an early draft and made very helpful comments. Lane Crothers in political science, Larry Youngs in history, and Logan Miller in archaeology each read earlier drafts. Lane said, "This book screams out for a preface," and he was right. He was also the one who asked me about a research question, thus helping me rethink the entire book. Larry and Logan were tremendously reassuring that, from their disciplinary perspectives, I was on track, and that they both learned a lot. The librarians at Milner Library at Illinois State University, including Chad Kahl, Elizabeth Babin, Vanette Schwartz, and Grace Allbaugh, have been valuable colleagues and good friends. I enjoyed working with Jill Thomas on the maps in this book, and I am grateful for her considerable cartographic skill. Responding to my eleventh-hour email, David Swanson made very helpful suggestions on population pyramids, as did Maria O. Smith (although, sadly, population pyramids do not appear here). A research assistant, Elizabeth Deiters, read it all and made detailed, invaluable comments, leading to yet other improvements; another, Katarina Hamende, has been very helpful with chapter references. I appreciate Eric Boerngen's efforts for a book subvention and Colby Cilento's work in helping me with copyright permissions. And I am grateful for an amazing copy editor, Brooke Lindell, who made this a much, much less error-laden book.

It has been easy to work with the staff at Springer Nature. Esther Otten and Hendrikje Tuerlings, my original editor and production assistant, were supportive and patient. I also appreciate Shinjini Chatterjee, Helen van der Stelt, and Bernadette Deelen-Mans. Shinjini has displayed tremendous patience, which I have long needed and vastly appreciated. Ino Rossi's comments as a reviewer have made this a better book.

I dedicate this book to the love of my life, Lisa Nancy Dawn Flanagan. She likely does not realize how important her ideas have been in writing this book. More importantly, her personal influence has made me a much better person. Most importantly, it is a joy to spend life with her.

Nobody mentioned above can save me from the inevitable errors, mistakes, omissions, and other problems that appear in this book. Authors tend to issue blanket *mea culpas* at the end of Acknowledgments sections about these problems. Now I know why. If you see any, you have my apologies.

Competing Interests The author has no competing interests to declare that are relevant to the content of this manuscript.

Contents

Chapter 1
Introduction: Human Groups, Globalization, and the Long Run

Abstract In this chapter, I make the case for the three arguments of this book. First, nearly all social scientists discuss globalization in terms of collective actors such as states, organizations, and ethnic groups. I will argue that we must research world history by assuming that humans are "hypersocial," or simultaneously individualistic and group-oriented. Second, ideas from social evolutionism can help structure a long-term world history that analyzes human groups by showing that three major transformations in group organization—the Agricultural Revolution, the Urban Revolution, and the global Industrial Revolution—led to our current global social system. Finally, an analytic framework comparing societies geographically and chronologically reveals great cultural diversity and millennia-long patterns and trends that influence human societies today.

Keywords Human nature · Hypersociality · World history · Social evolution · Cultural diversity · Long-term history

Introduction

Globalization is the major trend of our time. Capital, physical commodities, and services flow through global markets. Economic migrants and political refugees move from poorer countries to rich ones, facing increasingly steep resistance. Rich countries continue to outsource production, and thus employment, to poorer ones. International releases for films, books, and video games are now standard. Scientific knowledge is also increasingly international. Finally, neither climate change nor microbes such as COVID-19 respect international borders.

Different social scientists study and theorize globalization differently. Many draw out the implications of the suffix "-ization," which implies a historical process, agreeing with Bryant (1994:11) that the "present … is what the past … has made it." Economists study economic growth within states and trade between them (Stigliz 2012), and they often blur the line between theorization of, and historical study of, economic growth (e.g., Aghion and Durlauf 2005; Maddison 2006). Political scientists analyze the history of the current international state system and increasingly

T. C. Burr, *Societies Past and Present*, Emerging Globalities and Civilizational Perspectives, https://doi.org/10.1007/978-3-031-91532-1_1

study comparative politics and international relations together (Hobson 2009). Historical sociologists often examine the advent of "modernity" in Europe and North America over the last five centuries (Wood and Williamson 2007) or the growth of the capitalist world-economy in that period (Wallerstein 1974). Historians have long gone further back to study empires and civilizations; now world history is a well-established field of research and teaching (Stearns 2011; World History Association 2016). Archaeologists examine how small groups of hunter-gatherers in the distant past grew and changed over time into larger groups and, eventually, the civilizations that historians study (Scarre 2018).

I see two basic problems with globalization studies. First, while world prehistorians and world historians provide a foundation for starting in the distant past to understand the present—it was, after all, a single continuous process—separate disciplinary studies of historical globalization analyze only certain aspects, thus fragmenting our understanding. Second, the conceptual foundations of these interdisciplinary studies are inadequate in helping us to better understand the process. Different scholars have different assumptions about who and what they are studying. Most economists explicitly rely on the assumptions of the "rational-choice" theory (Kirchgässner 2008); many political scientists and some archaeologists join them. This stance is methodologically individualistic, though scholars often aggregate the results. Some anthropologists and archaeologists rely on the equally individualistic but biologically based assumptions of evolutionary psychology, or EP (Cosmides and Tooby 1992). Sociologists often resist making any assumptions about humans explicit, rejecting both rational choice and EP as hopelessly reductionist, unrealistic, and unjust (Rose and Rose 2000), though some historical sociologists draw selectively on both paradigms (Lenski 2005; Sanderson 2007; Chase-Dunn and Lerro 2014). Historians rarely explain their assumptions about people (Gaddis 2001).

Still, researchers' assumptions always shape their analyses. Globalization scholars constantly refer to collective actors or formally organized human groups because globalization usually operates through them, not primarily through individual action. Many political economy studies take collective actors for granted, regularly referring to states, societies, corporations, international organizations, and ethnic groups. Organizational and new institutional economists, following rational-choice assumptions, claim to explain how collective entities appear and operate (Olson [1965] 1971; Williamson 1985; Furubotn and Richter 2005), but they disregard much of human life, including the intense group loyalties or violent conflicts around them.

A world history that studies collective actors would help, ideally one founded on realistic assumptions of how human groups operate. How do separate individuals unite into a single entity, a collective actor? I will argue that the most useful assumptions about human groups come from a group of interdisciplinary scholars who find group formation rare in animals. They argue that humans are *eusocial* (Wilson 2012), *hypersocial* (Hrdy 2009), or *ultrasocial* (Turchin 2016) in contrast to most mammals—that is, we are simultaneously "groupish" and individualistic. Analyzing human groupishness explicitly would, I believe, help us understand globalization much better.

How do we organize the complex story of how the first human groups became societies and then one global social system over the millennia? I suggest that concepts of social evolution—the increase in size and complexity of human groups, not the evolution of human bodies—provide a balance of simplicity and complexity to trace this story. Specifically, three social revolutions (Childe [1936] 1966) or transformations (Sanderson 1995), all geographically incomplete, led to our current global social system. A millennia-long Agricultural Revolution in some places led hunter-gatherers in small camps to a more sedentary lifestyle in larger villages. In the Urban Revolution—again only in a few places—groups concentrated people, power, and resources in cities that ruled large areas of land. We now live in an ongoing global Industrial Revolution that in 250 years has transformed some societies and changed all.

Each of these transformations has taken many lifetimes to occur, and along the way, societies became greatly different from each other (Marcus 2008). Still, humans today are fundamentally the same as their hunter-gatherer ancestors. Thus, despite massive changes, there have been many continuities. Some, such as high mortality and fertility, lasted for most of history; others, such as economic and gender inequalities, started some seven millennia ago; and there has been a very long accumulation of knowledge, skills, and practices throughout. I will trace these deep continuities alongside the revolutionary changes.

In this chapter, I review scholarly research to support these three arguments: Humans are groupish and individualistic; three social transformations created our global world; and despite great cultural diversity from those transformations, strong continuities still influence us. The intellectual resources exist to support these claims, though they have not yet been brought together.

Assumptions About Human Nature

Researchers often neglect to detail their assumptions about humans, yet all scholarship is rooted in those assumptions. Shared sets of assumptions are *paradigms* (Kuhn [1962] 2012); there are at least three in the social sciences. Rational-choice theory and EP are types of methodological individualism (MI), in that they consider the attributes of individual people; a third epistemological approach, constructionism, assumes that humans operate in groups. I will argue that all three are inadequate to write a world history of groups and that hypersociality is a better foundation for such research, one that has yet to be fully fleshed out.

Paradigm has varied scholarly meanings. Kuhn ([1962] 2012) defined it as a puzzle or exemplar problem that natural science researchers could analogize to other problems. He and Masterman (1970) recognized that, instead, social scientists often defined paradigms as sets of ontological assumptions about people; while they objected to this usage, social scientists continued to use this definition (Dogan 2001; Ritzer 2005; Donmoyer 2008). Synonyms include school of thought, conceptual framework, general approach, conceptual vocabulary, or family of theories. Unlike some natural sciences, the social sciences regularly operate on multiple, competing paradigms (Dogan 2001).

MI is a metaparadigm; its adherents assume that individual persons are the main unit of analysis in explaining social phenomena (Voss and Abraham 2000; Heath 2020). One influential MI paradigm in the social sciences is the rational actor model. Rational-choice scholars assume individuals to be self-oriented, acting only on their own preferences and interests (Kirchgässner 2008). "Rational" means that people fulfill their preferences with actions "that are efficient and effective for achieving" them (Olson [1965] 1971:65). They prefer outcomes to actions that they find useful; thus, scholars analyze "utility" functions—rankings of preferences from most to least important (Ross 2019)—but tend to ignore how individuals form their utility functions. They assume people try to obtain maximum benefits from their utilities, they interpret social interaction as exchange, and they assume that aggregate outcomes of individual choices lead to spontaneous social order (Kirchgässner 2008; Ross 2019).

Three rational-choice approaches help researchers analyze interacting individuals. First, "games" are experiments of interactions between two or more people ("players"). Experimenters assume that players can apply interaction outcomes to their utility functions, can trace action sequences to those outcomes, and thus can select preferred actions. The Prisoner's Dilemma is the most famous, but there are many games (Ross 2019), some involving three or more players. Second, Olson ([1965] 1971) applied rational-choice assumptions to groups, finding that individuals have incentives to get a "free ride" by letting others do group tasks but reaping the same benefits. This problem grows as group size increases, leading to group coercion. Third, economists analyze how different individuals' interests create organizational problems and how solving such problems generates costs (Williamson 1985; Furubotn and Richter 2005).

While researchers from some disciplines use rational-choice thinking, others reject it. Many political scientists, some cultural ecologists in anthropology, and a few sociologists use rational-choice thinking; other political scientists attack it fiercely (Green and Shapiro 1994; Friedman 1996; Voss and Abraham 2000; Wittek 2013). Economists have long been modifying these assumptions since they recognized that rationality is imperfect (Simon 1997). Behavioral economists elaborate on this by analyzing the imperfections of human cognition, such as the use of heuristics, inequality aversion, norm conformity, and moods and emotions (Baddeley 2017; Thaler 2018; see McChesney 2013 for a vigorous rebuttal). Others suggest that the rational actor is "close to being a social moron" (Sen 1977:36). Economics students become more selfish than members of the public (Frank and Gills 1993), leading to accusations of unrealism (Ross 2019). Outside critics attack the approach for its lack of attention to cultural variability (Urbina and Ruiz-Villaverde 2019), especially in how it embodies cultural assumptions from "WEIRD" (Western, educated, industrialized, rich, and developed) countries (Blanton 2016; Ross 2019). Economic experiments in culturally diverse populations show great variation in outcomes, refuting its predictions (Ross 2019). Others argue that game theorists' assumptions "are unrealistically extreme" since they prevent players from communicating preferences or making agreements as they do in real life (Chacon and Mendoza 2017:425). Psychologists argue that rational-choice theorists ignore emotions, which "powerfully, predictably, and pervasively influence decision making" (Lerner et al. 2015:802, though see Baddeley 2017:Chap. 1). Finally, though it is "downright contrary to the principles of hard-line rational-choice theory," one game theorist concluded that separate individuals can join together and act as a single unit (Bacharach 2002:90).

Another MI paradigm came from biology. In the 1970s, some psychologists applied neo-Darwinian principles of genetics to humans (Wilson 1975). In EP, developed by the 1990s (Barkow et al. 1992), scholars argue that human brains evolved hundreds of separate "modules" that allow people to engage in specific tasks and a "domain-general" area of the brain that coordinates modules. EP scholars emphasize the importance of passing on one's genes (reproduction) in influencing human choices. Thus, many of them study differences between male and female brains and their consequent choices since bodies provide different constraints (Symons 1979; Baron Cohen 2003). Such researchers also assume that genes are "selected for" purely individually, not on a group basis.

Criticism of EP has been even more hostile than that applied to rational choice. While even critics admit that it is not racist (Rose and Rose 2000), the long history of social horrors being legitimized by eugenics has sparked worries that *any* discussion of humans as organic creatures will fall back into racism (Blanton 2016). The focus on reproduction has led critics to argue that EP ideas rest on European cultural assumptions of gender roles (Fausto-Sterling 2000). They say the supposed "sex contract," in which females exchange sex with men for protection and material resources, positions women as passive and men as active, when in fact women constantly make active choices about sex and reproduction (Hrdy 2009). Many brain scientists argue that "neuroplasticity," the brain's capacity to rewire itself through new physical habits, leads them to reject the idea of male and female brains (Joel and Vikhanski 2019; Rippon 2019).

Anthropologists and sociologists often criticize both MI approaches, yet they provide few alternatives. Anthropologists writing about "human nature" seem ambivalent (Fuentes et al. 2010), while many sociologists reject the very concept (Leahy 2017; Turner and Machalek 2018). Many in both groups argue that people are born as "blank slates" with no inborn programming, ready to be purely cultural creatures. They often cite research on feral children, those rare people who did not experience socialization by other people, who were usually neglected and isolated (Dombrowski et al. 2011:87). These children experience severe cognitive, language, emotional, and motor deficits. For them, "socialization will always be necessary" (Dombrowski et al. 2011:89). In contrast, EP scholars strongly criticize the developmental blank-slate metaphor (e.g., Pinker 2002), arguing that human minds have innate traits on which socialization operates.

The emphasis on socialization and enculturation takes its clearest form in social constructionism in sociology and political science, post-processualism in archaeology, and post-structuralism in literary criticism and philosophy (Gergen 2015; Iggers 2012; Woolf 2019). This family of epistemological approaches to humanness rests on group-oriented, though implicit, ontological assumptions. Constructionists argue that culture is so important among humans that it defines reality for groups; reality is, in effect, what groups say it is. People influence each other in terms of how they understand the world, so groups mentally "construct" those understandings:

> Whenever people describe reality … they are speaking from a particular standpoint … To be sure, *there is something*, but when you try to describe what that *something* is, you will inevitably rely on some tradition of sense making (Gergen 2015:5).

"Tradition," of course, is a property of groups. Thus, individuals in communities shape their beliefs, morals, norms, and language by influencing each other in interactions (Elder-Vass 2012).

This approach acknowledges the role of culture in human affairs and the diversity of cultural traditions around the world. The drawbacks of its most extreme versions are obvious. If cultural beliefs filter perception so much that there is only "something" that varies exclusively by the beholder(s), the logical end point is solipsism: Only the consciousness of the perceiver exists (Elder-Vass 2012). Likewise, if concepts shape perceptions so that the material world is only what a group says it is, this makes the world effectively unknowable, so research is pointless (Danermark et al. 2019). Most constructionists are *not* extreme, but as constructionism is primarily epistemological, its practitioners seem to avoid stating their ontological assumptions (Nightingale and Cromby 2002:704–705), making accusations of extremism possible. This approach also ignores findings tested against the physical world that have repeatedly been turned into practical objects (Ede and Cormack 2012). The lack of materialist assumptions makes all versions of constructionism only partly useful. Some try to remedy this, but they resort to critical realism to accomplish it (Nightingale and Cromby 2002; Elder-Vass 2012).

I find these three paradigms inadequate, as they are so partial. MI tends to be implemented in more "positivist" or nomothetic ways, assuming that similarities between humans are most important; Simpson (1945) called nomothetic scholars "lumpers": people who lump together social phenomena. Methodological individualists take parsimony to excess. EP scholars posit a person unconsciously obsessed with reproduction, while the "rational" actor is a half-human: selfish yet unemotional, calculating, a-cultural, and even lacking friends (Kirchgässner 2008). While economists are revising these assumptions (Baddeley 2017; Thaler 2018), that revision indicates a deeper problem. Constructionism is a strand within idiographic thinking, adopted by "splitters" who emphasize local, unique particulars in social life (Simpson 1945; Mjøset 2001). Constructionists' neglect of ontology can make them sound like they think people are bodiless thought-creatures, a stance—especially in its blank-slate version—subject to criticism (Pinker 2002) or occasionally parody (Sokal 1996). Another social science theory that has been elaborated into a paradigm, Marxism, is as social as constructionism, but in its own way, it is as unrealistically reductionist as rational choice or EP.

Overall, I think we need psychologically realistic assumptions about people (Ross 2019). Neglecting to explicitly state one's assumptions about humans creates an "explanatory vacuum" (Blanton 2016:55). Despite their refusals to state assumptions about people, sociologists nevertheless repeatedly demonstrate that they make such assumptions (Friesen 2015; Leahy 2017; Turner and Machalek 2018). The few who explicitly state them (Nolan and Lenski 2011; Chase-Dunn and Lerro 2014) seem to add assumptions about individual humans only to support the analysis of societies as systems. In contrast, I prefer an inductive approach: stating my ontological assumptions of individual humans and the smallest groups and then working outward to societies.

Some evolutionary scholars have developed a new paradigm that does just that, but it needs further development. It assumes humans are both individualistic and group-oriented, materialistic and cultural, and rational and emotional, taking a "both-and" stance, not an "either-or" approach. It has a relatively thin definition of culture, something social scientists could develop. As mentioned, this paradigm has different names: hypersociality, eusociality, or ultrasociality (Hrdy 2009; Wilson 2012; Turchin 2016). All use intensifiers to argue that humans are more group-oriented than most animals. Though these scholars all take a biocultural approach, they are diverse and eclectic, not at all united. Some emphasize how biological evolution led to humans depending on enculturation, citing feral children research (Sterelny 2012; Henrich 2015; Turner and Machalek 2018; Cartmill and Smith 2022). Others focus on how humans have a relatively standardized physical structure with many "biopossibilities," i.e., ones that give us a wide range of "creaturely capacities" that need enculturation (Bergesen 2005; Prinz 2014; quotes, Willey 2016:555; Tomasello 2019). A few continue to insist on the value of rational-choice thinking but reject methodological individualism (Gintis 2017); others focus on how emotion as a capacity deriving from ape emotionality became "supercharged" in humans (Lents 2016; Turner and Machalek 2018:348). As emotions guide most human behaviors (Tomasello 2019), many social phenomena make no sense unless we analyze them (Turner 2007). While this is still blank-slate thinking, the "slate" in this metaphor is limited by its physical structure, which has only some plastic capacities.

All hypersocial scholars argue that these capacities make humans able to form groups. "Eusocial" animals live together in a geographic space in multigenerational groups with an altruistic division of labor, a definition clearly applying to humans (Wilson 2012). Some argue that people are "inherently disposed" to live in groups (Tuomela 2013:10; see also Bacharach 2002; Read 2012; Tomasello 2019). Others argue that humans can form groups within larger groups within yet-larger groups *ad infinitum* (Gamble et al. 2014) or list the great diversity of those group forms (Turner and Machalek 2018). Humans share cooperative breeding—raising children both individually and collectively—with only two other primates, neither of them great apes (Hrdy 2009). Human groups learn collectively and cumulatively (Richerson and Boyd 2005; Sterelny 2012; Henrich 2015; Gintis 2017); cumulative cultural innovations over the last ten millennia made super-groups possible (Turchin 2016). These scholars do not assert mindless conformity, recognizing that humans are groupish *and* individualistic (Richerson and Boyd 2005; Stanish 2017; Turner and Machalek 2018). Some emphasize how difficult collective action is (Turchin 2016);

others point out that individual and group interests sometimes converge (Blanton 2016). Still, all agree that humans are simultaneously individualistic and groupish, a situation that creates tension.

To me, this paradigm is superior to the methodological individualism of rational choice or EP, and it is also superior to constructionism, which prevents us from talking about humans in general. Hypersociality is a "both-and" strategy that allows us to understand general patterns *and* specific cultures: Humans really are selfish, materialistic, and rational, *and* they are altruistic, cultural, and emotional—depending on circumstances. While some social scientists have contributed to this paradigm (Henrich 2015; Blanton 2016; Stanish 2017), most hypersocial scholars are not social scientists and cannot "explain the dynamics of social structures and their cultures" (Turner and Machalek 2018:249). I agree, so I will develop this paradigm further—produce my own version of it—by integrating it with social science concepts (Chap. 2).

Social Evolution as World History: The Deep Roots of Global Society

How do we apply this paradigm to world history? Basically, we trace changes in human groups from the deep past to the present. There are many nomothetic (general, universalizing) and idiographic (particularistic) social science approaches, so I will draw on and synthesize elements of both. The general (nomothetic) organizing framework I will use is called "social evolutionism" since it directly answers my research question: How did the descendants of small groups of hunter-gatherers from thousands of years ago develop today's global social system that contains billions of people? (I will discuss the idiographic approach below.) Many social scientists reject social evolutionism, yet its reputation derives mostly from its pre-1900 versions. I will explain how, in the last 80 years, scholars have stripped it of its mistaken, even repellent, characteristics and revamped it. "Social evolution" now means only the expansion and elaboration of human groups—nothing more.

But what evolves? As "society" is the central concept of this book—the "unit of analysis" in methodological terms—a provisional definition is warranted. Sztompka's definition (1994) is useful. First, any group is a set of relations between individuals, or between groups of individuals. Second, "social life is lived in time," so relations consist of interactions that change. Thus, *society* is "a multi-level, intermeshed set of processes" (Sztompka 1994:41 and 56). Following these ideas, I suggest that a society is a set of relations between multiple groups, and the members of these groups regard (voluntarily or coercively) this bundle of relationships as a single entity. The intergroup relations are usually more stable than relations between individuals, yet the former also change over time. In this way, any society is, so to speak, a moving target.

Societies grow larger or smaller depending on those interpersonal and intergroup processes. Social evolution is "the appearance of new forms of social or sociopolitical organization" (Marcus 2008:252). Over the last 15 millennia, many—but not all—human groups have grown, their organization becoming more elaborate. Social evolutionism—the study of the growth and elaboration of societies—exists mainly in archaeology and secondarily in sociology, often relying on ethnography (e.g., Sanderson 2001; Johnson and Earle 2000; Lenski 2005; Flannery and Marcus 2012). Forms around the world seem to resemble each other, so scholars often categorize societies by "types" (Carneiro 2003; Graeber and Wengrow 2021 object).

Many social scientists regard social evolutionism with hostility, previously for excellent reasons but recently from misunderstanding. The approach started without value judgments in the eighteenth century. However, most nineteenth-century European scholars were explicitly teleological and racist: Social evolution has a goal, and Europeans had achieved it. A few social evolutionists objected to this racism, but most others ignored them. After 1900, ethnographic anthropologists explicitly rejected social evolutionism, and it faded from sociology and archaeology. It was revived in the 1960s, stripped of its teleology and racism (Carneiro 2003 offers a history). While some still argue that liberal democracy is the best form of society (Fukuyama 2011, 2014), most social evolution scholars believe that individuals and groups have goals but that the process of social evolution has none (Carneiro 2003; Sanderson 2007) because of unintended consequences (Tang 2020). Now, one cannot find even hidden racism in social evolutionism (e.g., Johnson and Earle 2000; Wenke and Olszewski 2007; Flannery and Marcus 2012; Scarre 2018). Social "devolution," or shrinkage and simplification, is also possible (Carneiro 2003, though Tang 2020 demurs). Finally, while previous social evolutionists argued that larger and more complex societies were better, researchers now believe that people's well-being often deteriorated in more "evolved" societies (Cohen and Crane-Kramer 2007; Scott 2017; Graeber and Wengrow 2021).

Tracing the evolution of human societies relies on evidence scattered across the humanities and social sciences, mostly in archaeology and history. Archaeologists writing world prehistories usually start with analysis of the earliest, smallest societies and trace how they grew, stopping several thousand years ago for Afro-Eurasia and several hundred years ago for the Americas (e.g., Wenke and Olszewski 2007; Flannery and Marcus 2012; Fagan and Duranni 2014; Scarre 2018). World historians usually start analysis with a single chapter on those developments, tracing human history to today (e.g., McNeill and McNeill 2003; Roberts and Westad 2013; Strayer 2014) and emphasizing "civilizations" or large-scale societies (Stearns 2011:Chap. 5). Scholars in these two disciplines tend to stay out of each other's way, partly because the nature of their evidence differs: Archaeologists generally analyze physical objects such as movable "artifacts" (e.g., tools) and immovable "features" (e.g., buildings), while historians primarily read written documents (Historical archaeologists rely on both types of data.).

The two disciplines could complement each other, forming the backbone of a study to which others contribute. Historical social scientists mostly study the last few hundred years to understand the world today (e.g., Wallerstein 1974 and Thomas et al. 1987 in sociology; Watson 1992, Buzan and Little 2000, and Kaufman et al. 2007 in political science; Maddison 2006 and Aghion and Durlauf 2005, 2014 in economics; Wolf 1982 in anthropology), often in a theoretical vein. A few go much further back, studying large-scale systems (Abu-Lughod 1989; Frank and Gills 1993; Modelski and Thompson 1996; Chase-Dunn and Hall 1997; Korotayev et al. 2006).

Some scholars have been drawing on all of these studies to tell the long-term story of humans. The discipline of "Big History" is now several decades old, combining evidence from the natural sciences with human world history. Big Histories start with the Big Bang and trace the development of elements, stars, and planets; the cooling and formation of the earth; the beginning of life; the evolution of plants and animals; the appearance of mammals; and human history (Christian 2011; Spier 2015; Rodrigue et al. 2015). Both within and outside of this movement, scholars have developed what I would call Big Human History, in two approaches. The first is a chronologically structured, extremely interdisciplinary story (Mann 1986; Cook 2003; McNeill and McNeill 2003; Mazoyer and Roudart 2006; Modelski et al. 2008; Christian 2008, 2011; Carmack 2013; Brooke 2014; Chase-Dunn and Lerro 2014; Fukuyama 2011, 2014; Harari 2015; Kelly 2016; Faulkner 2018; Manning 2020; Szostak 2021). The other assumes this story and analyzes aspects of human life in commentaries (Diamond 1997; Shryock and Smail 2011; Turchin 2016; Vince 2019; Bregman 2019).

Why do we need a new account? Many of the existing approaches do not discuss social evolution; those that do usually discuss different kinds of societies separately (e.g., Nolan and Lenski 2011; Flannery and Marcus 2012; Chirot 2012). Furthermore, only a few (e.g., Chase-Dunn and Lerro 2014) attempt a multilevel analysis—i.e., large-scale social systems, societies within them, and groups within societies—but they are always left incomplete. I will attempt a consistent multilevel analysis, tracing how different kinds of societies related to each other over time.

Most historians and some social scientists object to the "types of societies" thinking typical of social evolutionism. Some idiographic scholars find categories of phenomena expressed as types to be flawed and ultimately illegitimate (e.g., Fernández-Armesto 2001:14). Even some social evolutionists resist them (e.g., Yoffee 2005; Earle 2017; Graeber and Wengrow 2021). In practice, most social evolutionists tend to use a few categories of societies since they are useful "shorthand" (Earle 2017:234). Others defend typologies (Lewellen 2003; Carneiro 2003; Marcus 2008). Carneiro goes furthest, arguing that typologies reflect *stages* in social evolution. As societies grow and elaborate, they develop similar sets of additional traits that constitute a new stage. Further growth and complexity lead to an additional cluster of similar traits, marking another stage. These are "the big steps in evolution" (Carneiro 2003:157).

While this thinking often disgusts idiographic scholars, I think they miss something. A literary analogy might help, using Jane Austen's works as an example. Filmmakers regularly adapt her works relatively directly; *Pride and Prejudice* ([1813] 1991) became a film in 1940, two television series (1980 and 1995), and another film in 2005. At least two film or television adaptations of each of her other novels also exist. However, others have created loose adaptations or "translations," placing them in cultural contexts very different from that of Regency England. Idiographic humanities scholars recognize how filmmakers transferred the understructure of plot and characters of *Pride and Prejudice* to different cultural contexts, such as *Bridget Jones's Diary* (Maguire 2001), *Bride and Prejudice* (Chadha 2004), *The Lizzie Bennet Diaries* (Su and Dunlap 2012–2014), and many others. There is also the brilliant rethinking of Austen's *Emma* ([1815] 2012) into *Clueless* (Heckerling 1995). William Shakespeare's works have also been fodder for translations; (e.g., *Macbeth* became Kurosawa's *Throne of Blood* [Kurosawa 1957]); readers may think of others.

Just as one can transfer a story's plot and characters to different cultural contexts, I see the stages in social evolution as the understructures of different kinds of societies, ones implemented in greatly different cultural forms (Marcus 2008; Feinman 2012). This analogy is limited: In the human past, societies in different regions developed similar understructures independently of each other rather than being diffused from one origination point and adapted locally. Still, larger and more complex patterns generally appeared in the same broad sequence in different places, thus suggesting Carneiro's (2003) stages. While he says that "evolution can be *both* unilinear and multilinear at the same time" (2003:235, original italics), I would rephrase it: Social evolution is uniserial but multilinear. That is, societies that do evolve move through one series of basic stages, but they follow quite varied cultural "lines" through them (Marcus 2008). Importantly, moving from one stage to another is not automatic, *pace* nineteenth-century social evolutionism.

I suggest that these stages result from cognitive capacities that all humans share. People can follow interactions between only a few others (Johnson 1982). As humans can "nest" groups within groups (Gamble et al. 2014; Earle 2017:236), once any group grows beyond these cognitive limits, members mentally and organizationally divide the larger group into subgroups. This enables them to continue to trace individual interactions within each subgroup, and now they can trace group–group interactions. This process can continue, and has continued, to create groups within groups, creating a hierarchy of nested groups. To me, this process creates broadly similar understructures of societies as ever-larger stages; however, the cultural specifics of nested groups vary tremendously.

How many stages are there? Service (1962) suggested four basic ones; scholars have disputed their number and names ever since (e.g., Fried 1975; Sanderson 1995; Kelly 2000; Lenski 2005; Flannery and Marcus 2012; Hayden 2014). In practice, archaeologists usually still use four basic stages. Others recognize that societies in the last five centuries have taken new forms (Sanderson 1995; Chase-Dunn and Lerro 2014). I will use six terms: *band*, *village*, *polygroup*,[1] and *state* up to 1900 CE, and *developed state* and *developing state* since then.

---

[1] After learning how diverse early complex societies were (Chap. 5), I rejected "chiefdom" for this neologism.

As this is a global history of societies, I use and combine the concept of globalization with social evolution concepts in varied ways. Many social scientists seem to think of globalization as occurring from the 1970s onward (e.g., Stigliz 2012; McMichael 2012; Dicken 2015). I reject this definition in favor of attention to the *longue durée* (Boivin and Frachetti 2018), so I identify two fundamental globalizations. At one point, *Homo sapiens sapiens* lived in small groups of mobile foragers only in sub-Saharan Africa; under such circumstances, talk of globalization or social evolution is nonsensical. Thus, while it was not transformative itself—it produced no new kind of society—the first globalization, humans migrating from sub-Saharan Africa to all continents, was the foundation for all other human social changes. Once I trace this process, I discuss the processes leading to new stages in social evolution as "revolutions" (Childe [1936] 1966; Sanderson 1995 called them "social transformations"). In the first transformation—the Agricultural Revolution, which began as the first globalization was ending—a few groups of mobile foragers gradually became independent villagers based on farming or herding; most humans globally continued mobile foraging, though village life spread geographically. Millennia later, in the Urban Revolution, leaders formed multivillage social units—at first polygroups (together, chiefdoms and their "analogues"), then states and empires, and eventually larger social systems. This process ultimately led to one large trade/interaction system in Afro-Eurasia and another in the Americas, the two isolated from each other. From the late 1400s, two processes occurred simultaneously: Europeans gradually, and usually violently, joined the two separate massive interaction systems into one even larger global system, and Europe developed a new stage in social evolution. Thus, I see this second globalization, which I also call the global Industrial Revolution, as the "culmination of processes that have unfolded over millennia" (Boivin and Frachetti 2018:2). The last few decades have seen an intensification of a very long process. The second globalization was, consequently, vastly different from the first one, so the relationship between globalization and industrialization is strenuously debated (see Chap. 8).

Each term is a synecdoche. The Agricultural Revolution started with domestication, but it also involved the development of villages, a demographic transition, and the appearance of warfare. The Urban Revolution spawned not only cities but also states and empires, conquest, new technologies, women's subordination, book religions, and permanent inequalities. Likewise, the global Industrial Revolution may have centered on industry, but it also involved print, science, capitalism, democracy, education, secularization, and gender changes—and it is still incomplete. Also, I am using these terms in their most expansive meanings. For instance, some economic historians argue that the Industrial Revolution occurred in Britain between 1760 and 1830 (though this is vigorously debated). From the perspective of ten to twenty millennia, however, the *global* Industrial Revolution of the last 250 years is still unfinished (Table 1.1).

These terms in particular, and "stage" thinking in general, obscure the messy reality of social evolution. For instance, by "villages" one can mean concentrations of complex hunter-gatherers, mobile camps of pastoralists, and settled farming villages. Thus, I do not use these terms as a mental straightjacket; rather, each is a loose shorthand—again, a set of "understructures"—only to get readers started.

Table 1.1 Transformations and social types

Major change	Resulting social form
Human physical evolution	Foraging bands
First globalization	No new social form; humans on all major land masses
Agricultural revolution	Independent villages surrounded by bands
Urban revolution	Polygroups and states surrounded by villages and bands
Industrial revolution/second globalization	Global system encompassing developed and developing states

Furthermore, such groups were always interlinked, and I will trace the systems those links sustained. Archaeologists discuss "peer-polity interaction spheres" (Renfrew 1986): Any "polity" (political unit or group) interacted with other, similar polities (its peers, such as similarly sized villages) within a large region through trade, intermarriage, or warfare. Political scientists call one subform of this a "state system." In contrast, sociologists talk about "world-systems" (Chase-Dunn and Hall 1997): The *world* in this phrase denotes a social world or system unto itself, not the entire planet.[2] Unlike peer-polity systems, world-systems are unequal and organized by a geographic division of labor. Bands, villages, polygroups, and states always existed within these larger systems; today, we exist within one global social system. Every such system tended to encompass and put into interaction different types of societies.

As mentioned, this multilevel account of transformations starting well over ten millennia ago will be primarily descriptive and only secondarily explanatory (theoretical). Even historians denigrate "mere" description, yet as Megill says, "Upon descriptions, explanations arise" (2007:103). Still, after tracing each social transformation, I review various explanations for these changes, then analyze the new social form. Along the way, I suggest *ad hoc* explanations rooted in the cognitive-emotional capacities of humans, which are seeds for later explanations. The descriptions of transformations, the explanatory reviews, and the scattered suggestions will help me sketch out an agenda for a theory of human history in the final chapter.

Overall, then, scholars have developed a conceptual apparatus to understand the expansion and elaboration of human groups in the last fifteen (or so) millennia, with its variety and its devolutions. This process is one way to understand world history; indeed, Yoffee (2005) repeatedly insists that social evolution *is* world history. These scholars no longer see "larger and more complex" as better; many recognize how life worsened in some ways as societies evolved. Most recognize a vast diversity within a few basic forms of society. Thus, the second and main argument of this book is that while each of these revolutions produced a new type of society, all three were and are incomplete geographically. The first two social revolutions never influenced all humans in their times; the third *affects* all humans but does not fully *include* them.

[2] Hence, the seemingly trivial hyphen is in fact necessary to distinguish a "world-system" from "the world."

Long-Term Continuities: Societies as Multidimensional Groups

Scholars of social evolution tend to break societies down into "parts," relying on metaphors such as arenas, areas, aspects, spheres, sectors, or dimensions of social life (Johnson and Earle 2000; Sanderson and Alderson 2005; Nolan and Lenski 2011). This makes possible an idiographic complement to the nomothetic framework of stages; it allows the comparison that shows the cultural diversity that developed over time. In each chapter, an analysis of such aspects will reveal how diverse types of societies can be. However, comparing societies along these aspects chapter after chapter allows us to see not only what changed but also what remained broadly the same. We will see that there have been long continuities throughout human history.

To support both assertions—cultural diversity and long-term continuities—I analyze societies in terms of seven aspects. Five are cultural-institutional, and two are material, both influenced by cultural patterns. (I define *culture* and *institutions* in the next chapter.) The first is the economy, including the three successive processes of production, distribution, and consumption. Subsistence production (or cultural ecology) involves extracting resources from the earth and transforming them (Ellen 1994; Lenski 2005). The question of "who gets what, when, how" (Lasswell [1936] 1950) studies the distribution of that produce. Consumption is the literal or figurative use of that produce (Carrier 1996). The second aspect, kinship, concerns both group organization and group survival over time. Anthropologists define it as "social relationships that usually coincide with biological ones" (Velioti-Georgopolous 2006:1369). I will analyze forms of marriage (monogamy, polygamy, and so on), postmarital residence (with whom does the new couple reside?), and descent systems (how people are related biologically and socially—traced primarily through the father or mother, or through both—or more complex systems). Sexual practices and gender roles also vary greatly (Jankowiak 1997; Pasternak et al. 1997). Politics is the third aspect of human groups to analyze. It includes power, defined as the capacity to accomplish, or the achievement of one's will against social resistance (Weber 1946); strategies for achieving group cohesion (loyalty, unity); leadership structures; and intergroup relations. Fourth, human groups have always been unequal in at least four ways. Individual differences and age are universal inequalities; gender and economic inequalities developed over time. The fifth aspect, institutions that help people manage ideas, include belief-systems and rituals to embody and express them; esthetic practices (image-making, music, dance, and stories) and practitioners (shamans and priests); systems of morality, or how people should treat one another; and the imposition of ideas on the physical world, "material culture," or technologies (Bourdieu [1979] 1984; Oudshoorn and Pinch 2003; Skibo and Schiffer 2008). One material aspect of societies is the relationship between humans and their physical environment, including its influence on groups and groups' influence on it (Michael Dougherty, personal communication). It also includes understanding the environment as "pool" from which to draw useful physical resources

(Ostrom et al. 2002) or "sink" into which we can discard used-up resources (i.e., pollution) (Ponting 2007). Demography, or the analysis of populations, is the final aspect (Abercrombie et al. 2000). This includes analysis of how many people are in a population at each age; births and deaths, usually per year; migration, or movement into and out of groups; and density, or the number of people per given unit of land (Chamberlain 2006).

This framework has one obvious disadvantage and several advantages. It is clearly artificial; people do not live their lives in "aspects" but are influenced by many factors simultaneously, especially in smaller-scale societies. Still, several academic disciplines are devoted to only one aspect or another (e.g., economics) because this sort of analysis can help us understand societies better by providing a conceptual vocabulary. Consequently, I will apply this framework of seven aspects to each social type, and I will also use it in discussions of change.

How does change happen? Philosophers debate the issue of free will versus determinism; in sociology, the question appears as "structure versus agency." The metaphor of *structure* helps us understand how institutions both enable and constrain human action: if one should do something, one often should not do other things. The English word *agency* comes from the Latin *agentem*, "effective, powerful," and appears in the French verb *agir*, "to act." Institutions constrain people, who nevertheless may not follow them. Thus, sociologists have discussed structure and agency—constraining social facts and individual social action as Durkheim and Weber put it, respectively—for over a century (Danermark et al. 2019). Some hypersocial scholars argue that cultural accumulation occurs so gradually that even "great" people contribute only a little to group life (Richerson and Boyd 2005; Henrich 2015), yet they also recognize that those changes add up. Were human beings perfectly imitative, no social change would happen. Instead, people think for themselves, sometimes following individual urges or new ideas rather than group traditions. Usually, a group punishes them, and they conform; when some resist, conflict occurs. This can and does change group institutions.

How people live out these seven aspects in their societies has changed tremendously in the last twenty millennia. Still, the tendency to conform and the physical limits of human capabilities have generated long continuities. Using these seven aspects repeatedly will help us see them. Some are ongoing patterns that have remained stable over the long term. Others are cumulative trends leading to long-term changes. One demographic pattern has been high mortality and fertility, lowering only in the last three centuries (Boone 2002; Livi-Bacci 2017). A demographic trend has been the ongoing growth in the human population; consequently, the number and size of groups and intergroup systems have also increased (Livi-Bacci 2017). One economic trend has been the slow, gradual expansion of trade systems, up to and including today's global economy (Korotayev et al. 2006). Two political trends include the increasing formalization of leadership structures, though this varies greatly (Lewellen 2003), and the growth in political interaction systems (Buzan and Little 2000). Original human groups were economically equal, and men and women seem to have shared equal power. Over the last seven to ten millennia, economic inequality has consistently grown and is now global (Flannery and Marcus

2012; Maddison 2006; Piketty 2014; Scheidel 2017), while gender inequality started roughly at the same time and has increased concomitantly (Lerner 1986; Bonvillain 2001; Therborn 2006). Scholars debate the long-term human influence on the physical world—some arguing that groups have often stayed in balance with it (Ostrom et al. 2002; Brooke 2014), others claiming that humans have had a destructive long-term effect (Chew 2001; Ponting 2007); few have disputed the latter argument for the last two centuries. Finally, human idea management through physical objects started in sign systems and images. The sequential development of information-storage systems, especially writing, the printing press, and science (Eisenstein 1983; Schmandt-Besserat 1996; Drori et al. 2003; McClellan and Dorn 2015; von Petzinger 2016), has resulted in the tremendous explosion of popular culture in film, television, music, video games, and social media today.

While it is most obvious in what I call "idea management," global cultural diversity developed early across all seven aspects. Archaeologists and historical linguists have shown that by the time humans populated the entire planet, specific cultural traditions had developed in different regions; this culturally regionalizing tendency intensified as populations increased. With the development of states came a special kind of cultural region, the "civilization." Unlike many scholars of the past, I do not equate *state* and *civilization*. Instead, I see civilizations as the long-term result of successive states in a given region; thus, states gave rise to civilizations and then existed within them. As the discussion in Chap. 6 will show, not all agree with this definition, but I will explain it more fully there. In Chap. 8, I will discuss the cultural phenomenon that I argue is gradually incorporating all cultural regions and civilizations: the global ecumene.

Following these long-term trends is as important as tracing social transformations. I get the sense that most people who think about the past think *only* about the past, rarely attempting to project past patterns and trends into the future (Fig. 1.1).

While people do try to forecast the future, most seem to have a very short time horizon, perhaps a matter of years. In contrast, archaeologists (Marcus 2008), Big Historians (Rodrigue et al. 2015), and even one postcolonial scholar (Robbins 2016) appreciate the conceptual and knowledge limits of one human lifetime, or even the period of European dominance, emphasizing instead much longer trends. We live within these trends; seeing how long they have lasted, or have started to reverse, helps us understand our situation within any long future (Fig. 1.2).

Since predicting the future is difficult, the line is dashed; no trend is guaranteed to continue. However, disregarding long-lasting patterns and trends blinds us to

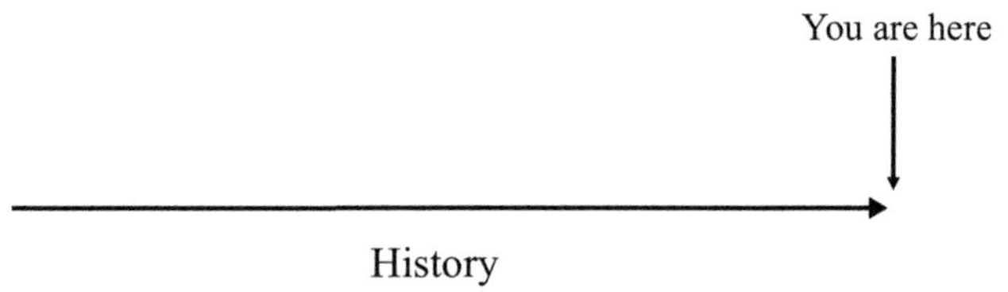

Fig. 1.1 Most people's perception of the past

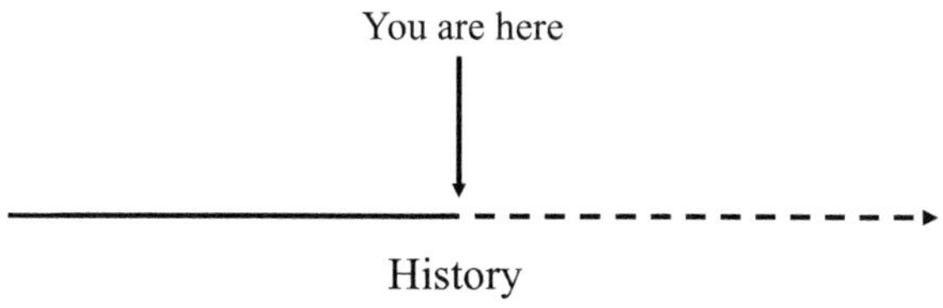

Fig. 1.2 Long term trends over time

possibilities; we could focus on "big picture" trends even if specific events are unknowable. For instance, COVID-19 took most people by surprise, but epidemiologists predicted some sort of global pandemic for decades. Likewise, many governments now worry about the convergence of falling birthrates, aging populations, and strained public retirement systems. Finally, in Chap. 11 I will suggest several possible futures by imaginatively extending the long-term patterns I have identified throughout the book.

Conclusion

This book is more social science than history, not only because I partly rely on a positivist framework—social evolution—but also because I am concerned with making sure readers understand how the past helps us understand the present. Many aspects of every kind of past society still exist today; studying the first humans and tracing both the transformations and the continuities will help us understand our current situation. Historians usually object to this approach as often leading to anachronism, "the willful or inadvertent misunderstanding of the past by applying standards or interpretations from outside the immediate era, context, or milieu under study," and a "less polite term for it is *presentism,*" applying the standards or interpretations of the present (Armitage 2023:47, original italics). Despite historians' hostility to presentism, Armitage partly defends it: some of its forms should be avoided, while others can help us understand history. Historians themselves cannot avoid their own position in the present in their choice of questions, their conceptual frameworks, and the audiences to whom they write (Armitage 2023:50). Furthermore, as "the present has nowhere to come from but the past" (p. 60), then "narrative presentism," or tracing the past into the present, is a plausible strategy; its main pitfall is teleological thinking. Though they are partly incompatible with narrative presentism, critical and strategic presentisms, which call for attention to possible alternatives to how history in fact worked out (Pp. 60–61) can help one avoid teleology.

These kinds of presentism make sense. I fully recognize that I analyze the past through Western, social science, twenty-first-century conceptual frameworks. A Chinese scholar would likely use different frameworks to address my questions—if they were even the same questions; an indigenous scholar would *very* likely use different frameworks and have different questions. Also, as discussed, the sheer scope of this book requires collapsing details into large-scale "facts," creating an ongoing opportunity for anachronisms to creep in. Still, I think that trying to understand the past on its own terms helps us better understand the present; keeping in mind that people were "making it up as they went along," not knowing what would happen, is much better than assuming historical directionality. Social evolution did go in a particular direction, but it could have gone in many others; figuring out why it happened in *this* way is part of any future theoretical agenda, but it must rest on an appreciation of contingency and unpredictability.

Given this presentist approach, in the concluding chapter, I address the implications of all three of the arguments of this book in reverse order. Implications of the long-term patterns and trends identified here for the future will probably be most interesting to the average reader. Likewise, few realize just how much of the past lives on in the present around the globe, and in such tremendously uneven ways. There are still hunting and gathering peoples scattered around the world, though beset by immense pressures. Most of us literally eat the results of the Agricultural Revolution daily; many still live like agropastoral villagers. Societies reshaped by the Urban Revolution laid their new patterns on some of those villagers, sometimes in small, regional "polities" but more often in expanding states and empires that conquered, then reworked social patterns. While the global Industrial Revolution influenced everyone on this planet and is remaking many societies today, some peoples want more of this influence, while many others reject it. Finally, there is an enormous amount of research on "who humans are," but few have pulled together a synthesis of this research. Most scholars seem to lean either toward a "there is no human nature" position, or they lean the other way to "human nature is relatively fixed." An appreciation that humans have a great deal in common as humans *and* that we differ tremendously, both individually and macrosocially, has been rare within academia. This situation is changing; it is the basic position of many of those who adhere to "hypersocial" tenets—including me. This line of research has underappreciated implications for human well-being; again, I will only sketch out beginnings of such a research program on this issue in the conclusion. In any case, I will start on substantiating those three arguments—many long-lasting continuities, three major social revolutions, and humans being hypersocial—by addressing the last one in the next chapter.

References

Abercrombie, Nicholas, Stephen Hill, and Bryan S. Turner. 2000. *The Penguin Dictionary of Sociology*. London: Penguin Books.

Abu-Lughod, Janet. 1989. *Before European Hegemony: The World System A.D. 1250–1350*. New York: Oxford University Press.

Aghion, Philippe, and Steven N. Durlauf. 2005. Handbook of Economic Growth. In *Handbooks in Economics 22*, vol. IIA and IIB. Amsterdam and Boston: Elsevier/North-Holland.

Aghion, Philippe, and Steven N. Durlauf. 2014. Handbook of Economic Growth. In *Handbooks in Economics 22*, vol. IIA and IIB. Amsterdam and Boston: Elsevier/North-Holland.

Armitage, David. 2023. In Defense of Presentism. In *History and Human Flourishing*, ed. D. McMahon, 44–69. New York: Oxford University Press.
Austen, Jane. [1813] 1991. *Pride and Prejudice*. New York: Knopf.
Austen, Jane. [1815] 2012. *Emma*. George Justice, ed. New York: W.W. Norton.
Bacharach, Michael. 2002. In *Beyond Individual Choice: Teams and Frames in Game Theory*, ed. Natalie Gold and Robert Sugden. Princeton and Oxford: Princeton University Press.
Baddeley, Michelle. 2017. *Behavioural Economics: A Very Short Introduction*. New York: Oxford University Press.
Barkow, Jerome H., Leda Cosmides, and John Tooby. 1992. *The Adapted Mind: Evolutionary Psychology and the Generation of Culture*. New York: Oxford University Press.
Baron Cohen, Simon. 2003. *The Essential Difference: The Truth about the Male and Female Brain*. New York: Basic Books.
Bergesen, Albert. 2005. Culture and Cognition. In *The Blackwell Companion to the Sociology of Culture*, ed. M. Jacobs and N. Hanrahan, 35–47. New York: Basil Blackwell.
Blanton, Richard E., with Lane F. Fargher. 2016. *How Humans Cooperate: Confronting the Challenges of Collective Action*. Boulder: University Press of Colorado.
Boivin, Nicole, and Michael D. Frachetti, eds. 2018. *Globalization in Prehistory*. New York: Cambridge University Press.
Bonvillain, Nancy M. 2001. *Women and Men: Cultural Constructs of Gender*. Upper Saddle River, NJ: Prentice Hall.
Boone, James L. 2002. Subsistence Strategies and Early Human Population History: An Evolutionary Ecological Perspective. *World Archaeology* 34 (1): 6–25.
Bourdieu, Pierre. [1979] 1984. *Distinction: A Social Critique of the Judgement of Taste*. Translated by Richard Nice. Cambridge, MA: Harvard University Press.
Bregman, Rutger. 2019. *Humankind: A Hopeful History*. Translated by Erica Moore and Elizabeth Manton. New York: Little, Brown and Company.
Brooke, John L. 2014. *Climate Change and the Course of Global History: A Rough Journey*. New York: Cambridge University Press.
Bryant, Joseph M. 1994. Evidence and Explanation in History and Sociology: Critical Reflections on Goldthorpe's Critique of Historical Sociology. *The British Journal of Sociology* 45 (1): 3–19.
Buzan, Barry, and Richard Little. 2000. *International Systems in World History: Remaking the Study of International Relations*. Oxford: Oxford University Press.
Carmack, Robert M. 2013. *Anthropology and the Global System: From Tribes to the Modern World System*. Lanham, MD: Altamira.
Carneiro, Robert L. 2003. *Evolutionism in Cultural Anthropology: A Critical History*. Cambridge, MA: Westview Press.
Carrier, James G. 1996. Consumption. In *Encyclopedia of Social and Cultural Anthropology*, ed. Alan Barnard and Jonathan Spencer. London and New York: Routledge.
Cartmill, Matt, and Fred H. Smith. 2022. *The Human Lineage*. 2nd ed. Hoboken, NJ: Wiley Blackwell.
Chacon, Richard J., and Ruben G. Mendoza, eds. 2017. *Feast, Famine, or Fighting? Multiple Pathways to Social Complexity*. New York: Springer.
Chadha, Gurinder. 2004. *Bride and Prejudice*. Los Angeles: Miramax Films.
Chamberlain, Andrew. 2006. *Demography in Archaeology*. Cambridge: Cambridge University Press.
Chase-Dunn, Christopher, and Thomas D. Hall. 1997. *Rise and Demise: Comparing World-Systems*. Boulder, CO: Westview Press.
Chase-Dunn, Christopher, and Bruce Lerro. 2014. *Social Change: Globalization from the Stone Age to the Present*. Boulder, CO: Paradigm.
Chew, Sing C. 2001. *World Ecological Degradation: Accumulation, Urbanization, and Deforestation, 3000 B.C.–A.D. 2000*. Walnut Creek, CA: AltaMira Press.
Childe, V. Gordon. [1936] 1966. *Man Makes Himself*. 4th ed. Bungay, Suffolk: Fontana.
Chirot, Daniel. 2012. *How Societies Change*. 2nd ed. Thousand Oaks, CA: Sage.

Christian, David. 2008. *This Fleeting World: A Short History of Humanity*. Great Barrington, MA: Berkshire.

Christian, David. 2011. *Maps of Time: An Introduction to Big History*. 2nd ed. Berkeley: University of California Press.

Cohen, Mark Nathan, and Gillian M. M. Crane-Kramer. 2007. Editors' Summation. In *Ancient Health: Skeletal Indicators of Agricultural and Economic Intensification*, ed. M. N. Cohen and G. M. M. Crane-Kramer, 320–343. Gainesville, FL: University Press of Florida.

Cook, Michael. 2003. *A Brief History of the Human Race*. New York: W.W. Norton.

Cosmides, Leda, and John Tooby. 1992. Cognitive Adaptations for Social Exchange. In *The Adapted Mind: Evolutionary Psychology and the Generation of Culture*, ed. J. H. Barkow, L. Cosmides, and J. Tooby, 163–228. New York: Oxford University Press.

Danermark, Berth, Mats Ekström, Ch. Jan, and Karlsson. 2019. *Explaining Society: Critical Realism in the Social Sciences*. 2nd ed. New York: Routledge.

Diamond, Jared. 1997. *Guns, Germs, and Steel: The Fate of Human Societies*. New York: W.W. Norton.

Dicken, Peter. 2015. *Global Shift: Mapping the Changing Contours of the World Economy*. 7th ed. New York: The Guilford Press.

Dogan, Mattei. 2001. Paradigms in the Social Sciences. In *International Encyclopedia of the Social and Behavioral Sciences*, ed. N. Smelser and P. Baltes, 11023–11027. Amsterdam: Elsevier.

Dombrowski, Stefan, Karen L. Gischlar, and Martin Mrazik, with Fred W. Greer III. 2011. "Feral Children." Pp. 81–94 in *Assessing and Treating Low Incidence/High Severity Psychological Disorders of Childhood*. New York: Springer.

Donmoyer, Robert. 2008. Paradigm. In *The SAGE Encyclopedia of Qualitative Research Methods*, ed. L. M. Given, 595–604. Thousand Oaks, CA: Sage.

Drori, Gili S., John W. Meyer, Francisco O. Ramirez, and Evan Schofer. 2003. *Science in the Modern World Polity: Institutionalization and Globalization*. Stanford, CA: Stanford University Press.

Earle, Timothy K. 2017. Chiefs, Chieftaincies, Chiefdoms, and Chiefly Confederacies: Power in the Evolution of Social Systems. In *Chiefdoms: Yesterday and Today*, ed. R. L. Carneiro, L. E. Grinin, and A. V. Korotayev, 233–256. Clinton Corners, NY: Eliot Werner.

Ede, Andrew, and Lesley B. Cormack. 2012. *A History of Science in Society: From Philosophy to Utility*. 2nd ed. Toronto: University of Toronto Press.

Eisenstein, Elizabeth L. 1983. *The Printing Revolution in Early Modern Europe*. Cambridge: Cambridge University Press.

Elder-Vass, Dave. 2012. *The Reality of Social Construction*. New York: Cambridge University Press.

Ellen, Roy. 1994. Modes of Subsistence: Hunting and Gathering to Agriculture and Pastoralism. In *Companion Encyclopaedia of Anthropology: Humanity, Culture and Social Life*, ed. T. Ingold, 197–225. London: Routledge.

Fagan, Brian M., and Nadia Duranni. 2014. *People of the Earth: An Introduction to World Prehistory*. 14th ed. New York: Pearson.

Faulkner, Neil. 2018. *A Radical History of the World*. London: Pluto Press.

Fausto-Sterling, Anne. 2000. Beyond Difference: Feminism and Evolutionary Psychology. In *Alas, Poor Darwin: Arguments Against Evolutionary Psychology*, ed. H. Rose and S. Rose, 209–228. New York: Harmony Books.

Feinman, Gary M. 2012. Comparative Frames for the Diachronic Analysis of Complex Societies: Next Steps. In *The Comparative Archaeology of Complex Societies*, ed. M. E. Smith, 21–43. New York: Cambridge University Press.

Fernández-Armesto, Felipe. 2001. *Civilizations: Culture, Ambition, and the Transformation of Nature*. New York: Touchstone.

Flannery, Kent, and Joyce Marcus. 2012. *The Creation of Inequality: How Our Prehistoric Ancestors Set the Stage for Monarchy, Slavery, and Empire*. Cambridge, MA: Harvard University Press.

Frank, Andre Gunder, and Barry Gills, eds. 1993. *The World System: Five Hundred Years or Five Thousand?* London: Routledge.

Fried, Morton H. 1975. *The Notion of Tribe*. Menlo Park, CA: Cummings Publishing.
Friedman, Jeffrey. 1996. *The Rational Choice Controversy: Economic Models of Politics Reconsidered.* New Haven, CT: Yale University Press.
Friesen, Bruce K. 2015. *Moral Systems and the Evolution of Human Rights*. New York: Springer.
Fuentes, Agustín, Jonathan Marks, Tim Ingold, Robert Sussman, Patrick V. Kirch, Elizabeth M. Brumfiel, Rayna Rapp, et al. 2010. On Nature and the Human. *American Anthropologist* 112 (4): 512–521.
Fukuyama, Francis. 2011. *The Origins of Political Order: From Prehuman Times to the French Revolution*. New York: Farrar, Straus and Giroux.
Fukuyama, Francis. 2014. *Political Order and Political Decay: From the Industrial Revolution to the Globalization of Democracy*. New York: Farrar, Straus and Giroux.
Furubotn, Eirik, and Rudolf Richter. 2005. *Institutions and Economic Theory: The Contribution of the New International Economics*. 2nd ed. Ann Arbor: University of Michigan Press.
Gaddis, John Lewis. 2001. In Defense of Particular Generalization: Rewriting Cold War History, Rethinking International Relations Theory. In *Bridges and Boundaries: Historians, Political Scientists, and the Study of International Relations*, ed. C. Elman and M. F. Elman, 301–326. Cambridge, MA: The MIT Press.
Gamble, Clive, John Gowlett, and Robin Dunbar. 2014. *Thinking Big: How the Evolution of Social Life Shaped the Mind.* London: Thames & Hudson.
Gergen, Kenneth J. 2015. *An Invitation to Social Construction*. 3rd ed. Thousand Oaks, CA: Sage.
Gintis, Herbert. 2017. *Individuality and Entanglement: The Moral and Material Bases of Social Life*. Princeton, NJ: Princeton University Press.
Graeber, David, and David Wengrow. 2021. *The Dawn of Everything: A New History of Humanity*. New York: Farrar, Straus, and Giroux.
Green, Donald P., and Ian Shapiro. 1994. *Pathologies of Rational Choice Theory: A Critique of Applications in Political Science*. New Haven, CT: Yale University Press.
Harari, Yoal Noah. 2015. *Sapiens: A Brief History of Humankind.* New York: HarperCollins.
Hayden, Brian. 2014. *The Power of Feasts: From Prehistory to the Present*. New York: Cambridge University Press.
Heath, Joseph. 2020 (2005). "Methodological Individualism." Stanford Encyclopedia of Philosophy. https://plato.stanford.edu/entries/methodological-individualism/
Heckerling, Amy. 1995. *Clueless*. Hollywood: Paramount Home Video.
Henrich, Joseph. 2015. *The Secret of Our Success: How Culture Is Driving Human Evolution, Domesticating Our Species, and Making Us Smarter*. Princeton and Oxford: Princeton University Press.
Hobson, John M. 2009. Comparative Politics and International Relations. In *The SAGE Handbook of Comparative Politics*, ed. T. Landman and N. Robinson, 175–189. Thousand Oaks, CA: Sage.
Hrdy, Sarah Blaffer. 2009. *Mothers and Others: The Evolutionary Origins of Mutual Understanding*. Belknap: Cambridge, MA.
Iggers, Georg C. 2012. *Historiography in the Twentieth Century: From Scientific Objectivity to the Postmodern Challenge, with a New Epilogue by the Author*. Middletown, CT: Wesleyan University Press.
Jankowiak, William. 1997. Gender and Sex Roles. In *The Dictionary of Anthropology*, ed. T. Barfield, 217–220. Malden, MA: Blackwell.
Joel, Daphna, and Luba Vikhanski. 2019. *Gender Mosaic: Beyond the Myth of the Male and Female Brain*. New York: Little, Brown Spark.
Johnson, Gregory A. 1982. Organizational Structure and Scalar Stress. In *Theory and Explanation in Archaeology*, ed. C. Renfrew, 389–421. New York: Academic Press.
Johnson, Allen W., and Timothy K. Earle. 2000. *The Evolution of Human Societies: From Foraging Group to Agrarian State*. 2nd ed. Stanford, CA: Stanford University Press.
Kaufman, Stuart J., Richard Little, and William C. Wohlforth, eds. 2007. *The Balance of Power in World History*. New York: Palgrave Macmillan.

Kelly, Raymond C. 2000. *Warless Societies and the Origin of War*. Ann Arbor: The University of Michigan Press.
Kelly, Robert L. 2016. *The Fifth Beginning: What Six Million Years of Human History Can Tell Us about Our Future*. Berkeley: University of California Press.
Kirchgässner, Gebhard. 2008. *Homo Economicus: The Economic Model of Behaviour and Its Applications in Economics and Other Social Sciences*. New York: Springer.
Korotayev, Andrey, Artemy Malkov, and Daria Khaltourina. 2006. *Introduction to Social Macrodynamics: Compact Macromodels of the World System Growth*. Moscow: URSS.
Kuhn, Thomas S. [1962] 2012. *The Structure of Scientific Revolutions*. 4th ed. With an introductory essay by Ian Hacking. Chicago and London: The University of Chicago Press.
Kurosawa, Akira. 1957. *Throne of Blood*. Brandon Films.
Lasswell, Harold D. [1936] 1950. Politics: Who Gets What, When, How. New York: Peter Smith.
Leahy, Terry. 2017. *Humanist Realism for Sociologists*. London and New York: Routledge.
Lenski, Gerhard E. 2005. *Evolutionary-Ecological Theory: Principles and Applications*. Boulder, CO: Paradigm.
Lents, Nathan H. 2016. *Not So Different: Finding Human Nature in Animals*. New York: Columbia University Press.
Lerner, Gerda. 1986. *The Creation of Patriarchy*. New York: Oxford University Press.
Lerner, Jennifer S., Ye Li, Piercarlo Valdesolo, and Karim S. Kassam. 2015. Emotion and Decision Making. *Annual Review of Psychology* 66:799–823.
Lewellen, Ted C. 2003. *Political Anthropology: An Introduction*. 3rd ed. Westport, CT: Praeger.
Livi-Bacci, Massimo. 2017. *A Concise History of World Population*. 6th ed. Chichester: Wiley-Blackwell.
Maddison, Angus. 2006. The World Economy. In *A Millennial Perspective and Volume 2: Historical Statistics*, vol. 1. Paris: OECD Publishing.
Maguire, Sharon. 2001. *Bridget Jones's Diary*. Burbank, CA: Miramax Home Entertainment.
Mann, Michael. 1986. *The Sources of Social Power*, A History of Power from the Beginning to A.D. 1760. Vol. I. New York: Cambridge University Press.
Manning, Patrick. 2020. *A History of Humanity: The Evolution of the Human System*. New York: Cambridge University Press.
Marcus, Joyce. 2008. The Archaeological Evidence for Social Evolution. *Annual Review of Anthropology* 37:251–266.
Masterman, Margaret. 1970. The Nature of a Paradigm. In *Criticism and the Growth of Knowledge*, ed. I. Lakatos and A. Musgrave, 59–90. New York: Cambridge University Press.
Mazoyer, Marcel, and Laurence Roudart. 2006. *A History of World Agriculture: From the Neolithic Revolution to the Current Crisis*. Translated by J. H. Membrez. New York: Monthly Review Press.
McChesney, Fred S. 2013. Behavioral Economists: Old Wine in Irrelevant New Bottles? *Supreme Court Economic Review* 21 (1): 43–76.
McClellan, James E., III, and Harold Dorn. 2015. *Science and Technology in World History: An Introduction*. 3rd ed. Baltimore: Johns Hopkins University Press.
McMichael, Philip. 2012. *Development and Social Change: A Global Perspective*. 5th ed. Thousand Oaks: SAGE Publications.
McNeill, J. R. R., and William H. McNeill. 2003. *The Human Web: A Bird's-Eye View of World History*. New York: W. W. Norton.
Megill, Allan. 2007. *Historical Knowledge, Historical Error: A Contemporary Guide to Practice*. With contributions by Steven Shepard and Phillip Honenberger. Chicago and London: The University of Chicago Press.
Mjøset, Lars. 2001. Theory: Conceptions in the Social Sciences. In *International Encyclopedia of the Social and Behavioral Sciences*, ed. N. Smelser and P. Baltes, 15641–15647. Amsterdam: Elsevier.

Modelski, George, and William R. Thompson. 1996. *Leading Sectors and World Powers: The Coevolution of Global Politics and Economics*. Columbia: The University of South Carolina Press.
Modelski, George, Tessaleno Devezas, and William R. Thompson. 2008. *Globalization as Evolutionary Process: Modeling Global Change*. New York: Routledge.
Nightingale, David J., and John Cromby. 2002. Social Constructionism as Ontology: Exposition and Example. *Theory and Psychology* 12 (5): 701–713.
Nolan, Patrick, and Gerhard Lenski. 2011. *Human Societies: An Introduction to Macrosociology*. 11th ed. Boulder, CO: Paradigm.
Olson, Mancur. [1965] 1971. *The Logic of Collective Action: Public Goods and the Theory of Groups*. Cambridge, MA: Harvard University Press.
Ostrom, Elinor, Thomas Dietz, Nives Dolšak, Paul C. Stern, Susan Stonich, and Elke U. Weber, eds. 2002. *The Drama of the Commons*. Washington, DC: National Academy Press.
Oudshoorn, Nelly, and Trevor Pinch, eds. 2003. *How Users Matter: The Co-Construction of Users and Technologies*. Cambridge, MA: The MIT Press.
Pasternak, Burton, Carol R. Ember, and Melvin Ember. 1997. *Sex, Gender, and Kinship: A Cross-Cultural Perspective*. Upper Saddle River, NJ: Prentice-Hall.
Piketty, Thomas. 2014. *Capital in the Twenty-First Century*. Translated by A. Goldhammer. Cambridge, MA: Harvard University Press.
Pinker, Steven. 2002. *The Blank Slate: The Modern Denial of Human Nature*. New York: Penguin.
Ponting, Clive. 2007. *A New Green History of the World: The Environment and the Collapse of Great Civilizations*. New York: Penguin.
Prinz, Jesse J. 2014. *Beyond Human Nature: How Culture and Experience Shape the Human Mind*. New York and London: W.W. Norton.
Read, Dwight W. 2012. *How Culture Makes Us Human: Primate Social Evolution and the Formation of Human Societies*. Walnut Creek, CA: Left Coast Press.
Renfrew, Colin. 1986. Introduction: Peer Polity Interaction and Socio-Political Change. In *Peer Polity Interaction and Socio-Political Change*, ed. C. Renfrew and J. F. Cherry, 1–18. New York: Cambridge University Press.
Richerson, Peter J., and Robert Boyd. 2005. *Not by Genes Alone: How Culture Transformed Human Evolution*. Chicago and London: The University of Chicago Press.
Rippon, Gina. 2019. *Gender and Our Brains: How New Neuroscience Explodes the Myths of the Male and Female Minds*. New York: Pantheon Books.
Ritzer, George. 2005. Paradigm. In *Encyclopedia of Social Theory*, ed. G. Ritzer, 543–544. Thousand Oaks, CA: Sage.
Robbins, Bruce. 2016. Prolegomena to a Cosmopolitanism in Deep Time. *Interventions* 18 (2): 172–186.
Roberts, J. M., and Odd Arne Westad. 2013. *A History of the World*. Oxford: Oxford University Press.
Rodrigue, Barry, Leonid Grinin, and Andrey Korotayev. 2015. From Big Bang to Galactic Civilizations: A Big History Anthology. In *The Ways that Big History Works: Cosmos, Life, Society and Our Future*, vol. III. Delhi: Primus Books.
Rose, Hilary, and Stephen Rose. 2000. In *Alas, Poor Darwin: Arguments Against Evolutionary Psychology*, ed. H. Rose and S. Rose. New York: Harmony Books.
Ross, Donald. 2019 (1997). "Game Theory." Stanford Encyclopedia of Philosophy. https://plato.stanford.edu/entries/game-theory/
Sanderson, Stephen K. 1995. *Social Transformations: A General Theory of Historical Development*. Cambridge, MA: Blackwell.
Sanderson, Stephen K. 2001. *The Evolution of Human Sociality: A Darwinian Conflict Perspective*. Lanham, MD: Rowman & Littlefield.
Sanderson, Stephen K. 2007. *Evolutionism and Its Critics: Deconstructing and Reconstructing an Evolutionary Interpretation of Human Society*. Boulder, CO: Paradigm Publishers.
Sanderson, Stephen K., and Arthur Alderson. 2005. *World Societies: The Evolution of Human Social Life*. Boston: Pearson.

Scarre, Chris, ed. 2018. *The Human Past: World Prehistory and the Development of Human Societies*. 4th ed. London: Thames & Hudson.
Scheidel, Walter. 2017. *The Great Leveler: Violence and the History of Inequality from the Stone Age to the Twenty-First Century*. Princeton and Oxford: Princeton University Press.
Schmandt-Besserat, Denise. 1996. *How Writing Came About*. Austin: University of Texas Press.
Scott, James. 2017. *Against the Grain: A Deep History of the Earliest States*. New Haven, CT: Yale University Press.
Sen, Amartya K. 1977. Rational Fools: A Critique of the Behavioral Foundations of Economic Theory. *Philosophy and Public Affairs* 6 (4): 317–344.
Service, Elman. 1962. *Primitive Social Organization: An Evolutionary Perspective*. New York: Random House.
Shryock, Andrew, and Daniel Lord Smail, eds. 2011. *Deep History: The Architecture of Past and Present*. Berkeley: University of California Press.
Simon, Herbert A. 1997. *Administrative Behavior: A Study of Decision-Making Processes in Administrative Organizations*. 4th ed. New York: The Free Press.
Simpson, George Gaylord. 1945. *The Principles of Classification and a Classification of Mammals*. New York: American Museum of Natural History.
Skibo, James M., and Michael B. Schiffer. 2008. *People and Things: A Behavioral Approach to Material Culture*. New York: Springer.
Sokal, Alan D. 1996. Transgressing the Boundaries: Toward a Transformative Hermeneutics of Quantum Gravity. *Social Text* 14 (1–2): 217–252.
Spier, Fred. 2015. *Big History and the Future of Humanity*. Malden, MA: Blackwell.
Stanish, Charles. 2017. *The Evolution of Human Co-operation: Ritual and Social Complexity in Stateless Societies*. New York: Cambridge University Press.
Stearns, Peter N. 2011. *World History: The Basics*. London and New York: Routledge.
Sterelny, Kim. 2012. *The Evolved Apprentice: How Evolution Made Humans Unique*. Cambridge, MA: The MIT Press.
Stigliz, Joseph. 2012. *The Price of Inequality: How Today's Divided Society Endangers Our Future*. New York: Norton.
Strayer, Robert W. 2014. *Ways of the World: A Brief Global History with Sources*. 2nd ed. Boston: Bedford/St. Martin's.
Su, Bernie, and Margaret Dunlap. 2012–2014. *The Lizzie Bennet Diaries*. Agreeable Entertainment.
Symons, Donald. 1979. *The Evolution of Human Sexuality*. New York: Oxford University Press.
Szostak, Rich. 2021. *Making Sense of World History*. New York: Routledge.
Sztompka, Piotr. 1994. *The Sociology of Social Change*. Cambridge, MA: Blackwell.
Tang, Shiping. 2020. *On Social Evolution: Phenomenon and Paradigm*. New York: Routledge.
Thaler, Richard H. 2018. From Cashews to Nudges: The Evolution of Behavioral Economics. *American Economic Review* 108 (6): 1265–1287.
Therborn, Göran. 2006. *Between Sex and Power: Family in the World, 1900–2000*. London and New York: Routledge.
Thomas, George M., John W. Meyer, Francisco O. Ramirez, and John Boli. 1987. *Institutional Structure: Constituting State, Society, and the Individual*. Newbury Park, CA: Sage.
Tomasello, Michael. 2019. *Becoming Human: A Theory of Ontogeny*. Cambridge, MA: Harvard University Press.
Tuomela, Raimo. 2013. *Social Ontology: Collective Intentionality and Group Agents*. New York: Oxford University Press.
Turchin, Peter. 2016. *Ultrasociety: How 10,000 Years of War Made Humans the Greatest Cooperators on Earth*. Chaplin, CT: Beresta Books.
Turner, Jonathan. 2007. *Human Emotions: A Sociological Theory*. London: Routledge.
Turner, Jonathan H., and Richard Machalek. 2018. *The New Evolutionary Sociology: Recent and Revitalized Theoretical and Methodological Approaches*. New York and London: Routledge.
Urbina, Dante A., and Alberto Ruiz-Villaverde. 2019. A Critical Review of *Homo Economicus* from Five Approaches. *American Journal of Economics and Sociology* 78 (1): 63–93.

Velioti-Georgopolous, Maria. 2006. Kinship and Descent. In *Encyclopedia of Anthropology*, ed. H. J. Birx, 1369–1371. Thousand Oaks, CA: Sage.
Vince, Gaia. 2019. *Transcendence: How Humans Evolved Through Fire, Language, Beauty, and Time*. New York: Basic Books.
Von Petzinger, Genevieve. 2016. *The First Signs: Unlocking the Mysteries of the World's Oldest Symbols*. New York: Atria Books.
Voss, Thomas, and Martin Abraham. 2000. Rational Choice Theory in Sociology: A Survey. In *The International Handbook of Sociology*, ed. S. R. Quah and A. Sales, 50–83. London: Sage.
Wallerstein, Immanuel. 1974. The Rise and Future Demise of the Capitalist World-System: Concepts for Comparative Analysis. *Comparative Studies in Society and History* 16 (4): 387–415.
Watson, Adam. 1992. *The Evolution of International Society: A Comparative Historical Analysis*. London and New York: Routledge.
Weber, Max. 1946. In *From Max Weber: Essays in Sociology*, ed. H. H. Gerth and C. Wright Mills. New York: Oxford.
Wenke, Robert J., and Debora I. Olszewski. 2007. *Patterns in Prehistory: Humankind's First Three Million Years*. New York: Oxford University Press.
Willey, Angela. 2016. Biopossibility: A Queer Feminist Materialist Science Studies Manifesto, with Special Reference to the Question of Monogamous Behavior. *Signs: Journal of Women in Culture and Society* 41 (3): 553–577.
Williamson, Oliver E. 1985. *The Economic Institutions of Capitalism*. New York: Free Press.
Wilson, Edward O. 1975. *Sociobiology: The New Synthesis*. Belknap: Cambridge, MA.
Wilson, Edward O. 2012. *The Social Conquest of Earth*. New York and London: W.W. Norton.
Wittek, Rafael. 2013. Rational Choice Theory. In *Theory in Social and Cultural Anthropology: An Encyclopedia*, ed. R. J. McGee and R. L. Warms, 688–690. Thousand Oaks, CA: Sage.
Wolf, Eric R. 1982. *Europe and the People Without History*. Berkley, CA: University of California Press.
Wood, William R., and John Williamson. 2007. Comparative Historical Sociology. In *21st Century Sociology: A Reference Handbook*, ed. C. D. Bryant and D. L. Peck, 118–128. Thousand Oaks, CA: Sage.
Woolf, Daniel. 2019. *A Concise History of History: Global Historiography from Antiquity to the Present*. New York: Cambridge University Press.
World History Association. 2016. "World History Programs." Retrieved August 11, 2016. http://www.thewha.org/resource-links/resources-for-graduate-students/graduate-programs/
Yoffee, Norman. 2005. *Myths of the Archaic State: Evolution of the Earliest Cities, States and Civilizations*. New York: Cambridge University Press.

Chapter 2
Our Common Humanity: Hominin Evolution, the Reasonable Actor, and Groups

Abstract Humans evolved over several million years. While many hominin species appeared in this period, only *Homo sapiens sapiens* survives today. Scholars studying biological evolution debate why and how hominin evolution led to this result, discussing language, cooperative breeding, multilevel intentionality, cultural imitation and cumulation, groups-within-groups, and human ontogenesis. A conceptual model of individual humans analyzes physical bodies, emotional capacities and motivations, and mental capabilities, many of which make us group-capable. Relationships between individuals, or ongoing interactions, give rise to social networks. Bounded networks, or groups, always operate through institutions, or cultural patterns for divisions of labor. Groups face problems such as defection, free-riding, groupthink, interpretive heterogeneity, and inequality. Moral codes, rooted in value-based norms, help reduce these problems, as do leadership and commitment devices, but conflict still occurs. Intergroup relations often support alliances, but these can also break out into destructive conflicts. Groups change both normally, through turnover, and unusually, through internal conflict and external problems. To apply this model to world history, we must use secondary historical sources carefully.

Keywords Hominin evolution · Hypersociality · Cultural imitation · Cultural learning · Collective action problems · Institutions · Conflict

Introduction

Here, I explicitly develop my assumptions about human beings. To improve upon the hypersocial paradigm, I add social science ideas, somewhat like Henrich (2015). Most models of "human nature" are methodologically individualist, but if humans are groupish, then we must also understand how groups operate (or not) to address what is human. I develop two conceptual models, one of the "reasonable" human and the other of groups of these humans. Scholars occasionally (Sibley 1953; Toulmin 2001; Grossman et al. 2020) use the "reasonable actor" from North American legal thinking to question the rational actor model. While rational-choice

T. C. Burr, *Societies Past and Present*, Emerging Globalities and Civilizational Perspectives, https://doi.org/10.1007/978-3-031-91532-1_2

scholars obviously assume people have reasons for their behavior, I think reasonableness depicts humans in a more psychologically realistic way compared to rationality (Grossman et al. 2020). That is, I suggest that there are many more possible reasons for human behavior than only calculations based on individual selfishness. While Moran (2003) focuses on reasonableness itself, rejecting the modeling of reasonable persons, I disagree. In discussing individual reasonable humans, I will repeatedly use the tripartite analytical framework of mental, emotional, and physical aspects of individuals common in psychology (Yzerbyt and Demoulin 2010).

First, I trace hominin evolution and review theories of that evolutionary path. Next, I develop a conceptual model of reasonable human individuals. Third, I discuss aspects of individuals that enable them to live in groups. Fourth, I develop a simple conceptual model of groups from these individual traits, drawing on many social science disciplines. Fifth, I briefly address methodological issues. Finally, I describe how each chapter is organized.

Hominin Evolution: How and Why

As discussed, many social scientists object when biologists theorize human evolution, but I have not seen social scientists reject evolution *in toto*; they object to its purported results. I start with the evolution of "hominins," a category including all species of *Australopithecus* and *Homo* (Cartmill and Smith 2022:83), and review how they became us. Paleoanthropologists use three lines of evidence to do this: fossilized skeletons, DNA evidence from them, and archaeological evidence around them. Scholars increasingly add a fourth type, comparison to great apes (Schick and Toth 2015; Shelby Putt, personal communication). The following brief sketch provides a foundation for discussion of why this evolution happened, and its result: *Homo sapiens sapiens*.

Mammalian evolution led to hominin evolution. Mammals appeared over 200 million years ago (mya). Unlike dinosaurs, mammals were warm-blooded, furry, and long-snouted, giving birth to live young, whom they suckled (Mears 2015). After most dinosaurs disappeared around 65 mya, mammal species proliferated. Primates evolved around 30 mya, and monkeys 25 mya. Brain growth led to longer gestational periods. Great apes evolved next—orangutans 12 mya and gorillas 8 mya. Paleoanthropologists estimate that the "Last Common Ancestor" (LCA) of chimpanzees, bonobos, and hominins existed between 4 and 7 mya (Gamble et al. 2014; Schick and Toth 2015; Cartmill and Smith 2022). Their life stages consisted of infancy, a juvenile period, and adulthood (Nowell 2021).

In the next few million years, twenty different species of hominins evolved. There is no consensus on "the number, distinctiveness, classification, or behavior" of these species, given the considerable "in-between" skeletal evidence (Cartmill and Smith 2022:139). Thus, the following account is tentative. Partly bipedal species such as *Sahelanthropus tchadensis* or *Orrorin tugenensis* evolved first, then *Ardipithecus ramidus*, which was fully bipedal (Gamble et al. 2014; Schick and

Toth 2015; Cartmill and Smith 2022). Bipedal species, including various *Australopithecus* species and *Paranthropus*, evolved with reduced incisors, larger brains, and sophisticated tool use (Schick and Toth 2015). Various *Australopithecus* species seem to have been the ancestors of all *Homo* species, who appeared before 2 mya and clearly coexisted in time and place with *Australopithecus* (Cartmill and Smith 2022:188). *Homo* seems to have had a fourth life stage, childhood, between infancy and the juvenile period (Nowell 2021).

Paleoanthropologists have found skeletons that they group into various *Homo* species. While they have categorized them variously as *ergaster, rudolfensis, habilis*, and *erectus*, there is no consensus, so some—not all—group them into Habilines, Ergasters, and Erectines (Cartmill and Smith 2022). However, all of their brains were larger than those of Australopithecines. *Homo erectus* existed for over one million years, in that time developing larger brains, smaller jaws and teeth, better tools, longer walking distances, and longer childhoods (Nowell 2021). They started by using "Oldowan"-style tools for hunting and cutting meat, and the "Acheulean" industry (tool kit or set of tools) appeared around 1.76 mya. Their mild sexual dimorphism (a difference in size between average females and average males) is associated with monogamy.

Researchers originally assumed *erectus* evolved in Africa, but some argue that Erectines evolved in western Asia and moved into East Asia and Africa (Bellwood 2013; Groves 2015; Cartmill and Smith 2022). Some think that "human populations across Eurasia, from … around 770 Kya [thousands of years ago] down to the present, have always been connected by continent-spanning patterns of migration and gene flow" (Cartmill and Smith 2022:383). Only some argue that *Homo heidelbergensis* then evolved; because of the "in-between" skeletons, others disagree—though all agree they were no longer Erectines (Klein 2013:112–113; Cartmill and Smith 2022). Heidelberg brain sizes ranged from 1200 to 1500 cubic centimeters; they controlled fire, used wooden spears, hunted collectively, and ranged across Eurasia (Gamble et al. 2014; Groves 2015).

From these Heidelberg or in-between species, Neandertals (some scholars now omit the "h" from "Neanderthal") evolved in western Eurasia; Denisovans developed in eastern Asia; and our ancestors evolved in Africa. Since two identical populations need about 280,000 generations to become separate species (Cartmill and Smith 2022:468), and these three populations were far more closely related than that, some think all three were subspecies of *Homo sapiens*—that is, *Homo sapiens neandertalensis*, *Homo sapiens denisova*, and *Homo sapiens sapiens*. Increasingly, scholars think that *neandertalensis* and *sapiens* interbred along much of the Mediterranean for many millennia (Cartmill and Smith 2022; Li et al. 2024). Researchers know little about Denisovans but think Neandertals had full language, hunted animals collectively and gathered plants, played music, and buried their dead (Cartmill and Smith 2022). Most scholars recognize Sapiens as a distinct group before 200 kya. Their groups, geographic ranges, and fertility rates were all greater than those of Neandertals. Both subspecies had longer childhoods, with a new, fifth stage of development, the adolescent growth spurt. Some think that Neandertal child development was shorter than that of Sapiens (Nowell 2021:15,

32); others argue that Sapiens' development was more social and thus cognitively different (Henrich 2015); and others see no cognitive difference between the two (Cartmill and Smith 2022).

Researchers also argue that hominins increasingly relied on cultural behaviors and lived in ever-larger groups; they debate why, usually comparing us to great apes. All great ape females marry out, as do male gorillas and orangutans, but male chimpanzees and bonobos do not. Both sexes of all four species are promiscuous, so paternity is unknown. Mothers help children and male siblings help each other, so the "only permanent social structure among great apes … is the larger, more inclusive community" (Turner and Machalek 2018:297). Female bonobos form alliances and are thus female-dominant, unlike male-dominant chimpanzees. Bonobo and human females are sexually active throughout their cycles and do not display ovulation (Hrdy 2009). Sexual dimorphism varies: male gorillas are twice as big as females, while female bonobos are 80 percent as large as males. All intelligent mammals, including "dolphins, whales, elephants, [great apes,] and humans … are … highly emotional" (Turner and Machalek 2018:307).

Explanations of physical evolution among hominins vary. Some think a shift from olfactory to visual dominance made life in trees easier for all apes (Turner and Machalek 2018). Most agree that climate-driven recession of forests forced apes from jungles to open fields, leading to bipedalism. This led to greater predation by megafauna, encouraging tighter groups (Gamble et al. 2014; Cartmill and Smith 2022). All agree that bipedalism freed hands for manipulating tools. Sophisticated tools such as wooden spears and flaking technologies allowed greater use of all resources, expanding the range of *Homo* environments. A few argue that the most important change was an improved shoulder joint that allowed *Homo erectus* to throw, improving hunting and defense (Turchin 2016). Many think a recursive, feedback-driven process between hominin thinking and learning, physical environments, and genetic evolution started around 2 mya (Richerson and Boyd 2005; Sterelny 2012; Henrich 2015; Cartmill and Smith 2022), a process called "gene-culture co-evolution." Control of fire occurred by 780 kya and probably earlier (Cartmill and Smith 2022), allowing cooked food. This reduced jaw and abdomen size, making room for larger brains. In turn, *Homo* could be fully upright, accelerating gene-culture coevolution. Some think that all group adults, not just parents, treated children as apprentices by starting with easy tasks and providing more difficult ones later (Sterelny 2012), including quarrying rocks, making tools, hunting, gathering, fire-making, infant- and childcare, and trade. Others emphasize how cumulative, collective, intergenerational learning accelerated gene-culture coevolution (Richerson and Boyd 2005; Henrich 2015).

Both apprentice learning and cultural accumulation depended on several changes. Emotions "amplified" through hominin species, so Sapiens developed "supercharged" emotions compared to all apes (Gamble et al. 2014:152; Turner and Machalek 2018:307). This allows relationship flexibility, both when living in close groups and when handling strangers. Increased emotionality allowed increased cognitive complexity (Turner and Machalek 2018); research shows that damage to emotional capacities compromises decision-making, while "a continued ability to express feelings" aids reasoning (Damasio 1994:xii).

Cumulative learning seems to have driven two cognitive changes. One was the development of language, which some argue evolved from emotion-expressing gestures (Turner and Machalek 2018). The other change was "mentalizing," i.e., understanding others' internal states, especially their intentions. This means each human has a "theory of mind"—knowing that others have similar mental and emotional capacities but different ideas and feelings (Gamble et al., 2014; Blanton 2016). Apes also seem to have theory of mind but the human version is more complex. For instance, one person can know that a second person has an opinion about a third person's beliefs; this is "third-order" intentionality, and most adults can understand fifth-order intentionality. This allows coordination of intentions, the foundation for collective action. Both changes allowed increased sociality, and some argue that brain size and community size coevolved: Larger communities required bigger brains, leading to larger communities. This also made possible forming larger groups from smaller ones (Gamble et al. 2014).

Alloparenting, or nonparental childcare, also furthered gene-culture coevolution. Great apes do not alloparent, though some primates (tamarins and marmosets) do (Hrdy 2009). Great ape mothers keep infants to themselves for months, while human mothers share and retrieve their babies from birth (Hrdy 2009). Also, there is more variation in alloparenting among *sapiens* fathers "than in all other primates combined" (Hrdy 2009:162). Researchers suggest different paths to alloparenting. In a sex-contract vein, some say hominin males started to protect and provision certain females more than others, eventually developing pair-bonds and getting to know their children well. When males fostered ties between their children and their mothers and brothers, they effectively created families (Henrich 2015). Others argue for the "grandmother hypothesis." Child mortality for new mothers in all primate species is much higher than for experienced mothers (Hrdy 2009). Perhaps extended postmenopausal lifespan allowed females to teach their daughters (and sometimes sons) childcare, reducing child mortality. Scholars argue that this change came before pair-bonding (Hrdy 2009; Nowell 2021). Both sex-contract and grandmothering researchers agree that alloparenting appeared fully among Erectines.

Another debated explanation for gene-culture coevolution is intergroup competition. Many espouse it (Richerson and Boyd 2005; Wilson 2012; Bowles and Gintis 2013; Gamble et al. 2014; Henrich 2015); others disagree (Hrdy 2009; Sterelny 2012; Haas and Piscitelli 2013; Turchin 2016). Henrich (2015) argues that there are five types of intergroup competition: violence, differential survival, migration, faster reproduction, and cultural transmission. Most focus on how intergroup violence drove hominin evolution; few address the other types.

These debates about what made "hominins" into us—Sapiens—over millions of years help us understand human capacities now. Bipedality and brain size; jaw, shoulder, and abdominal changes; learning, imitation, and cumulation of knowledge; language; group size, mentalizing, and groups-within-groups; alloparenting; and intergroup competition all help guide a conceptual model of individual, group-capable humans, and human groups.

Our Common Humanity: Individuals

In the next three sections, I develop this conceptualization. I offer something that I hope is reasonably balanced, though it must be imperfect, being so early in development.

We start with the most obvious: our bodies. Like all mammals, each human has a head, torso, and four limbs. Our eyes face forward, giving us depth perception. As bipeds, humans have two legs and two arms; our hands have fine motor skills and coordinate with our eyes. We compensate for weakness compared to most mammals with complex tools. We reproduce sexually, with female and male bodies. Humans are mildly sexually dimorphic; average adult females are about seven-eighths as tall as average adult males, who have stronger upper bodies. We eat omnivorously, cooking plants and animals to predigest them.

Motivations put this organic structure into action. They are "drives, needs, [and] urges … that shape [the body's] responses to stimuli" (Hewitt 2007:114; Bargh et al. 2010:269). Outside stimuli reveal these inner impulses; people also consciously set goals, desired states of affairs. However, each person "comprises many different, often conflicting, motives" (Bargh et al. 2010:292). Researchers divide motivations into distal (underlying) and proximate (immediate) ones; both can be conscious or unconscious. Biologists suggest that two major distal motivations are physical and genetic survival (reproduction). Each distal motive becomes a proximate one in a specific situation: survival manifests as hunger or thirst, reproduction as sexual desire, and so on. Other distal motivations include developing relationships with other humans and belonging in groups (Baumeister 2005) and curiosity—a "restlessness" for discovery and experimentation (Belsey 2006:123–125; Richerson and Boyd 2005). The capacity to work consciously toward goals seems to have developed late in hominin evolution (Bargh et al. 2010). Some research supports Maslow's proposed hierarchy of needs (1943), but relationships can support positive subjective well-being even when basic needs are not met (Tay and Diener 2011).

Emotions help us understand our motivations. They consist of four components: an outside stimulus, an internal feeling (physiological response), external (facial or bodily) signs, and a cognitive appraisal of how the world meets expectations (or not). Researchers disagree on whether appraisals create physical feelings or whether feelings lead to cognitive judgments (Keltner and Lerner 2010:328). *Emotion* describes intense, short-term feelings with specific objects; sentiments and moods are longer-term, less intense, and more diffuse feelings (Keltner and Lerner 2010; Ben-Ze'ev 2013). Scholars debate how universal emotions are. Ekman (2007) argued that people across cultures recognize similar facial expressions, while others find more variability. Ekman suggested that fear, anger, sadness, happiness, disgust, and surprise are recognized, while others suggest even fewer universal emotions: "aggression, helplessness, startle and wanting" (Prinz 2014:266). Many agree that people combine and elaborate basic emotions into complex ones, such as shame, guilt, or alienation (Turner 2007; Prinz 2014), and that "in practice, cultures construct quite different emotions" (Keltner and Lerner 2010:334; see Rosaldo [1989] for more on commonly shared human emotions and cultural differences). Cultural rules govern how people express emotions.

Psychologists argue that emotions have many purposes. They turn distal motivations into proximate ones for behavior (Baumeister 2005). We focus emotions more often on people than on things (Manstead 2007). We share emotions with others; with ongoing interaction, people develop strong emotions about others, creating relationships (Turner 2007; Keltner and Lerner 2010). Scholars talk about "ties" or "bonds" between people (e.g., Granovetter 1973). While we physically hold hands or hug, we are "connected" mostly emotionally and mentally. Losing the person and the relationship creates painful emotions; good relationships improve well-being.

Emotions support thinking. They influence cognitive processes such as attention, judgment, decision-making, and attitudes toward others (Keltner and Lerner 2010; Lerner et al. 2015). Most researchers agree that people understand the world through "concepts" or mental representations; when something is not immediately present, we *re*present it mentally through concepts (Medin et al. 2000). Concepts help us categorize people, actions, events, and objects in the world. Each person has many concepts, which are structured hierarchically (Henrich 2015). Larger categories (e.g., animal) contain smaller ones (mammal, then dog, then pit bull). Conceptual hierarchies intertwine: "dog" is linked to other concepts such as hunting, obedience, and play (Medin et al. 2000). Each person likely has hundreds of conceptual hierarchies, so their connections suggest a web-like conceptual structure in each person's mind (Geertz 1973). Memory stores concepts for use in two ways. We have three versions of long-term memory: episodic memory (past events); semantic memory (information stored in concepts); and procedural memory (physically embodied in behaviors) (Turkington and Harris 2001). Short-term "working" memory consists of presently conscious thoughts relying on long-term memory.

Concepts provide the "building blocks" of beliefs—propositional statements about the world that people hold true (Dretske 2005:85). We develop beliefs to explain our physical and social worlds to ourselves to make "the world a predictable and controllable place" (Yzerbyt and Demoulin 2010:1027). Concept-supported beliefs help humans think faster and better than animals do. Cognition, mentally working with information encoded in concepts, occurs in short-term memory. Activities include "decision-making, planning, sequencing, temporal tagging … the updating, maintenance, and integration of information," problem-solving, making judgments, and imagination (Coolidge and Wynn 2009:36–39). These require living mentally in the past or future (Bargh et al. 2010). Much human cognition is unconscious, yet "Unconscious processes are of paramount importance for human functioning" (Dijksterhuis 2010:299). *Pace* rational-choice assumptions, conscious decision-making often leads to suboptimal (regretted) decisions (Dijksterhuis 2010:254). Incubation or unconscious pondering supports creativity, and distal motivations are often unconscious in proximate situations, as is perception.

Human cognition is quite imperfect (Simon 1997; Baumeister 2005; Gilovich and Griffin 2010). Emotions and moods heavily influence cognition. People often use heuristics (mental shortcuts) to develop conclusions. Culture also influences cognition, sometimes leading to false conclusions. Thus, people can hold false beliefs and deceive others (Baumeister 2005); they actively seek evidence to confirm their beliefs, often ignoring disconfirming evidence (Alcock 2018). This occurs for three reasons: Belief retention helps the person avoid group rejection, it prevents existential crises (if beliefs are false, the world makes no sense, which creates fear), and it prevents self-perception of gullibility and consequent shame (Kahan et al. 2013; Alcock 2018).

Despite talk of "mentalizing," we are imprisoned in our own minds. We transcend this by using signs to represent concepts to others. The meaning of each sign is the concept(s) one is trying to indicate. Signs are natural (darkening skies signifying a storm) or social, involving facial and bodily gestures and sounds in animals and humans (Lents 2016). Humans also use cultural signs called "symbols" (Danesi 2008). All human groups have language, a system of symbols expressing shared concepts. Language and other symbol systems (e.g., gestures and clothing) allow us to understand others' internal experiences more than any other animal does.

Theory of mind requires consciousness of self as separate from others, so humans can also think reflexively (i.e., about themselves). The "self-concept" consists of all mental representations of oneself (Owens 2006; Swann and Bosson 2010). These include personal and social identities, categorizations based on characteristics shared with others. Culture influences self-concepts—in some societies, people see themselves as bounded, fixed selves; in others, they understand themselves as different in different contexts (Heine 2010). Self-concept influences self-esteem (emotional self-appraisal) (Baumeister 2005). People compare themselves to others, often as being "above" or "below" others on some characteristic (Locke 2014:18). This "social comparison" heavily influences the self-concept and thus self-esteem. People generally want to feel good about themselves, which motivates them to improve their behavior with others.

All these capacities are potential; socialization (enculturation) activates and develops them. Babies are born helpless, with only a nipple-rooting instinct. They become *people* as others interact with them as they grow; inborn raw capacities develop as children learn concepts, beliefs, sign systems, emotional styles, and behaviors. Enculturation starts within families and continues with peers and others. Socialization styles vary around the world, sometimes letting children observe and imitate and other times involving conscious parental teaching (West 2006; Birney and Sternberg 2011; Chen and Rubin 2011).

This conceptual model is only a start since individuals vary. For instance, while there are two physical versions of humans, female and male, reproductive biology research shows that these two categories greatly oversimplify reality: "While most people are in most … respects exclusively male or exclusively female, it is not true of everybody" (Dea 2023:2). Specifically, "intersex" people vary from the two basic versions. About 1.7 percent of humans fit the most inclusive definition of intersexuality. However, most conditions under this definition are not physically obvious; only "about .02 per cent of the population" has anatomically visible intersexual characteristics (Dea 2023:110). Thus, until the last few decades, all societies detected intersex babies only occasionally, likely dealing with them in an *ad hoc* way, and thus they probably defined only two sexes. Thus, I will use "male" and "female" or "men" and "women" throughout.

Still, the sexes are more similar than different. One sex contributes vastly more to reproduction; otherwise, differences are a matter of averages. Intelligence testing shows no difference between average intelligence in males and females; average differences in visual and verbal abilities are slight (Lips 2005). In same-sex small-group experiments, both sexes show equal tendencies toward dominance; in mixed-sex groups, males dominate only slightly; and dominance patterns vary culturally (Lips 2005). Paternal alloparenting varies so much across societies that the sex contract is not universal (Hrdy 2009). Also, both sexes strategize sexually and take initiative in mating, sometimes in different ways (Dunbar 2008; Hrdy 2009; Starkweather and Hames 2012).

Individuals vary in other ways, including balance, hand-eye coordination, strength, and other physical abilities (Smith et al. 2015) and in levels and types of intelligence (Flynn 2012). Other inborn variations include Down's syndrome, autism, *situs inversus*, and others (Buller 2005; Howlin et al. 2011). Some have strong, internally felt tendencies such as attraction to the same sex or transgenderism (Patterson and D'Augelli 2012; Dea 2023). In a single generation, such variations seem rare and thus "abnormal" (Buller 2005). I think, however, that as such variations always appear in every generation, they simply exhibit normal variability among humans.

Our Common Humanity: Groupish Individualists

This abstract conceptualization of "human" is only partial. Again, the blank-slate metaphor is accurate, but incomplete. Humans at birth are mostly "blank," but social scientists ignore the "slate's" characteristics. There are two more important aspects of individuals: Internal human capacities that, as children develop, make them capable of forming and living in groups, and the likelihood that human beings are the most individualistic animals on earth.

Through comparative experiments on chimpanzee, bonobo, and human young, Tomasello (2019) shows how thinking with concepts, understanding others' emotional and mental states, sharing intentionality, and imagining past and future all develop in human childhood to make children group-capable. Aware of the WEIRD problem, he also compares children cross-culturally. He repeatedly asserts the importance of socialization inferred from feral children research: Human children's "unique maturational capacities would be inert without a sociocultural context within which to develop" (2019:7, 8, 52, 132, 187, and 314). Human children focus much more on others than apes do and imitate more (p. 89), and to learn culture, human childhoods are much longer than those of any great ape (Pp. 26–27).

Within the overall ontogenetic (maturational) path, all human children develop social capabilities at broadly the same age. Young babies take turns exhibiting emotions with caretakers; by 10 months, they share intentions with others; they start helping others at 14 months; and they imagine what others are thinking at 2 years (Tomasello 2019:52–54). After age three, they protest defection, act out of sympathy and obligation, and understand social norms (Pp. 174, 208, 220, 225). In line with Asch's (1956) findings, three-year-olds feel pressure to conform; they also initiate third-party enforcement of norms. By age four, humans coordinate with others better than adult chimpanzees do (Pp. 144, 175, 256), and "from three to five years of age, young children become ever more group-minded" (Tomasello 2019:254). Tomasello thinks that children learn the dual structure of individuals-in-groups on their own, not from adults. By age five, children understand culturally specific "good" and "bad" reasons for behaviors and can spontaneously develop unified groups of three or more with strangers, including creating and enforcing group rules (p. 260). By age six, children understand others' beliefs, coordinate with others, give reasons for their thinking and behavior, and engage in collective decision-making (p. 161). Ultimately, in a group, children "understand themselves to be operating interdependently as a single agent" (Tomasello 2019:342).

Tomasello's career-long study of children (2019) provides much evidence that we are strongly "groupish." Ironically, hypersocial scholars also imply that humans are highly individualistic. We have a well-developed sense of self, the self-concept. We share "the cognitive awareness of one's existence" (Read 2012:146) with "great apes, dolphins, probably whales, and elephants" (Turner and Machalek 2018:422). Mirror experiments show that human children exhibit self-awareness at 18 months of age, 6 months earlier than chimpanzees do (Read 2012). To live in groups, children learn self-regulation (Tomasello 2019). This intensifies self-attention, allowing the self-control that helps one avoid shame, guilt, and embarrassment with others (Turner and Machalek 2018). Thus, there is abundant evidence that humans are both strongly group-oriented *and* quite individualistic, probably more so than any other animal. While different cultures develop different self-concepts (Heine 2010; Prinz 2014), talk of "individualist" and "collectivist" cultures ignores the reality that supposedly individualistic Westerners are clearly enmeshed in social obligations, networks, and conformity, while purportedly collectivist East Asian individuals clearly think for themselves.

This seeming paradox—that humans are simultaneously more groupish and more individualistic than all other apes—is key for understanding world history. It sets up the potential for conflict, exactly what some hypersocial scholars argue (Richerson and Boyd 2005; Wilson 2012). Some claim that public debates about collective and individual influences mirror the individualism inherited from the LCA and the "supercharged" interpersonal emotions supporting groupishness (Turner and Machalek 2018:323). Others assert that these opposing tendencies depend on circumstances; Tuomela (2013) argues that humans can think in both "I-mode" and "we-mode," while Stanish (2017:15) argues that people are "simultaneously ego directed and prosocial under the appropriate conditions." Certainly, individual and collective interests sometimes align (Blanton 2016). Thus, I suggest that each individual is in *tension*, not necessarily in conflict.

Our Common Humanity: Groups

This tension in each person between individual desire and group conformity helps us understand the following conceptualization of human groups. It applies mostly to the earliest *Homo sapiens sapiens* groups (before 100 kya) and only partly to "small groups" in modern societies. While ontologically only individuals exist (Tuomela 2013), a "group" is not a mere aggregate of individuals; instead, individuals coordinate their actions so completely that the group acts as a single unit (Bacharach 2002; Tuomela 2013; Tomasello 2019). This suggests some constructionist thinking: How do individuals form and maintain a group-as-unitary-actor, especially with individualistic tendencies? We must expect that this both succeeds and fails, which raises other questions: What breaks groups apart? How do they handle that? How do groups deal with other groups? Most importantly, how do groups change?

This section answers these five questions in five subsections, building on the sections above. The first analyzes group formation and unity. The second addresses conflict and disunity. The third considers unity-maintenance strategies. A fourth analyzes intergroup relations, and the last one examines group change. Early on, I define *culture* carefully. Blanton (2016) derisively calls mathematically oriented scholars of hominin evolution "biomathematicians," an understandable criticism. Though discussing gene-culture coevolution, their definition of culture is thin; many treat it only as information and norms (e.g., Turchin 2016), perhaps—as Blanton implies—to make it quantifiable. (Again, Henrich [2015] is an exception.) I think a "thick" analysis of culture would help us better understand societies; Geertz (1973) famously advocated thick description, but this must rest on a similarly comprehensive definition of culture, one I try to develop.

Cohesion: Group Unity

How do individuals construct a group? People divide survival and other tasks through patterns of taken-for-granted behaviors, or institutions.

The building block of groups is the dyad, a relationship ("tie" or "bond") between two individuals. Relationships develop through interaction, or "social processes involving face-to-face encounters" (Elliot 2006:304–305). One person acts toward another, who then responds to the first, each aware of the other. Thus, *inter*action rests on contingent actions and responses: looking at and talking to each other, perhaps touching, and learning about each other. The process embodies reciprocity,[1] or turn-taking (Elliot 2006). Some see interactions as a series of exchanges of attention, listening, effort, objects, and/or touching (Cook 2000).

[1] When I was struggling with social exchange theory, the social psychologist Diane Felmlee asked me: "Do you want to be in an unequal relationship?" The question answered itself, clarifying my understanding of reciprocity.

Relationships activate emotions. People tend to respond reciprocally to actions—positive responses to positive actions and negative to negative (Fehr and Gächter 2000), a tendency which manifests in early childhood (Tomasello 2019). These actions tend to stimulate equivalent emotions (positive or negative) in the other (Gamble et al. 2014). Some relationships are intense, with deep caring about the other's well-being, while others are "instrumental," concerned only with benefits to self; and yet, others are a mix of these two types (Leary 2010).

From dyadic relations, people develop groups. Each person has multiple dyadic ties; two people tying to a third or more create a social network (Baumeister 2005), so in all interactions, each person must keep the broader network in mind. For instance, person A must manage her tie to person B while being aware that she and B both know persons C, D, and E, who have their own networks—and so on beyond those. Ultimately, humanity is one massive network (McNeill and McNeill 2003). When a part of this network devotes its actions to a given purpose, divides tasks between individuals, and knows that others do not share their purpose, they form a group. Thus, I define a "group" as a social network with a mental boundary around it, devoted to a purpose, with interdependent members (Hackman and Katz 2010).

Individual actions and interactions within groups depend upon what we call culture. Crucially, culture is not simply information but *ways to store* information. Using concepts to categorize objects in the world, people learn how the world operates and develop beliefs (relations between concepts); multiple, interrelated beliefs constitute belief-systems, mostly assumed rather than explicit. Individuals use shared concepts, beliefs, and belief-systems to process and store information, to learn more, and to guide their behaviors in the physical world and toward each other.

As implied, beliefs generate emotions, especially when they are challenged. Values are deeply held, emotionally salient, and semiconscious principles derived from beliefs. They imply behaviors, but they are so broad that people apply them differently in different situations. (Values research is cross-culturally comparative, highly quantitative, and fragmented: compare, e.g., Hofstede et al. 2010 to Inglehart 2018). Morals are more specific conceptions of right and wrong than values are, and they generate norms, behavioral expectations in given situations (Read 2012), or informal, implicit scripts. Morality is mostly about how to treat other people (Haidt and Kesebir 2010), while values are broader (e.g., honesty, respect for nature). Some norms (informal scripts, procedures, traditions, and practices) are not moral, but instrumental or practical. With physical repetition, norms become neurologically embodied—they become skills.

People communicate in interactions by exchanging signs, especially symbols. One person's use of a set of signs activates information stored in concepts and beliefs in the other person. Said differently, the signs person A uses *mean* something to persons B (or C, D, and so on). Meaning goes beyond concepts and beliefs; signs usually communicate emotionally laden values and morals. People also develop meanings (emotion-laden concepts) for physical objects, and behaviors also send messages; following a group norm indicates agreement with group beliefs and morals. With ongoing interaction, groups develop sign systems, from which they draw and use specific signs in given situations. Language is only the most obvious sign system; physical objects (clothing, tools, decorations) and gestures also have their own sign systems.

Thus, (1) shared human sign systems communicate (2) shared conceptual and belief-systems and information, (3) shared values and morals, and (4) shared norms (behavioral scripts). These four elements taken together constitute "culture." While this definition is artificial, since we live within a mix of nearly automatic ideas, assumptions, symbols, meanings, connotations, knowledge, and procedures, it can still help people better understand culture. Furthermore, elements of culture are often self-contradictory: Explicitly stated beliefs and deeper, semiconscious assumptions may be at odds; assumed values may not match putative moral behaviors; and individuals may share most beliefs, values, and behaviors with each other, but not all. Thus, "culture" is never perfectly uniform within any group; the larger the group, the more it varies across that group.

Enhanced theory of mind combined with culture makes possible deeper, more complex, and subtler interactions between humans than between animals, which also makes the creation of groups possible. A common culture helps people share thoughts, experience similar emotions, and coordinate actions, and it enables our uniquely powerful manipulation of the physical world. Over time, culture changes: Trial and error reveal good-enough ways to do things, which spread through imitation. The childhood tendency to imitate those around them becomes the adult tendency to imitate the most prestigious (i.e., the most competent in some way) or to copy the most frequent practice around themselves (Henrich 2015). Over time, knowledge accumulates as group members share their physical and social experiences through common concepts and beliefs (Henrich 2015).

Groups must carry out tasks to survive physically. People learned long ago that repeated practice of a task increases skill in that task. If individuals practice different tasks, their specialization increases skill even more, though specific skills are limited. Scholars call task specialization and its learning "the division of labor" (Smith [1776] 1937). Reciprocity—trading the results of labor—helps both individuals, and the division of labor also creates a division of information (Henrich 2015) since each individual has specialized, often embodied, knowledge. Furthermore, two people learning two distinct but interdependent tasks produce much more than twice as much as two people doing both tasks. This multiplying effect is known as "economy of scale" (Stanish 2017); larger groups are many times more productive than smaller ones. Finally, groups, not merely individuals, can divide labor between them.

Institutions formalize, somewhat, the division of labor. There are many definitions of institutions (Immergut 2011). Since some define them as "blueprints" for collective behaviors (Portes 2010), I define them as "cultural patterns for a division of labor." They consist of roles and relations between roles (Portes 2010; Blanton 2016). Roles are culturally specific definitions of positions, or task-sets. Relations are culturally specific reciprocal obligations between roles. For instance, bodies, not institutions, define reproductive roles (gestation and birth); beyond this, culture defines behavior (parenting or alloparenting). In some hunter-gatherer groups, men only hunt and women only gather, while other groups allow great flexibility; and in all, roles rest on specific concepts of "women" and "men." For instance, some groups culturally accept individual expression of internally felt tendencies such as same-sex attraction or transgenderism and allow (or occasionally mandate) associated practices; others suppress them (Neill 2009; Dea 2023).

Institutions are also practical. "Labor" is not only economic; it consists of action taken to achieve any goal. Each institution satisfies a need of the people in the group (Tuomela 2013). Economic institutions enable physical survival; kinship institutions help people make and raise new people, define relationships, and organize groups; and so on. They embody the cumulative learning of groups to avoid destructive situations (Poteete et al. 2010; Tuomela 2013; Ross 2019). They also ensure that economies of scale are built into group functioning (Stanish 2017) and often embody a "costless threat" to exclude people from those returns to scale if they do not conform (Willer et al. 2017). Finally, in embodying cumulative learning, institutions simplify cognition "by making activities routine and computationally easier" (Tuomela 2013:215).

Shared institutions help groups live in physical environments. Multiple forms of reciprocity between people lead to interdependence, which "integrates" the group (Polanyi 1957). Scholars call the resulting condition cohesion, solidarity, loyalty, or unity; these consist of mental agreement on principles and actions, positive emotions toward other group members, and behavioral coordination (Shotola 2000; Gamble et al. 2014) in which people can think as "we" or "I" depending on circumstances (Bacharach 2002; Tuomela 2013). Groups are mental constructions based on "people's capacity to imagine and pretend things, compose wholes of them, and entertain counterfactual contents in one's mind" (Tuomela 2013:49). Members think of the group as a single thing, an *entity* (Yzerbyt and Demoulin 2010). Entitativity (being an entity) is a variable: Groups are more or less cohesive, so they require ongoing maintenance. These efforts create trust between members, which improves cohesion and group success. The group's cultural concepts, including roles, shape individuals' self-concepts, though group membership is only one identity among many for any individual. Still, group identities within individuals make coordination between them easier; one knows one's role, others' roles, and the reciprocal relations between them, and one feels one's membership emotionally (Tuomela 2013).

Conflict: Group Disunity

Despite accepting group purposes and assuming group cultural patterns, people remain individuals. They vary randomly on characteristics such as intelligence, ambition, emotional perceptiveness, creativity, physical strength, and altruism. While many people help others, even at a cost (Fehr and Gächter 2000; Hrdy 2009; McAuliffe et al. 2015), experiments in more individualistic WEIRD societies show that 20–30 percent of people are nonreciprocating, i.e., selfish; the percentage might be lower elsewhere. Still, ethnographers argue that there are selfish people in every group, who reciprocate only under pressure (e.g., Henrich 2015:309).

These variations confirm the tension between self-interested and group orientations discussed above. Rational-choice scholars talk about individual interests as an obstacle to forming groups (Olson [1965] 1971), but what are "interests"? Before 1800, the word originally meant "the totality of human aspirations." After, its meaning narrowed to "economic advantage" (Hirschman 1977:32). Some define it by distinguishing between needs and wants, with an emphasis on needs (Dowding 2011), while others loosely equate it with "fundamental goals and values" (Swedberg 2005:104). The word also connoted rational recognition of ongoing desires rather than momentary "passions" (Hirschman 1977). Still, I define interests as mental (rational) conclusions about perceived paths to fulfilling desires (emotions) in given situations. As institutions shape situations and institutions are culturally constructed, interests are at least partly constructed, so they could be misperceived (Dowding 2011). If a situation changes, interests may change; underlying desires probably do not. It is usually in one's interest to get along with others, contribute to group survival, receive group benefits, and enjoy relationships, yet people have multiple group memberships, out-group friendships, lust, dislikes, and desires for curiosity and glory (self-esteem needs). Thus, multiple individual desires (motivations) (Bargh et al. 2010) will sometimes lead to individual interests that are different from group interests. Sometimes, momentary passions do overwhelm interests, so people momentarily disobey moral strictures; other times, people rationally and calculatedly betray their groups to fulfill their desires.

Two of these actions are free-riding and defection. Free-riding (social loafing) involves not doing one's expected work but benefiting from the group's work. Defection (betrayal) involves making promises and not keeping them—either lying, or promising honestly, then changing one's mind. Both rest on deception—not only hiding information, as other primates do, but actively representing things that do not exist. Free-riding is culturally variable; in some societies, people punish free-riding *and* its opposite, "social striving" (working more than others for the group's benefit) (Heine 2010), as it over-obligates others to the strivers.

There are three other potential problems. Individuals can perceive situations differently from each other, partly from personality differences and partly from differing kinds of ties to other groups (multiple memberships, out-friendships, etc.). This "interpretative heterogeneity" may increase group creativity, but it can reduce cooperation (Bardhan and Dayton-Johnson 2002). Larger groups have more interpretive heterogeneity, so they unify less easily than smaller groups do (Ostrom et al. 2002; Bardhan and Dayton-Johnson 2002). The opposite problem is "groupthink" or hyperconformity. Asch's classic studies of conformity (1956) support arguments that humans imitate more than they innovate (Richerson and Boyd 2005; Henrich 2015). However, conformity is situational. In insulated, high-stress situations, individuals may think nearly identically, ignoring different interpretations (Hollingshead et al. 2005). Responding to a quickly changing situation becomes difficult, narrowing the conceptual range of solutions, thus preventing the creativity that new situations often require. If groupthink reaffirms old solutions to new situations, disaster may ensue.

The final problem is inequality. Social hierarchies are endemic among primates. Those with "higher" status have more "social respect, recognition, importance, and prestige" (Fiske 2010:941) and experience increased survival, safety, and improved reproductive opportunities. They also enjoy enhanced self-esteem: "a person who occupies one position is [supposedly] 'better than' a person who occupies another" (Lovaglia et al. 2005:146). Among humans, there may be two types of hierarchy, dominance and prestige (Henrich 2015). Dominance involves intimidation and coercion through aggressive attention-seeking; people only imitate them through fear. People achieve prestige through high competence in some area of life; when they add self-deprecating and generous actions, people imitate them more, deferring to them to learn more (Henrich 2015). Prestige has the same material and emotional benefits as dominance because creating a positive social environment eventually has indirect benefits for the prestigious. The dominant, however, tend to create negative social environments.

Dominance often involves maintaining group cohesion for the dominator's benefit; perhaps it combines defection and free-riding. Some anthropologists suggest that another normal individual variation is dominance-seeking; individuals who combine this trait with intellectual talent and a willingness to work are "upstarts" or "aggrandizers" (Boehm 1999; Hayden 2014). Aggrandizers seem to represent the extreme of these characteristics; the other extreme would be relaxed, unambitious egalitarians. Most people would fall somewhere in between. So far, only Henrich (2015) implies that competence-based, prosocial prestige is part of aggrandizing. Perhaps combining competence, generosity, and (hidden) ambition might create less opposition.

Defection, free-riding, interpretive heterogeneity, groupthink, and domination may create conflict, at bottom a disagreement about the combination of interdependence and deprivation (De Dreu 2010). Disagreement only matters if people depend on each other; conflict emerges if some think others are depriving them of an expected outcome. Emotions motivate actions during conflict; extreme emotions intensify it. For instance, being dominated is "materially unpleasant [and] also emotionally painful" (Križan and Smith 2014:63), creating resentment (anger, indignation, and contempt combined) if unfair; depression if considered fair; and envy either way (Križan and Smith 2014). Conflicts can split groups into factions, or two or more opposing social networks. Some aggrandizers use deception to split groups and set factions against each other to gain power. This always involves mental "information" (deception) to stoke negative emotions, especially fear, about the other faction.

Preventing, Reducing, and Ending Conflict

Conflict threatens the interdependence of group members, and therefore everyone's survival, so groups try to prevent, reduce, or end it through three strategies: moral systems, leadership, and commitment devices. All require managing one's own and others' emotions (De Dreu 2010). Once conflict breaks out, factions have four choices: fight, negotiate or problem solve, withdraw, or submit (De Dreu 2010).

Morality rests in part on the human tendency to third-party punishment of inequity that develops in three-year-olds (Tomasello 2019:265). It consists of "inside-the-head psychological mechanisms and outside-the-head cultural products" that "suppress or regulate selfishness [to] make cooperative social life possible" (Haidt and Kesebir 2010:818, 800). Group members share values, or semiconscious assumptions, about what is desirable or proper across situations (Karp 2000); they usually rank them from more to less important. Morality turns values into judgments and then behaviors. Some aspects of moral cognition are conscious (moral reasoning), while others are unconscious (moral intuition); the latter produce more moral judgments (Haidt and Kesebir 2010). Punishments include gossip, i.e., monitoring for and talking about selfish behaviors, which can damage the subjects' reputations so that others refuse to interact with them (Henrich 2015). Other punishments include verbal abuse, physical violence, and material deprivation. Alternatively, rewards encourage prosocial behavior. Cross-cultural comparative studies of conformity show that group members follow group norms more than general norms, an effect "larger for moral norms than for non-moral norms" (Haidt and Kesebir 2010:819). When individuals' moral judgments are similar, they activate group identity, increasing cohesion.

Leadership also reduces or ends conflict. This involves someone mobilizing others to act toward a collective goal. It is always social since "wherever there are leaders, there must be followers" (Hogg 2010:1189). Decisions reached by consensus are usually successful but time-consuming; a leader deciding for the group is more efficient but riskier. There are no inherent leaders: different situations require different personal characteristics. Competence-based prestige "forms a foundation for leadership" (Henrich 2015:118), but people can achieve it through dominance. Leadership creates inequality; concepts can legitimate it, thus creating "authority" (Shotola 2000) to mitigate this problem, but ending conflict by fiat can create resentment.

Third, groups create cohesion by requiring commitment actions. Practices such as tattoos or painful or terrifying (and then relieving) ceremonies can demonstrate loyalty to the group and raise the cost of defection (Sterelny 2012; Henrich 2015). These are another way to ensure that group identity becomes part of an individual's self-concept (Abrams et al. 2005). Also, collective rituals based on common, often supernatural, beliefs combined with music and dancing "create strong personal ties, deep trust, and a profound sense of group solidarity" (Henrich 2015:160).

Once conflict breaks out, groups can fight, sometimes using violence. This is costly because losers lose more than winners win (Baumeister 2005), creating lasting resentment. Revenge (negative reciprocity) may force the conflict to spiral (De Dreu 2010). When a fight stalemates and a path to resolution exists, negotiation is possible. While this saves resources, health, and lives, conflict can spark creativity, break groupthink, and recombine groups (De Dreu 2010). If loss is likely, one side may withdraw or submit, accepting the emotional consequences.

Intergroup Relations

The groups-within-groups capacity (Gamble et al. 2014) means that groups always relate to other groups. Individuals have at least two reasons for intergroup interaction. Any small group may lack enough marriage partners or physical resources. Intergroup marriages create intergroup kin relations, children and reciprocal alloparenting cement them, and repetition produces a larger kin group (Read 2012). Repeated trade can create trust and material interdependence.

Humans tend to judge other groups by their standards, so outsiders' practices may spark anger or disgust. Ethnocentrism is the tendency to evaluate others through one's own cultural assumptions (Hebl and Madera 2007), and "In-group favoritism accompanied by out-group prejudice is one of the most well-documented phenomena in all of contemporary social psychology" (Tomasello 2019:252). At first, groups have little accurate information about each other, so they may develop stereotypes that produce negative judgments, or prejudices (De Dreu 2010). Circumstances change the power of ethnocentrism. For instance, two different groups competing for land will feel more negatively about each other. Since intergroup conflict increases group cohesion, "groups tend to be far more noncooperative than individuals," so intergroup conflict is more likely than interpersonal conflict (De Dreu 2010:1001), and negative moral judgments tend to escalate conflict. This can lead to "vicarious retribution": If someone injures a member of another group, that group may retaliate against the entire first group, not against individuals. This often starts a spiral of retributions—a feud (De Dreu 2010). While stalemate, dominance, and withdrawal may solve feuds, superordinate goals, such as a common threat, may de-escalate them, perhaps allowing both to form one group (Sherif 1958).

Group Change

Conflict is, of course, a kind of change within or between groups. Two basic forms of change exist. Typical change involves demographic turnover: births, deaths, and migration in or out. The planet also imposes its daily and annual cycles on the physical environment of groups in semipredictable ways. "Innovative drift," or individual incremental changes accumulating over time, also occurs (Richerson and Boyd 2005). Atypical change is unpredictable but still takes recognizable forms. If "aggrandizers" exist, they are born occasionally and randomly. As they mature, they may disrupt the group, and other conflicts occur randomly (Arrow et al. 2005; Kelly 2013). Groups can merge, assimilate to each other, or separate into factions. Feuds may lead to the group leaving an area, and annual supplies of plants and animals are never fully predictable.

I think this abstract sketch applies to all human groups. All have belief-systems storing information, language, institutions, and morality. All have individuals who defect or free-ride; they are all subject to interpretive heterogeneity, groupthink, or inequalities, and thus experience conflict. All groups have multiple, interconnected subgroups, all deal with other groups, and all experience normal turnover and periods of unusual change.

I constructed these visions of individual humans and the groups that they form to make my assumptions explicit because I believe that all humans do have some things in common and because I believe that understanding those commonalities helps us understand how societies are structured, how they operate (or not), and how they change. In the following chapters, I will suggest additional human capacities on an *ad hoc* basis and discuss them more fully in the conclusion. However, how does one turn the abstractions above into that world history?

Evidence and Methods

Historians regard discussions of method as "unnecessarily displaying ductwork" (Gaddis 2001:301). Comparative-historical social scientists also avoid it (Lange 2013). In contrast, I think I owe readers an explanation of how I researched, organized, and wrote this book, especially given its multidisciplinary approach. This includes epistemology, or underlying research approaches; how "institutions" become five of the "aspects" of society; families of theories of social change; what the available evidence is from the past, and how to infer what that past was from that evidence; and dating conventions.

The two broad stances to researching human societies, the nomothetic (generalizing) approach or the idiographic (particularizing) attitude, seem mutually exclusive. Both have problems—excessive parsimony or ignorance of obvious similarities. However, comparative history tends to follow a "both-and" epistemological stance, acknowledging similarities and differences among humans (Lange 2013:7, 18). The method "transcends" the nomothetic-idiographic division, constituting "its own approach" (Arthur 2011:176; Bryant 1994). Thus, I organize the book overall through the social evolution framework, while I organize each chapter partly through the aspects-of-societies approach. This allows comparison by place, sustaining the particularistic argument of cultural diversity, and by time, supporting the continuities argument.

This also applies to theories of social change. While explanations of social transformations differ (e.g., specific theories of the Agricultural and Urban Revolutions are dissimilar), many scholars draw on one of two basic paradigms, structural-functionalism or conflict theory, to develop specific theories. Total rejection of functionalist theories is silly: If societies completely cease to function, everyone dies. If human groups exist, they are somewhat functional. However, conflict scholars ask a crucial question that structural-functionalists never seem to ask (but should): Functioning primarily for whom—or everyone, or for an elite? A society can function "enough" to keep most alive in misery, while the few flourish. Thus, many scholars explain how a few achieve permanent dominance. Most debates follow structural-functionalist or conflict lines. My "both-and" bias should be obvious: Solving a serious functional problem is an excellent way to gain power. Once installed permanently, elites can betray this deal and free-ride, though overdoing free-riding has consequences.

What evidence did I use to trace this story? While the past no longer exists, it left physical traces, the primary data of archaeologists and historians, which are equivalent to experimental observations, participant observation notes, or survey or interview answers. Archaeologists use fossils (remnants of bodies), artifacts made by humans (e.g., tools), and features embedded in landscapes (e.g., buildings) (Gamble 2015; Sutton 2019), while historians use one kind of artifact, written documents (Rahikainen and Fellman 2012). Inference, or drawing conclusions about what we cannot observe from what we can observe, is the bread and butter of researchers; how do these scholars infer what the past was like? Archaeologists examine specific attributes of fossils, artifacts, and features, such as raw materials, production techniques, and wear use. Archaeologists also use DNA evidence, reconstructed languages, and chemical and isotopic evidence, comparing these against each other and against archaeological evidence (Kaiser et al. 2012; Campbell 2013; Gamble 2015). Historians apply principles of "source criticism" (Rahikainen and Fellman 2012; Rampolla 2018; Storey 2021) to documents, including the examination of the "what, who, when, where, how, and why" of each document's production; analysis of the document's content; and comparison to other documents of the time. Since so little survives the passage of time, both disciplines always question the representativeness of each piece of evidence. Both analyze collections of their respective kinds of evidence in context to make inferences about the past.

Archaeologists use consciously what historians tend to assume: ethnographic inference. Based on the principle that fundamental human physical and mental processes, and the operations of physical environments, remain similar across time (Meltzer 2009; Gamble 2015; Woolf 2019), archaeologists use anthropological reports to make analogies from groups in the recent past to groups of the distant past. Using present groups as analogs of past groups is risky (Kelly 2013), so archaeologists use this cautiously and in limited ways (Wobst 1978). I will use this technique in Chaps. 3, 4, and 5, but (I hope) even more cautiously than archaeologists do.

Archaeologists and historians report their data and inferences in articles and books. Historians call these "secondary" sources; they constitute extended interpretations of primary data. World historians and historical social scientists regularly use secondary sources as data (Lustick 1996; Adas 2012; Lange 2013; Myrdal 2016), something that most historians (e.g., Weinstein 2012) and some sociologists (Goldthorpe 1991) question or even reject since the evidence is filtered through another scholar's mind: "facts" may be only interpretations. Having previously relied on primary sources to write comparative history, I understand this criticism. Still, others suggest distinguishing between interpretation and "reportage" of basic facts; understanding interpretive stances of the historians one uses; and "triangulating" or comparing secondary sources (Bryant 1994; Lustick 1996; Arthur 2011). Also, my experience in analyzing primary sources showed that understanding them required contextualizing them within evidence from secondary ones, i.e., trusting secondary sources somewhat. Here I have used mostly secondary sources, but I also relied on "tertiary" sources, prehistories and histories synthesizing large quantities of secondary sources, for which Myrdal (2016) advocates in world history. Again, my primarily descriptive approach focused my attention on the reportage of basic facts; these have often consisted of summarized facts or empirical generalizations.

To tell this story, I used the multiple dating schemes of different disciplines. So far, I have used "mya" and "kya;" archaeologists also use "before the present" (BP), which is not identical to kya, since "the present" is defined as January 1, 1950. Still, in most cases, it is close enough, as most archaeologists offer estimated ranges of dates. Radiocarbon dating has to be "calibrated" to different amounts of carbon in the atmosphere each year, so I use only calibrated dates. Historians used "BC" or Before Christ and A.D. or *Anno Domini* ("the year of our Lord," not "after death," as undergraduates often think); now they use BCE and CE (Before the Common Era and the Common Era). I will use kya and BP in Chaps. 3 and 4, BP and BCE in Chap. 5, and BCE and CE in Chap. 6; all dates after that are CE. Since archaeologists use different names for sequences in different regions, I have avoided all named sequences (such as "Palaeolithic").

Analyzing the past on its own terms requires care. For instance, I will sometimes say "what is now" for place names (e.g., "what is now Canada"). I will not use terms like *preagricultural*, *preindustrial*, *nonstate*, or other terms that assume that agriculture, industry, and states would develop until after they have developed in the narrative. Since Chaps. 3–6 deal with most of human history while Chap. 7 deals with the period 1450 CE to about 1910 CE, and Chaps. 8–10 analyze the last 115 years or so, readers may conclude that this derives from a social science presentist bias, the better survival of more recent sources, or Eurocentrism. Perhaps—or perhaps it is because the world's population has grown greatly since 1800 and many times since 1900, so there are vastly more people, doing a great deal more, with more complex interconnections. *There is so much more to say* about the last two centuries than about any previous period. Finally, I will remind readers (probably too often) that, rather than marching steadily toward some glorious goal, people in the past were usually "making it up as they went along" within a flow of unpredictable events—something we are, I think, still doing.

Conclusion

I made my assumptions about humans clear in this chapter: We are both individualistic and groupish, with many ways to organize our groups, split them apart, put them back together, change them, and grow them. I have also explained my methods in researching and writing this history. The following story is not "the truth," given the limitations of what has survived from the past and my own conceptual frameworks, but it is as true as I have been able to make it.

Readers should know and look for the four discursive strategies, or modes of analysis, that I used in organizing the historical Chaps. 3–10). To help readers, I have identified these sections with standardized headings throughout (to see how, refer to the Table of Contents). The four modes are narrative analysis, theoretical analysis, contextual analysis, and social pattern analysis. I wrote these separately for Chaps. 3–6, while I mix them somewhat in Chaps. 7–10.

Each mode accomplishes something different. When included, narrative analysis—under the heading "Telling the Tale"—is always chronological. In these sections, I tell a basic story of the development and usually geographic expansion of a new social form. These are meant to be simple and brief. Theoretical analysis, always under the heading "Explaining … Debates," almost always appears right after narrative analysis, reviewing debates about why the story played out the way it did. Thus, it is always a review of various theories—of the human diaspora (Chap. 3), of the Agricultural Revolution (Chap. 4), of the Urban Revolution (Chaps. 5–6), and of the global Industrial Revolution (Chap. 8). As stated above, groups always linked themselves to other groups; thus, the third kind of analysis—under "The Larger Context" in Chaps. 4–6 and under "Global Context" in Chap. 7—addresses the social context in which each kind of society existed. Only Chap. 3 omits this because of the tight focus on nearly contextless conditions, what I will call the "global human frontier." Finally, the fourth type of analysis applies the seven "aspects"—five institutional and two material—to societies under the heading "Social Patterns," although this is combined with contextual analysis under subheadings in Chap. 7. Chapters 8, 9, and 10 consist exclusively of social pattern analysis.

References

Abrams, Dominic, Michael A. Hogg, Steve Hinkle, and Sabine Otten. 2005. The Social Identity Perspective on Small Groups. In *Theories of Small Groups: Interdisciplinary Perspectives*, ed. M. S. Poole and A. B. Hollingshead, 99–138. Thousand Oaks, CA: Sage.

Adas, Michael. 2012. Comparative History and the Challenge of the Grand Narrative. In *A Companion to World History*, ed. D. Northrup, 227–243. Malden, MA: Wiley-Blackwell.

Alcock, James E. 2018. *Belief: What It Means to Believe and Why Our Convictions Are So Compelling*. Amherst, NY: Prometheus Books.

Arrow, Holly, Kelly Bouas Henry, Marshall Scott Poole, Susan Wheelan, and Richard Moreland. 2005. Traces, Trajectories, and Timing. In *Theories of Small Groups: Interdisciplinary Perspectives*, ed. M. S. Poole and A. B. Hollingshead, 313–368. Thousand Oaks, CA: Sage.

Arthur, Mikaila Mariel Lemonik. 2011. The Neglected Virtues of Comparative-Historical Methods. In *New Directions in Sociology: Essays on Theory and Methodology in the 21st Century*, ed. I. Zake and M. DeCesare, 172–192. London: McFarland.

Asch, Solomon E. 1956. Studies of Independence and Conformity. I. A Minority of One Against a Unanimous Majority. *Psychological Monographs: General and Applied* 70 (9): 1–70.

Bacharach, Michael. 2002. In *Beyond Individual Choice: Teams and Frames in Game Theory*, ed. Natalie Gold and Robert Sugden. Princeton and Oxford: Princeton University Press.

Bardhan, Pranab, and Jeff Dayton-Johnson. 2002. Unequal Irrigators: Heterogeneity and Commons Management in Largescale Multivariate Research. In *The Drama of the Commons*, ed.

E. Ostrom, T. Dietz, N. Dolšak, P. C. Stern, S. Stonich, and E. U. Weber, 87–112. Washington, DC: National Academy Press.
Bargh, John A., Peter M. Gollwitzer, and Gabrielle Oettingen. 2010. Motivation. In *Handbook of Social Psychology*, ed. S. T. Fiske, D. T. Gilbert, and G. Lindzey, 5th ed., 268–316. Hoboken, NJ: Wiley.
Baumeister, Roy. 2005. *The Cultural Animal: Human Nature, Meaning, and Social Life*. New York: Oxford University Press.
Bellwood, Peter. 2013. *First Migrants: Ancient Migration in Global Perspective*. Malden, MA: Wiley Blackwell.
Belsey, Catherine. 2006. Biology and Imagination: The Role of Culture. In *Human Nature: Fact and Fiction*, ed. R. H. Wells and J. McFadden, 111–127. New York: Continuum.
Ben-Ze'ev, Aaron. 2013. Emotions. In *Encyclopedia of Philosophy and the Social Sciences*, ed. B. Kaldis, 242–244. Thousand Oaks, CA: SAGE.
Birney, Damian P., and Robert J. Sternberg. 2011. The Development of Cognitive Abilities. In *Developmental Science: An Advanced Textbook*, ed. M. H. Bornstein and M. E. Lamb, 353–388. New York: Psychology Press.
Blanton, Richard E., and with Lane F. Fargher. 2016. *How Humans Cooperate: Confronting the Challenges of Collective Action*. Boulder: University Press of Colorado.
Boehm, Christopher. 1999. *Hierarchy in the Forest: The Evolution of Egalitarian Behavior*. Cambridge, MA: Harvard University Press.
Bowles, Samuel, and Herbert Gintis. 2013. *A Cooperative Species: Human Reciprocity and Its Evolution*. Princeton, NJ: Princeton University Press.
Bryant, Joseph M. 1994. Evidence and Explanation in History and Sociology: Critical Reflections on Goldthorpe's Critique of Historical Sociology. *British Journal of Sociology* 45 (1): 3–19.
Buller, David J. 2005. *Adapting Minds: Evolutionary Psychology and the Persistent Quest for Human Nature*. Cambridge, MA: The MIT Press.
Campbell, Lyle. 2013. *Historical Linguistics: An Introduction*. Edinburgh: Edinburgh University Press.
Cartmill, Matt, and Fred H. Smith. 2022. *The Human Lineage*. 2nd ed. Hoboken, NJ: Wiley Blackwell.
Chen, Xinyin, and Kenneth H. Rubin, eds. 2011. *Socioemotional Development in Cultural Context*. New York: The Guilford Press.
Cook, Karen S. 2000. Social Exchange Theory. In *Encyclopedia of Sociology*, ed. E. F. Borgatta and R. J. V. Montgomery, vol. 4, 2nd ed., 2669–2676. New York: Macmillan.
Coolidge, Frederick L., and Thomas Wynn. 2009. *The Rise of Homo sapiens: The Evolution of Modern Thinking*. Chichester: Wiley-Blackwell.
Damasio, Antonio. 1994. *Descartes' Error: Emotion, Reason, and the Human Brain*. New York: G. P. Putnam.
Danesi, Marcel. 2008. *Of Cigarettes, High Heels, and Other Interesting Things: An Introduction to Semiotics*. 2nd ed. New York: Palgrave Macmillan.
De Dreu, Carsten K. W. 2010. Social Conflict: The Emergence and Consequences of Struggle and Negotiation. In *Handbook of Social Psychology*, ed. S. T. Fiske, D. T. Gilbert, and G. Lindzey, 5th ed., 983–1023. Hoboken, NJ: Wiley.
Dea, Shannon. 2023. *Beyond the Binary: Thinking About Sex and Gender*. 2nd ed. Peterborough, Ontario: Broadview Press.
Dijksterhuis, Ap. 2010. Automaticity and the Unconscious. In *Handbook of Social Psychology*, ed. S. T. Fiske, D. T. Gilbert, and G. Lindzey, 5th ed., 228–267. Hoboken, NJ: Wiley.
Dowding, Keith. 2011. Interests. In *Encyclopedia of Power*, 349–351. Thousand Oaks, CA: Sage.
Dretske, Fred. 2005. Belief. In *P. 85 in The Oxford Companion to Philosophy*, ed. T. Honderich. New York: Oxford University Press.
Dunbar, Robin. 2008. Kinship in Biological Perspective. In *Early Human Kinship: From Sex to Social Reproduction*, ed. N. J. Allen, H. Callan, R. Dunbar, and W. James, 131–150. Oxford: Blackwell.

Ekman, Paul. 2007. *Emotions Revealed: Recognizing Faces and Feelings to Improve Communication and Emotional Life*. 2nd ed. New York: Owl.
Elliot, Anthony. 2006. Interaction. In *Cambridge Dictionary of Sociology*, ed. B. S. Turner, 304–305. New York: Cambridge University Press.
Fehr, Ernst, and Simon Gächter. 2000. Fairness and Retaliation: The Economics of Reciprocity. *Journal of Economic Perspectives* 14 (3): 159–181.
Fiske, Susan T. 2010. Interpersonal Stratification: Status, Power, and Subordination. In *Handbook of Social Psychology*, ed. S. T. Fiske, D. T. Gilbert, and G. Lindzey, 5th ed., 941–982. Hoboken, NJ: Wiley.
Flynn, James R. 2012. *Are We Getting Smarter? Rising IQ in the Twenty-First Century*. New York: Cambridge University Press.
Gaddis, John Lewis. 2001. In Defense of Particular Generalization: Rewriting Cold War History, Rethinking International Relations Theory. In *Bridges and Boundaries: Historians, Political Scientists, and the Study of International Relations*, ed. C. Elman and M. F. Elman, 301–326. Cambridge, MA: The MIT Press.
Gamble, Clive. 2015. *Archaeology: The Basics*. 3rd ed. London and New York: Routledge.
Gamble, Clive, John Gowlett, and Robin Dunbar. 2014. *Thinking Big: How the Evolution of Social Life Shaped the Mind*. London: Thames & Hudson.
Geertz, Clifford. 1973. *The Interpretation of Cultures*. New York: Basic Books.
Gilovich, Thomas D., and Dale W. Griffin. 2010. Judgment and Decision Making. In *Handbook of Social Psychology*, ed. S. T. Fiske, D. T. Gilbert, and G. Lindzey, 5th ed., 542–588. Hoboken, NJ: Wiley.
Goldthorpe, John H. 1991. The Uses of History in Sociology: Reflections on Some Recent Tendencies. *British Journal of Sociology* 42 (2): 211–230.
Granovetter, Mark S. 1973. The Strength of Weak Ties. *American Journal of Sociology* 78 (6): 1360–1380.
Grossman, Igor, Richard P. Elbach, Jacklyn Koyama, and Qalsar B. Sahi. 2020. Folk Standards of Sound Judgment: Rationality Versus Reasonableness. *Science Advances* 6 (2): 1–14.
Groves, Colin. 2015. Hominin Migrations before *Homo sapiens*: Out of Africa—How Many Times? In *The Global Prehistory of Human Migration*, ed. P. Bellwood, 18–25. Malden, MA: Wiley Blackwell.
Haas, Jonathan, and Matthew Piscitelli. 2013. The Prehistory of Warfare: Misled by Ethnography. In *War, Peace, and Human Nature: The Convergence of Evolutionary and Cultural Views*, ed. D. P. Fry, 168–190. New York: Oxford University Press.
Hackman, J. Richard, and Nancy Katz. 2010. Group Behavior and Performance. In *Handbook of Social Psychology*, ed. S. T. Fiske, D. T. Gilbert, and G. Lindzey, 5th ed., 1208–1251. Hoboken, NJ: Wiley.
Haidt, Jonathan, and Selin Kesebir. 2010. Morality. In *Handbook of Social Psychology*, ed. S. T. Fiske, D. T. Gilbert, and G. Lindzey, 5th ed., 797–832. Hoboken, NJ: Wiley.
Hayden, Brian. 2014. *The Power of Feasts: From Prehistory to the Present*. New York: Cambridge University Press.
Hebl, Michelle R., and Juan M. Madera. 2007. Ethnocentrism. In *Encyclopedia of Social Psychology*, ed. R. F. Baumeister and K. D. Vohs, 315. Thousand Oaks, CA: Sage.
Heine, Steven J. 2010. Cultural Psychology. In *Handbook of Social Psychology*, ed. S. T. Fiske, D. T. Gilbert, and G. Lindzey, 5th ed., 1423–1464. Hoboken, NJ: Wiley.
Henrich, Joseph. 2015. *The Secret of Our Success: How Culture Is Driving Human Evolution, Domesticating Our Species, and Making Us Smarter*. Princeton and Oxford: Princeton University Press.
Hewitt, John P. 2007. *Self and Society: A Symbolic Interactionist Social Psychology*. 10th ed. New York: Allyn and Bacon.
Hirschman, Albert O. 1977. *The Passions and the Interests: Political Arguments for Capitalism before Its Triumph*. Princeton, NJ: Princeton University Press.

Hofstede, Geert, Gert Jan Hofstede, and Michael Minkov. 2010. *Cultures and Organizations: Software of the Mind.* London: McGraw-Hill.
Hogg, Michael. 2010. Influence and Leadership. In *Handbook of Social Psychology*, ed. S. T. Fiske, D. T. Gilbert, and G. Lindzey, 5th ed., 1166–1207. Hoboken, NJ: Wiley.
Hollingshead, Andrea B., Gwen M. Wittenbaum, Paul B. Paulus, Randy Y. Hirokawa, Deborah G. Ancona, Randall S. Peterson, Karen A. Jehn, and Kay Yoon. 2005. A Look at Groups From the Functional Perspective. In *Theories of Small Groups: Interdisciplinary Perspectives*, ed. M. S. Poole and A. B. Hollingshead, 21–62. Thousand Oaks, CA: Sage.
Howlin, Patricia, Tony Charman, and Mohammad Ghaziuddin, eds. 2011. *The SAGE Handbook of Developmental Disorders.* Thousand Oaks, CA: Sage.
Hrdy, Sarah Blaffer. 2009. *Mothers and Others: The Evolutionary Origins of Mutual Understanding.* Belknap: Cambridge, MA.
Immergut, Ellen M. 2011. Institutions and Institutionalism. In *International Encyclopedia of Political Science*, ed. B. Badie, D. Berg-Schlosser, and L. Morlino, 1203–1207. Thousand Oaks, CA: Sage.
Inglehart, Ronald F. 2018. *Cultural Evolution: People's Motivations are Changing, and Reshaping the World.* New York: Cambridge University Press.
Kahan, Dan M., Ellen Peters, Erica Cantrell Dawson, and Paul Slovic. 2013. "Motivated Numeracy and Enlightened Self-Government." Social Science Research Network. Retrieved December 29, 2016. doi: https://papers.ssrn.com/sol3/papers.cfm?abstract_id=2319992
Kaiser, Elke, Joachim Burger, and Wolfram Schier. 2012. *Population Dynamics in Prehistory and Early History: New Approaches Using Stable Isotopes and Genetics.* Boston: De Gruyter.
Karp, David R. 2000. Values Theory and Research. In *Encyclopedia of Sociology*, ed. E. F. Borgatta and R. J. V. Montgomery, vol. 5, 2nd ed., 3212–3227. New York: Macmillan.
Kelly, Robert L. 2013. *The Lifeways of Hunter-Gatherers: The Foraging Spectrum.* New York: Cambridge University Press.
Keltner, Dacher, and Jennifer S. Lerner. 2010. Emotion. In *Handbook of Social Psychology*, ed. S. T. Fiske, D. T. Gilbert, and G. Lindzey, 5th ed., 317–352. Hoboken, NJ: Wiley.
Klein, Richard. 2013. Hominin Dispersals in the Old World. In *The Human Past: World Prehistory & the Development of Human Societies*, ed. C. Scarre, 3rd ed., 84–123. London: Thames & Hudson.
Križan, Zlatan, and Richard H. Smith. 2014. When Comparisons Divide: Invidious Emotions and Their Social Control. In *Communal Functions of Social Comparison*, ed. Z. Križan and F. X. Gibbons, 60–94. New York: Cambridge University Press.
Lange, Matthew. 2013. *Comparative-Historical Methods.* Los Angeles: Sage.
Leary, Mark R. 2010. Affiliation, Acceptance, and Belonging. In *Handbook of Social Psychology*, ed. S. T. Fiske, D. T. Gilbert, and G. Lindzey, 5th ed., 864–897. Hoboken, NJ: Wiley.
Lents, Nathan H. 2016. *Not So Different: Finding Human Nature in Animals.* New York: Columbia University Press.
Lerner, Jennifer S., Ye Li, Piercarlo Valdesolo, and Karim S. Kassam. 2015. Emotion and Decision Making. *Annual Review of Psychology* 66:799–823.
Li, Liming, Troy J. Comi, Rob F. Bierman, and Joshua M. Akey. 2024. Recurrent Gene Flow Between Neanderthals and Modern Humans Over the Past 200,000 Years. *Science* 385:158–168.
Lips, Hilary. 2005. *Sex & Gender: An Introduction.* 5th ed. Boston: McGraw-Hill.
Locke, Kenneth D. 2014. Agency and Communion in Social Comparisons. In *Communal Functions of Social Comparison*, ed. Z. Križan and F. X. Gibbons, 11–38. New York: Cambridge University Press.
Lovaglia, Michael, Elizabeth A. Mannix, Charles D. Samuelson, Jane Sell, and Rick K. Wilson. 2005. Conflict, Power, and Status in Groups. In *Theories of Small Groups: Interdisciplinary Perspectives*, ed. M. S. Poole and A. B. Hollingshead, 139–184. Thousand Oaks, CA: Sage.
Lustick, Ian. 1996. History, Historiography, and Political Science: Multiple Historical Records and the Problem of Selection Bias. *The American Political Science Review* 90 (3): 605–618.

Manstead, Antony S. R. 2007. Emotion. In *Encyclopedia of Social Psychology*, ed. R. F. Baumeister and K. D. Vohs, 286–291. Thousand Oaks, CA: Sage.
Maslow, Abraham. 1943. A Theory of Human Motivation. *Psychological Review* 50 (4): 370–396.
McAuliffe, Katherine, Jillian J. Jordan, and Felix Warneken. 2015. Costly Third-Party Punishment in Young Children. *Cognition* 134 (1): 1–10.
McNeill, J. R., and William H. McNeill. 2003. *The Human Web: A Bird's-Eye View of World History*. New York: W. W. Norton.
Mears, John A. 2015. "The Meaning of Humankind's Emergence: From the Largest Possible Perspective." Pp. 243–254 in *From Big Bang to Galactic Civilizations: A Big History Anthology*. Vol. II: Education and Understanding: Big History Around the World, edited by B. Rodrigue, L. Grinin, and A. Korotayev. Delhi: Primus Books.
Medin, Douglas L., Julia Beth Proffitt, and Hillarie C. Schwartz. 2000. Concepts: An Overview. In *Encyclopedia of Psychology*, ed. A. E. Kazdin, vol. II, 242–245. New York: Oxford University Press.
Meltzer, David J. 2009. *First Peoples in a New World: Colonizing Ice Age America*. Berkeley: University of California Press.
Moran, Mayo. 2003. *Rethinking the Reasonable Person: An Egalitarian Reconstruction of the Objective Standard*. New York: Oxford University Press.
Myrdal, Janken. 2016. On Source Criticism in World History. In *Methods in World History: A Critical Approach*, ed. A. Jarrick, J. Myrdal, and M. W. Bondesson, 45–84. Lund: Nordic Academic Press.
Neill, James. 2009. *The Origins and Role of Same-Sex Relations in Human Societies*. Jefferson, NC and London: McFarland & Company.
Nowell, April. 2021. *Growing Up in the Ice Age: Fossil and Archaeological Evidence of the Lived Lives of Plio-Pleistocene Children*. Philadelphia: Oxbow Books.
Olson, Mancur. 1971 (1965). *The Logic of Collective Action: Public Goods and the Theory of Groups*. Cambridge MA: Harvard University Press.
Ostrom, Elinor, Thomas Dietz, Nives Dolšak, Paul C. Stern, Susan Stonich, and Elke U. Weber, eds. 2002. *The Drama of the Commons*. Washington, DC: National Academy Press.
Owens, Timothy J. 2006. Self and Identity. In *Handbook of Social Psychology*, ed. J. DeLamater, 205–232. New York: Springer.
Patterson, Charlotte J., and Anthony R. D'Augelli, eds. 2012. *Handbook of Psychology and Sexual Orientation*. New York: Oxford University Press.
Polanyi, Karl. 1957. The Economy as Instituted Process. In *Trade and Market in the Early Empires: Economies in History and Theory*, ed. K. Polanyi, C. M. Arensberg, and H. W. Pearson, 243–269. Chicago: Henry Regnery Company.
Portes, Alejandro. 2010. *Economic Sociology: A Systematic Inquiry*. Princeton, NJ: Princeton University Press.
Poteete, Amy R., Marco A. Janssen, and Elinor Ostrom. 2010. *Working Together: Collective Action, the Commons, and Multiple Methods in Practice*. Princeton and Oxford: Princeton University Press.
Prinz, Jesse J. 2014. *Beyond Human Nature: How Culture and Experience Shape the Human Mind*. New York and London: W.W. Norton.
Rahikainen, Marjatta, and Susanna Fellman. 2012. On Historical Writing and Evidence. In *Historical Knowledge: In Quest of Theory, Method, and Evidence*, ed. S. Fellman and M. Rahikainen, 5–44. Newcastle upon Tyne: Cambridge Scholars Publishing.
Rampolla, Mary Lynn. 2018. *A Pocket Guide to Writing History*. 9th ed. Boston: Bedford/St. Martin's.
Read, Dwight W. 2012. *How Culture Makes Us Human: Primate Social Evolution and the Formation of Human Societies*. Walnut Creek, CA: Left Coast Press.
Richerson, Peter J., and Robert Boyd. 2005. *Not by Genes Alone: How Culture Transformed Human Evolution*. Chicago and London: The University of Chicago Press.
Rosaldo, Renato. 1989. *Culture and Truth: The Remaking of Social Analysis*. Boston: Beacon Press.

Ross, Donald. 2019 (1997). "Game Theory." Stanford Encyclopedia of Philosophy. https://plato.stanford.edu/entries/game-theory/

Schick, Kathy and Nicholas Toth. 2015. "Big History and the Major Stages of Human Evolution." *From Big Bang to Galactic Civilizations: A Big History Anthology*. Vol. III: The Ways that Big History Works: Cosmos, Life, Society and Our Future, edited by B. Rodrigue, L. Grinin, and A. Korotayev. Delhi: Primus Books.

Sherif, Muzafer. 1958. Superordinate Goals in the Reduction of Intergroup Conflict. *American Journal of Sociology* 63 (4): 349–356.

Shotola, Robert. 2000. Small Groups. In *Encyclopedia of Sociology*, ed. E. F. Borgatta and R. J. F. Montgomery, 2610–2622. New York: Macmillan.

Sibley, W. M. 1953. The Rational Versus the Reasonable. *The Philosophical Review* 62 (4): 554–560.

Simon, Herbert A. 1997. *Administrative Behavior: A Study of Decision-Making Processes in Administrative Organizations*. 4th ed. New York: The Free Press.

Smith, Adam. [1776] 1937. *An Inquiry into the Nature and Causes of the Wealth of Nations*, edited by E. Cannan. New York: Modern Library.

Smith, Thomas Jay, Robert A. Henning, Michael G. Wade, and Thomas Fisher. 2015. *Variability in Human Performance*. Boca Raton, FL: CRC Press.

Stanish, Charles. 2017. *The Evolution of Human Co-operation: Ritual and Social Complexity in Stateless Societies*. New York: Cambridge University Press.

Starkweather, K. E., and R. Hames. 2012. A Survey of Non-Classical Polyandry. *Human Nature* 23 (2): 149–172.

Sterelny, Kim. 2012. *The Evolved Apprentice: How Evolution Made Humans Unique*. Cambridge, MA: The MIT Press.

Storey, William Kelleher. 2021. *Writing History: A Guide for Students*. 6th ed. New York: Oxford.

Sutton, Mark Q. 2019. *Archaeology: The Science of the Human Past*. New York: Routledge.

Swann, William B., Jr., and Jennifer K. Bosson. 2010. Self and Identity. In *Handbook of Social Psychology*, ed. S. T. Fiske, D. T. Gilbert, and G. Lindzey, 5th ed., 589–628. Hoboken, NJ: Wiley.

Swedberg, Richard. 2005. *Interest*. New York: Open University Press.

Tay, Louis, and Ed Diener. 2011. Needs and Subjective Well-Being Around the World. *Journal of Personality and Social Psychology* 101 (2): 354–365.

Tomasello, Michael. 2019. *Becoming Human: A Theory of Ontogeny*. Cambridge, MA: Harvard University Press.

Toulmin, Stephen. 2001. *Return to Reason*. Cambridge, MA: Harvard University Press.

Tuomela, Raimo. 2013. *Social Ontology: Collective Intentionality and Group Agents*. New York: Oxford University Press.

Turchin, Peter. 2016. *Ultrasociety: How 10,000 Years of War Made Humans the Greatest Cooperators on Earth*. Chaplin, CT: Beresta Books.

Turkington, Carol, and Joseph R. Harris. 2001. *The Encyclopedia of Memory and Memory Disorders*. 2nd ed. New York: Facts on File.

Turner, Jonathan. 2007. *Human Emotions: A Sociological Theory*. London: Routledge.

Turner, Jonathan H., and Richard Machalek. 2018. *The New Evolutionary Sociology: Recent and Revitalized Theoretical and Methodological Approaches*. New York and London: Routledge.

Weinstein, Barbara. 2012. The World Is Your Archive? The Challenges of World History as a Field of Research. In *A Companion to World History*, ed. D. Northrop, 63–78. Malden, MA: Wiley-Blackwell.

West, Barbara. 2006. Socialization. In *Encyclopedia of Anthropology*, ed. H. J. Birx, vol. 5, 2097–2099. Thousand Oaks, CA: Sage.

Willer, David, Pamela Emanuelson, Yamilette Chacon, and Richard J. Chacon. 2017. How Chiefdom and Early State Social Structures Resolve Collective Action Problems. In *Feast, Famine or Fighting? Multiple Pathways to Social Complexity*, ed. R. J. Chacon and R. G. Mendoza, 417–452. Cham: Springer.

Wilson, Edward O. 2012. *The Social Conquest of Earth*. New York and London: W.W. Norton.
Wobst, H. Martin. 1978. The Archaeo-Ethnology of Hunter-Gatherers or the Tyranny of the Ethnographic Record in Archaeology. *American Antiquity* 43 (2): 303–309.
Woolf, Daniel. 2019. *A Concise History of History: Global Historiography from Antiquity to the Present*. New York: Cambridge University Press.
Yzerbyt, Vincent, and Stéphanie Demoulin. 2010. Intergroup Relations. In *Handbook of Social Psychology*, ed. S. T. Fiske, D. T. Gilbert, and G. Lindzey, 5th ed., 1024–1083. Hoboken, NJ: Wiley.

Chapter 3
The First Globalization: Foraging Bands on the Global Human Frontier, 100,000–12,000 BP

Abstract Modern humans evolved in Africa. Over many millennia, they migrated throughout the planet, sometimes through rapid migration, die-offs, and return migration, and sometimes through cautious movement. Population growth, empty land, conflict avoidance, curiosity, and adaptability seem to have caused this global dispersal. These human groups subsisted on foraging—hunting, gathering, and fishing. Evidence increasingly shows that the sexual division of labor was extremely flexible; some argue there was no division. Groups often shared food collectively and used objects practically and symbolically. Kinship tended toward arranged marriages, bilocal postmarital residence, and bilateral descent, again with flexibility. Groups ("bands") averaged around 25 people, probably part of interband communities. Leadership was rare, and group membership was loose, flexible, and constantly recombining. Most inequality was derived from differences in age and personal qualities, while economic inequality was nonexistent, and a rough gender equality existed. Forager ideologies were animistic; most groups had shamans and regular rituals. Women usually bore five or six children, but roughly half of those died by age 20. Scholars debate whether warming trends or humans themselves caused the extinctions of megafauna.

Keywords Human diaspora · Prehistoric hunter-gatherers · Foragers · Bands · Pleistocene · Migration

Introduction

We have seen how hominins evolved, becoming three kinds of humans. Prehistorians discuss "archaic" humans, Neandertals and Denisovans, and "modern" humans—us (Kuhn and Steiner 2018; Cartmill and Smith 2022 object to this categorization). Here I trace how modern humans migrated from Africa globally, incorporating archaic humans along the way; analyze aspects of that movement; and speculatively describe their way of life, applying the foregoing analysis of groups. In this chapter, I make three arguments. First, as discussed, I call this intergenerational journey the "first globalization." It was not transformational; just by itself, it did not lead to a

T. C. Burr, *Societies Past and Present*, Emerging Globalities and Civilizational Perspectives, https://doi.org/10.1007/978-3-031-91532-1_3

new type of society. It was merely geographic expansion. Still, out of all the hominins that evolved across Afro-Eurasia, only one species, *Homo sapiens sapiens*, made it to the Americas, and only with humans on all continents can we even talk about "global" processes at all. Thus, this first globalization was the foundation for all subsequent developments. Second, prehistorians studying this period debate the human cognitive capacities developed, and this debate helps extend the analysis of humans in Chap. 2. Finally, I argue that groups on what I call "the global human frontier" were probably organized as simple foraging bands. It was group organization that was simple; individual people were as complex as ever. Many call this a stereotype, arguing that hunter-gatherers were quite varied (Kelly 2013a). I agree, but I think that on the frontier, they broadly fit this stereotype.

This construct, the "global human frontier," is explicitly anachronistic but useful. Modern humans moving across the planet, generation after generation, must have known that some lands had people and others did not, but probably did not think in planetary terms; instead, they simply survived where they were and moved to uninhabited lands when it made sense. The "frontier" concept results from our after-the-fact overview of this process. While I will sometimes use the more experience-oriented phrase "edge of human habitation," the global human frontier was rarely an edge; instead, it was probably a zone some miles (kilometers) wide within which people moved back and forth, its width changing with different circumstances (Feuer 2016). Finally, meetings of archaic and modern humans make the "human" frontier idea a bit fuzzy.

The limited available evidence restricts what we know, so I supplement archaeological evidence with ethnographic, linguistic, and genetic evidence. Some archaeologists directly infer ethnographically (e.g., Peoples et al. 2016; Stewart et al. 2020), but others acknowledge that while living foragers today "operate under the same principles" as those did in the past (Kelly 2013a:237), they only use them to generate hypotheses that they test with archaeological evidence (see Lemke 2018b for a good example). Inferring specific details of life for groups on the global human frontier with secondary and tertiary sources is effectively impossible; I infer only the most basic principles of social organization, and *only on the frontier itself*, not after it moved on (for that, see Chap. 4). I think such organization was broadly similar during this journey, but specific beliefs, knowledge, kinship structures, and so on varied greatly across diverse physical environments over millennia. I use the secondary and tertiary ethnographic literature (e.g., Binford 2001; Kelly 2013a; Steadman 2009; etc.); use secondary and tertiary archaeological sources to refine these findings; and speculate carefully on what life was like for prehistoric foragers alone in their environments (Wobst 1978; Kelly 2013a). The frontier posed the extreme: in one direction, there were no people at all. Archaeologists seem to ignore some ethnographic findings (e.g., Ensor 2013 argues this about kinship), so I aim only for plausibility. I lay out this picture in three sections: history, debate, and institutional analysis.

Telling the Tale: Humans Populate the Planet

Modern humans evolved either in east-central or southern Africa. They went north out of Africa, encountering and mating with archaic humans, who went extinct either through absorption, out-competition, or warfare with modern humans, or all three. Modified modern humans then went east through South Asia to Australia, north into Central Asia, or northwest into Europe. East Asians continued into northeastern Siberia and went through Beringia to the Americas. This intergenerational journey populating all continents took 50 (or 90) millennia to complete, ending sometime between 20,000 and 10,000 years ago. This first "globalization" of humanity was not perfectly global; some areas remained uninhabited.

Archaeologists developed three arguments about the timing of this process (Kaifu et al. 2015; Cartmill and Smith 2022). Some argue that modern humans moved from Africa to southwest Asia, then South Asia around 100 kya. Others suggest that they left Africa around the time of the Toba volcano super-eruption of 74 kya, following the southern coastlines. Finally, some argue that modern humans "spread to eastern Asia after 50,000 years ago" (Kaifu et al. 2015:544; also, Cartmill and Smith 2022:457). I present evidence for the first and third timelines.

Evidence supports an African origin for modern humans. The earliest "sophisticated technology and personal ornamentation... [appeared] earlier in Africa than elsewhere in the world" (Kuykendall and Heyerdahl-King 2014:28). Genetically, all modern humans are descended from one set of ancestors who lived 60,000 years ago (Brooke 2014; Groves 2015). Both linguistic and genetic human diversity are greatest in Africa and narrowest in South America (Atkinson 2011; Kuykendall and Heyerdahl-King 2014), so the "currently accepted chronology" is that modern humans originated in east Africa (Cartmill and Smith 2022:418–419), though some think southern Africa (Wood 2019:23); numerous fossilized footprints in southern Africa from 153 to 71 kya are "overwhelmingly likely" to be from *sapiens* (Helm et al. 2023:63).

Scholars debate when modern humans left Africa. Geneticists have identified mitochondrial DNA (only from mothers) from modern humans in European Neandertals from over 200,000 years ago (Posth et al. 2017:6), and Y chromosomes (only from fathers) from modern humans among European Neandertals between 300,000 and 100,000 years ago (Petr et al. 2020), supporting assertions that Neandertals and Sapiens interacted across the Mediterranean (Cartmill and Smith 2022). Archaeological evidence shows that modern humans left Africa around 100,000 BP in either "a brief range expansion" (Groves 2015:23) or the first main exodus (Wood 2019). They definitely left Africa by 60 kya (Kaifu et al. 2015), across the mouth of the Red Sea into Arabia, up the Nile Valley and across the Sinai Peninsula, or both (Gamble 2013). Many argue for a coastal movement into Iran and Pakistan, "hopping" between river mouths and then going up river valleys (Mithen 2020; Wood 2019), though they might have had boats (Erlandson 2017). Genetic evidence places humans in India 70–50 kya (Gamble 2013; Wood 2019), but the oldest modern human skeleton found in that location dates only from 31 kya (Petraglia and Boivin 2014).

They encountered Neandertals in Central Asia (Gamble 2013:256). Upon (re) meeting them, modern humans interbred with them (Stoneking and Harvati 2015; Cartmill and Smith 2022), so all modern humans outside sub-Saharan Africa were a little Neandertal.[1] Perhaps this caused "extinction by assimilation" of Neandertals (Bellwood 2013:38) since modern humans may have "demographically overwhelmed the local archaic peoples" with larger groups (Cartmill and Smith 2022:466). They reached Southeast Asia and its islands, then a peninsula prehistorians call "Sunda," either around 70,000 or 55,000 BP; sites exist from 45,000 BP (O'Connor and Bulbeck 2014). Some think modern humans reached New Guinea and Australia, then joined in one continent that prehistorians call "Sahul," by 68,000 BP (Clarkson et al. 2017); most think they arrived by 40 kya (Gamble 2013; Stoneking and Harvati 2015; Mithen 2020; Cartmill and Smith 2022). Genetic evidence demonstrates many separate crossings, not one movement (Bellwood 2013), that probably required "the use of boats for deep-sea fishing" (Davidson 2014:378), which were common "in a wide area of the western Pacific" (Kaifu et al. 2015:554; O'Connor 2015).

Modern humans went northwest into Europe by 43–45 kya. They may have encountered more Neandertals (Stoneking and Harvati 2015). Archaeologists debate whether "there was conflict, interbreeding, acculturation, or simply careful avoidance" (Mithen 2020:386), though European Neandertals may have already been extinct (Pettitt 2014; Cartmill and Smith 2022). The earliest potentially modern human tool assemblages were found in Bulgaria and Hungary, dated to 44–46,000 BP; modern humans reached England by 42,000 BP (Gamble 2013) and southern Germany and northern Spain by 40,000 BP (Jochim 2011). Sites exist in Ukraine and southern Russia from before 30 kya, and a mix of various "tool kits" (standard sets of tools, or "industries") existed from Poland to Spain. The last potentially Neandertal assemblage found is from 28,000 BP in southern Spain. After that, all sites found have modern human industries.

Archaeologists disagree about when modern humans reached northern Asia. Some suggest three major routes: through the Fertile Crescent and the Caucasus into Russia; from India through the Himalayas into Central Asia, reaching southern Siberia by 45 kya; or up the Pacific Coast from Sunda through what is now the Philippines, then one large island, by 67,000 BP (Wood 2019). Sites with Denisovan skeletal material have been found in southern Siberia, in Tibet, and in Laos (Demeter et al. 2022). Two of these three routes put modern humans in contact with Denisovans (Stoneking 2012) which produced another assimilation, so there was not much of a "frontier" in this era. Once through these lands, slightly modified modern humans moved in several directions. The evidence seems to show "fully anatomically modern humans in East Asia" near what is now Beijing by 42–39,000 BP (Barnes

[1] Today all humans, including sub-Saharan Africans, have some Neandertal genes (Chen et al. 2020).

2015:67), and "the arrival of modern humans" in what is now Japan "by 38,000 cal BP" (Kaifu et al. 2015:40). While some argue that modern humans arrived by 100 kya, most scholars agree with the ~40 kya dating, though there is much debate (Cartmill and Smith 2022:432–433). Finally, modern humans "reached the Arctic [northern Siberia] by 33,000 cal BP" (Kaifu et al. 2015:543; map, 546).

The latest ice age forced reverse migration; ice recession allowed return. The last global spread of glaciers, called the Last Glacial Maximum (LGM), peaked around 20,000 years ago (Mithen 2020), and humans retreated out of northern Eurasia. Then the planet warmed, polar ice melted, global climate became wetter, and the oceans rose. This separated Britain from Europe, New Guinea and Tasmania from Australia, and Siberia from Alaska. Global landmass shrunk by about 8% (Meltzer 2009). Intermediate warming and cooling periods (Brooke 2014) led to back-and-forth remigration. For instance, humans repopulated the East European Plain around 20,000 kya (Hoffecker 2005) but returned to northern Europe from Iberia by 12.9 kya (Bellwood 2013). They also retreated from and remigrated into Siberia. During the LGM, northeastern Siberia was not glaciated, but "fuel, shelter, and food were significantly limited," making survival difficult (Graf 2015:511). Some archaeologists think that humans "more or less permanently colonized far northeastern Asia only after 18,000 years BP" (Meltzer 2009:20, 189).

Siberians migrated into the Americas, but scholars debate the path into North America. Tool kits in prehistoric Siberia, Alaska, and Canada are similar; American dental evidence shows a Siberian origin; and genetic evidence supports this conclusion (Meltzer 2009). At the time, "Beringia," a stretch of land 600 miles wide north to south mostly free of glaciers, connected northeastern Asia and northwestern North America (Wood 2019). Two ice sheets to the southeast blocked the way into North America; between them was an ice-free but inhospitable corridor (Wood 2019). Glaciers retreated after 14,500 BP, so this corridor widened and became habitable (Meltzer 2009). The Pacific Coast was ice-free from Alaska to what is now Washington state between 16,000 and 14,500 BP; it reglaciated and deglaciated again by 13,400 BP.

Scholars debate when humans moved beyond Beringia into the Americas (Pauketat and Sassaman 2020) because the types of evidence do not match. Genetic evidence shows that humans "arrived in Beringia... sometime between 20,000 and 40,000 years ago" (Pauketat and Sassaman 2020:148; see also Mithen 2020:395). Some argue that humans arrived in North America before the LGM; cranial evidence allows for, but does not demonstrate conclusively, this possibility (Kornfield and Politis 2014). Bluefish Caves in northwest Yukon had human-worked mammoth bone 24,000 years ago (Pauketat and Sassaman 2020). Researchers found footprints in the south-central United States seemingly similar to today's human footprints dated to 23,000 BP (Bennett et al. 2021). Others claim that the Chiquihuite Cave in northern Mexico has remains of a human group from at least 19,000 BP (Ardelean et al. 2020).

Understanding eastward expansion in the Americas affects the understanding of arrival time. Some archaeologists think that people moved in a general eastward "wave of advance" (Pauketat and Sassaman 2020:123), but others think that they followed river valleys eastward. DNA samples "hint at rapid population growth and dispersal of humans throughout the Americas between about 14,500 and 13,000 years ago" (Fagan 2019:56). Some evidence supports an earlier timeline, including sites in Pennsylvania (16–19,000 BP), Virginia (15,000–17,000 BP), South Carolina (16,000 BP), and Texas (15,500 BP), but many "reserve judgment" on the dating for these sites (Pauketat and Sassaman 2020:138–142). Archaeologists used to see the Clovis industry in North America (13,200–12,900 BP) as the first American arrivals; now, some think Clovis was a fast-spreading cultural tradition (Gamble 2013; Pauketat and Sassaman 2020).

At what point humans entered South America obviously influences when we think they came to North America. Most archaeologists accept human habitation at Monte Verde in southern Chile by 14,500 BP (Mithen 2020). Some argue that a coastal route from Beringia into North America was not likely until 14,700 BP, before Monte Verde (Mithen 2020), yet taking only 200 years to get to Monte Verde through different environments was impossible (Mithen 2020). This points to an immediate post-LGM human arrival in the Americas. Others argue that the now-submerged coastal route was deglaciated by 16,000 BP (Pauketat and Sassaman 2020), providing a more plausible 2000-year period to get to Monte Verde. Some researchers claim to have found human artifacts in Brazil in the early LGM, well before 20,000 BP (Pansani et al. 2023). If others confirm this, a general rethinking of the timing might be required. Still, humans had certainly arrived in Tierra del Fuego by 12,500 BP (Mithen 2020), and there are 37 different sites or layers within sites in South America dated to between 11,000 and 10,000 years ago (Moore 2014), demonstrating a rapid movement through the continent and the end of the global human frontier.

The broad outlines of the first globalization are clear, but the timing is not. Modern humans originated in southern or eastern Africa and reached southwestern Asia by 100,000 BP. Either then or around 60,000 BP, they headed north and east. They moved into South Asia, Southeast Asia, and Australia before 60,000 BP or by 45,000 BP. Modern humans moved into Europe by 45,000 BP and far northern Asia by 33,000 BP; they retreated southward during the LGM, then returned. People may or may not have arrived in the Americas before the LGM, but they certainly lived in southern South America fourteen millennia ago.

Explaining Why and How Humans Populated the Planet: Debates

Archaeologists constantly analyze why most prehistoric social changes happened, but for this first, massive expansion, there are only scattered subtheoretical discussions, not true debate (e.g., Meltzer 2009; Brooke 2014; Kaifu et al. 2015; Mithen 2020; Wood 2019; only Gamble 2013 approaches a full treatment). Populating the entire planet required millennia, so it was obviously intergenerational: Over a lifetime, any one person moved only a little. While reasons for moving varied greatly by time and place, general questions are still worth posing. First, were humans in any given place "pushed" out by population pressure or "pulled" by environmental richness and curiosity? Second, when were humans cognitively and behaviorally "modern"? This raises questions about flexibility and curiosity as adaptive traits. Further issues include attitudes affecting the speed of migration; a coast-and-rivers strategy or inland movement; and responses to climate change.

Why did modern humans populate the entire planet? Archaeologists analyze motivations through the classic push-or-pull debate common in migration studies (Lemke 2018a): Do problems in an area "push" people out, or do other, nicer areas "pull" them in? The clearest push would have been population growth beyond food resources, called "population pressure" (Kelly 2013b). Population growth occurred, so population pressure probably did as well. While some argue that foragers respond to population pressure by fighting (see below), most scholars argue that foraging bands consistently respond to too-few resources by moving away (Meltzer 2009; Hrdy 2009; Kelly 2013b), and on the frontier, there was always empty land.

On the other hand, empty land plus human curiosity might have pulled people in. Biomass (total available plants and animals) on the frontier was as high as humans ever experienced, seven times greater than now (Bar-On et al. 2016:6508). Some contend that since plant and animal densities and distributions fluctuated, human curiosity helped people locate new foods (Richerson and Boyd 2005; Meltzer 2009). Many hunter-gatherers know ranges of 300,000 km^2, usually far larger than their home ranges (Kelly 2013a). Thus, curiosity coupled with uninhabited land could have been a strong pull factor. Push and pull factors may also have worked together.

Curiosity raises another question: When did we develop our current package of mental capabilities, i.e., become "cognitively modern"? Since we can only infer these capacities from artifacts and since humans did not display all intellectual functions in any one place, scholars debate cognitive modernity in two positions. Some argue for a creative "explosion" because of dramatic improvement in blade technology, the appearance of art objects, and the settlement of Australia through the use of sea travel over many thousands of years. They think this occurred around 70,000 years ago or later (Brooke 2014; Mithen 2020). Others maintain that personal adornment, rock art, grave goods, iconic representation, and materials preparation all show that humans were cognitively modern by 100 kya, and archaeological changes demonstrate a gradual evolution involving cumulative learning based on those capacities (Henrich 2015; Von Petzinger 2016; Wood 2019; see also Kuykendall and Heyerdahl-King 2014; Kaifu et al. 2015).

This debate overlaps with discussions of archaic humans. Human skeletal evidence is often missing at sites where obviously human, but either archaic or modern, tools are found. Tool assemblages before 50,000 BP were simple and uniform across Eurasia, while tool kits after are consistently more sophisticated and varied (Kaifu et al. 2015; Kuhn and Steiner 2018). Some archaeologists conclude that the earlier assemblages indicate archaic humans and the later tool kits modern humans; others think modern humans produced some earlier assemblages (Kuykendall and

Heyerdahl-King 2014; Kaifu et al. 2015; Kuhn and Steiner 2018). Cognitive modernity is thus difficult to infer, so curiosity is also.

Cognitive modernity matters because humans had to learn many new environments to populate the planet. Foragers usually need hundreds of years to learn a specific environment thoroughly (Richerson and Boyd 2005; Meltzer 2009; Kelly 2013a; Mithen 2020). Thus, migrating humans needed several cultural traits: general knowledge of plants and animals; geological knowledge to find raw materials for tools; adaptable social organization; willingness to learn and to share information; and good wayfinding skills (Meltzer 2009)—and curiosity.

Questions around why humans populated all continents also raise the question of how they did it. Scholars suggest two strategies for migrating into an uninhabited area, called either "cautious" and "bold" (Meltzer 2009), or "contented" (usually older foragers) and "adventurous" (usually younger) (Wood 2019). The cautious, contented strategy would have involved slow expansion outward due to population growth and thus increasing pressure on land. Bold, adventurous movement would have consisted of migrating quickly into an area, using up food, and moving on. As discussed below, all foraging bands existed within larger communities, providing allies and mates. Bold movement might have been risky because it isolated a band from these groups, but it might have paid off in easy living.

A surprisingly small number of people could have moved successfully. Mathematical simulations showed that even under varying conditions, a group of 30 people with sufficient females could increase in size over several centuries (McArthur et al. 1976), making a bold strategy viable. Older foragers today are better at all survival tasks (hunting, gathering, and childcare) than younger foragers (Hrdy 2009; Kelly 2013a; Henrich 2015); prehistoric older foragers were likely similar. Were the "adventurous" always young, or was bringing experienced older people a better survival strategy in new environments? I think cautious movement happened most often under two conditions: along north-south axes and at abrupt environmental transitions. Scholars suggest that environments varied greatly from south to north (Kaifu et al. 2015; Mithen 2020; Kuhn and Steiner 2018), implying less variation from west to east. Examples of abrupt transitions include where the North Gangetic Plain meets the Himalayas in Asia, the ocean between Sunda and Sahul, and the Rockies meeting the Great Plains in North America. Extended scouting makes more sense in these cases.

Evidence of cautious strategies exists. Repopulating Europe after the LGM seems to have happened in two phases. Archaeologists have found what they interpret as overnight camps. People "likely... traveled north during the summer, returning to southern base-camps to report on what they had seen." This lasted "about 500 years, or twenty generations." Cautious indeed! Next, "groups of families relocated their base-campus to live permanently in the north" (Mithen 2020:389, all quotations). Exploratory scouting strongly implies an awareness that moving into unfamiliar, uninhabited lands was risky. Other scholars contend that some north-south migrations were quick. Brooke (2014) argues for fast northward Eurasian expansion, but the unstable Pleistocene climate forced die-offs. This implies repeated migrations. Meltzer argues that the corresponding expansion in the Americas was also fast, reaching Tierra del Fuego in southern South America "within 1,000 years of leaving Alaska" (Meltzer 2009:5), but the arrival debate (discussed above) raises questions about this argument.

The overall migration strategy probably modified travel speed. Many scholars conclude that humans followed a coast-and-river strategy (Petraglia and Boivin 2014; Pettitt 2014; Kaifu et al. 2015; Mithen 2020; Wood 2019). Relying on "rich marine resources" (Wood 2019:28), people moved along the coasts from one river mouth to another for fresh water, then moved inland up rivers. Other scholars argue for an inland route following lakes into Central Asia (Petraglia and Boivin 2014; Kaifu et al. 2015). Deglaciating ocean rise covered coasts several miles inland, so the evidence is buried, but the strategy still makes sense. Figure 3.1 implies a single wave of humans at all points on this journey, but a coast-and-rivers strategy implies more locally complex decisions along the way (Pettitt 2014).

Limited evidence restricts what archaeologists can infer about why and how humans populated the planet. When humans became cognitively "modern"—developing curiosity, cumulative learning, and boldness or caution in movement—is still debated, and archaic and modern humans may both have had these capacities. Certainly, people on the global human frontier constantly learned about plants, animals, land, and climate. Population growth and environmental richness may have pushed or pulled human groups along. People may have moved mostly along coasts,

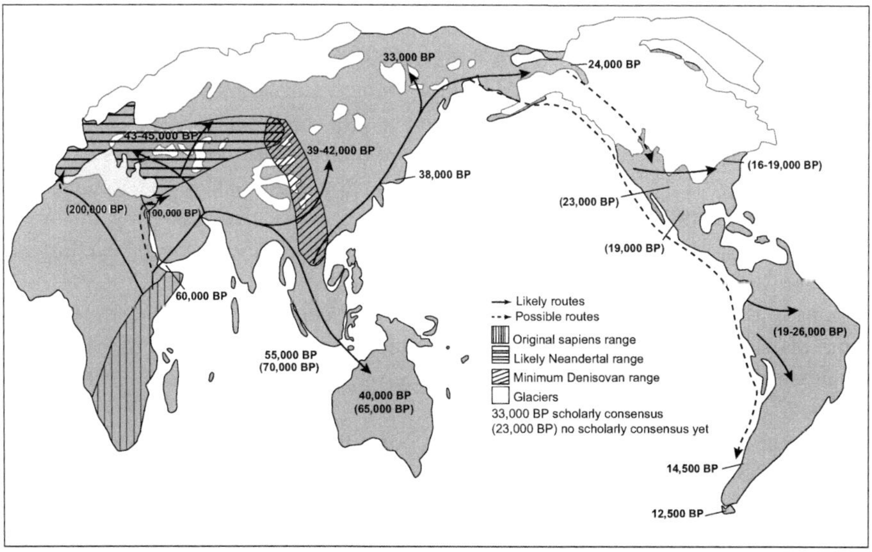

Fig. 3.1 Global Human Dispersal. *Sources*: Meltzer (2009), Jochim (2011), Bellwood (2013), Gamble (2013), Davidson (2014), Kuykendall and Heyerdahl-King (2014), Moore (2014), Petraglia and Boivin (2014), Groves (2015), Kaifu et al. (2015), Stoneking and Harvati (2015), Clarkson et al. (2017), Mithen (2020), Posth et al. (2017), Wood (2019), Ardelean et al. (2020), Pauketat and Sassaman (2020), Petr et al. (2020), Bennett et al. (2021), Cartmill and Smith (2022) and Demeter et al. (2022)

then inland along rivers, or maybe they used lakes. Migration speed depended on factors such as bringing elders along or not; the newness or familiarity of environments; and the availability of fresh water.

Social Patterns of Foraging Bands, or "Simple Hunter-Gatherers"

Despite these variations, groups on the global human frontier—and *only* on the frontier—seem to have been organized similarly. Understanding this first way of life is important in itself and important for understanding what came later. The following account relies on ethnographic and archaeological evidence modified by genetic and linguistic evidence with some speculation.

Band Economies

Prehistoric bands produced for "subsistence," or physical survival, through foraging. They made tools, including clothing and shelter, to do this and shared or traded much of what they foraged, consuming it both practically and symbolically.

Foraging consists of taking wild foods from the environment, including hunting, gathering, and fishing. Land was dramatically more productive than now (Bar-On et al. 2016), except during the LGM. People hunted individual animals, collectively attacked herds or drove them over cliffs, and caught small game in nets and pit traps (Kaifu et al. 2015; O'Connor 2015). Collective hunting has more success than individual hunting (Kelly 2013a). Since most megafauna were dangerous, it made even more sense on the frontier. By 60 kya, humans were also gathering a "broad-spectrum" of plants (Liu and Chen 2012; Pettitt 2013). In Southwest Asia, humans ate over 100 species of plants, including seeds, leaves, fruits, wild cereals, and legumes (Bar-Yosef 2014), and human sites in central Thailand and in South America show many plant remains (O'Connor and Bulbeck 2014; Moore 2014). Neandertals and Sapiens both seem to have ground grains into flour in Italy around 40 kya (Lippi et al. 2023).

Scholars debate the division of labor in prehistoric foraging bands. Many previously thought that men hunted and women gathered among modern humans and maybe among Neandertals (Price and Feinman 2010). Others argue that Neandertals focused mostly on large game: Women and children chased megafauna, and men brought them down. Modern humans supposedly developed the sexual division of labor in Africa, arguing that it was learned, not "natural." The specialization and complementary nature of such a division would have given modern humans a survival advantage over archaic humans (Kuhn and Stiner 2006, Kuhn and Steiner 2018).

Lately, however, others object. Re-analysis of ethnographies shows great flexibility in labor patterns. In some societies, men and women divide labor rigidly; in others, there is more flexibility; and in others, women hunt as men do (Owen 2005; Kuhn and Stiner 2006; Kelly 2013a). Individual women learn to hunt sometimes in unusual circumstances, sometimes in regular ones (Kuhn and Stiner 2006; Jarvenpa and Brumbach 2014). Re-analysis of forager ethnographies from 1888 to 2018 CE shows that women actively hunted in 79% of societies—often large game, bringing infants in slings and children on foot (Anderson et al. 2023). Prehistorically, female skeletons were buried throughout the Americas with big-game hunting tools just after the frontier disappeared; scholars estimate that women did 30–50% of the hunting (Haas et al. 2020). Skeletal wear and traumas related to repetitive motion are similar between the sexes in both Neandertals and Sapiens (Lacy and Ocobock 2024), and females are physiologically more adapted to long-distance "persistence" hunting (hours-long chases of prey) (Ocobock and Lacy 2024). Some think that the stereotypical division appeared in hunter-gatherer groups only in the last few centuries (Owen 2005).

The sequence of tasks helps us understand hunting further. Turning a live animal into food involves tracking, pursuit, killing, carrying, and butchering. Archaeologists call such a sequence of operations the *chaîne opératoire* (Sellet 1993). Such chains vary by global region and locally within them, but they always involve both female and male labor (Brumbach and Jarvenpa 2006). In the tropics, people rely more on gathering plants; near the poles, they eat much more meat, gained from hunting. In a rigid division, men would have provided more calories in the Arctic, women more in the tropics, yet even in the Arctic today, women butcher meat to prevent spoilage (Brumbach and Jarvenpa 2006; Mitchell 2009; Jarvenpa and Brumbach 2014). On the frontier, the effort and danger involved in taking down massive animals would have been wasted without butchering.

Humans also developed tool-making sequences. These consisted of quarrying raw materials, creating generalized "cores," and from those cores, making scrapers, blades, needles, and arrow- and spearheads. Hominins started working with rock; humans later developed multimaterial tools from rock, wood, antlers, bone, marine shells, ivory, and animal hide. They seem to have developed clothing between 170 and 83 kya (Toups et al. 2011).

As discussed, any division of labor only works with reciprocal relations. Contemporary foraging groups share food in what scholars call "generalized reciprocity": When a hunter kills an animal, everyone in the group gets some meat. Hunters share large animal kills more than small ones since hunters cannot eat it all; obligating everyone else is smarter than wastage. Only a few hunts succeed, and some hunters fail for a while. Thus, repayment sometimes takes a while; receiving meat after a kill obligates each hunter to share meat when successful. It becomes "generalized" when hunters must give meat to everyone every time.

However, sharing practices vary among modern foragers and probably varied among prehistoric ones. First, "most food in most foraging societies remains within the family that procured it" (Kelly 2013a:149). Beyond families, meat is shared much more often than plant food; authors debate whether people share plants or not

(Read 2012; Kelly 2013a). Some Inuit groups have "seal-partners" in the winter, the two sharing cuts of meat only between themselves (Read 2012). Also, unmarried men share the most; distant kills are less shared than those close to camp; and sharing increases in variable or risky environments, lessening in reliable ones (Kelly 2013a). Poor hunters may contribute in other ways, such as gathering firewood.

Sharing increases survival. Among the Aché, foragers in South America, food-sharing results "in an 80 percent increase in nutritional status and a nearly three-fold increase in the predictability and regularity of daily food intake" (Bogin 2001:161). Prehistoric foragers very likely shared in this way. Also, collective hunting requiring multiple attackers might instead generate the idea that everyone deserves a piece of the kill. Fish might have been shared rarely.

Reciprocity takes other forms. Seal-partner hunting, mentioned above, is called "balanced" reciprocity; another form is what we call "trade," usually in terms of reciprocal gift giving. Prehistoric foragers traded objects over hundreds of miles. Researchers have found ostrich-egg jewelry from around 33,000 years ago in what is now Lesotho, in what was an ostrich-less region, from hundreds of kilometers away (Stewart et al. 2020). Archaeologists found flint tools in Russian sites quarried 150–300 km away; shells in Austria from the Mediterranean and Black Sea; and Pyrenean stone, and Atlantic and Mediterranean shells at inland French sites. Obsidian traveled between islands in what is now Japan. In North America, objects made of obsidian, shell, and copper moved hundreds of miles from their natural sources (Barham and Mitchell 2008; Price and Feinman 2010; Jochim 2011; Pettitt 2013; Kaifu et al. 2015).

Ethnographies help us understand archaeological findings. Each member of the!Kung regularly gives gifts to kin in other bands, later receiving gifts of equivalent value (Flannery and Marcus 2012), so each person develops a personal network. Everyone exchanges gifts, and each network links to all others. Often, a receiver gives an object to another, creating "down the line" trade; repeated giftings or individuals traveling long distances to trade can move objects hundreds of miles. Finally, some members barter, which is the immediate and "direct exchange [of objects] of equivalent value" (Stanish 2017:132). On the frontier, some environments might have lacked raw materials for tools; trade networks could help groups cross geologically poor spaces to better places.

Prehistoric foragers consumed objects both practically and symbolically. Food, drink, and clothing had obvious practical uses. Prehistoric foragers also sometimes constructed houses (Price and Feinman 2010; Pettitt 2013). Interestingly, nearly every utilitarian object is decorated, and many objects have only ritual value (Price and Feinman 2010; Pettitt 2013). Prehistoric peoples made and used necklaces, bracelets, figurines of humans, and pendants of shells and ivory. They often used red ochre (iron ore) to decorate corpses in graves (Pettitt 2013). The large panels of prehistoric art in European caves may have been spiritually oriented (Pettitt 2013). Collective consumption now includes, and may have then included, religious rituals such as healings and initiations. Ritual objects include musical instruments—drums, horns, and rattles—and masks (Hayden 2003). Some argue that foragers commonly hold feasts (Flannery and Marcus 2012), while others argue that they did only seasonally or from windfalls (Hayden 2014).

Prehistoric foraging bands survived by gathering plants and hunting animals and sharing them with each other. They consumed products both practically and symbolically.

Kinship in Bands

Kinship institutions were and are simple. On the frontier, I think kinship tended toward monogamous marriage, multilocal residence, and bilateral descent. I also think that people were flexible. Kinship defines communities of interlinked bands today, and probably then as well. Inferring prehistoric kinship patterns is difficult from archaeological evidence (Ensor 2013), so I will infer patterns from ethnographic, linguistic, and genetic evidence.

Marriage among foragers today is usually with distant kin and monogamous. People are related to nearly everyone they know, so they usually marry beyond their first cousins or cousins' children. A bold migration strategy might have led to first-cousin marriage; cautious groups would have had more potential mates. Parents tend to arrange first marriages; people arrange their own later ones. Polygyny is rare among foragers today, polyandry even rarer (Kelly 2013a), though informal polyandry is rather common (Starkweather and Hames 2012) and likely existed in early human history. Simplicity aids mobility, and monogamy is simple. Still, Inuit marriage types suggest wide possibilities: love marriages, polygynous marriages, polyandrous marriages, and comarriages, in which two men marry two women (Flannery and Marcus 2012).

Ethnographically, hunter-gatherer sexual practices were even looser. Many North American, Australian, and sub-Saharan African forager groups accepted casual adolescent same-sex sexual relations, expecting the adolescents to grow out of them, though some groups in Africa had little homosexuality. Most married a member of the opposite sex; a few maintained same-sex sexual relations (Neill 2009). Across South America, there was a third gender consisting of males who engaged in the passive role in sex with other men. The "two-spirit" person was a common institution in North and South America and Siberia. There were both male and female two-spirits (Neill 2009).

Contemporary hunter-gatherers also have divorce. Among the!Kung and the Agta, most divorces occurred in the first 5 years of marriage, being rare thereafter (Howell 1979; Headland 1987). Divorce is twice as common as widowhood. Most divorced women remarry, while many widows do not. In both hunter-gatherer groups, the couple accomplishes both marriage and divorce by stating the fact and then carrying out the subsequent actions of living together, consummating the marriage, and raising children, or separating and living apart. This was likely for prehistoric foragers.

Several factors influence postmarital residence. Nearly all foraging bands require the groom's service to the bride's family (Flannery and Marcus 2012), requiring premarital residence. Cross-culturally, women do most newborn and small infant care (Pasternak et al. 1997). With toddlers, childcare varies more, so mothers are more mobile. Alloparenting increases child survival, and grandmothers and sisters are common allomothers (Hrdy 2009). Father alloparenting varies partly by hunting strategy; longer-distance hunters do less childcare. Patrilocal foragers (living with the groom's family) tend to know specific regions, limiting them geographically. Matrilocality (living with the bride's family) tends to familiarize hunters with animal tendencies, opening them geographically. Fishing is common among patrilocal bands; long-distance hunting bands are often matrilocal. Bilocality occurs in unstable environments with varying resources (Layton 2008). Child socialization requires exposure to different environments, so parents might have lived sequentially with different groups (Binford 2001).

Marital residence probably varied more on the prehistoric frontier. Anthropologists previously assumed patrilocality (Service 1962). Others argue that prehistoric foragers were likely matrilocal (Knight 2008; Opie and Power 2008) because of paternity uncertainty, strong relationships between brothers and sisters, and grandmotherly infant care. Also, "genetic data on sub-Saharan African hunter-gatherers indicate a matrilocal residential bias" (Knight 2008:80). In some bands today, couples first live with the groom's parents and move to the bride's mother at the first pregnancy (Hrdy 2009). Kelly (2013a) argues that bilocality is an ethnographic stereotype, yet his own data (p. 226) show that people in at least half of all bands are multilocal, and many others make various "exceptions." Others argue that people prefer living with siblings over parents (Dyble et al. 2015). Marital locality on the frontier was probably flexible, with a bilocal bias. Boldly moving bands probably had fewer options than cautious ones.

Descent among ethnographically studied foraging bands tends toward bilaterality: Children treat both their mother's and father's relatives as relatives. Three-quarters of modern foragers have bilateral, ambilineal, or double-descent systems (Kelly 2013a). Patrilineal bands, in which the father's "line" matters more, do exist but are rare; matrilineal systems are rarer. Prehistoric descent is difficult to know, but a reconstructed history of the Khoisan language family shows that "bilateral descent appears to have been the original rule" (Ehret 2008:225), so at least some foraging bands on the frontier were bilateral. Knight (2008) argues for an overall matrifocal bias.

Many scholars argue that any band is "a continuously shifting combination of" members, not a permanent, cohesive group (Binford 2001:465–466; Dyble et al. 2015; Henrich 2015). The preference of living with siblings and alternation between the spouse's families (Dyble et al. 2015) means that modern bands are a mix of strangers and kin. Any given person sees half of their band as close kin, a quarter are distant kin, and a quarter are strangers. Constant interband movement introduces people to new people, building social networks.

Foraging bands today frequently interact with other bands within communities of a few hundred people. Kin categories help people identify relationships: Direct kin relations, indirect ones, or fictive ties. These larger communities are the source for constant recombination of residential groups (bands) (Read 2012), and they probably existed on the prehistoric frontier. A band of 25 people isolated in new territory is at risk; staying in touch with their larger communities, however tenuously, made more sense. Still, some bold, isolated groups probably existed.

Foraging bands on the global human frontier probably had kinship patterns broadly similar to those today. They likely existed within larger communities of a few hundred kin with whom they traded, married, and visited. Marriage was usually monogamous and arranged variously. Residence was likely relatively fluid, subject to visits, new children, and the vagaries of environmental richness. Descent was probably most often bilateral, but flexible.

Band Politics

Constant interaction, band turnover, and collective action are all political, but frontier foraging bands have and probably had few political institutions. They are (and were) simply and loosely organized, usually making decisions by consensus; leadership is, and probably was, episodic, not permanent; and there was violence but seemingly no war.

Several factors shaped group cohesion on the prehistoric frontier. Constant recombination might lower cohesion, and people do not always get along. Risky moves into new territory would require trustworthy companions and thus cohesion, as would material interdependence and emotional ties. Opting out of food-sharing is not merely free-riding; it violates cultural rules of kinship which structure people's identities (Read 2012).

Decision-making among foraging bands now takes two forms. Most decisions result from consensus after time-consuming discussions (Johnson and Earle 2000), but groups sometimes have leaders. Generally, it is "episodic," appearing only during certain situations, based on personal qualities. Some argue that leaders mostly handle interband relations (Binford 2001:352); others say that leaders develop only in specific situations requiring quick decision-making and rapid coordination (Lewellen 2003). For instance, the Shoshone had a "rabbit boss" who organized the annual rabbit hunt: Everyone chased rabbits into nets, then clubbed and skinned them. Otherwise, the rabbit boss was just another group member (Johnson and Earle 2000). The !Kung have permanent headmen or headwomen who can only persuade or lead by example (Nolan and Lenski 2011). Both possibilities likely operated on the frontier.

Scholars very contentiously debate how violent prehistoric bands were and whether this violence ever amounted to warfare. Neandertals were certainly violent: "[P]ractically every reasonably complete Neandertal skeleton has at least one healed injury" (Smith 2017:37), mostly in the upper body. These are consistent with what scholars have concluded about Neanderthal hunting tactics involving close-in stabbing, something that is quite dangerous with large animals. Many other injuries are on top of the skull, which falls rarely create; appear on the left side where a right-handed assailant would strike; and are sometimes healed and sometimes unhealed. These are best explained as interpersonal violence (Smith 2017). Prehistoric and modern humans were, and are, just as violent. Homicide rates among foraging bands today are high, occurring among the "vast majority" of hunter-gatherer groups (Keeley 1996:30–31). Homicides occur between 4 and 15 years apart today (Kelly 2000, 2013a). Most prehistoric Sapiens skeletons have injuries similar to the Neandertal ones (Gat 2017).

Whether this amounted to intergroup violence or "war" depends on definitions. Scholars define war as lethal conflict between independent polities (LeBlanc and Register 2003; LeBlanc 2014); as violence between two foraging bands (Boehm 2013); and as "the killing of any member of the enemy group" (Kelly 2000:5). Types of violence within and between bands include a group executing a disruptive and violent member (Boehm 1999; Kelly 2000); one man killing another man in a different band (Boehm 2013); and ongoing feuds or pitched battles between groups (Kelly 2000; Boehm 2013). Kelly argues that these gradations matter: among the most simply organized foragers—likely on the frontier—the tendency to bilateral kinship orients individuals only toward other individuals, not toward groups, which creates individual, not group, liability for killing (Kelly 2000). The likelihood of multilocal residence and consequent group fluidity suggests that group cohesion was high enough for survival but too low to support intergroup violence.

Many factors support this conclusion. Ethnographically, leaving rather than fighting is a typical foraging band strategy to reduce tensions (Kelly 2000; Hrdy 2009; Meltzer 2009; Boehm 2013; Kelly 2013b), and the prehistoric global human frontier was exactly when people could mostly easily move. Surviving skeletal evidence shows no organized, collective violence for most of human prehistory; it only appears around 12 kya (Haas and Piscitelli 2013) when the main global frontier was gone. Furthermore, scholars estimate the human population at half a million to a few million around the entire world at the LGM (Haas and Piscitelli 2013), suggesting an extremely sparsely populated world, and intergroup violence is rare or non-existent at very low population densities (Kelly 2000).

The debate is complicated by assumptions about human nature and potential exceptions. Some assume that "aggressive hair-trigger responses to threats, the desire to win, and the desire for revenge" would be highly adaptive prehistorically (LeBlanc 2014:31). Others think that on the frontier, "hostility toward strangers would have been decidedly disadvantageous" and people were probably able to "recognize even strangers as friends" (Meltzer 2009:67). I tend to agree with those who say that people are predisposed to both peace and violence (Gat 2017; Kissel and Kim 2019). For survival, cooperation was necessary, but not getting along with some people is also normal (Kelly 2013a). Since "big-game hunting and warfare are essentially the same adaptation" (LeBlanc 2014:31), they were ready for it when conditions led them into fighting with other groups. Two situations for this might have existed. First, southward retreats during the LGM might have caused population pressure in specific locales (Boehm 2013). Second, inter-subspecies violence was possible. A Neandertal skeleton from 50,000–75,000 BP had an injury that researchers think might have been made with a thrown spear. If Neandertals did not throw, a modern human might have killed them (Smith 2017:40–43), maybe in a group-to-group fight. Still, the lack of skeletal evidence, the ease of leaving, and the fluidity of groups made intergroup violence much less likely than interindividual violence, so I conclude provisionally that warfare did not exist on the global human frontier.

Foraging band politics on the prehistoric frontier were likely informal. They were usually consensus-based. Leaders probably appeared only for specific tasks. There was plenty of interpersonal violence, but violence between groups seems not to have occurred.

Band Inequalities

Both power and inequalities within bands were limited. Among foragers today, and probably in the past, there are four foundations for inequalities: age, personal characteristics, economic resources, and sex and gender. Inequalities always exist in foraging bands (Kelly 2013a), but there are and were no *patterned* inequalities except age.

As everyone grows older, age is the most physical foundation for inequality. Prehistorically, "children comprised at least 40 to 65% of the population" (Nowell 2021:1). They had some influence, but adults obviously had more. Furthermore, physical abilities increase up to a point, then gradually decay, but emotional wisdom and mental knowledge always increase from ongoing experience and reflection. Thus, middle-aged men and women have the most power in bands today (Van den Berghe 1973; Flannery and Marcus 2012), and likely prehistorically.

A second foundation for inequality was individual characteristics, such as "skills, intelligence, [and] athletic abilities" (Binford 2001:467). In a group of 25 people, individual differences matter; people mentally rank others by their personal characteristics. Some scholars think hunting is the most important skill (Kelly 2013a), but tracking skills, plant finding, peacemaking, public speaking, and age-related experience are also important group resources. The most competent, usually those in middle age with the most skill and as-yet unreduced physical capabilities, would likely have the most influence and prestige (Henrich 2015).

As mentioned, ethnographers note that occasionally people wanting to dominate—"upstarts" or "aggrandizers"—appear randomly in bands (Boehm 1999; Hayden 2014). Since this causes conflict, it is dangerous for foragers (Hayden 2003); it would have been more dangerous when migrating boldly. Thus, bands develop traditions to limit aggrandizers. Most people want neither dominance nor subordination, so they "form a moral community [and] develop an egalitarian ethos" through sheer numbers (Boehm 1999:12). Different groups have different strategies. The!Kung institutionalize insults and discounting to maintain equality. The Inuit have more humor-based traditions. The Hadza and many Australian groups also use such tactics (Flannery and Marcus 2012; Kelly 2013a), suggesting that aggrandizers appear regularly. Bands also engage in "demand sharing," constantly requesting gifts of food and other objects, and punish those who are stingy (Kelly 2013a). When these tactics fail, groups increase the harshness of their sanctions, up to and including execution (Boehm 1999; Kelly 2000).

Third, the simplest bands are economically egalitarian and very likely were prehistorically. They enjoy only occasional and therefore "opportunistic" material surpluses, such as a beached whale or an unusually lush year (Thompson and Moore 2015), but this unpredictability cannot support economic inequality. Also, landownership institutions probably did not develop on the frontier; treating land as a common resource made more sense because food availability was probably uncertain in new landscapes until people learned them.

Finally, sex and gender are contentious issues in hunter-gatherer studies. Many ethnographers argue that men have more power in band societies: There is a "widespread tendency for male hunters to actively dominate their mates within the domestic unit" (Boehm 1999:62). Others emphasize how men provide meat and fats, giving them slightly more power and prestige (Kelly 2013a). Many disagree. Arguments that early human groups were matrifocal suggest the power of mothers (Knight 2008), and a need for negotiations between men and women about postmarital residence implies basic gender equality (Dyble et al. 2015). Endicott (1999) argues three points. First, many groups value foods other than meat, such as fruits or fats. Second, men tend to decide men's issues, and women decide women's issues. While fathers often arrange daughters' marriages, mothers decide whether the marriage occurs. Sometimes the oldest male leads a group, yet other times there are "natural leaders," including women. Third, bands do not spiritually devalue women; many tell stories of powerful goddesses. Finally, burials show no differences between children, women, and men (Lacy and Ocobock 2024).

While ethnographers debate gender (in)equalities in bands, most archaeologists simply seem to assume sexual inequality (Bolger 2006); a growing minority question this (Jarvenpa and Brumbach 2014). Critics say that archaeologists treat the gendered division of labor as "a list of things that women cannot do, should not do, or are prohibited from doing by men," and assume that pregnant and nursing women are relatively immobile (Brumbach and Jarvenpa 2006:505, 506). Chatters (2014) argues that skeletal evidence shows that Paleoindian males dominated their wives with harsh but nonlethal violence; this pattern disappeared over time as women "domesticated" them. In contrast, Snow (2013) argues that most handprints in cave art were female, suggesting that women engaged in important behaviors. Finally, the main objection to ethnographic inference—that today's foragers are surrounded by larger, inegalitarian societies (Wobst 1978)—suggests that some male hunter-gatherers may have learned gender inequality from others; it may not have existed prehistorically.

Geographic variations influence the debate. As mentioned, perhaps men bring in more calories in Arctic regions, women more in tropical regions, and contributions are mixed in temperate zones (Binford 2001). Others say that men provide half of all calories in the tropics and more elsewhere (Kelly 2013a). Either way, the moving prehistoric frontier might have made gender equality a dynamic condition with men having more power in the north, while gender relations equalized as people moved south. Chatters's (2014) argument about domestic violence supports this. Yet if women had their own weapons (Haas et al. 2020), this could be mistaken.

Overall, prehistoric foraging bands on the global human frontier might have had the fewest inequalities in human groups ever because the only universal inequality was age. Otherwise, economic resources providing power were minimal, personal characteristics varied, and inequality between women and men seems to have been minor or nonexistent in foraging bands.

Band Ideas

As a reminder, culture is not merely "information." People use concepts not only to store information; they construct belief-systems to make sense of the world. They also care deeply about some beliefs, and they usually act on their beliefs and morals. Again, current beliefs and practices are our best working model for prehistoric band belief-systems. Anthropologists group forager belief-systems under the term "animism." Rituals and spiritual leaders help foragers implement their beliefs daily. Their material culture included art, decoration, and tools.

Animism is a general term; each group has its own beliefs and practices. Commonalities among foragers include believing that "there is a vital force that imbues everything in creation" (Hayden 2003:52), so all elements of the natural world are alive and interconnected (Peoples et al. 2016): People, plants, animals, mountains, rivers, the weather, and even stars. Thus, eating plants and animals is a sacred activity; sickness is lost vital force; and connection with this vital force creates euphoric states. Most foragers also have two general concepts, the multilayered universe and the ordering of time. One common layering is the living world, the upper world, and the lower world. Many, but not all, foraging peoples believe in reincarnation—the human spirit leaving one body at death and being reborn into another (Flannery and Marcus 2012). Creation myths usually suggest two states, one past and one present. The past was less formed and incomplete. Among African Bushmen, the "First People" were part animal and part human; an act of transformation separated them into animals and humans (Guenther 1999). Aboriginal Australians talk about a "Dreamtime" in the past, one that blends with the present.

This sense of interrelatedness helps foragers manage physical resources. For instance, foragers can transplant plants, yet often refuse because they feel the plants "belong" spiritually to a given region. If animals are not hunted, they will feel that humans do not depend on them; appropriate prayers must accompany the hunt. Foragers use night skies for navigational, but also for religious and calendrical, purposes, and often have solstice ceremonies. When the ancestors moved through the landscape, they created trees, water holes, and rock formations, so "the landscape exists in the resident's mind as a mental map of lines and places that make up the past, present, and future of existence" (Steadman 2009:92).

Most foragers act out their belief-systems in rituals producing ecstatic states of consciousness. These are "a direct connection with the sacred forces of the universe" (Hayden 2003:7). They enter trances or sing, dance, or beat drums. These rituals create strong emotional bonds between people; sharing ideologies and emotions commits people "to ideals that no earthly external conditions appear capable of altering" (Hayden 2003:31). All groups have certain individuals who are skilled in interceding with sacred forces, usually called "shamans" (though each group has its own term). Shamans have many roles: healer, ceremony-leader, problem solver, diviner of the future, sorcerer, pharmacologist, and storyteller (Steadman 2009; Peoples et al. 2016). In rituals, shamans use masks, drums, and rattles to enter ecstatic states, to mediate between the levels of the universe. They rarely use hallucinogenic drugs to enter trances, instead using "induced starvation, thirst, physical exhaustion, [and] sleep deprivation" (Hayden 2003:33). During trances, shamans transform into animals, usually local animals, such as bears in North America, jaguars in South America, and antelopes in southern Africa. They then travel across the land or into other spiritual realms, usually to heal the sick or improve hunting.

Shamanic traditions vary widely. For instance, "among the Bushmen of South Africa, many people take on this role" (Hayden 2003:48). In other groups, only a few are shamans. It is a part-time role; mostly, they are just group members. In most bands, shamanism is gender-specific, usually male (Guenther 1999), though some groups have female shamans. Many Native American groups believe that the Supreme Being is neither male nor female, so there is a "strong association between two-spirits and shamanism" (Neill 2009:38). Two-spirit people are often considered powerful; shamans defer to them in some ceremonies. Foragers often believe that shamans help their communities, but others think they make people sick or kill them magically.

How does this apply to the prehistoric frontier? Through linguistic reconstruction based on forager groups from around the world, Peoples et al. (2016) argue that the elements of forager spirituality appeared in a sequence. Animism appeared first, then belief in a single creator deity. Following that, ideas of the afterlife and the institution of shamanism formed (perhaps after humans left Africa), and finally, rituals that promote "high levels of cooperation" took hold (Peoples et al. 2016:262).

Morality also supports cooperation. It is impossible to reconstruct specific prehistoric systems of morality, but ethnographic inference helps us make general statements. Within families, the incest taboo is effectively universal (Pasternak et al. 1997:104). While foragers privilege self and family, they also foster group-oriented values, including generosity, and place importance on "being humble in demeanor, along with being truthful" (Boehm 2012:35), which supports egalitarianism. Foragers also monitor others' behavior through gossip and punish violations through stern advice, mocking, group shaming, ostracism, expulsion, and occasionally execution. Boehm (2012) argues strongly that prehistoric foragers would have had similar moral systems but also argues that forager morality is flexible in practice. Some argue that foragers do (and did) not worship spirits so much as "treated them as one would acquaintances" (Friesen 2015:29), and that beliefs about spirits were separate from moral norms in dealing with people.

"Art" for foragers is the ability to make images and is "integral to the business of living" (Lewis-Williams 2014:625), not a special practice. Enough has survived that ethnographic inference is less helpful. Beads were found in Blombos and Sibudu caves in southern Africa dating from 77,000 BP (Lombard 2013) and from later periods in Greece and Russia (Smail et al. 2011). Humans also sculpted human figurines in bone, clay, stone, antler, and wood (Derevianko et al. 2014). They made music: Researchers found (and played replicas of) 40 different flutes that have been found across Europe. Art, music, and dance may have "evolved as a package that committed sacred lore to memory more effectively than any lecture" (Flannery and Marcus 2012:63).

Famously, prehistoric humans made cave art. It started when humans entered Europe (Von Petzinger 2016), and cave and rock art is found "on all of the inhabited continents" (Steadman 2009:63; Von Petzinger 2016; Brumm et al. 2021; MorcoteRíos et al. 2021). Humans made three types of images: Humans, therianthropes (human-animal hybrids), and geometric shapes. A common image in Europe and Australia (Steadman 2009; Von Petzinger 2016) is the hand-stencil, formed by putting one's hand with fingers outstretched on a cave wall, then blowing pigment at it; lifting the hand away reveals a hand shape. Snow (2013) concludes that women made more of these stencils than men did, though some disagree (Hughes 2013).

Humans may also have communicated with each other through cave signs. Von Petzinger (2016) found that the same geometric signs appeared in caves from Portugal to Russia; two-thirds of these signs appeared in the earliest sites (i.e., on the frontier). She categorized them into a 32-sign typology, commenting, "This is a surprisingly small number... [in] an entire continent and a time span of 30,000 years" (Von Petzinger 2016:163). Most signs fit this system, and she argues that this was a system of signs with which European bands communicated across time and space.

People also made a wide range of tools. They include the spear-thrower (atlatl), which doubles the thrown distance; the bow and arrow; rope; poison for spear and arrow tips; nets for fishing and fishhooks; eyed needles for sewing sophisticated clothing; huts and igloos for shelter; boats for travel and seahunting; slings for baby carrying and containers, both important for plant gathering; and torches (Price and Feinman 2010).

Archaeological sites and reconstructive archaeology show evidence of child socialization. Knapping stones takes years to learn, best started when young. Many sites from this era have "debitage" or leftovers from knapping that show errors in raw materials choices and in techniques (Nowell 2021). Both girls and boys likely learned knapping, probably starting as play. Children were also involved in making cave art, sometimes as novice artists (Nowell 2021).

Ethnographic inference and archaeological evidence suggest that as prehistoric foragers migrated globally, they were probably animistic; likely had shamans and regular rituals; made fixed and portable images; and fashioned musical instruments, weaponry, clothing, and buildings, generating great diversity as they moved through different physical environments.

Band Demography

Prehistoric frontier foragers lived difficult lives because their mortality rate was so high, which spurred high fertility. The total global population was extremely small, so human density on the planet's land surface was low, even when the frontier ended. Bands averaged about 25 people and were always part of larger communities. Mortality, fertility, group composition, and population density all influence migration strategies now, and probably on the frontier.

Death was a constant presence in the human past. Up to about 300 years ago, between a third and a half of all people died before age 20 (Livi-Bacci 2017). Among foragers today, "50–60 percent of children survive to age fifteen (of those, 60–70% reach age forty-five)" (Kelly 2013a:200). In a sample of 17 modern hunter-gatherer peoples, mortality at age 15 was almost 50% (Volk and Atkinson 2013). Prehistoric evidence supports these historical and ethnographic mortality trends. Paleodemography, the study of prehistoric populations, is contentious because the evidence is messy and difficult to analyze (Konigsberg and Herrmann 2006). "Sexing" skeletons is easier for adults than for children, individual variation in growth stages makes determining age at death difficult, and environmental factors influence bones. Representative skeletal samples are difficult to find. Paleodemographers still think that ethnographically described forager mortality is a good model for prehistoric forager mortality (Konigsberg and Herrmann 2006).

Most deaths worldwide through the life course followed a common pattern. Mortality was high for infants, from genetic defects or childbirth problems; high but declining for children; lowest around puberty; higher in early adulthood; and high again in late adulthood (Livi-Bacci 2017). Figure 3.2 plots mortality by age among three populations of contemporary foragers that follows the historical pattern. Others think mortality rises more slowly from adolescence to age 40. Accelerating mortality after that age still means that "at least one-fourth of the population is likely to live as grandparents for 15–20 years" (Gurven and Kaplan 2007:331). Other research broadly confirms this overall pattern (Séguy et al. 2008:88, 95), though it varies by population.

Grief at others' deaths seems universal among humans. Studies of primates, human infants, and many cultures show that people exhibit similar facial expressions and bodily gestures after losing an intimate relation (Horwitz 2020), but mourning, the public expression of grief, varies culturally (Rosenblatt 2020). Some societies support long, public mourning periods; others allow only a few days, though many people continue grieving for a while in private (Rosenblatt 2020; Stearns 2020). Also, children, economically dependent on others, grieve both the person and the loss of material support. Beliefs and normative rituals help people cope with grief (Scheper-Hughes 1993; Stearns 2020). Grieving, constantly renewed by new deaths, may also explain the high level of violence among prehistoric and modern foragers (Rosaldo 1989).

There is extensive ethnographic evidence of grief among foragers at someone's death (Woodburn 1982). Forager teens feel the deaths of their kin keenly (Hewlett

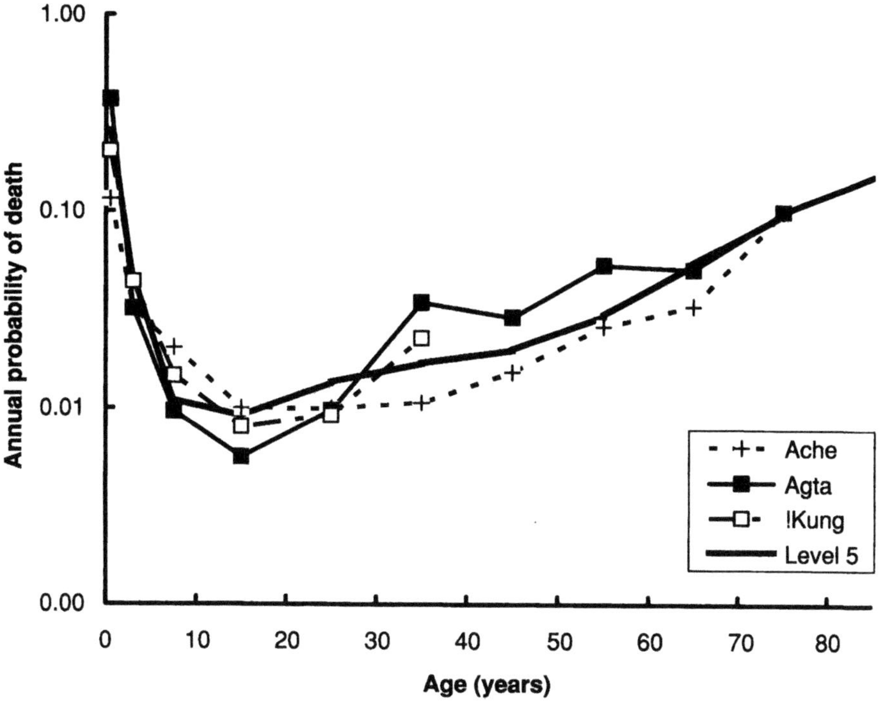

Fig. 3.2 Death rates at different ages, Hunter Gatherers. *Source*: Chamberlain (2006:64)

2005). Being with other people, something particularly easy in bands, consoles them. Children worry about survival after their parents die; relatives can reassure them (Woodburn 1982). Evidence from many societies suggests the most debilitating grief when losing a child, but mothering behaviors among foragers only provide indirect evidence of it (Fouts and Silverman 2015). The belief in reincarnation (Flannery and Marcus 2012) may be a grief-management strategy for the loss of an infant.

The high mortality rate also likely led to high fertility. Forager women devote tremendous resources to fertility, parenting, and alloparenting that men do not. Fertility leaves no archaeological record, so paleodemographers estimate prehistoric fertility rates. Estimates of the total fertility rate (TFR) of foragers today range from around 4–6 children (Chamberlain 2006; Pennington 2001; Boone 2002; Kelly 2013a). Foraging band women average a fecund period of about 20 years; they try to space births about 4 years apart since "children...increase a woman's energy needs *and* decrease her ability to meet those needs" (Kelly 2013a:191–192). Actual mean spacing is about 3.3 years. Some argue that starvation episodes lower body fat and create amenorrhea (interrupted menstrual cycles), reducing fecundity (Nolan and Lenski 2011). Others say that the marginal diet, combined with constant hard physical labor, means that women's bodies may "know" when they cannot sustain a fetus (Kelly 2013a:198). Many societies have postbirth taboos against sexual intercourse (Pasternak et al. 1997), and perhaps STI-induced infertility lowers births among foragers (Pennington 2001). Finally, infanticide controls effective fertility. A too-short birth interval creates a grim choice: Risk a demonstrably viable toddler for the benefit of a completely untested newcomer, or kill a new baby (Hrdy 2009)? Female infanticide is rare and mostly Arctic. Birth-spacing infanticide is even rarer, reflecting perceived alloparenting availability (Kelly 2013a:212).

Mortality and fertility shape population growth. If TFR was almost six, and half survived to age 20, then almost three children per woman survived to reproduce; this is almost a 50% increase per generation, a doubling within 60 years. Thus, some argue that humans engaged in "explosive" expansion across Eurasia (Kaifu et al. 2015:544). Higher environmental biomass might have lowered childhood mortality, weaning might have been easier, birth intervals were perhaps a little shorter (raising average TFR), and infanticide was maybe unnecessary. Most paleodemographers accept these factors, yet still think that the human population grew very slowly. The global population around 35–30,000 BP "did not exceed several hundred thousand" (Livi-Bacci 2017:24) and grew only 0.008% per year—or less than 0.1% per every thousand years—doubling around every 8400 years. Estimates of the global human population around 10–13,000 BP, at the end of the frontier, range from less than half a million to eight million (Haas and Piscitelli 2013; Livi-Bacci 2017). As humans dispersed around the globe, the Pleistocene climate varied wildly, with more than twenty warming spikes and at least six deep cold epochs (Brooke 2014). Cycles of rapid growth and catastrophic shrinkage due to cold-induced famine and STI-induced infertility may have hindered long-term population growth (Pennington 2001; Boone 2002). This low global population meant a sparse global density. Estimates of prehistoric global density range from a minimum of 0.04 people per square kilometer and a maximum of 0.1–0.115 km^2 (Haas and Piscitelli 2013). Density was lowest on the frontier, would have varied depending on any given region's biomass, and would have changed over time. Even now, most simple foraging bands have a density below one person per square kilometer (Kelly 2013a), so a band would live on a territory of at least 25 square kilometers (almost ten square miles).

Age-specific mortality and fertility rates combined to create a typical age distribution. Paleodemographers argue that all populations before the last few centuries were subject to normal "attritional" mortality plus occasional mortality catastrophes (Bocquet-Appel and Bacro 2008). Thus, prehistoric forager populations would usually have consisted of many children, some young adults, fewer middle-aged adults, and very few elderly people (Chamberlain 2006; Gurven and Kaplan 2007; Nowell 2021), but whatever mortality catastrophes occurred could change this general pattern for some generations geographically and chronologically.

The simplest foraging bands average around 25 persons. Why? In small group decision-making, individuals cannot cognitively handle interactions between more than six people (Johnson 1982). Since men usually decide men's issues, and women decide women's issues (Endicott 1999; Binford 2001), up to six reproductive couples form each band's core, and children and elderly people bring groups to the average range. Also, a group of roughly 25 people best balances the variance in food returns from the land with local resource depletion of that land (Kelly 2013a). Thus, most prehistoric frontier bands probably averaged 25 people. Archaeologically, sometimes groups were smaller: At one site in North America, two bands met, one having 14–18 people and the other 13–17 people (Flannery and Marcus 2012).

As mentioned, bands existed in larger communities. Family is the basic building block, bands are a larger group, multiple bands unite periodically (Binford 2001), and these groups exist within larger kin-communities, called "tribes" or "societies" (Read 2012), of 350 to 500 people comprised of more than 10 bands—the boundaries of which seem based on always knowing who is related to whom. These larger groups constituted social networks through which resources flowed; having relationships with distant people helped even out lean years. Also, the larger community would meet social needs—mating, friendship, alloparenting—that bands could not.

Simple foraging bands are usually mobile, following a few basic patterns. "Fission-fusion" organization involves living in "microbands" (of roughly 25 people) for most of the year and gathering in "macrobands" in the shortest and richest periods. In these meetings, trade, marriages, information exchanges, and ceremonies all occur. Repeated annual meetings create long-lasting, strong relationships, increasing survival for everyone facing variable environments instead of "family members or friends who... only come around when they are in dire need [and thus] are bitterly resented" (Hayden 2003:29). Such gatherings usually number less than 100 people (Binford 2001), always below the largest community size. Some Paleoindian groups engaged in fission-fusion patterns, such as those in what is now northern Colorado (Meltzer 2009). Longer settlement at rivers, lakes, or oceans might have occurred.

Different movement strategies affected relations with the land. Today, people in bands camp for a few days in one spot, making out-and-back foraging trips (Kelly 2013a). When they reach diminishing food returns, they move camp. They usually move annually within a "range," a specific region, arriving at the same place at the same time each year (Kelly 2013a; Sutton and Anderson 2010). Furthermore, "women's foraging should normally determine when and where camp is moved" (Kelly 2013a:102), as gathering returns and ranges are smaller than hunting ones. Portable housing setup and takedown time, terrain difficulty, distance to the next camp, how much to carry (including children), and the riskiness of the next area all influence moving. Bands on the frontier probably had ranges, perhaps expanding them over time or through scouting.

What camps looked like varied widely. Some argue that prehistoric peoples lived in caves "in all parts of the world" (Bellwood 2013:83); others say that caves had "bat colonies, big unfriendly bears, meager lighting, damp walls, and poor ventilation" of fire smoke (Von Petzinger 2016:80), so they more likely lived in cave entrances and rock overhangs. They also have and had mobile shelters. Siberian and North American peoples have versions of the conical tipi; other temporary shelters included huts, igloos, and lean-tos (Meltzer 2009; Bar-Yosef 2014; Derevianko et al. 2014; Moore 2014; O'Connor and Bulbeck 2014).

These patterns recall questions about movement on the frontier. Each group had few allies and fewer enemies, and a way to escape them. Did groups regularly move out of contact with their communities, or did only the bold and adventurous form one new band and send messages back? Was scouting on the frontier consistent, or did it occur more on north-south axes or at abrupt transitions, with bold, rapid movement along west-east lines? Are "contented and cautious" and "bold and adventurous" only Western research categories, or are they both common attitudes among all humans?

Finally, as they moved, humans continued to evolve physically. As people moved northward in Asia, their skin lightened, darkening again as they moved southward in the Americas. Those who moved into colder climates became shorter and stockier to conserve body heat; the reverse happened into warmer climates. Hair straightened, eyes developed a variety of shapes, teeth became differently shaped, and internal changes, such as new immunities, occurred (Binford 2001; Hoffecker 2005; Nolan and Lenski 2011; Cartmill and Smith 2022). Despite this millennia-long biological diversification, modern humans remained one species.

Population patterns in foraging bands included high mortality; almost six births per woman; many children, some adults, and a few elders in each group; an average group size of 25 people within larger communities; regular, episodic mobility; and low population density.

Foragers in Their Environments

The land affected every aspect of prehistoric foraging life. How much did humans influence it? Prehistoric people used fire to alter landscapes and domesticated dogs. Still, scholars debate why so many megafauna (for instance, mammoths, saber-toothed tigers, and giant sloths) went extinct. They offer three answers: Extinctions were mostly driven by climate change; they were mostly caused by humans; or both factors led to them. There is no consensus.

Humans had two clear impacts on their environment during the Pleistocene era. Foragers seem to have learned that setting brushfires increased the fertility of the soil, fostering plant growth that fed megafauna (Flannery and Marcus 2012; Stiner and Feeley-Harnik 2011). Humans also domesticated wolves; hunting with dogs greatly increases the amount of meat procured (Derr 2011). Nobody knows how this happened, or whether it happened once or several times in different places (Bergstrom et al. 2020). The earliest dog archaeological finds are 31.7 kya (in what is now Belgium), then 29.9 kya (Siberia), and 26–27 kya (Czechia).

A larger issue is the late Pleistocene megafaunal extinctions. Thirty-five genera of mammals vanished from the North American Pleistocene landscape by 10,800 BP (Meltzer 2009), constituting "70–80% of the megafauna genera in the Americas" (Barnosky et al. 2016:856). Similarly severe extinctions happened in Europe and Australia; they were moderate in northern Asia and North Africa, and mild in sub-Saharan Africa and southern Asia (Sandom et al. 2014:2 [map]). As scholars agree that the extinctions occurred roughly simultaneously with the global human diaspora and with climate change, explaining these extinctions is difficult.

Some argue that climate change caused the extinctions. The shift from the dry, cold, and variable Pleistocene climate to the stable and warm Holocene era around 12,000 BP involved melting glaciers. Oceans rose and rainfall increased, heavily affecting plant patterns around the world. Basically, megafauna went extinct because their food sources disappeared (Price and Feinman 2010; Stewart et al. 2021). Indeed, climate-driven change in eastern Beringia seems to have triggered megafaunal extinctions between 13,500 and 10,000 BP, yet megafauna existed in "remote pockets" of land long after humans arrived (Murchie et al. 2021:11).

Others disagree. The "overkill" hypothesis (Martin 1973) suggests that as humans moved into Eurasia and North America, they caused megafauna extinctions. The hunting was easy since animals were unfamiliar with humans who hunted prey animals, not predators, who then died off. On the frontier, there was less concern for conserving resources, as movement was possible. Many agree. There are cases of foragers overhunting today (Kelly 2013a). Others assert that humans were the "primary cause of Pleistocene megafaunal extinction in Australia" (Van der Kaars et al. 2017). Humans pressured populations of tortoises, fish, and land mammals so that the average size of each animal shrank (Stiner and Feeley-Harnik 2011). Most analyses are continental; Sandom et al. (2014) analyzed extinctions globally, concluding that humans caused them. Bartlett et al. (2016) likewise argue that humans had the dominant role in these extinctions across the planet, except in Asia, but climate change may have helped.

Other researchers doubt the overkill hypothesis. Many hunter-gatherers purposely avoid over-foraging to leave future resources (Kelly 2013a). Some argue that there is "little archaeological evidence to support" human-caused megafauna extinction in Australia (Hiscock 2015:58), and others say that "woolly mammoth... and naumann's elephant... co-existed with humans on the Japanese Archipelago for at least 7000–12,000 years," making overkill unlikely (Kaifu et al. 2015:548). Others agree for North America and eastern Beringia (Meltzer 2009; Murchie et al. 2021; Stewart et al. 2021).

Some argue that both factors contributed to these extinctions. For instance, Broughton and Weitzel (2018) argue that in North America, humans caused extinctions in some areas, climate did in other areas, and in one region, both operated. However, they seem to be in the minority; most argue primarily for one cause or the other. The debate continues.

Conclusion

In fact, debates continue for most of these issues. I have argued three points in this chapter, tracing debates around them throughout. First, modern humans, *Homo sapiens sapiens*, migrated from Africa into Eurasia and then into the Americas. I call this the first globalization; it was the most basic and foundational one, putting humans on all continents. Yet scholars endlessly debate its details: Did Sapiens originate in East Africa, Southern Africa, or across the entire continent (Chap. 2)?

Did a few of them leave Africa 100 or 60 millennia ago? And did they arrive in South America before, during, or after the LGM?

Second, figuring out when "cognitive modernity" developed—when our ancestors developed our mental-emotional capacities—would help us understand when and how the first globalization occurred. All capacities discussed in Chap. 2 helped humans disperse globally: small groups within larger ones, culture, cumulative learning, emotional bonds, institutions, and morality all played a part. However, people also had to be curious enough to learn about widely varying environments. Did they move quickly or slowly into them? We do not yet know.

Finally, I used ethnographic inference to develop what I hope is a plausible picture of Sapiens' lives on the "global human frontier." People seem to have lived in mobile groups of 25 people or so in an ever-changing collection of kin and friends, connected to many other such groups. They lived from wild plants and animals, sharing food, with only a little leadership, rough equality, and a high average mortality and fertility. Separating "aspects" of groups seems artificial, but analysis shows their deep interrelations. Foraging required constant movement on the land, keeping group sizes small and leading to informal politics, economic equality, low or nonexistent gender inequality, and flexible residence patterns. Subsistence practices affected mortality, in turn affecting fertility and influencing—but, it seems, not determining—gender roles. Conflict on the edge of habitation seems to have been minimal, or groups handled it by moving away. Each group had its own beliefs, rituals, kinship systems, food, foraging strategies, and childcare practices, different from those of other foragers, both geographically and chronologically.

Focusing on human groups on the frontier has been useful, but those groups were always a small minority of humans. What happened to the vast majority who remained behind?

References

Anderson, Abigail, Sophia Chilczuk, Kaylie Nelson, Roxanne Ruther, and Cara Wall-Scheffler. 2023. The myth of man the hunter: Women's contribution to the hunt across ethnographic contexts. *PLoS One* 18 (6): e0287101. https://doi.org/10.1371/journal.pone.0287101.

Ardelean, Ciprian F., Lorena Becerra-Valdivia, Mikkel Winther Pedersen, Jean-Luc Schwenninger, Charles G. Oviatt, Juan I. Macías-Quintero, Joaquin Arroyo-Cabrales, et al. 2020. Evidence of human occupation in Mexico around the last glacial maximum. *Nature* 584:87–108.

Atkinson, Quentin D. 2011. Founder effect model of language expansion from Africa. *Science* 332 (6027): 346–349.

Barham, Lawrence, and Peter Mitchell. 2008. *The first Africans: African archaeology from the earliest toolmakers to Most recent foragers*. New York: Cambridge University Press.

Barnes, Gina L. 2015. *The archaeology of East Asia: The rise of civilization in China, Korea, and Japan*. Havertown: Oxbow Books.

Barnosky, Anthony, Emily L. Lindsey, Natalia A. Villavicencio, Enrique Bostelmann, Elizabeth A. Hadly, James Wanket, and Charles R. Marshall. 2016. Variable impact of late-quaternary Megafaunal extinction in causing ecological state shifts in north and South America. *Proceedings of the National Academy of Sciences* 113 (4): 856–861.

Bar-On, Yinon M., Rob Phillips, and Ron Milo. 2016. The biomass distribution on earth. *Proceedings of the National Academy of Sciences* 115 (25): 6506–6511.

Bartlett, Lewis J., David R. Williams, Graham W. Prescott, Andrew Balmford, Rhys E. Green, Anders Eriksson, Paul J. Valdes, Joy S. Singarayer, and Andrea Manica. 2016. Robustness despite uncertainty: Regional climate data reveal the dominant role of humans in explaining global extinction of late quaternary Megafauna. *Ecography: Pattern and Diversity in Ecology* 39 (2): 152–161.

Bar-Yosef, Ofer. 2014. Upper Palaeolithic hunter-gatherers in Western Asia. In *The Oxford handbook of the archaeology and anthropology of hunter-gatherers*, ed. V. Cummings, P. Jordan, and M. Zvelebil, 252–278. New York: Oxford University Press.

Bellwood, Peter. 2013. *First migrants: Ancient migration in global perspective*. Malden: Wiley Blackwell.

Bennett, Matthew R., David Bustos, Jeffrey S. Pigati, Kathleen B. Springer, Thomas M. Urban, Vance T. Holliday, Sally C. Reynolds, et al. 2021. Evidence of humans in North America during the last glacial maximum. *Science* 373:1528–1531.

Bergstrom, Anders, Laurent Frantz, Ryan Schmidt, Erik Ersmark, Ophelie Lebrasseur, Linus Girdland-Flink, Audrey T. Lin, et al. 2020. Origins and genetic legacy of prehistoric dogs. *Science* 370:557–564.

Binford, Lewis R. 2001. *Constructing frames of reference: An analytical method for archaeological theory building using ethnographic and environmental data sets*. Berkeley: University of California Press.

Bocquet-Appel, Jean-Pierre, and Jean-Noël Bacro. 2008. Estimation of an Age Distribution with Its Confidence Intervals Using an Iterative Bayesian Procedure and a Bootstrap Sampling Approach. In *Recent Advances in Palaeodemography: Data, Techniques, Patterns*, ed. J.-P. Bocquet-Appel, 63–82. Dordrecht, The Netherlands: Springer.

Boehm, Christopher. 1999. *Hierarchy in the Forest: The evolution of egalitarian behavior*. Cambridge: Harvard University Press.

Boehm, Christopher. 2012. *Moral origins: The evolution of virtue, altruism, and shame*. New York: Basic Books.

Boehm, Christopher. 2013. The biocultural evolution of conflict resolution between groups. In *War, peace, and human nature: The convergence of evolutionary and cultural views*, ed. D. P. Fry, 315–340. New York: Oxford University Press.

Bogin, Barry. 2001. *The growth of humanity*. New York: Wiley–Liss.

Bolger, Diane. 2006. Gender and human evolution. In *Handbook of gender in archaeology*, ed. S. M. Nelson, 453–501. New York: AltaMira Press.

Boone, James L. 2002. Subsistence strategies and early human population history: An evolutionary ecological perspective. *World Archaeology* 34 (1): 6–25.

Brooke, John L. 2014. *Climate change and the course of global history: A rough journey*. New York: Cambridge University Press.

Broughton, Jack M., and Elic M. Weitzel. 2018. Population reconstructions for humans and Megafauna suggest mixed causes for north American Pleistocene extinctions. *Nature Communications* 9:1–12.

Brumbach, Hetty Jo, and Robert Jarvenpa. 2006. In *Handbook of gender in archaeology*, ed. S. M. Nelson, 503–535. New York: AltaMira Press.

Brumm, Adam, Adhi Agus Oktaviana, Basran Burhan, Budianto Hakim, Rustan Lebe, Jian-xin Zhao, Priyatno Hadi Sulistyarto, et al. 2021. Oldest cave art found in Sulawesi. *Science Advances* 7:1–2.

Cartmill, Matt, and Fred H. Smith. 2022. *The human lineage*. 2nd ed. Hoboken: Wiley Blackwell.

Chamberlain, Andrew. 2006. *Demography in archaeology*. Cambridge: Cambridge University Press.

Chatters, James C. 2014. Wild-type colonizers and high levels of violence among Paleoamericans. In *Violence and warfare among hunter-gatherers*, ed. M. W. Allen and T. L. Jones, 70–96. Walnut Creek: Left Coast Press.

Chen, Lu, Aaron B. Wolf, Wenqing Fu, Liming Li, and Joshua M. Akey. 2020. Identifying and Interpreting Apparent Neanderthal Ancestry in African Individuals. *Cell* 180 (4): 677–687.

Clarkson, Chris, Zenobia Jacobs, Ben Marwick, Richard Fullagar, Lynley Wallis, Mike Smith, Richard G. Roberts, et al. 2017. Human occupation of northern Australia by 65,000 years ago. *Nature* 547:306–310.

Davidson, Ian. 2014. Hunter-gatherers in Australia: Deep histories of continuity and change. In *The Oxford handbook of the archaeology and anthropology of hunter-gatherers*, ed. V. Cummings, P. Jordan, and M. Zvelebil, 368–404. New York: Oxford University Press.

Demeter, Fabrice, Clément Zanolli, Kira E. Westaway, Renaud Joannes-Boyau, Philippe Duringer, Mike W. Morley, Frido Welker, et al. 2022. A middle Pleistocene Denisovan molar from the Annamite chain of northern Laos. *Nature Communications* 13:2557–2569.

der Kaars, Van, Gifford H. Sander, Chris S. M. Miller, Ellyn J. Turney, Dirk Nürnberg Cook, A. Joachim Schönfeld, Peter Kershaw, and Scott J. Lehman. 2017. Humans rather than climate the primary cause of Pleistocene Megafaunal extinction in Australia. *Nature Communications* 8:14142.

Derevianko, Anatoly P., Sergei V. Markin, and Andrei V. Tabarev. 2014. The Palaeolithic of northern Asia. In *Of the Oxford handbook of the archaeology and anthropology of hunter-gatherers*, ed. V. Cummings, P. Jordan, and M. Zvelebil, 310–327. New York: Oxford University Press.

Derr, Mark. 2011. *How the dog became the dog: From wolves to our best friends*. New York: The Overlook Press.

Dyble, M., G. D. Salali, N. Chaudhury, A. Page, D. Smith, J. Thompson, L. Vinicius, R. Mace, and A. B. Migliano. 2015. Sex equality can explain the unique social structure of hunter-gatherer bands. *Science* 348 (6236): 796–798.

Ehret, Christopher. 2008. Reconstructing ancient kinship in Africa. In *Early human kinship: From sex to social reproduction*, ed. N. J. Allen, H. Callan, R. Dunbar, and W. James, 200–231. Malden: Blackwell Publishing.

Endicott, Karen L. 1999. Gender relations in hunter-gatherer societies. In *The Cambridge Encyclopedia of hunters and gatherers*, ed. R. B. Lee and R. Daly, 411–419. New York: Cambridge University Press.

Ensor, Bradley E. 2013. *The archaeology of kinship: Advancing interpretations and contributions to theory*. Tucson: The University of Arizona Press.

Erlandson, Jon. 2017. Coastlines, marine ecology, and maritime dispersals in human history. In *Human dispersal and species movement: From prehistory to the present*, ed. N. Boivin, R. Crassard, and M. D. Petraglia, 147–163. New York: Cambridge University Press.

Fagan, Brian M. 2019. *Ancient North America: The archaeology of a continent*. New York: Thames and Hudson.

Feuer, Bryan. 2016. *Boundaries, Borders and Frontiers in archaeology: A study of spatial relationships*. Jefferson: McFarland & Company.

Flannery, Kent, and Joyce Marcus. 2012. *The creation of inequality: How our prehistoric ancestors set the stage for monarchy, slavery, and empire*. Cambridge: Harvard University Press.

Fouts, Hillary N., and Lisa S. Silverman. 2015. Parenting and environmental risk: An examination of child loss and maternal involvement among Bofi foragers in Central Africa. *Human Nature* 26:73–88.

Friesen, Bruce K. 2015. *Moral systems and the evolution of human rights*. New York: Springer.

Gamble, Clive. 2013. *Settling the earth: The archaeology of deep human history*. New York: Cambridge University Press.

Gat, Azar. 2017. *The causes of war and the spread of peace: But will war rebound?* New York: Oxford University Press.

Graf, Kelly E. 2015. Modern human response to the last glacial maximum in Siberia. In *Emergence and diversity of modern human behavior in Paleolithic Asia*, ed. Y. Kaifu, M. Izuho, T. Goebel, H. Sato, and A. Ono, 506–532. College Station: Texas A&M University Press.

Groves, Colin. 2015. Hominin migrations before *Homo sapiens*: Out of Africa—How many times? In *The global prehistory of human migration*, ed. P. Bellwood, 18–25. Malden, MA: Wiley Blackwell.

Guenther, Mathias. 1999. From Totemism to shamanism. In *The Cambridge Encyclopedia of hunters and gatherers*, ed. R. B. Lee and R. Daly, 426–433. New York: Cambridge University Press.

Gurven, Michael, and Hillard Kaplan. 2007. Longevity among hunter-gatherers: A cross-cultural examination. *Population and Development Review* 33 (2): 321–365.

Haas, Jonathan, and Matthew Piscitelli. 2013. The prehistory of warfare: Misled by ethnography. In *War, peace, and human nature: The convergence of evolutionary and cultural views*, ed. D. P. Fry, 168–190. New York: Oxford University Press.

Haas, Randall, James Watson, Tammy Buonasera, John Southon, Jennifer C. Chen, Sarah Noe, Kevin Smith, Carlos Viviano Llave, Jelmer Eerkens, and Glendon Parker. 2020. Female hunters of the early Americas. *Science Advances* 6:1–10.

Hayden, Brian. 2003. *Shamans, sorcerers and saints: A prehistory of religion.* Washington: Smithsonian.

Hayden, Brian. 2014. *The Power of feasts: From prehistory to the present.* New York: Cambridge University Press.

Headland, Thomas N. 1987. Kinship and social behavior among Agta Negrito hunter-gatherers. *Ethnology* 26 (4): 261–280.

Helm, Charles W., Andrew S. Carr, Martin G. Lockley, Hayley C. Cawthra, Jan C. De Vynck, Mark G. Dixon, Willo Stear, and Guy H. H. Thesen. 2023. Dating the Pleistocene Hominin Ichnosites on South Africa's cape south coast. *Ichnos: An International Journal for Plant and Animal Traces* 30 (1): 49–68.

Henrich, Joseph. 2015. *The secret of our success: How culture is driving human evolution, domesticating our species, and making us smarter.* Princeton: Princeton University Press.

Hewlett, Bonnie L. 2005. Vulnerable lives: The experience of death and loss among the aka and Ngandu adolescents of the Central African Republic. In *Hunter-gatherer childhoods: Evolutionary, developmental, & cultural perspectives*, ed. B. S. Hewlett and M. E. Lamb, 322–342. New York: Transaction.

Hiscock, Peter. 2015. Early Old World migrations of *Homo sapiens*: Archaeology. In *The global prehistory of human migration*, ed. P. Bellwood, 38–48. Malden: Wiley Blackwell.

Hoffecker, John F. 2005. *A prehistory of the north: Human settlement of the higher latitudes.* New Brunswick: Rutgers University Press.

Horwitz, Allan V. 2020. The medicalization of grief. In *Exploring grief: Towards a sociology of sorrow*, ed. M. H. Jacobsen and A. Petersen, 171–187. New York: Routledge.

Howell, Nancy. 1979. *Demography of the Dobe !Kung.* New York: Academic Press.

Hrdy, Sarah Blaffer. 2009. *Mothers and others: The evolutionary origins of mutual understanding.* Cambridge: Belknap.

Hughes, V. (2013). Were the first artists mostly women? *National Geographic.* http://news.nationalgeographic.com/news/2013/10/131008-women-handprints-oldest-neolithic-cave-art/

Jarvenpa, Robert, and Hetty Jo Brumbach. 2014. Hunter-gatherer gender and identity. In *The Oxford handbook of the archaeology and anthropology of hunter-gatherers*, ed. V. Cummings, P. Jordan, and M. Zvelebil, 1243–1265. New York: Oxford University Press.

Jochim, Michael. 2011. The upper Paleolithic. In *European prehistory: A survey*, ed. S. Milisauskas, 67–124. New York: Springer.

Johnson, Gregory A. 1982. Organizational structure and scalar stress. In *Theory and explanation in archaeology*, ed. C. Renfrew, 389–421. New York: Academic Press.

Johnson, Allen W., and Timothy K. Earle. 2000. *The evolution of human societies: From foraging group to agrarian state.* 2nd ed. Stanford, CA: Stanford University Press.

Kaifu, Yōsuke, Masami Izuho, and Ted Goebel. 2015. Modern human dispersal and behavior in Paleolithic Asia: Summary and discussion. In *Emergence and diversity of modern human behavior in Paleolithic Asia*, ed. Y. Kaifu, M. Izuho, T. Goebel, H. Sato, and A. Ono, 535–566. College Station: Texas A&M University Press.

Keeley, Lawrence H. 1996. *War before civilization.* New York: Oxford University Press.

Kelly, Raymond C. 2000. *Warless societies and the origin of war.* Ann Arbor: The University of Michigan Press.

Kelly, Robert L. 2013a. *The Lifeways of hunter-gatherers: The foraging Spectrum*. New York: Cambridge University Press.
Kelly, Robert L. 2013b. From the peaceful to the warlike: Ethnographic and archaeological insights into hunter-gatherer warfare and homicide. In *War, peace, and human nature: The convergence of evolutionary and cultural views*, ed. D. P. Fry, 151–167. New York: Oxford University Press.
Kissel, Marc, and Nam C. Kim. 2019. The emergence of human warfare: Current perspectives. *American Journal of Biological Anthropology* 168 (S67): 141–163.
Knight, Chris. 2008. Early human kinship was matrilineal. In *Early human kinship: From sex to social reproduction*, ed. N. J. Allen, H. Callan, R. Dunbar, and W. James, 61–82. Malden: Blackwell Publishing.
Konigsberg, Lyle W., and Nicholas P. Herrmann. 2006. The Osteological evidence for human longevity in the recent past. In *The evolution of human life history*, ed. K. Hawkes and R. R. Paine, 267–306. Santa Fe: School of American Research Press.
Kornfield, Marcel, and Gustavo G. Politis. 2014. Into the Americas: The earliest hunter-gatherers. In *The Oxford handbook of the archaeology and anthropology of hunter-gatherers*, ed. V. Cummings, P. Jordan, and M. Zvelebil, 405–433. New York: Oxford University Press.
Kuhn, Steven, and Mary Steiner. 2018. The antiquity of hunter-gatherers revisited. In *Foraging in the past: Archaeological studies of hunter-gatherer diversity*, ed. A. K. Lemke, 247–268. Louisville: University Press of Colorado.
Kuhn, Steven L., and Mary C. Stiner. 2006. What's a mother to do? The division of labor among Neandertals and modern humans in Eurasia. *Current Anthropology* 47 (6): 953–980.
Kuykendall, Kevin L., and Isabelle S. Heyerdahl-King. 2014. Modern human origins in Africa: A review of the fossil, archaeological, and genetic perspectives on early *Homo sapiens*. In *The Oxford handbook of the archaeology and anthropology of hunter-gatherers*, ed. V. Cummings, P. Jordan, and M. Zvelebil, 214–251. New York: Oxford University Press.
Lacy, Sarah, and Cara Ocobock. 2024. Woman the Hunter: The Archaeological Evidence. *American Anthropologist 126* (1): 19–31.
Layton, Robert. 2008. What can ethnography tell us about human social evolution? In *Early human kinship: From sex to social reproduction*, ed. N. J. Allen, H. Callan, R. Dunbar, and W. James, 113–127. Malden, MA: Blackwell Publishing.
LeBlanc, Steven A. 2014. Forager warfare and our evolutionary past. In *Violence and warfare among hunter-gatherers*, ed. M. W. Allen and T. L. Jones, 26–48. Walnut Creek, CA: Left Coast Press.
LeBlanc, Steven A., and Katherine E. Register. 2003. *Constant battles: The myth of the peaceful, Noble savage*. New York: St. Martin's Press.
Lemke, Ashley K. 2018a. Hunter-gatherers and archaeology. In *Foraging in the past: Archaeological studies of hunter-gatherer diversity*, ed. A. K. Lemke, 3–18. Louisville: University Press of Colorado.
Lemke, Ashley K. 2018b. Underwater archaeology and Archaeo-ethnology of prehistoric hunter-gatherers: Examining the role of ethnography in prehistoric forager research beneath the north American Great Lakes. In *Foraging in the past: Archaeological studies of hunter-gatherer diversity*, ed. A. K. Lemke, 49–75. Louisville: University Press of Colorado.
Lewellen, Ted C. 2003. *Political anthropology: An introduction*. 3rd ed. Westport, CT: Praeger.
Lewis-Williams, J. D. 2014. Art for the living. In *The Oxford handbook of the archaeology and anthropology of hunter-gatherers*, ed. V. Cummings, P. Jordamn, and M. Zvelebil, 625–642. New York: Oxford University Press.
Lippi, Mariotti, Biancamaria Aranguren Marta, Simona Arrighi, Davide Attolini, Stefano Benazzi, Francesco Boschin, Silvia Florindi, et al. 2023. New evidence of plant food processing in Italy before 40ka. *Quaternary Science Reviews* 312 (July): 108161. https://doi.org/10.1016/j.quascirev.2023.108161.
Liu, Li, and Xingcan Chen. 2012. *The archaeology of China: From the late Paleolithic to the early bronze age*. New York: Cambridge University Press.

Livi-Bacci, Massimo. 2017. *A concise history of world population*. 6th ed. Chichester: Wiley/Blackwell.

Lombard, Marlize. 2013. Hunter-gatherers in southern Africa before 20,000 years ago. In *The Oxford handbook of African archaeology*, ed. P. Mitchell and P. Lane, 367–386. New York: Oxford University Press.

Martin, Paul S. 1973. The discovery of America. *Science* 179 (4077): 969–974.

McArthur, Norma, I. W. Saunders, and R. L. Tweedie. 1976. Small population isolates: A microsimulation study. *The Journal of the Polynesian Society* 85 (3): 307–326.

Meltzer, David J. 2009. *First Peoples in a New World: Colonizing ice age America*. Berkeley: University of California Press.

Mitchell, Peter. 2009. Hunters and gatherers. In *The Oxford handbook of archaeology*, ed. B. Cunliffe, C. Gosden, and R. A. Joyce, 411–444. New York: Oxford University Press.

Mithen, Steven. 2020. Peopling the world. In *The Oxford handbook of archaeology*, ed. B. Cunliffe, C. Gosden, and R. Joyce, 2nd ed., 371–408. New York: Oxford University Press.

Moore, Jerry D. 2014. *A prehistory of South America: Ancient cultural diversity on the least known continent*. Boulder: University of Colorado Press.

MorcoteRíos, Gaspar, Francisco Javier Aceituno, José Iriarte, Mark Robinson, and Jeison L. Chaparro-Cárdenas. 2021. Colonisation and early peopling of the Colombian Amazon during the late Pleistocene and the early Holocene: New evidence from La Serranía La Lindosa. *Quaternary International* 578:5–19.

Murchie, Tyler J., Alistair J. Monteath, Matthew E. Mahony, George S. Long, Scott Cocker, Tara Sadoway, Emil Karpinski, et al. 2021. Collapse of the mammoth-steppe in Central Yukon as revealed by ancient environmental DNA. *Nature Communications* 12:1–18.

Neill, James. 2009. *The origins and role of same-sex relations in human societies*. Jefferson, NC and London: McFarland & Company.

Nolan, Patrick, and Gerhard Lenski. 2011. *Human societies: An introduction to macrosociology*. 11th ed. Boulder: Paradigm Press.

Nowell, April. 2021. *Growing up in the ice age: Fossil and archaeological evidence of the lived lives of Plio-Pleistocene children*. Philadelphia: Oxbow Books.

O'Connor, Sue. 2015. Crossing the Wallace line: The maritime skills of the earliest colonists in the Wallacean archipelago. In *Emergence and diversity of modern human behavior in Paleolithic Asia*, ed. Y. Kaifu, M. Izuho, T. Goebel, H. Sato, and A. Ono, 214–224. College Station: Texas A&M University Press.

O'Connor, Susan, and David Bulbeck. 2014. *Homo sapiens* societies in Indonesia and south-eastern Asia. In *The Oxford handbook of the archaeology and anthropology of hunter-gatherers*, ed. V. Cummings, P. Jordan, and M. Zvelebil, 346–367. New York: Oxford University Press.

Ocobock, Cara, and Sarah Lacy. 2024. Woman the hunter: The physiological evidence. *American Anthropologist* 126 (1): 7–18.

Opie, Kit, and Camilla Power. 2008. Grandmothering and female coalitions: A basis for matrilineal priority? In *Early human kinship: From sex to social reproduction*, ed. N. J. Allen, H. Callan, R. Dunbar, and W. James, 168–186. Malden, MA: Blackwell Publishing.

Owen, Linda R. 2005. *Distorting the past: Gender and the division of labor in the European upper Paleolithic*. Tübingen: Kerns Verlag.

Pansani, Thais R., Briana Pobiner, Pierre Gueriau, Mathieu Thoury, Paul Tafforeau, Emmanuel Baranger, Águeda V. Vialou, et al. 2023. Evidence of artefacts made of Giant sloth bones in Central Brazil around the last glacial maximum. *Proceedings of the Royal Society B* 290:2023–0316.

Pasternak, Burton, Carol R. Ember, and Melvin Ember. 1997. *Sex, gender, and kinship: A cross-cultural perspective*. Upper Saddle River: Prentice-Hall.

Pauketat, Timothy R., and Kenneth E. Sassaman. 2020. *The archaeology of ancient North America*. New York: Cambridge University Press.

Pennington, Renee. 2001. Hunter-gatherer demography. In *Hunter-gatherers: An interdisciplinary perspective*, ed. C. Panter-Brick, R. H. Layton, and P. Rowley-Conwy, 170–204. Cambridge: Cambridge University Press.
Peoples, Hervey C., Pavel Duda, and Frank W. Marlowe. 2016. Hunter-gatherers and the origins of religion. *Human Nature* 27:261–282.
Petr, Martin, Mateja Hajdinjak, Fu Qiaomei, Elena Essel, Hélène Rougier, Isabelle Crevecoeur, Patrick Semal, et al. 2020. The evolutionary history of Neanderthal and Denisovan Y chromosomes. *Science* 369:1653–1656.
Petraglia, Michael D., and Nicole Boivin. 2014. *Homo sapiens* societies: South Asia. In *The Oxford handbook of the archaeology and anthropology of hunter-gatherers*, ed. V. Cummings, P. Jordan, and M. Zvelebil, 328–345. New York: Oxford University Press.
Pettitt, Paul. 2013. The rise of modern humans. In *The human past: World prehistory and the development of human societies*, ed. C. Scarre, 3rd ed., 124–173. London: Thames & Hudson.
Pettitt, Paul. 2014. The European upper Palaeolithic. In *The Oxford handbook of the archaeology and anthropology of hunter-gatherers*, ed. V. Cummings, P. Jordan, and M. Zvelebil, 279–309. New York: Oxford University Press.
Posth, Cosimo, Christoph Wißing, Keiko Kitagawa, Luca Pagani, Laura van Holstein, Fernando Racimo, Kurth Wehrberger, et al. 2017. Deeply divergent archaic mitochondrial genome provides lower time boundary for African gene flow into Neanderthals. *Nature Communications* 8:16046–16055.
Price, T. Douglas, and Gary Feinman. 2010. *Images of the past*. 6th ed. New York: McGraw Hill.
Read, Dwight W. 2012. *How culture makes us human: Primate social evolution and the formation of human societies*. Walnut Creek: Left Coast Press.
Richerson, Peter J., and Robert Boyd. 2005. *Not by genes alone: How culture transformed human evolution*. Chicago: The University of Chicago Press.
Rosaldo, Renato. 1989. *Culture and truth: The remaking of social analysis*. Boston: Beacon Press.
Rosenblatt, Paul C. 2020. Diversity in human grieving: Historical and cross-cultural perspectives. In *Exploring grief: Towards a sociology of sorrow*, ed. M. H. Jacobsen and A. Petersen, 37–51. New York: Routledge.
Sandom, Christopher, Søren Faurby, Brody Sandel, and Jens-Christian Svenning. 2014. Global late quaternary Megafauna extinctions linked to humans, not climate change. *Proceedings of the Royal Society B* 281:1–9.
Scheper-Hughes, Nancy. 1993. *Death without weeping: The violence of everyday life in Brazil*. Berkeley: University of California Press.
Séguy, Isabelle, Luc Buchet, Arnaud Bringé, Magali Belaigues-Rossard, Paul Beurnier, Nadège Couvert, and Carole Perraut. 2008. Model life tables for pre-industrial populations: First application in Palaeodemography. In *Recent advances in Paleodemography: Data, techniques, patterns*, ed. J. Bocquet-Appel, 83–118. Dordrecht: Springer.
Sellet, Frédéric. 1993. Chaîne Opératoire: The concept and its applications. *Lithic Technology* 18 (1–2): 106–112.
Service, Elman. 1962. *Primitive social organization: An evolutionary perspective*. New York: Random House.
Smail, Daniel Lord, Mary Stiner, and Timothy K. Earle. 2011. Goods. In *Deep history: The architecture of past and present*, ed. A. Shryock and D. L. Smail, 219–241. Berkeley: University of California Press.
Smith, Martin. 2017. *Mortal wounds: The human skeleton as evidence for conflict in the past*. Barnsley: Pen & Sword Military.
Snow, Dean R. 2013. Sexual dimorphism in European upper Paleolithic cave art. *American Antiquity* 78 (4): 746–761.
Stanish, Charles. 2017. *The Evolution of Human Co-operation: Ritual and Social Complexity in Stateless Societies*. New York: Cambridge University Press.
Starkweather, K. E., and R. Hames. 2012. A survey of non-classical polyandry. *Human Nature* 23 (2): 149–172.

Steadman, Sharon R. 2009. *The archaeology of religion: Cultures and their beliefs in worldwide context*. Walnut Creek: Left Coast Press.

Stearns, Peter N. 2020. Grief in modern history: An Ongoing evolution. In *Exploring grief: Towards a sociology of sorrow*, ed. M. H. Jacobsen and A. Petersen, 21–36. New York: Routledge.

Stewart, Brian A., Yuchao Zhao, Peter J. Mitchell, Genevieve Dewar, James D. Gleason, and Joel D. Blum. 2020. Ostrich eggshell bead strontium isotopes reveal persistent macroscale social networking across late quaternary southern Africa. *Proceedings of the National Academy of Science* 117 (12): 6453–6462.

Stewart, Mathew, W. Christopher Carleton, and Huw S. Groucutt. 2021. Climate change, not human population growth, correlates with late quaternary Megafauna declines in North America. *Nature Communications* 12 (965): 1–15.

Stiner, Mary C., and Gillian Feeley-Harnik. 2011. Energy and EcoSystems. In *Deep history: The architecture of past and present*, ed. A. Shryock and D. L. Smail, 78–102. Berkeley: University of California Press.

Stoneking, Mark. 2012. Genetic evidence concerning the origins and dispersals of modern humans. In *Causes and consequences of human migration: An evolutionary perspective*, ed. M. H. Crawford and B. C. Campbell, 11–20. New York: Cambridge University Press.

Stoneking, Mark, and Katerina Harvati. 2015. Early Old World migrations of *Homo sapiens*: Human biology. In *The global prehistory of human migration*, ed. P. Bellwood, 26–37. Malden, MA: Wiley Blackwell.

Sutton, Mark Q., and E. N. Anderson. 2010. *Introduction to cultural ecology*. 2nd ed. New York: Rowman & Littlefield.

Thompson, Victor D., and Christopher R. Moore. 2015. The sociality of surplus among late archaic hunter-gatherers of coastal Georgia. In *Surplus: The politics of production and the strategies of everyday life*, ed. C. T. Morehart and K. DeLucia, 245–266. Boulder: University of Colorado Press.

Toups, Melissa A., Andrew Kitchen, Jessica E. Light, and David L. Reed. 2011. Origin of clothing lice indicates early clothing use by anatomically modern humans in Africa. *Molecular Biology and Evolution* 28 (1): 29–32.

Van den Berghe, Pierre. 1973. *Age and sex in human societies: A biosocial perspective*. Belmont: Wadsworth.

Volk, Anthony A., and Jeremy A. Atkinson. 2013. Infant and child death in the human environment of evolutionary adaptation. *Evolution and Human Behavior* 34 (3): 182–192.

Von Petzinger, Genevieve. 2016. *The first signs: Unlocking the mysteries of the world's oldest symbols*. New York: Atria Books.

Wobst, H. Martin. 1978. The Archaeo-ethnology of hunter-gatherers or the tyranny of the ethnographic record in archaeology. *American Antiquity* 43 (2): 303–309.

Wood, Barry. 2019. Beachcombing and coastal settlement: The Long migration from southern Africa to Patagonia—The greatest journey ever made. *Journal of Big History* III (4): 19–46.

Woodburn, James. 1982. Social dimensions of death in four African hunting and gathering societies. In *Death and the regeneration of life*, ed. M. Bloch and J. Parry, 187–210. New York: Cambridge University Press.

Chapter 4
The Agricultural Revolution: Group Diversification, Domestication, and Expansion, 33,000–7000 BP

Abstract After humans reached all continents, they slowly filled in empty areas. Groups in some places grew larger and more settled. Many foragers began to collect, manage, and store wild foods. After the Last Glacial Maximum, the climate became warmer and wetter, and in some regions, groups domesticated plants and animals. Villagers relied on collecting, horticulture, or herding. Groups started to organize themselves according to descent, increasing group cohesion. Villages developed multiple, informal leaders and warfare. Economic and gender inequalities emerged in a few areas. Spiritual beliefs often remained animistic, while some groups started emphasizing their deceased ancestors. Population grew more quickly, some groups became more sedentary, and environmental degradation increased. Villagers developed more intense and longer-distance trade networks. These new practices spread from their origins to new areas.

Keywords Villages · Domestication · Horticulture · Pastoralism · Agriculture · Warfare · Population growth

Introduction

If social organization on the human frontier was broadly similar in the global diaspora, it started diversifying for those remaining behind—the vast majority. Millennia before the frontier disappeared in South America, groups permanently living across Afro-Eurasia sometimes grew larger, sedentary, and more complex (Lemke 2018), some living in villages of permanent buildings. When the climate became warmer and wetter after 12,000 BP, a few groups around the world started to intervene in plant and/or animal lives rather than merely gathering and hunting them, a process that slowly "domesticated" plants and animals. By 7000 BP, most of these groups mixed foraging and domesticating plants and animals; a few relied fully on farming.

Scholars call this new way of life—farming domesticated plants and herding domesticated animals—the Agricultural Revolution. Some object to calling a change lasting thousands of years a "revolution" (Jain 2006; Ponting 2007; Graeber and Wengrow 2021). While this is fair, context defines suddenness. After over two

T. C. Burr, *Societies Past and Present*, Emerging Globalities and Civilizational Perspectives, https://doi.org/10.1007/978-3-031-91532-1_4

million years of hominin foraging, developing agriculture in three thousand years *was* sudden (Price and Bar-Yosef 2011: S172), and the consequences of domestication were truly revolutionary: "[P]eople in sedentary villages … lived in fundamentally different ways than their Late Pleistocene ancestors" (Cohen 2011: S274). Many people's lives worsened compared to their foraging ancestors (Cohen and Crane-Kramer 2007; Scott 2017). In this chapter, I argue that this was indeed revolutionary since changing subsistence practices changed many aspects of social life. Groups grew larger, kin organization elaborated, leadership forms increased in number, warfare or intergroup violence seems to have started, cultural variation increased around the world, and regional networks grew.

Despite these changes, many deep patterns continued, and new ones started. Most people around the world remained foragers or started collecting and storing wild foods. Trade systems expanded, and people invented new forms of storage. Mortality and fertility both increased, with fertility increasing faster. People continued clustering, though clusters increased in size and sometimes were less mobile. Everyone remained animist, but many venerated dead ancestors. Patterned inequalities beyond age started, and humans had deeper impacts on their environments.

Both the changes and continuities depended on, and revealed, human capacities. People creatively responded to new conditions with new techniques; what worked became part of cumulative cultural learning, which diversified around the world. The nested-group capacity enabled group sizes to increase, and individuals joined multiple groups. People also developed more elaborate institutions; this and ethnocentrism seem to have supported increased warfare, while intergroup institutions helped increase the geographic size of trade regions.

To trace these changes and continuities, I start by outlining how forager groups became more complex and diverse and how and where domestication happened. I then lay out theories of why domestication occurred. Then I analyze the new social patterns. Finally, I trace out geographically larger patterns: how farming spread, how regional systems developed, and how foragers and farmers related.

Telling the Tale: The Development of Domesticating Societies

While hunter-gatherer groups on the global human frontier probably fit the stereotype of simple, egalitarian bands, hunter-gatherers today vary tremendously, well beyond this stereotype (Kelly 2013; Lemke 2018). The growth and elaboration of forager groups lasted millennia, reversing during the Last Glacial Maximum (LGM) and starting again after the climate warmed. After another, briefer cooling and warming, humans in many places started farming, increasing the variety of groups around the world.

Only a few people lived on the global human frontier. When humans first moved into a region, they learned its environment, and the population usually grew. Nearly everyone stayed in the region permanently, moving within their annual range; only a few eventually moved farther out, thus moving the frontier. The process of most staying and a few moving ultimately populated all continents; obviously, this required population growth. Still, the global human frontier did not end when people reached southern South America; instead, people probably lived in "a patchily inhabited world in which large, uninhabited areas separated locations" of foragers (Binford 2001:453). Slowly, people filled in these "patchy" areas—say, lands between river valleys—gradually closing the now-scattered frontier. Given how long global dispersal took, this process of "infill" (Lemke 2018) began in Afro-Eurasia long before people arrived in South America.

As infill occurred, foragers' mobility varied greatly (Kelly 2013). If a region was abundant with plants and animals, being sedentary for weeks or months was easier, so foragers finding these rich regions established settlements that lasted months, even years, and stored food. From 33,000 to 25,000 BP, people in Europe lived under rock shelters for months at a time (Hayden 2014a). The Chauvet cave, with spectacular art, housed large groups (Von Petzinger 2016). From 25,000 to 19,000 BP, there were sedentary settlements in Eastern Europe (Hayden 2014a), with large, multiroom structures made of wood, skin, and mammoth bone (Pettitt 2014). Most sites had storage pits, drying racks, basketry, and textiles. Scholars call foragers who store "collectors" and settled groups who practiced collecting "complex hunter-gatherers."

These patterns ended, then reappeared as the global climate changed. The LGM ended about 20,000 years ago. The planet then warmed for several thousand years, warming more after 14,600 BP. Glaciers melted, monsoon systems started, the tropics became lusher, and the Central Asian tundra steppe retreated. From 12,900 to 11,700 BP, global temperatures cooled again in the "Younger Dryas"; then the earth warmed again and has remained warm since (Brooke 2014). Scientists call this period of warm climate the Holocene Era. Glaciers melted more, sea levels rose, the atmosphere became more humid, and carbon dioxide (which plants breathe) increased. While varying over time, the Holocene climate has been more stable than the Pleistocene (Brooke 2014; Cummings 2014).

After the LGM, humans intensified broad-spectrum foraging. This was a shift from eating long-lived plants and animals with few offspring to species with short lives who "reproduce in incalculably large numbers" (Wenke and Olszewski 2007:240). They invented new technologies oriented to "small and faster prey: nets, traps, bows and arrows, baskets" (Brooke 2014:140) and smaller and finer stone tools, or "microliths" (Cummings 2014). Most foragers remained mobile, but complex hunter-gatherers reappeared, either as "transhumant," moving annually between a few permanent settlements, or settling permanently. People returned to northern Europe, producing the spectacular Lascaux cave art. In Eastern Europe, "[d]welling structures employing large numbers of mammoth bones increase in number" (Pettitt 2014:292). These settlements were larger than foraging camps and may have been unequal economically (Hayden 2014a). In Chinese caves, archaeologists found pottery, the first in the world (21,700–19,300 BP) (Barnes 2015:93). Pottery indicates sedentism beyond normal forager camping; tools accompanying pots also indicate increased sedentism (Hayden 2014a). Sedentary hunter-gatherer villages appeared elsewhere in China (Cohen 2011). In Southwest Asia around 15,000 BP, people started settling seasonally or permanently in villages of over 100 people. Many used sickles for harvesting and mortars and pestles to grind grains, living in stone huts (Watkins 2018).

Villages of broad-spectrum collectors increased in size and number around the world in the Holocene. In Southwest Asia, settled hunter-gatherer villages of 18–22 people existed between 17,400 and 15,000 BP. From 13,500 to 11,600 BP, villages of 60 people developed. In the next thousand years, village populations averaged 330 people. After 10,000 BP, they averaged 760 people, growing to over three thousand by 8700 BP. By 8000 BP, they had scattered back to smaller villages (Brooke 2014; Watkins 2018). Hunter-gatherer villages with houses and storage also developed in East Asia (Cohen 2011). In Japan, villages had prestige items, differential burials, pit buildings, and ventilated hearths (Barnes 2015). Pottery appeared in the then-grassy Sahara "in the ninth/tenth millennia BP." The oldest were from Niger (tenth millennium BP) and were "richly decorated" (Barich 2013:446). Semisedentary societies, with incipient inequalities, also developed there and in the Nile Valley. Coastal peoples in southern Brazil settled around 8000 BP, fishing and collecting shells and plants (Moore 2014). One southern European village had "prestige goods, feasting, ancestral or totemic sculptures and burials, permanent architecture, large-sized settlement, and inequalities in burials" before 5000 BP (Hayden 2014a:650).

People also responded to climate warming economically. They developed "management" techniques to supplement hunting, gathering, and fishing (Graeber and Wengrow 2021), including "land clearance, planting, sowing, weeding, harvesting, soil drainage and irrigation," and burning (Harris 2007:25). Others in Southwest Asia, East Asia, Libya, and Egypt started managing herd animals (Cohen 2011; Zeder 2011; Marshall and Weissbrod 2011). Still-mobile South American peoples also started managing plants (Piperno 2011). These supplementary practices lasted for centuries in some places and for millennia in others. Some archaeologists call mixed foraging/collecting and managing people "in-between" societies (Barnes 2015:117).

Cumulative learning led eventually to increased interventions. One tactic was selectivity: choosing larger seeds from plants, killing the more obstreperous animals, and breeding the more docile animals (Graeber and Wengrow 2021). Repeating selectivity over millennia, humans gradually developed woolier, meatier, and more docile animals and bigger and more harvestable plants (Hayden 2014a). Although they combined hunting, gathering, fishing, herding, and cultivating for millennia, repeated selections ultimately led to "domesticated" species, those dependent on humans for reproduction. In turn, relying mostly or totally on domesticated foods meant that managers became farmers (Bar-Yosef 2011: S182).

These processes took several millennia to become transformative, so any individual saw little change. Probably nobody knew they were doing something revolutionary, though some think humans made changes for broader purposes (Stiner and Feeley-Harnik 2011). Different definitions of "domestication" make accurate dating difficult (compare, e.g., Price and Bar-Yosef 2011 with Larson et al. 2014). Also, since research is ongoing, scholarly understandings can alter quickly (Zeder 2011):

While scholars looked for "centers" of domestication, they now analyze entire regions (Barker 2009; Harris 2007; Price and Bar-Yosef 2011; Zeder 2011).

The process started at the end of the Younger Dryas (Larson et al. 2014: Table S1, p. 4). Southwest Asians started managing pigs by 11,500 BP, wheat and barley by 11,250 BP, domesticating all three by 9000 BP, and more plants and animals by 8000 BP. Mesoamericans and South Americans began domesticating squash by 10,000 BP; Mesoamericans began domesticating maize by 9000 BP, and South Americans coca by 8000 BP. East Asian peoples were domesticating pigs by 8500 BP and millet by 8000 BP. South Asians and North Africans started domesticating cattle by 8000 BP and 7750 BP, respectively, and managed other animal species. (As "infill" took time, foragers did not live in all the land portrayed in Fig. 4.1.)

Scholars are unsure of the exact sequence of plant domestication, and even less sure about animal domestication. This included three elements: constraining animals' movement, regulating their breeding, and controlling their feeding. Wolves may have helped domesticate themselves by scavenging from humans, but herd animals make these three tasks easier. Management may have started with guarding a herd, then penning it, and finally killing off the difficult animals, thus breeding for docility. Others suggest that people domesticated animals by capturing baby animals, raising them, and breeding those who got along better with humans (Stiner and Feeley-Harnik 2011). Globally, people consistently domesticated only five herd animals: Goats, sheep, pigs, cows, and horses. Eventually, they also domesticated "two kinds of camels, donkeys, llamas, reindeer, water buffalo, yaks, and Bali cattle" (Brown 2007:78), only some of which are true herd animals. Women may have domesticated plants and men domesticated animals, but given the debate about this division of labor, perhaps not.

The earliest domestications occurred independently of each other. Africans and South Asians may have borrowed from other regions, but evidence of independent

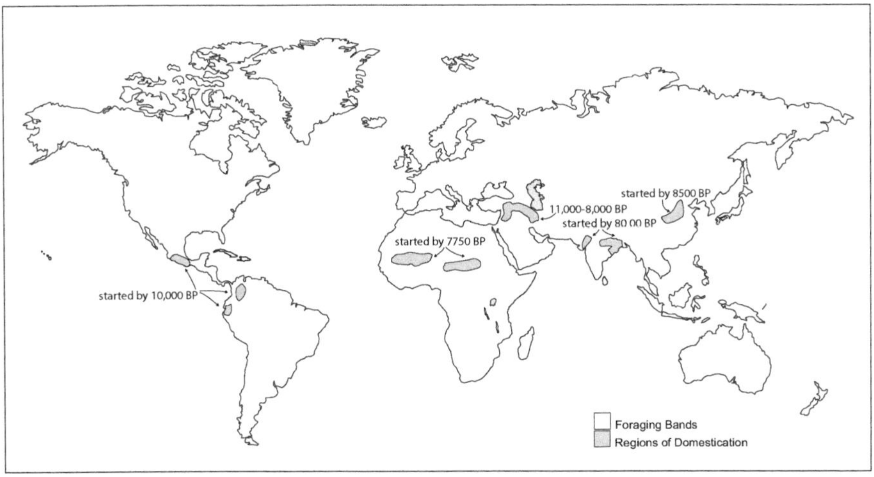

Fig. 4.1 Regions of first domestication. Source: Larson et al. (2014); Houston (2018)

domestication in both regions keeps appearing (Fuller 2011; Marshall and Weissbrod 2011). Around the world in this period, most people remained mobile, broad-spectrum foragers; fewer engaged in subsistence management; even fewer settled in villages; and domesticators were the rarest.

Explaining Settlement and Domestication: Debates

As management and domestication started in "at least seven widely dispersed and ecologically disparate regions of the world" in "near synchroneity" after many millennia of no change (Piperno 2011: S464), scholars ask the obvious question: Why? The debate ranges across "genetics, archaeobotany, zooarchaeology, geoarchaeology, and archaeology" (Larson et al. 2014:6139) in two camps. Some argue for general explanations, in structural-functionalist or conflict veins. Others reject general theories, arguing for unique explanations for specific places.

The debate starts with complex hunter-gatherers. Some (Hayden 2014a) say that attempts at aggrandizing led to hunter-gatherer complexity. Some foragers found environments lush enough to reduce their mobility, allowing storage. Burials demonstrate both unequal wealth between lineages and regularly produced surpluses: "prestige objects, rich graves, middens indicative of some sedentism, and storage" occur in some places and not others (Hayden 2014a:645–646). Others strongly disagree, arguing that ceremonial centers, such as Göbekli Tepe in Anatolia and those in Eastern Europe "from Kraków to Kiev," functioned as sites for trade and cohesion; there were "monuments and magnificent burials, but little else to indicate the growth of ranked societies" (Graeber and Wengrow 2021:90, 92). Something like seasonal foraging-band fission-fusion strategies seemed to occur on a massive scale, and ethnographers have observed seasonal changes in social organization (Graeber and Wengrow 2021:107–111).

Studying complex hunter-gatherers provides only a limited explanation; some never started food management, much less domestication, while some mobile foragers started cultivating and domesticating (Piperno 2011). Still, groups that managed and domesticated plants and animals broadly increased in average size and decreased in average mobility. Thus, scholars debate population theories of domestication. Perhaps circumscribed population growth led people to innovate (Boserup 1965). Yet others claim that population growth spurs innovation by having more minds working the same problem and more interactions between those minds (Brooke 2014:104–105). In these theories, groups are usually already sedentary in lush, productive areas and managing resources, such as river valleys like the Tigris and Euphrates area, the Nile Valley, and the Yangzi (Brooke 2014). While few sites show population density leading to food shortages, people could have seen ecological overuse happening, and they could have transitioned quickly (Mazoyer and Roudart 2006). Still, sometimes populations grew first and then groups started farming and tending animals, while other times they started domestication first and then their populations grew (Barker 2009).

Others argue that population growth in complex hunter-gatherers changed political strategies. As land was more productive, kin groups owned the land, so kin groups became more important, and sharing now occurred only within them (Hayden 2014b). Eventually, some kin groups had more resources than others, so economic inequality and competition increased. Feasting demonstrated group wealth; another group's failure to hold a better feast demonstrated its weakness, so pressures to top each other created upwardly spiraling competitive feasting cycles (Hayden 2014b), increasing food productivity. Increased population meant increased interconnections between groups; settled villages allowed people to regularly find trade partners (Bogucki 1999). People traded unusual, even "luxury," foods and drugs regionally (Anderson 2017), perhaps increasing food productivity (Hayden 2003).

Other scholars theorize domestication as risk management. Local annual variations in food supplies created risk for foragers, so relying on a broad spectrum of foods reduced risk and increased options. After a while, people emphasized and stored only a few plants and animals, increasing the risk again. To make that food supply more reliable, people started managing some foods, which ultimately led to domesticating them (Piperno 2011).

Many argue against general explanations. Southwest and East Asians domesticated both plants and animals at roughly the same time (Zeder 2011; Barnes 2015); mobile Africans domesticated only animals at first, domesticating plants millennia later (Marshall and Weissbrod 2011); and mobile South Americans domesticated only plants at first (Piperno 2011). These differences suggest unique explanations resting on local conditions. At one conference, scholars proffered "available protodomesticates, human sedentism, higher population density, resource abundance, geographic and/or social constraints, processing and harvesting technology, storage, and wealth accumulation" as causes of domestication (Price and Bar-Yosef 2011: S172).

Since domestication started only after the Younger Dryas, everyone might agree that climate change was a precondition for agriculture. Still, scholars cannot agree on why it happened in any given place or why it happened several times (Price and Bar-Yosef 2011).

Social Patterns of Prehistoric Villages

All scholars seem to agree that agriculture transformed human societies. Settling in villages created important changes; sedentism and domestication together created a true transformation. New distribution and consumption practices developed. Groups became larger and perhaps more cohesive, more often oriented to kinship. Warfare and permanent, but still informal, leadership developed. Gender and economic inequalities sometimes developed. More diverse and intense technologies and artistic styles emerged. Population growth accelerated; this and other changes increased the human impact on the land.

Village Economies

People in villages procured and processed economic resources in four ways—foraging, collecting, horticulture, or pastoralism—which they often combined. Foraging and collecting are on a continuum (Lane 2014), depending on the length of storage. Horticulture is subsistence gardening, and pastoral peoples herd animals. Reduced mobility seems to have changed distribution patterns, and having more belongings changed consumption.

Collecting required a larger, more complex division of labor—new *chaînes opératoires*. Trips to hunt and gather were fewer but longer. Groups dried and stored meat, fish, grains, legumes, and nuts and developed more specialized tools. In Southwest Asia after 20,000 BP, people continued hunting while collecting and storing plants and seeds (Watkins 2018). In the Sahara and Sahel, people started using ceramics—first for water, then for storage—while foraging (Barich 2013). By 10,000 BP, "the production of ceramics burgeoned across East Asia" (Barnes 2015:97) for storing and cooking. Peoples in western South America emphasized seafood after 11,000 BP, moving seasonally between coast and mountains; they later established "more permanent settlements along the coast" and engaged in marine resource collection (Moore 2014:107). With increasing aridity around 9000 BP, mobile North American peoples in the Great Basin began "using semi-permanent base camps with basic storage facilities" (Fagan 2019:150).

Mixed foraging-collecting strategies shaded into casual, then active, plant and animal management, then domestication. Societies starting food production[1] relied primarily on horticulture, herding, or agropastoralism. Horticulture involves small-scale cultivation of plants using hand tools such as hoes, digging sticks, or shovels (Sutton and Anderson 2010). In one form, swidden ("slash and burn") horticulture, people cut down plants, burn them, and mix the ash into the soil. They claim a large area but use only some of it annually, moving from plot to plot every few years, to let most of the land regenerate its fertility (Sutton and Anderson 2010). People started this in central Mexico and Panama by 7000 BP (Piperno 2011). Other horticulturalists use floodplains; annual flooding drops sediment, and receding floods leave newly fertile fields. People seem to have done this at Çatalhöyük in Southwest Asia (Graeber and Wengrow 2021) and from the Levant to the Nile Valley after 7200 BP (Haaland and Haaland 2013). People started growing rice in bodies of water by 7000 BP in East Asia (Higham 2018). In Peru, people began to cut irrigation canals from streams to water fields (Piperno 2011).

Pastoralists herd animals. They are sometimes more mobile than foraging bands in moving their flocks between pastures. While people always used animals for "primary" products—meat, bone, and hide—in this period they started using animals for "secondary" products, such as milk, traction for transportation, manure,

---

[1] Previously, I called the act of turning wild plants and animals into food "production," but archaeologists use the term "food production" for farming in contrast to merely obtaining wild plants and animals.

food storage, and wool to weave into clothing. Scholars debate this shift (Sherratt 1981; Greenfield 2014). Humans domesticated dogs for secondary uses (hunting, traction, and protection), so some suggest that people domesticated herding animals for secondary uses (Shipman 2014). The process is also unclear: "[M]ilk exploitation occurred simultaneously with or quickly followed upon the domestication of some livestock" (Greenfield 2014:32). Specialized pastoralists developed in the Levant in the eighth millennium BP, and African pastoralists a little later. Many supplemented reliance on herds with hunting, gathering, fishing, and trade (Marshall and Weissbrod 2011; Honeychurch and Makarawicz 2016:346).

In the early Holocene, people commonly relied on mixed subsistence strategies; after some millennia, they started relying primarily on domesticated foods. Scholars call the reliance on domesticated plants and animals together "agropastoralism." This first appeared in Southwest Asia by 9000 BP. Across the region, peoples relied on goats, sheep, pigs, and cattle "alongside [fully domesticated] cereals and various pulses, including emmer and einkorn wheat, the naked wheats, barleys, pea, lentil, chickpea, flax, and bitter vetch" (Leppard 2022:240).

Changes in production methods changed distribution in at least two ways. Storage responded to seasonality: Producing food in one season and living off it in the next (Bogucki 1999). Ethnographically, village societies tend to share less between families than foraging bands do (Hayden 2014a; Lane 2014). Archaeologically, in Southwest Asia, some villages likely had collective storage by 11,000 BP; by 10,000 BP, there was increasing storage within individual houses, though collective hunting and feasting continued (Watkins 2018:218). Storage in East Asian villages after 10,000 BP might have been for entire villages or only for households, but households often contained upwards of 30 people (Barnes 2015:101).

By the early Holocene, predictable locations and population and production increases raised intervillage trade well above the level of interband trade (Bogucki 1999:155) in Southwest Asia (Watkins 2018); distances increased for many objects, especially those symbolizing membership in large exchange networks. Trade also increased in East Asia, especially on Honshu in Japan (Barnes 2015). Also, "[m]eeting in predetermined places is common" ethnographically (Stanish 2017:138); ritual sites in the early Holocene such as Göbekli Tepe made this possible (Watkins 2018). These sites probably helped people renew friendships and arrange marriages. Increased trade may have started a regional division of labor: different villages provided different goods, increasing interdependence. Ethnographically, raiding is an unusual but consistent form of exchange (Stanish 2017:111): One group stole goods from another, then the other stole from the first. It rarely appears archaeologically, but was likely.

Consumption practices also changed. Sedentism "allowed the accumulation of material goods" (Scarre 2018b:188), so surpluses helped the sheer number of objects to increase; villagers have much more food, clothing, shelter, and leisure products than bands do. Complex hunter-gatherers buried individuals in Sungir, northern Russia (roughly 30,000 BP), Dolní Věstonice, Moravia (26,000 BP), and in the Dordogne (16,000 BP), while wearing massive, lavish sets of jewelry made of shells, mammoth ivory, animal teeth and skins, and antlers. Each piece took hundreds of hours to make (Hayden 2014a; Graeber and Wengrow 2021). While archaeologists differ on how to interpret these burials, they show consumption practices well beyond those of foraging bands on the frontier. Indeed, there was a "blossoming of artistic traditions" in this era "in beads, pendants, bone pins, small sculptures, and wall paintings" (Smail et al. 2011:232). Origins of the materials indicate a wide exchange area, so these goods solidified social relationships.

Communal consumption continued, taking different forms. Feasts after hunting persisted in Southwest Asia. Some archaeologists think feasts maintained village cohesion (Watkins 2018:218), while others call them competitive political practices (Hayden 2014b). Perhaps they were both. Yet new consumption practices appeared. Many dwellings in Çatalhöyük contained wall decorations, such as wild cattle skulls. Even semisedentary patterns provided places to store things not being used, so more kinds of storage developed. People invented pottery several times, first in China by 19,300 BP (Barnes 2015), and then made baskets, houses, and storage buildings.

Overall, millennia of foraging, foraging-collecting, and management led to domestication in horticulture, pastoralism, and agropastoralism. Group reciprocity narrowed somewhat, and exchange networks became denser and larger. People had more, and more varied, objects.

Kinship in Villages

As subsistence production changed, so did kinship forms. Inferring kinship systems archaeologically requires ethnographic analogy (Ensor 2021). Archaeologists often infer unilineal descent in domesticating peoples (Bogucki 1999; Watkins 2018). Increased production and larger kin groups reduced incentives for sharing, so collectively owned land became more important; group boundaries may have firmed up. Here, I review ethnographies of kinship in villages, then analyze evidence for early Holocene villages in Europe and Southwest Asia.

The flexible kin systems of foragers fitted their mobile way of life; in villages, kin institutions seem to have responded to new conditions. Anthropologists often call descent groups "corporate" groups, from the Latin word *corpus* for "body," denoting the unity of the kin group. The most common type of descent in villages is unilineality—the descent "line" defines the group. Patrilineal descent goes from father to son to grandson; women marry into the line. Matrilineal descent goes from mother to daughter to granddaughter; men marry in. Patrilineality is most common ethnographically; matrilineality is less common; and other, more complex forms are even less common. Residence patterns match descent groups imperfectly: matrilineality usually goes with matrilocality, but not always, and patrilocality and patrilineality also usually coincide—but again, not always. Geographic residence near descent-group members helps groups that face internal conflict; when outsiders threaten, different descent groups may disperse throughout a village (Pasternak et al. 1997:265–266).

Lineages form the basis of larger descent groups. Anthropologists call these "segmentary" kin societies (Pasternak et al. 1997:268–271). Clans are sets of lineages in which all claim descent from a common ancestor (Pasternak et al. 1997; Flannery and Marcus 2012). Many villages have sets of interrelated clans. Two clan sets in a village are called "moieties" (from the French word *moitié*, or "half"), while three or more are known as "phratries" (Flannery and Marcus 2012). In warfare, sometimes smaller segments join to make larger ones.[2]

Gradually increased sedentism in the early Holocene may have led to these descent and residence patterns. Segmentary groups carefully define rules around exogamy and endogamy. Sometimes one must marry out of one's clan set; other times, first-cousin marriage becomes common since it keeps property in the family, and groups develop complex rules about which cousins to marry.[3] Parental involvement in arranging marriages probably increases. Dowry, granting the bride her inheritance on marriage, and bride price, groom labor for the bride's family, become more common, depending on circumstances.

Applying these findings archaeologically is difficult. Households gradually superseded individuals or married couples as the basic social unit (Stanish 2017:104–109). Cross-culturally, bilateral descent tends to produce "haphazard" arrangements of dwellings (Ensor 2021:163). Patrilineal-patrilocal residences tend to be small but numerous, in a cluster "formally-arranged around a small plaza" (Ensor 2021:161), while matrilocal-matrilineal houses tend to be larger but fewer in number. Archaeologists can combine these cross-cultural findings with ancient DNA and isotopic evidence to infer descent and residence (Ensor 2021:Chap. 2).

Kinship in this period started shifting to these patterns. Ethnographic inference indicates that Europeans before 7000 BP generally lived in bilateral kinship and bilocal residence systems (Ensor 2021), though some sites suggest increasing matrilocality over time in Central Europe with some patrilocality in Iberia (Scarre 2018a). Thus, groups seem to have remained flexible for millennia, moving slowly to unilineal, corporate descent groups. I found no other in-depth kinship studies outside Europe, but some argue that Southwest Asians had segmentary societies (Watkins 2008) and East Asian households often contained upwards of 30 people (Barnes 2015); perhaps they developed matrilineal-matrilocal kin structures.

Scholars also debate the gendered division of labor in village societies. Ethnographically, men tend to clear areas, women plant and tend crops, and both sexes harvest the crops; this varies by circumstances (Levy 2006). In some places, gender roles were fluid among collecting, horticultural, and pastoral societies. Levantine skeletons from just after the Younger Dryas "suggest a convergence in

[2] From what I can tell, archaeologists tend to treat all these structures as "segmentary," while ethnographers do not, since in different societies, segments (lineages, clans, etc.) have different obligations.

[3] Geneticists have shown that risks of cousin marriage are far lower than previously believed (Conniff 2003).

men's and women's overall muscle demands, with possible reductions in sexual labor divisions" in some places in the region, but not in others (Peterson 2010:253). Despite assumptions of male cattle tending, women have also tended cattle, been central to dairying, and have long made textiles from animal hair (Crabtree 2006). However, some places show a clear division of labor. In Iberian burials and rock art, males are much more associated with bows and arrows, female skeletons more with pottery (Cintas-Peña and García-Sanjuan 2019). In South Asia (10,000–7000 BP), rock art consistently portrays women "in less hazardous activities such as vegetable and food gathering or fish and rat catching" (Jain 2006:72). Whether these are accurate portrayals or merely ideals is open to interpretation.

Ethnographically, village societies experience sexual and gender fluidity. Many allow same-sex sexual relationships among adolescents, which sometimes do and sometimes do not continue into adulthood, even continuing with heterosexual marriage (Neill 2009:Chap. 2). Some require same-sex relationships among both boys and girls in initiation ceremonies, while others merely allow them. Archaeologically, there seems to be evidence that "in prehistoric northern Europe … shamans harnessed both male and female sexual potentials" (Voss 2008:325).

Inferring kinship patterns prehistorically is difficult. Europeans before 7000 BP seem to have retained bilateral descent and bilocal residence, yet elsewhere, trends toward uni-residential systems appear. Gendered divisions of labor seem varied among domesticating societies, some archaeologically showing a strong division and others retaining the flexibility of forager ancestors.

Village Politics

Changes in subsistence, sedentism, and kinship interacted to create two political changes. Leadership in villages is difficult to see archaeologically; ethnographic inference suggests that village leadership was somewhat institutionalized, achieved, informal, and heterarchical. Second, warfare developed; some argue that it was a function of increasing group cohesion.

Ethnographically, leadership in villages is usually "heterarchical"—there are always multiple leaders since individuals are "ranked in a number of different ways" (Crumley 1995:3), so people have different kinds of power. Leadership is institutionalized because positions always exist in people's minds; people achieve them by strenuous effort and others' acknowledgment (Flannery and Marcus 2012). Aspiring leaders create and lead social networks. This probably happened prehistorically for three reasons. First, villages are usually larger than bands, and formalization follows increases in group size and density (Lewellen 2003; Watkins 2018; McIntosh 1999 objects). Second, while frontier bands rarely needed to control land, settled peoples usually did; demographic growth increased pressures on common resources, which created common-pool resource problems that leadership could solve (Ostrom et al. 1994; Johnson and Earle 2000). Finally, spiritual megasites (see below) likely required coordination to build.

Forms of village leadership vary ethnographically and probably did prehistorically. Lineage or clan leaders may form councils. Some groups have "Big Men" or "Big Women." Aspirants do people favors to build a network of obligations, then call them in later, usually for third parties, which creates more obligations. These include organizing and funding collective projects, including "large-scale works like dams, weirs, and defense structures" (Johnson and Earle 2000:211) that benefit many people; ceremonial feasts for which Big Men exhort their followers to outcompete other's followers; and keeping "a retinue of warriors" in war-filled lands (Johnson and Earle 2000:212; Flannery and Marcus 2012) to defend nonwarriors, from whom they receive payments. Nonkinship associations called "sodalities" create links across lineages or villages. They include age sets, ancestor cults (Hayden 2003), or other secret societies. (The societies' existences are not secret; instead, they hold secret, ritual knowledge [Hayden 2018:7]). These societies often had ritual buildings ethnographically; archaeologically, several sites in Southwest Asia over 10,000 years ago had buildings that look much like men's association buildings. Ritual houses were also found at Göbekli Tepe and in other Mesopotamian centers (Flannery and Marcus 2012). There is some evidence of older village leaders, both male and female, in Central European burials (Milisauskas 2011).

Leaders sometimes manage relations between communities, which often existed in conflict. In this period, violence seems to have shifted from an individual to a group basis: War. While the most simply organized foragers ethnographically engaged only in individual-to-individual violence (Kelly 2000), "segmentary" societies—ones with unilineal, cohesive descent groups—started to hold entire groups responsible for their individual members' behaviors, creating "social substitutability" (Kelly 2000), which led to what psychologists call "vicarious retribution" (De Dreu 2010): If people in group A cannot kill a murder's perpetrator in group B, group A will kill someone else in group B. Reciprocal retribution—someone in group B then kills someone in group A—usually spirals into ongoing feuds (De Dreu 2010).

Social substitutability seemed to combined with other factors to generate warfare. Once the frontier moved on and population grew, groups started competing for land; infill eased this only somewhat. All scholars studying warfare (Kelly 2000; Boehm 2013; Kelly 2013; LeBlanc 2014; Gat 2017) agree that population pressure (population growth in a limited area, pressing on food resources) led to war; some (Boehm 2013; Gat 2017) argue that male-male competition for females also contributed, which others downplay (Kelly 2000, 2013). In addition, complex foragers and some villagers continued hunting and thus using weapons; the mortality rate remained high (see below), perhaps generating grief, rage, and fear (Rosaldo 1989; Turner 2007); and cultural differences between groups led to ethnocentrism (Hebl and Madera 2007).[4]

[4] Political scientists would probably mention anarchy as another condition, since heterarchical leadership meant no pacifying central power. This is anachronistic, as central leadership had not yet developed anywhere.

Ethnographically, warfare takes two forms. Raids and ambushes are common, especially at dawn, with few deaths per incident but a large cumulative death rate; massacres and pitched battles are rare (LeBlanc 2006). Prehistorically, evidence for war appears only in the Holocene. At Jebel Sahaba in Sudan around 13,000 years ago, individuals had multiple wounds (or "pincushioning"); more women and children than men were killed; and some had forearm parry fractures, all indicators of intergroup violence (Kelly 2000:149–150; Smith 2017:Ch. 4). Evidence of a massacre also exists in Nataruk, Kenya from around 10 kya (Smith 2017). Australian rock art from 10 kya also depicts collective violence. From then, archaeological evidence of war increasingly appears (Haas and Piscitelli 2013; Smith 2017), most frequently in Southwest Asia. Çatalhöyük had "maces and enlarged projectile points" from around 10 kya, and near it, there were "several clearly fortified locations" in the ninth millennium BP. By 7000 BP, "unambiguous indications of war become common" in the region (Ferguson 2006:483–484), including defensive walls (Flannery and Marcus 2012). There is evidence of massacres from the ninth millennium BP in southern Germany and Austria around 7000 BP. In two of these sites, "there were adult males, children of all ages, older adult females but no women of child-bearing age" buried (Smith 2017:101). It is likely that feuds and small raids also occurred (Smith 2017).

The conclusions above about kinship raise the question of whether segmented kinship or intergroup conflict developed first. Perhaps social substitutability and intergroup violence evolved together (Pasternak et al. 1997; Kelly 2000). The lack of cohesion beyond the clan or secret society made decisive collective action difficult. Villagers often created "buffer zones" of unplanted lands between villages, creating an early warning area for attacks (Kelly 2000; LeBlanc 2006). Fewer planted areas created more resource pressure, intensifying the stakes, so war often continued once it started. However, Kelly (2005) argues that peacemaking and nonviolent competition could make buffer zones unnecessary, opening up more land.

Again, this depends upon definitions. Most scholars cite, yet then ignore the implications of, the Kellys' separate (2000, 2013) analyses of hunter-gatherers. The existence of simply organized bands and complex foragers, and many in between, makes testing Kelly's (2000) definition of "warfare" crucial. Neither Boehm (2013) nor Gat (2017) distinguishes between kinds of forager groups, yet both infer war from this mix of groups. If violence did occur purely interpersonally in the simplest forager bands, then their conclusions are untenable. However, if Kelly (2000) is mistaken that social substitutability—true group-against-group thinking—led to vicarious retribution and war, then we might accept their conclusions, and those of LeBlanc (2014).

The debate also ignores women. Given the high mortality rate, gender egalitarianism, increasing evidence of female familiarity with weaponry (Haas et al. 2020; Anderson et al. 2023), complex relations with male relatives, and the stakes—the survival of self and children—women were likely involved. The evidence cited above that no women of childbearing age were found in at least one massacre, plus some ethnographic evidence of warfare starting because of "women willingly leaving their husbands, and joining other men" (Gat 2017:45), suggests that women's role in the origins of warfare is severely understudied. Were women behind-the-scenes fomenters of conflict, active combatants, peacemakers, victims, or in all four roles depending on conditions? This needs more research.

Villagers might also have developed what became a common coping mechanism. In dehumanization, "people deny humanness to others on the basis of their group membership, reserving humanness for their ingroup" (Haslam et al. 2013:38), characterizing outsiders as animals or objects (also, Smith 2011). Dehumanizing attitudes about out-groups are closely associated with willingness to commit violence against such groups (Haslam and Loughnan 2016); they are often found preceding mass atrocities (Vollhardt and Campbell-Obaid 2016:166–169). Some critics question dehumanization in general, arguing that those committing violence want their victims to experience very human emotions (Fiske and Rai 2015). Others argue that dehumanizers attribute negative but human characteristics to enemies (Over 2021). Yet others respond that this is how dehumanization works: ideologies express the idea that the out-group is a human group whose morality is wrong—less than human (Phillips 2022). Thus, it is an extension of ethnocentrism (Smith 2011). Finally, some argue that violence is in fact "morally motivated. Morality is about regulating social relationships, and violence is one way to regulate relationships" (Fiske and Rai 2015:xxii). In this view, villagers would have morally justified, to themselves, raids and massacres against other groups, usually as protection of self and group.

These cultural theories ignore the extremely high mortality of the past that generated grief, fear, and rage (Rosaldo 1989; Turner 2007). This may have helped turn ethnocentrism into dehumanization or moral violence. In ongoing warfare, tactical dehumanization may have become ideological. Ethnographies from Brazil and New Guinea show that many village-level peoples culturally dehumanize neighboring peoples as part of attacking them (Smith 2011). Alternatively, intergroup violence may have generated moral codes of defending one's group violently. Either way, there would have been a cultural component to incipient warfare.

Overall, then, leadership in villages seems to have become more institutionalized than in foraging bands. It seems that intergroup violence or war consisted of feuds and raids founded in social substitutability and perhaps resulting in dehumanizing ideologies.

Inequalities in Villages

The increase in intergroup violence derived partly from increased inequality, which exists in villages ethnographically and perhaps began prehistorically. The rough gender equality on the global human frontier also began to change but only in a few places, and only minimally.

Economic inequality seems to have started with sedentism and storage. People accessed solar energy stored in plants and animals by work—putting their life energy into collecting, farming, or herding to produce food and other objects. In mobile bands, stored objects were rare; in villages, they accumulated for trade or use. As products of their own life energy, people treated objects as scarce, hard-won resources for survival; if someone took them in a raid, a counterraid (and anger) made sense. In many places, people developed a surplus through the accumulation of belongings, and harvests may not have provided much food above subsistence.

Scholars disagree about when economic inequality started after the frontier moved on. Wealth differences between complex hunter-gatherer families may have appeared before the LGM in Europe (Hayden 2014a). Burial goods became more elaborate, perhaps differentiating status (Bogucki 1999). In some sites in Russia, nearly all burials had grave goods; only some had prestige goods, which may have reflected inequalities or personal qualities (Scarre 2018b). Elsewhere, villages of peoples with "in-between" subsistence strategies show some, though minimal, wealth inequalities. Some houses in Çatalhöyük had more objects than others had, perhaps due to generational accumulation and resulting prestige (Graeber and Wengrow 2021:222). Other Southwest Asian villages had larger and smaller houses, and more and fewer goods (Bogucki 1999; Watkins 2018; Graeber and Wengrow 2021). Still, to 7000 BP, the region shows no serious economic inequalities. There were likewise only minimal economic inequalities in East Asia (Barnes 2015; Higham 2018) and none anywhere else.

At first, gender equality seems to have continued. Villages early in the transition to farming in Southwest Asia seem gender egalitarian (Peterson 2010). Both sexes contributed to production, rituals included women at all levels, and neither burials nor spatial analyses suggest inequality. At Çatalhöyük, "the female form was a special focus of ritual attention," and the masculine and feminine domains were kept apart in imagery (Graeber and Wengrow 2021:220). In Iberia, for a few millennia after 8000 BP, male and female skeletons in some burials show no real differences in dietary intake, occupation, residence patterns, burial postures or grave goods, or pathologies, except for physical trauma (Cintas-Peña and García-Sanjuan 2019). After 7000 BP in China, skeletal and burial evidence show similar diets and diet-related pathologies between males and females and similarities in grave goods (Dong et al. 2017).

Later, real gender inequality seems to have begun. Formal burial was accorded to Iberian males much more often than to females (Cintas-Peña and García-Sanjuan 2019). In what is now the Czech Republic, burial evidence from before 7000 BP shows that males ate more meat than some females, males "were buried with more numerous and more varied grave goods than females," and grave goods suggest a strongly gendered division of labor (Masclans Tatorre et al. 2021:267). Female burials indicate a much wider variability, so the evidence is equivocal. Ethnographically, many villages have both men's and women's houses, but there were usually more men's houses, and they were often more important (Flannery and Marcus 2012).

Overall, then, economic inequalities appeared in some areas, especially among complex hunter-gatherers, yet most villagers around the world seem to have remained egalitarian. Gender inequalities began in a few places, but most still seemed gender egalitarian.

Village Ideas

Living in larger groups also spurred human creativity, as more minds interacting meant more ideas generated (Henrich 2015). New beliefs and practices about ancestors developed, and technologies elaborated in the new social and material situation.

Spiritual beliefs in villages were mostly similar to those in foraging bands. People still seem to have lived with beliefs in pervasive supernatural forces. Some villagers extended this to emphasize ancestor veneration. Inferences from ethnographies suggest that such veneration was "absent in early humans" (Peoples et al. 2016:272). Ancestor veneration appears more in complex hunter-gatherer groups, occurring frequently in horticultural ones (Nolan and Lenski 2011). As one's ancestors moved to the larger spirit world at death, they could influence other spirits animating the wind, water, and land near villages. Villagers may have appealed to these familiar, but now supernatural, authorities for fertility and spiritual guidance (Hayden 2003).

Burial practices clearly indicate more attention to ancestors. Archaeologists find many more burials among villagers than among foragers (Bogucki 1999; Jain 2006; Higham 2018; Watkins 2018). In Southwest Asia, some buried their kin under house floors. Others buried them, then exhumed the skull, leaving the body. Some re-formed the face on the skull with clay. Elsewhere, they reburied many skulls together, separately from the bodies, or buried their dead in house walls (Watkins 2018). In southeastern Europe, people buried the dead within settlements, and in Central Europe, they buried them just outside; South Asian villagers also buried people near their huts, and some East Asians first buried their relatives near houses, then in cemeteries nearby (Jain 2006; Milisauskas 2011; Higham 2018).

Of course, specific beliefs are unrecoverable. Still, social scientists interpret this emphasis on ancestors in two ways. First, it may have operated as social control. The current elderly would soon be "the ancestors" controlling the land's fertility; angering them while they were still alive would be foolish (Hayden 2003). Second, ancestor veneration tended to arise with land inheritance; burial might have been a way to stake a claim to the land. Both suggestions are plausible. In addition, high mortality meant that people regularly lost loved ones to death, so burial, exhumation, and reburial may have had emotional as well as spiritual purposes.

Spiritual leaders—shamans—continued into this period. Archaeologists think an elderly woman buried with potentially ritual objects in Southwest Asia before 12,000 BP was a shaman (Watkins 2018). A few burials in East Asia contain objects rare in the region; archaeologists have interpreted the people as ritual leaders (Higham 2018:234–235). Rock art in South Asia portrays people dancing or wearing masks; interestingly, such images appear far from residential areas, suggesting that making imagery itself might have been ritualistic (Jain 2006).

Ritual sites became important, starting with complex foragers. Sedentism allowed people in villages "to hold scheduled, calendrical rituals at the same every year" (Flannery and Marcus 2012:163). One village in Syria (11,600–10,800 BP) had ritual structures with collective storage facilities (Watkins 2018:212). Possibly thousands of hunter-gatherer-cultivators built Göbekli Tepe in Anatolia, using it for ceremonies for at least seven hundred years (Watkins 2018:217). Villages in East Asia later had ritual structures (Barnes 2015). There also seem to have been some ritual buildings in southeastern European villages by 7000 BP (Milisauskas 2011).

Morality systems also changed with domestication. Matrilineal-matrilocal groups have the fewest restrictions on people's sexual behavior, while patrilineal-patrilocal societies have much more (Pasternak et al. 1997). Since sharing between families reduced in villages (Hayden 2014a; Lane 2014), reciprocal obligations to kin increased. Also, pastoralists tended to develop a "culture of honor." Ethnographies of over 1000 pastoralist societies show that they have a strong tendency to punish unfair behavior through violence. Flocks are easier to steal than land is, so developing a reputation for retaliation should deter would-be thieves (Cao et al. 2021). This probably did not develop fully before 7000 BP, but this pattern seems to have begun then.

Once people started settling, the sign systems on the frontier (Von Petzinger 2016) became elaborated in "semasiographic" systems. These systems do not represent speech; instead, each sign, usually an image, directly represents a physical object (Sampson 1985; Powell 2012). Such systems were "widespread in small-scale societies" (Trigger 2004:61). Around 10,000 BP, Southwest Asians developed a system of clay tokens: cones, spheres, disks, and cylinders. Archaeologists found a tablet made later explaining the meaning of each counter, and it seems that "agriculture brought about a need for accounting" (Schmandt-Besserat 1996:102), though not all agree. Other systems seem to have grown out of rock art, becoming more conventional over time (Gelb 1952:Chap. 2), continuing into the present in the Americas, Europe, and Africa.

People continued making esthetic objects and images. By 25,000 BP, complex hunter-gatherers produced many "sculptured objects in the form of bone, ivory, and antler" in caves in southern Germany and "thousands of beads" in southwestern France (Hayden 2014a:646). The number and variety of rock art sites and objects increased to 7000 BP. Simple foragers may have made the earlier cave art in Europe (~30 kya), while complex hunter-gatherers may have produced later cave art, perhaps in secret societies (Hayden 2018). Rock art from southeastern Europe, Southwest Asia, and South Asia indicates dancing (Jain 2006:72–73; Milisauskas 2011:200); South Asian images also depict hunting, fishing, eating, and burials. Göbekli Tepe in Anatolia had massive stone structures shaped like the letter T, with carvings of animals, and small ceremonial areas surrounded by large ones (Watkins 2018:217). Much pottery in East Asia was highly decorated (Barnes 2015:98–99). Some figurines seem to indicate binary gender systems, while others indicate more mixed or diverse systems (Peterson 2010). Some European figurines have breasts and pelvic triangles, others have beards and penises, some have no sex characteristics, and a few have both breasts and penises (Milisauskas 2011:200–201), possibly portraying third genders. Most were of females; archaeologists suggest that they were "dolls, educational aids, personal ornaments, symbols of power, representations of people or ancestral figures, goddesses, pornographic objects," or healing devices (Milisauskas 2011:202).

Practical technologies also proliferated. Pottery and baskets appeared nearly everywhere domestication did. People around the world developed bows and arrows, axes, hammers, and other general tools. Pastoralists developed sieves and other vessels to handle milk; wagons; and cloth woven from animal fur (Bogucki 1999:227). Horticulturalists developed farming tools such as sickles, hoes, and small irrigation systems and cooking implements such as mortars and pestles (Watkins 2018). Other technologies included boats and oars, and woven cloth (Bogucki 1999). Where people settled seasonally or permanently, they built residential and other buildings that they maintained for decades or centuries in Southwest Asia, East Asia, and Europe.

Villagers elaborated their idea-lives over millennia, imposing them on the material world. The number and variety of beliefs, patterns for physical forms, and practices increased greatly after the Holocene started, perhaps because demographic growth generated more creativity.

Village Demography

Population growth accelerated in regions of plant and animal management, even more after full farming developed. Many groups alternated seasonally between mobility and sedentism; others settled fully. Furthermore, the sheer size of communities increased; paleodemographers debate this "agricultural demographic transition (ADT)" (Bocquet-Appel 2011). They think that both fertility and mortality increased but that fertility grew more.

All global population estimates from this period are rough. Some estimate the global population at around six million in 12,000 BP (Livi-Bacci 2012:25), while other estimates suggest growth from approximately seven million to 38 million from 11,000 to 5000 BP (Brooke 2014:259). These estimates incorporate much uncertainty; yet others estimate the global population at 10 million people around 10,000 BP (Barker 2009:448). To us, global population density was still extremely low, but it was much higher in richer-resource areas.

Archaeologists long assumed that horticulture led to sedentism, but they increasingly find that horticulture sometimes succeeded it by centuries. Sometimes residential sedentism increased logistical mobility (i.e., "commuting" to forage, collect, or farm) (Kelly 2013). In other places, settlement clearly occurred before horticulture, producing villages. Camps became longer-lasting, a trend reversible depending on conditions. However, settlements of complex hunter-gatherers only sometimes continued a tradition of village life, while farming societies in this period usually did, sometimes growing into towns.

Villages increased in size and number over time. For instance, settlements of 300–400 people appeared in Southwest Asia by 11,500 BP. By 10,000 BP, there were many villages of a few hundred people each across much of Southwest Asia. Village life was established in Japan by 11,000 BP and by 10,000 BP in northern China; then villages expanded across these regions (Barnes 2015). In Peru, hunter dwellings became larger and more permanent after 9800 BP (Moore 2014). Likewise, in India there are remains of "semi-permanent, hut-like structures" with "paved floors" from this period, and toward 7500 BP, houses became larger (Jain 2006:65).

The largest town in this period, Çatalhöyük in southern Anatolia, lasted from around 9100 to 8000 BP, with a population of between 5000 and 10,000 people at its peak of 8500 BP. Excavations show that houses were "built like rectangular boxes pushed together, the four walls of one touching the walls of the four adjacent houses" (Watkins 2018:220). Access to each house was through a rooftop door. No other settlement in the world was as large at the time.

Pastoral peoples remained nomadic, moving their animals with them. Herds constituted surplus storage: Pastoralists needed only a few animals each day, so they also traded them, used them as favors, and paid bride price with them. Still, pastoral nomadism was and is a variable (Smith 2005). Some "are on the move constantly"; others live in villages and grow plants; and others are partly nomadic and partly sedentary (Galaty and Johnson 1990; Smith 2005:5–6). The southern Sahara before 7000 BP was green; herders there were partly or fully nomadic (Smith 2005:61–62, 92). They might seem similar to foragers, but a group of people moving a herd made any pastoralist camp more like a mobile village than a forager camp.

Mortality increased in domesticating societies. Farming communities grew faster than foraging bands did (Bocquet-Appel 2011), and archaeologists debate why. The oldest argument held that domestication led to a more stable food supply, better nutrition, lowered mortality, and population growth (Livi-Bacci 2012). Actually, both sedentism and domestication had multiple effects. Reliance on only a few foods led to narrower diet variety and worsening nutrition, both increasing mortality; child mortality remained very high (Volk and Atkinson 2013). Sedentism created conditions for increased diseases and new parasites, also increasing mortality. While some argue that farming mortality "does not differ substantially" from its rate among foragers (Chamberlain 2006:67), most agree that mortality increased after domestication (Cohen and Crane-Kramer 2007; Bocquet-Appel 2011; Livi-Bacci 2012).

Population grew because fertility increased more than mortality did. Prehistoric and ethnographic data show that introducing full agriculture tends strongly to increase the total fertility rate (TFR) (Bocquet-Appel 2011) from birth-interval shortening. Why? As groups settled, women carried children less often and for shorter distances, thus expending overall less energy. Also, higher-calorie foods were more available, lowering the age of first menses (Bocquet-Appel 2011), and increasing the number and quality of weaning foods (Chamberlain 2006; Price and Feinman 2010), so populations grew (Bocquet-Appel 2011; Livi-Bacci 2012).

Fertility seems to have increased first; mortality increased after several generations. In some places in Europe, men lived slightly longer lives than women because many women from ages 15 to 30 died, probably in childbirth (Milisauskas 2011). Women also seem to have lived shorter lives in some Levantine areas (Peterson 2010). Problems for both sexes included skull bones becoming more porous; dental caries ("cavities") from increased carbohydrate consumption; imbalanced diets; and infectious diseases spreading more easily (Cohen and Crane-Kramer 2007). Despite better weaning foods, weaning became stressful, increasing childhood mortality. Sanitary problems increased, including rodents and water contamination (Bocquet-Appel 2011). Overall, then, researchers conclude that "health commonly declined through prehistory" over several millennia (Cohen and Crane-Kramer 2007:3443). Nevertheless, the age structure of villager populations was likely similar to that of foragers: There were many children, some young adults, fewer middle-aged adults, and very few elderly people. Some argue that this was the general shape of populations before the last few hundred years (Bocquet-Appel and Bacro 2008).

High mortality probably still influenced people's emotional lives. South Asian rock art shows one person burying four others, with two people nearby holding their faces in their hands (Jain 2006:73). The grief and fear underlying villagers' daily lives could have become rage needing an outlet (Rosaldo 1989; Turner 2007); combined with high population growth within limited land, increased mortality, and more cohesive groups, this makes warfare understandable.

As people started managing, then domesticating, their food sources, the global human population grew, and grew greatly in specific areas. Some groups gradually became partly sedentary, a few becoming fully sedentary. Life became harsher as both fertility and mortality increased. However, populations grew, as fertility increased more quickly than mortality did.

Villages in Their Environments

In this era, global climate change affected humans' physical environments around the world. Through the global diaspora, growing populations, and intensifying production, humans increasingly affected those environments, mostly in domesticating areas.

The shift from the Pleistocene to the Holocene climate eras was the backdrop to management and domestication. "Lasting for more than 4,000 years, the Early Holocene was the warmest period the earth had seen for at least 100,000 years" (Brooke 2014:133). It had two immediate effects. The massive global drop in biomass due to (human- or climate-caused) megafaunal extinctions (Stiner et al. 2011; Brooke 2014) was one; the other was the shift to broad-spectrum foraging around the world. Most megafaunal extinctions occurred in Australia, North America, and Siberia, none of which were the first regions of domestication (Brooke 2014).

The warmer, wetter climate helped domestication. In the early Holocene, "intense summer monsoon rains" fell "across mid-latitude Asia, northern Africa, and the tropical Americas" (Brooke 2014:150). More evaporation over the Mediterranean Sea increased rainfall in Southwest Asia, increasing regional fertility. Wild lowland plants gradually moved uphill, and the growing season expanded. The Yellow River in China increased its flow. Alternating seasonal droughts and rains increased upstream erosion of loess, a fertile soil, and seasonal floodplains downstream annually deposited loess, renewing soil fertility. The same pattern of annual rains, upstream erosion, and floodplains also made the Nile and Indus River valleys fertile. Mesoamerica and parts of South America also saw increased Holocene rainfall (Brooke 2014; Penna 2015:Chap. 3), without many circumscribed river floodplains, but with environmental diversity, including cold, temperate, and warm climates.

In turn, domesticating humans influenced the environment in two ways. First, farming degraded the land. Horticulturalists relied much more on plant foods than foragers did (Stiner and Feeley-Harnik 2011). Horticulture required clearing the natural system and creating an artificial habitat where humans grew plants and raised stock animals, destroying "natural balances and inherent stability of the original ecosystem" (Ponting 2007:68). Increased wind and rain eroded soil more, requiring extra manure and irrigation or long periods of fallow outside floodplains.

Second, deforestation became a problem. Humans started cutting down trees to clear land for farming; to stoke fires for making pottery, cooking, and burning limestone to plaster walls; and to build wooden structures, fence herds, and make tools. Deforestation occurred in some places (Ponting 2007; Barnes 2015; Headrick 2020), where it sometimes led to soil erosion, degrading soil fertility. When people ran out of wood—the walk was too far—they moved, abandoning their villages. This happened in what is now Jordan by 8000 BP (Headrick 2020). Archaeologists repeatedly mention other abandoned sites around the world. Most responded to environmental shifts beyond human control; some occurred from the centuries-long environmental effects of human activities.

Agriculture really was revolutionary. Farming and herding, storage and accumulation, increased trade, more cohesive groups, warfare, new inequalities, population growth, and environmental degradation formed a new way of life for humans.

The Larger Context of Domestication

The Agricultural Revolution also created larger social trends. Regional cultures developed; regional interaction areas grew; and farmers interacted with or replaced foragers.

Cultural Regionalization

The longer humans stayed in specific places, the more cultures diversified. Early on, people from Europe to Siberia had one style of tool manufacture, the "Aurignacian." Over time, different regional tool styles developed, such as the European "Gravettian" (Haas and Piscitelli 2013). People reached Southeast Asia, but Aurignacian-using people did not, so styles had changed. The Clovis point tradition seemingly diffused across North America; by 10,500 BP, it "had been supplanted by a variety of regionally distinctive projectile points" (Meltzer 2009:283). After hunter-gatherers populated the entire continent, large cultural regions developed (Fagan 2019:69), and later, style zones became smaller and more numerous (Meltzer 2009; Kornfeld and

Politis 2014). Similar style diversification occurred in South America (Moore 2014:125). African hunter-gatherers also developed regional cultural patterns (Connah 2018).

Cave art also illustrates growing regional cultural differences. European cave art has a distinctive style of big flowing lines. The oldest art there started around 40 kya, and the youngest is from about 15 kya (David 2017). Australian art started likely before the LGM (David 2017:207); it has long, angular figures and segmented details, styles which regionalized after the LGM (David 2017:211). Beyond styles, what cave artists represented varied greatly by region: "Unlike Australian and southern African art from the same broad period, cave art in Europe offers no depiction of landscape, no horizon, no vegetation, almost no depiction of human-animal interaction, almost no hunting scenes," and few people (Smith 2018:8).

Linguistic reconstruction shows that linguistic diversity started in Africa. As humans dispersed globally, "thousands of languages" came to exist (Fischer 2018:65). Linguistic historians have reconstructed language "families," some of which were established by 7000 BP, including the Niger-Congo languages of sub-Saharan Africa, the Afro-Asiatic languages of North Africa and Southwest Asia, the Sino-Tibetan family, and the Dravidian in South Asia. Many unrelated language families existed in North and South America, suggesting multiple migrations or other processes. Their geographic distribution today only partly reflects the likely geographic distributions of their ancestor languages in prehistory (Fischer 2018).

Eventually, even more local cultures developed. Many archaeologists argue strongly that specific adaptations to local environments broadly drove cultural differences (Fagan 2019:66–74). Others argue that some groups define themselves culturally against nearby groups, emphasizing differences in a process called "schismogenesis" (Watkins 2008:162; Graeber and Wengrow 2021). For instance, between 12,000 and 10,000 BP, the uplands of Southwest Asia had violent, male-oriented imagery; nearby lowlands had peaceful, female-oriented imagery (Graeber and Wengrow 2021:Chap. 6). As the original meaning of the imagery is lost to us, we cannot know that "being unlike them" is why the two groups were different, but it is plausible.

Thus, as humans traveled intergenerationally across the planet, their cultural systems—tool styles, artistic styles, and languages—diversified regionally.

Regional Trade Systems

Trade systems also grew geographically and demographically. Europeans had trade systems across several hundred kilometers well before domestication (Chapman 2008). Around 17,000 BP, European hunter-gatherers may have been getting rich from exporting meats, skins, and garments decorated "with fur, feathers, teeth, claws, and beads" (Hayden 2014a:651). Hunter-gatherers developed pottery independently several times; pottery spread over ten millennia so that by 7500 BP, some used it across Siberia, from the Baltic to the Pacific, others across large parts of the Sahara (Hommel 2018:22, map).

After domestication and population growth, exchange networks encompassed more people, increased the density of interconnections, and grew geographically larger. Over 11,000 years ago, people in the Levant and Mesopotamia traded obsidian over two or three hundred kilometers for practical purposes and up to 900 kilometers for symbolic ones, and they also traded marine shells, pottery, tokens, and basalt (Watkins 2008). A large-scale exchange network developed down the entire length of the Persian Gulf (over 1000 kilometers long) on both coasts, mostly conducted on watercraft between Mesopotamian domesticating peoples and Arabian hunter-gatherers (Carter 2018). By 7000 BP, the Levantine-Mesopotamian and Mesopotamian-Gulf systems seem joined, making much of Southwest Asia one trading area. To the east, the Jomon people in Japan lived in several small, neighboring ecological zones. By 7000 years ago, they traded pottery, jade, obsidian, and amber with each other (Barnes 2015:127–131).

The peer-polity interaction sphere concept (Renfrew 1986) describes the large Southwest Asian trade system before 7000 BP. It consisted of relatively equal and autonomous villages, i.e., polities, networked diffusely with each other (Watkins 2008; Carter 2018). One archaeologist expanded the *chaîne opératoire* concept from operations within a small group to the quarrying, trade, and use of obsidian across the Mediterranean. Around 7000 BP, this seems organized one way, but millennia later, it was organized differently: production sites changed, and one receiving site changed its obsidian sources several times (Tykot 2003). This geographic expansion of the *chaîne opératoire* concept parallels the "commodity chain" concept in sociology (Bair 2009), which archaeologists are beginning to use to understand exchange systems (Earle 2010), though without noting similarities to large-scale *chaînes opératoires*. Intensified foraging, management and domestication of food sources, and pottery production all probably created more complex sequences of raw materials extraction, manufacturing, distribution, and use, plus goods moving in reverse for exchange, but I found almost no studies tracing such sequences for this period.

Finally, ideas also moved throughout the Southwest Asian interaction sphere. Specific styles of projectile points diffused throughout the region after 12,000 BP, and as they changed over time, the new styles diffused. Core reduction procedures also moved through the region after 10,000 BP, as did architectural patterns and the postburial separation of skulls from bodies, although each locale practiced this in slightly different ways. Finally, the use of T-shaped monoliths and animal carvings at ceremonial centers became regional (Watkins 2008).

The Diffusion of Domesticating Societies

Arabian mobile hunter-gatherers trading with Mesopotamian farmers illustrates another trend. Most people globally remained foragers for many millennia and filled in empty areas. Since farming groups grew faster demographically, farming spread outside its sites of origin. Thus, foragers and farmers interacted in three ways: conversion, replacement, and trade.

Farming practices diffused geographically. People originally domesticated grains, and then sheep and goats, in southern Anatolia and the northern Levant. Farmers spread northwest into Anatolia, west to the island of Cyprus, southwest to the Red Sea, south through the Euphrates River valley, east to the Tigris River valley, and finally, south to the Persian Gulf (Watkins 2018), ultimately creating the regional interaction discussed above. Horticulture spread from Southwest Asia eastward into Central and South Asia by 9000 BP, moving into the Nile Valley by 7000 (Bellwood 2005; Barker 2009). Agropastoralism spread northwest to what is now Greece and Bulgaria between 9000 and 8500 BP. It moved further west to northern Italy, southern France, and eastern Spain by 7500 BP, reaching southern Germany, Ukraine, and Russia around 7000 BP (Milisauskas 2011). Likewise, farming "in northern China involved expansion east and west into sparsely occupied new lands" (Higham 2018:236). Rice cultivation moved up the Yangzi River valley between 10,000 and 8000 years ago. Domesticated plants in South America spread northward and westward; peanuts "appeared on the North Coast of Peru after 8500 BP and were carried north to Panama by 7600 BP" (Moore 2014:137). People in Mexico domesticated maize by 8700 BP, which "spread widely through the Americas ... reaching lower Central America by 7600 BP" (Moore 2014:139). Some in the Americas followed the same Afro-Eurasian sequence from foragers to cultivators to managers to domesticators; others started farming abruptly (Moore 2014:Chap. 5) (Fig. 4.2).

This gradualness or abruptness relates to how and whether agriculture diffused. It occurred in two ways. The evidence for Europe comes from archaeology, genetics, and linguistics and is fiercely debated (Bellwood 2005:Chap. 11). In the "cultural" or crop-migration model (Bellwood 2005; Manning 2013), farmers stayed put, sharing crops and techniques with neighbors, so farming techniques moved from population to population. In the "demic" or migration model (Bellwood 2005; Manning 2013), farmers themselves moved and brought crops and techniques with them. European archaeologists debate which model applies, though all agree "that farming was introduced by genetically distinct people" (Milisauskas 2011:158). Later, perhaps farmers used both methods (Milisauskas 2011; Manning 2013), depending on location, time, and circumstances. Some (Bellwood 2005; Manning 2013) believe that the intermarriage of foragers and farmers was likely. However, total and quick replacement occurred only rarely; in most places, it was piecemeal, creating a "mosaic-like" pattern of diffusion (Scarre 2018b:187). Thus, Eurasians were selective, some continuing partial foraging.

Archaeologists rarely ask whether domestication spread peacefully or violently, but this is changing. They rarely say "conquest" or "warfare," instead saying "replacement" or "displacement" (Bellwood 2005; Manning 2013). Cultural movement was likely peaceful, but farmers moving into new areas could have been violent. Some say that some foragers resisted farmers and herders moving into their lands (Brooke 2014:157–159); others argue that agriculture spread throughout Europe without warfare (Ferguson 2006). Greece has little evidence for indigenous

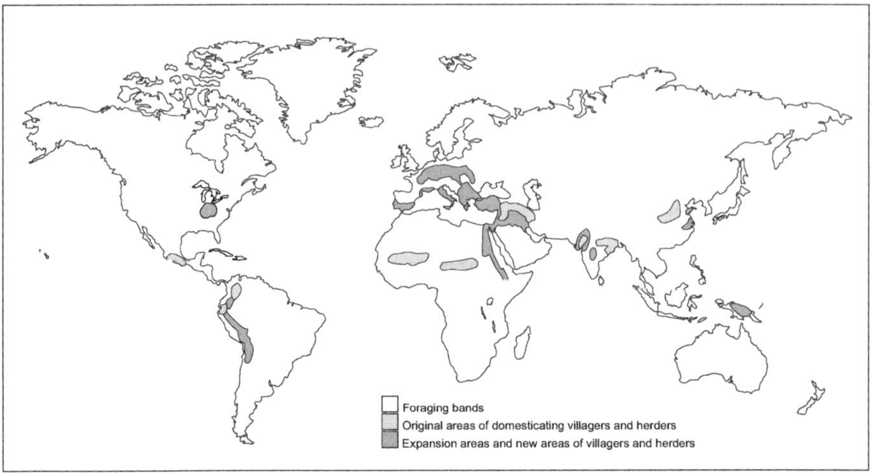

Fig. 4.2 Spread of agriculture by 7000 BP (5000 BCE). Sources: Black (2005); Bahn (2009); Houston (2018); Scarre (2018a)

hunter-gatherer populations where farmers first arrived (Scarre 2018a); perhaps agropastoralists engaged in infill.

However, foragers and farmers did live together. Foragers coexisted with farmers "in many parts of the world for *millennia*, without adopting agriculture" (Bellwood 2005:42, original emphasis), such as the aforementioned Arabian foragers and Mesopotamian farmers (Carter 2018). Indeed, during the first period of management, foragers blended gradually into "in-between" societies (Barnes 2015). Trade relations might have continued as some societies gradually adopted farming; rarely, farming peoples moved back to foraging (Bellwood 2005).

Thus, the Agricultural Revolution spurred large-scale geographic change. People responded to varied environments with diverse cultural patterns. Settled villagers expanded trade networks beyond those of foragers, building large-scale exchange systems often including foragers. Whether farming techniques or farmers themselves moved is debated, but farming clearly spread beyond its regions of origin, and sometimes farmers and foragers lived together.

Conclusion

Around 33,000 years ago, humans lived throughout Africa, across Afro-Eurasia, and on the continent of Sahul. As humans were moving into the Americas, some Afro-Eurasian foraging groups began to settle, collect food, and grow larger. The

LGM forced them southward; when the glaciers retreated and humans were in the Americas, some Southwest Asians again settled more. After the Younger Dryas, Southwest Asians, East Asians, northern Africans, and South Americans intensified their subsistence strategies. By 7000 BP, people lived in new ways in these regions, in South Asia, and in Mesoamerica. Foragers continued living in North America and most of South America; most of Africa; northern Europe; and Northern, Central, Southeast, and much of South Asia; and Australia.

The Agricultural Revolution was the first social transformation in human history. Again, despite denials that a change playing out over 3000 years was not a "revolution," something so new after two million years was certainly revolutionary. These people lived in a new way of life. They worked hard to grow food from the earth and to control animals, and they changed many species. Some reduced their sharing just as the number of their possessions increased. In some areas, kinship seems to have shifted to more rigid descent-based groups. Warfare clearly started, and dehumanizing or moral ideologies may have justified them. Economic and gender inequalities also seem to have begun in some places. Ancestry became more important spiritually, though in still-inscrutable ways, and people developed more technologies, most notably pottery. Some moved less and settled more. Mortality increased, but fertility increased more, so populations grew. Farmers' impact on the physical environment increased, sometimes degrading it locally. Their effective zones of interaction grew greatly, and they expanded geographically, converting, pushing out, or interacting with foraging peoples. These regional trade systems developed entire divisions of labor across them.

Still, there were many continuities. Trade systems expanded and elaborated. People still lived mentally in a supernatural world, adding ancestors to their concerns. Instead of clustering and settling for a few days, as their ancestors had, they stayed for weeks, months, years, or generations. Kinship remained central to relationships. Intergroup conflict seems to have grown out of interpersonal violence. Foragers had some environmental impact; villagers intensified it.

Human capacities—for learning and invention, for multiple and nested groups, and for deep bonds—made possible all these changes. However, the changes raise new questions. What was "the group" in these new societies? Was it the household, as that is where people lived most of their lives? If kinship sometimes became more rigid, did that define groups more strictly, i.e., were kin "segments," lineages or clans, sometimes next door, other times scattered throughout villages, the main group? Was the village itself the group, or was a "tribe" of villages in one region the group? The answer likely depended on specific local circumstances, but domestication expanded the range of possible groups around the world.

Domestication did not simply happen a few times and stop. It spread geographically, and more peoples in more places around the world developed similar practices equally independently. Full domestication then had further consequences.

References

Anderson, E. N. 2017. Agricultural Origins and Early Development. In *Routledge Handbook of World-Systems Analysis*, ed. S. J. Babones and C. Chase-Dunn, 39–41. New York: Routledge.

Anderson, Abigail, Sophia Chilczuk, Kaylie Nelson, Roxanne Ruther, and Cara Wall-Scheffler. 2023. The Myth of Man the Hunter: Women's Contribution to the Hunt Across Ethnographic Contexts. *PLoS One* 18 (6): e0287101.

Bahn, Paul, ed. [2000] 2009. *Atlas of World Archaeology*. London: Brown Reference.

Bair, Jennifer. 2009. Global Commodity Chains: Genealogy and Review. In *Frontiers of Commodity Chain Research*, ed. J. Bair, 1–34. Stanford, CA: Stanford University Press.

Barich, Barbara. 2013. Hunter-Gatherer-Fishers of the Sahara and the Sahel 12,000–4,000 years Ago. In *The Oxford Handbook of African Archaeology*, ed. P. Mitchell and P. Lane, 445–460. New York: Oxford University Press.

Barker, Graeme. 2009. Early Farming and Domestication. In *The Oxford Handbook of Archaeology*, ed. B. Cunliffe, C. Godsen, and R. A. Joyce, 445–483. New York: Oxford University Press.

Barnes, Gina L. 2015. *The Archaeology of East Asia: The Rise of Civilization in China, Korea, and Japan*. Havertown, PA: Oxbow Books.

Bar-Yosef, Ofer. 2011. Climatic Fluctuations and Early Farming in West and East Asia. *Current Anthropology* 52 (S4): S175–S193.

Bellwood, Peter. 2005. *First Farmers: The Origins of Agricultural Societies*. Malden, MA: Blackwell.

Binford, Lewis R. 2001. *Constructing Frames of Reference: An Analytical Method for Archaeological Theory Building Using Ethnographic and Environmental Data Sets*. Berkeley: University of California Press.

Black, Jeremy, ed. 2005. *The Atlas of World History*. London: Dorling Kindersley Limited.

Bocquet-Appel, Jean-Pierre. 2011. The Agricultural Demographic Transition During and After the Agriculture Inventions. *Current Anthropology* 52 (S4): S497–S510.

Bocquet-Appel, Jean-Pierre, and Jean-Noël Bacro. 2008. Estimation of an Age Distribution with Its Confidence Intervals Using an Iterative Bayesian Procedure and a Bootstrap Sampling Approach. In *Recent Advances in Palaeodemography: Data, Techniques, Patterns*, ed. J.-P. Bocquet-Appel, 63–82. Dordrecht, The Netherlands: Springer.

Boehm, Christopher. 2013. The Biocultural Evolution of Conflict Resolution Between Groups. In *War, Peace, and Human Nature: The Convergence of Evolutionary and Cultural Views*, ed. D. P. Fry, 315–340. New York: Oxford University Press.

Bogucki, Peter. 1999. *The Origins of Human Society. The Blackwell History of the World*. Malden, MA: Blackwell.

Boserup, Ester. 1965. *The Conditions of Agricultural Growth: The Economics of Agrarian Change under Population Pressure*. New York: Aldine Publishing.

Brooke, John L. 2014. *Climate Change and the Course of Global History: A Rough Journey*. New York: Cambridge University Press.

Brown, Cynthia Stokes. 2007. *Big History: From the Big Bang to the Present*. New York: New Press.

Cao, Yiming, Benjamin Enke, Armin Falk, Paola Giuliano, and Nathan Nunn. 2021. Herding, Warfare, and a Culture of Honor: Global Evidence. Working Paper No. 29250, National Bureau of Economic Research, Cambridge, MA.

Carter, Robert. 2018. Globalizing Interactions in the Arabian Neolithic and the 'Ubaid. In *Globalization in Prehistory: Contact, Exchange, and the "People Without History"*, ed. N. Boivin and M. D. Frachetti, 43–79. New York: Cambridge University Press.

Chamberlain, Andrew. 2006. *Demography in Archaeology*. Cambridge, UK: Cambridge University Press.

Chapman, John. 2008. Approaches to Trade and Exchange in Earlier Prehistory (Late Mesolithic-Early Bronze Age). In *Prehistoric Europe: Theory and Practice*, ed. A. Jones, 333–355. Chichester, UK: Wiley-Blackwell.

Cintas-Peña, Marta, and Leonardo García-Sanjuan. 2019. Gender Inequalities in Neolithic Iberia: A Multi-Proxy Approach. *European Journal of Archaeology* 22 (4): 499–522.
Cohen, Daniel Joel. 2011. The Beginnings of Agriculture in China: A Multiregional View. *Current Anthropology* 52 (S4): S273–S293.
Cohen, Mark Nathan, and Gillian M. M. Crane-Kramer. 2007. Editors' Summation. In *Ancient Health: Skeletal Indicators of Agricultural and Economic Intensification*, ed. M. N. Cohen and G. M. M. Crane-Kramer, 320–343. Gainesville: University Press of Florida.
Connah, Graham. 2018. Holocene Africa. In *The Human Past: World Prehistory and the Development of Human Societies*, ed. C. Scarre, 4th ed., 344–387. London: Thames and Hudson.
Conniff, Richard. 2003. Go Ahead, Kiss Your Cousin: Heck, Marry Her If You Want to. *Discover Magazine*, Aug. 1. http://discovermagazine.com/2003/aug/featkiss#.UuiJgWQo46g
Crabtree, Pam. 2006. Women, Gender, and Pastoralism. In *Handbook of Gender in Archaeology*, ed. S. M. Nelson, 571–592. New York: Rowman & Littlefield.
Crumley, Carole L. 1995. Heterarchy and the Analysis of Complex Societies. *Archeological Papers of the American Anthropological Association* 6 (1): 1–5.
Cummings, Vicki. 2014. Hunter-Gatherers in the Post-Glacial World. In *The Oxford Handbook of the Archaeology and Anthropology of Hunter-Gatherers*, ed. V. Cummings, P. Jordan, and M. Zvelebil, 435–455. New York: Oxford University Press.
David, Bruno. 2017. *Cave Art*. London: Thames & Hudson.
De Dreu, Carsten K. W. 2010. Social Conflict: The Emergence and Consequences of Struggle and Negotiation. In *Handbook of Social Psychology*, ed. S. T. Fiske, D. T. Gilbert, and G. Lindzey, 5th ed., 983–1023. Hoboken, NJ: John Wiley & Sons.
Dong, Yu, Chelsea Morgan, Yurii Chinenov, Ligang Zhou, Wenquan Fan, Xiaolin Ma, and Kate Pechenkina. 2017. Shifting Diets and the Rise of Male-Based Inequality on the Central Plains of China during Eastern Zhou. *Proceedings of the National Academy of Sciences* 114 (5): 932–937.
Earle, Timothy. 2010. Exchange Systems in Prehistory. In *Trade and Exchange: Archaeological Studies from History and Prehistory*, ed. C. D. Dillian and C. L. White, 205–217. New York: Springer.
Ensor, Bradley E. 2021. *The Not Very Patrilocal European Neolithic: Strontium, ADNA, and Archaeological Kinship Analyses*. Oxford: Archaeopress Publishing.
Fagan, Brian M. 2019. *Ancient North America: The Archaeology of a Continent*. 5th ed. New York: Thames and Hudson.
Ferguson, R. Brian. 2006. Archaeology, Cultural Anthropology, and the Origins and Intensification of War. In *The Archaeology of Warfare: Prehistories of Raiding and Conquest*, ed. E. N. Arkush and M. W. Allen, 469–523. Gainsville: University Press of Florida.
Fischer, Steven Roger. 2018. *A History of Language*. London: Reaktion Books.
Fiske, Alan Page, and Tage Shakti Rai. 2015. *Virtuous Violence: Hurting and Killing to Create, Sustain, End, and Honor Social Relationships*. New York: Cambridge University Press.
Flannery, Kent, and Joyce Marcus. 2012. *The Creation of Inequality: How Our Prehistoric Ancestors Set the Stage for Monarchy, Slavery, and Empire*. Cambridge, MA: Harvard University Press.
Fuller, Dorian Q. 2011. Finding Plant Domestication in the Indian Subcontinent. *Current Anthropology* 52 (S4): S347–S362.
Galaty, John G., and Douglas L. Johnson. 1990. Introduction: Pastoral Systems in Global Perspective. In *The World of Pastoralism: Herding Systems in Comparative Perspective*, ed. J. G. Galaty and D. L. Johnson, 1–31. New York: Guildford Press.
Gat, Azar. 2017. *The Causes of War and the Spread of Peace: But Will War Rebound?* New York: Oxford University Press.
Gelb, I. J. 1952. *A Study of Writing: The Foundations of Grammatology*. Chicago: The University of Chicago Press.
Graeber, David, and David Wengrow. 2021. *The Dawn of Everything: A New History of Humanity*. New York: Farrar, Straus, and Giroux.

Greenfield, Haskel J. 2014. Some Reflections on the Origins and Intensification of Dairying in the Archaeological Record. In *Animal Secondary Products: Domestic Animal Exploitation in Prehistoric Europe, the Near East and the Far East*, ed. H. J. Greenfield, 20–39. Havertown, PA: Oxbow Books.
Haaland, Randi, and Gunnar Haaland. 2013. Early Farming Societies Along the Nile. In *The Oxford Handbook of African Archaeology*, ed. P. Mitchell and P. Lane, 541–554. New York: Oxford University Press.
Haas, Jonathan, and Matthew Piscitelli. 2013. The Prehistory of Warfare: Misled by Ethnography. In *War, Peace, and Human Nature: The Convergence of Evolutionary and Cultural Views*, ed. D. P. Fry, 168–190. New York: Oxford U.P.
Haas, Randall, James Watson, Tammy Buonasera, John Southon, Jennifer C. Chen, Sarah Noe, Kevin Smith, Carlos Viviano Llave, Jelmer Eerkens, and Glendon Parker. 2020. Female Hunters of the Early Americas. *Science Advances* 6:eabd0310.
Harris, David R. 2007. Agriculture, Cultivation and Domestication: Exploring the Conceptual Framework of Early Food Production. In *Rethinking Agriculture: Archaeological and Ethnoarchaeological Approaches*, ed. T. Denham, J. Iriarte, and L. Vrydaghs, 16–35. Walnut Creek, CA: Left Coast Press.
Haslam, Nick, and Steven Loughnan. 2016. How Dehumanization Promotes Harm. In *The Social Psychology of Good and Evil*, ed. A. G. Miller, 2nd ed., 140–158. New York: The Guilford Press.
Haslam, Nick, Steven Loughnan, and Elise Holland. 2013. The Psychology of Humanness. In *Objectification and (De)Humanization. 60th Nebraska Symposium on Motivation*, ed. S. J. Gervais, 25–51. New York: Springer.
Hayden, Brian. 2003. *Shamans, Sorcerers and Saints: A Prehistory of Religion*. Washington, DC: Smithsonian.
Hayden, Brian. 2014a. Social Complexity. In *The Oxford Handbook of the Archaeology and Anthropology of Hunter-Gatherers*, ed. V. Cummings, P. Jordan, and M. Zvelebil, 643–662. New York: Oxford University Press.
Hayden, Brian. 2014b. *The Power of Feasts: From Prehistory to the Present*. New York: Cambridge University Press.
Hayden, Brian. 2018. *The Power of Ritual in Prehistory: Secret Societies and Origins of Social Complexity*. New York: Cambridge University Press.
Headrick, Daniel R. 2020. *Humans Versus Nature: A Global Environmental History*. New York: Oxford University Press.
Hebl, Michelle R., and Juan M. Madera. 2007. Ethnocentrism. In *Encyclopedia of Social Psychology*, ed. R. F. Baumeister and K. D. Vohs, 315. Thousand Oaks, CA: SAGE.
Henrich, Joseph. 2015. *The Secret of Our Success: How Culture Is Driving Human Evolution, Domesticating Our Species, and Making Us Smarter*. Princeton and Oxford: Princeton University Press.
Higham, Charles. 2018. East Asian Agriculture and Its Impact. In *The Human Past: World Prehistory and the Development of Human Societies*, ed. C. Scarre, 230–260. London: Thames and Hudson.
Hommel, Peter. 2018. What's the Point? Globalization and the Emergence of Ceramic-Using Hunter-Gatherers in Northern Eurasia. In *Globalization in Prehistory: Contact, Exchange, and the "People Without History"*, ed. N. Boivin and M. D. Frachetti, 15–42. New York: Cambridge University Press.
Honeychurch, William, and Cheryl A. Makarawicz. 2016. The Archaeology of Pastoral Nomadism. *Annual Review of Anthropology* 45:341–359.
Houston, Rob, ed. 2018. *History of the World, Map by Map*. New York: Dorling Kindersley.
Jain, V. K. 2006. *Prehistory and Protohistory of India: An Appraisal: Palaeolithic—Non-Harappan—Chalcolithic Cultures*. New Delhi: D.K. Printworld.
Johnson, Allen W., and Timothy Earle. 2000. *The Evolution of Human Societies: From Foraging Group to Agrarian State*. 2nd ed. Stanford, CA: Stanford University Press.

Kelly, Raymond C. 2000. *Warless Societies and the Origin of War*. Ann Arbor: The University of Michigan Press.

Kelly, Raymond C. 2005. The Evolution of Lethal Intergroup Violence. *PNAS* 102 (43): 15294–15298.

Kelly, Robert L. 2013. *The Lifeways of Hunter-Gatherers: The Foraging Spectrum*. New York: Cambridge University Press.

Kornfeld, Marcel, and Gustavo G. Politis. 2014. Into the Americas: The Earliest Hunter-Gatherers in an Empty Continent. In *The Oxford Handbook of the Archaeology and Anthropology of Hunter-Gatherers*, ed. V. Cummings, P. Jordan, and M. Zvelebil, 405–436. New York: Oxford University Press.

Lane, Paul J. 2014. Hunter-Gatherer-Fishers, Ethnoarchaeology, and Analogical Reasoning. In *The Oxford Handbook of the Archaeology and Anthropology of Hunter-Gatherers*, ed. V. Cummings, P. Jordan, and M. Zvelebil, 104–150. New York: Oxford University Press.

Larson, Greger, Dolores R. Piperno, Robin G. Allaby, Michael D. Purugganan, Leif Andersson, Manuel Arroyo-Kalin, Loukas Barton, et al. 2014. Current Perspectives and the Future of Domestication Studies. *Proceedings of the National Academy of Sciences* 111 (17): 6139–6146.

LeBlanc, Steven A. 2006. Warfare and the Development of Social Complexity: Some Demographic and Environmental Factors. In *The Archaeology of Warfare: Prehistories of Raiding and Conquest*, ed. E. N. Arkush and M. W. Allen, 437–468. Gainsville: University Press of Florida.

LeBlanc, Steven A. 2014. Forager Warfare and Our Evolutionary Past. In *Violence and Warfare Among Hunter-Gatherers*, ed. M. W. Allen and T. L. Jones, 26–48. Walnut Creek, CA: Left Coast Press.

Lemke, Ashley K. 2018. Hunter-Gatherers and Archaeology. In *Foraging in the Past: Archaeological Studies of Hunter-Gatherer Diversity*, ed. A. K. Lemke, 3–18. Louisville: University Press of Colorado.

Leppard, Thomas P. 2022. Process and Dynamics of Mediterranean Neolithization (7000-5500 BC). *Journal of Archaeological Research* 30 (2): 231–283.

Levy, Janet E. 2006. Gender, Heterarchy, and Hierarchy. In *Handbook of Gender in Archaeology*, ed. S. M. Nelson, 219–246. New York: AltaMira Press.

Lewellen, Ted C. 2003. *Political Anthropology: An Introduction*. 3rd ed. Westport, CT: Praeger.

Livi-Bacci, Massimo. 2012. *A Concise History of World Population*. 5th ed. Chichester, UK: Wiley/Blackwell.

Manning, Patrick, with Tiffany Trimmer. 2013. *Migration in World History*, 2nd ed. London and New York: Routledge.

Marshall, Fiona, and Lior Weissbrod. 2011. Domestication Processes and Morphological Change: Through the Lens of the Donkey and African Pastoralism. *Current Anthropology* 52 (S4): S397–S413.

Masclans Latorre, Alba, Penny Bickle, and Caroline Hamon. 2021. Sexual Inequalities in the Early Neolithic? Exploring Relationships Between Sexes/Genders at the Cemetery of Vedrovice Using Use-Wear Analysis, Diet and Mobility. *Journal of Archaeological Method and Theory* 28 (1): 232–273.

Mazoyer, Marcel, and Laurence Roudart. 2006. *A History of World Agriculture: From the Neolithic Age to the Current Crisis*. Translated by James H. Membrez. New York: Monthly Review Press.

McIntosh, Susan Keech. 1999. Pathways to Complexity: An African Perspective. In *Beyond Chiefdoms: Pathways to Complexity in Africa*, ed. S. K. McIntosh, 1–30. New York: Cambridge University Press.

Meltzer, David J. 2009. *First Peoples in a New World: Colonizing Ice Age America*. Berkeley: University of California Press.

Milisauskas, Sarunas. 2011. Early Neolithic, the First Farmers in Europe, 700–5500/5000 BC. In *European Prehistory: A Survey*, ed. S. Milisauskas, 2nd ed., 153–221. New York: Springer.

Moore, Jerry D. 2014. *A Prehistory of South America: Ancient Cultural Diversity on the Least Known Continent*. Boulder: University of Colorado Press.

Neill, James. 2009. *The Origins and Role of Same-Sex Relations in Human Societies*. Jefferson, NC and London: McFarland & Company.
Nolan, Patrick, and Gerhard Lenski. 2011. *Human Societies: An Introduction to Macrosociology*. Boulder, CO: Paradigm.
Ostrom, Elinor, Roy Gardner, and James Walker, eds. 1994. *Rules, Games, and Common Pool Resources*. University of Michigan Press: Ann Arbor.
Over, Harriet. 2021. Seven Challenges for the Dehumanization Hypothesis. *Perspectives on Psychological Science* 16 (1): 3–13.
Pasternak, Burton, Carol R. Ember, and Melvin Ember. 1997. *Sex, Gender, and Kinship: A Cross-Cultural Perspective*. Upper Saddle River, NJ: Prentice-Hall.
Penna, Anthony N. 2015. *The Human Footprint: A Global Environmental History*. Malden, MA: John Wiley & Sons.
Peoples, Hervey C., Pavel Duda, and Frank W. Marlowe. 2016. Hunter-Gatherers and the Origins of Religion. *Human Nature* 27:261–282.
Peterson, Jane D. 2010. Gender and Early Farming Societies. In *Handbook of Gender in Archaeology*, ed. S. M. Nelson, 537–570. New York: Rowman & Littlefield.
Pettitt, Paul. 2014. The European Upper Palaeolithic. In *The Oxford Handbook of the Archaeology and Anthropology of Hunter-Gatherers*, ed. V. Cummings, P. Jordan, and M. Zvelebil, 279–309. New York: Oxford University Press.
Phillips, Ben. 2022. 'They're Not True Humans': Beliefs about Moral Character Drive Denials of Humanity. *Cognitive Science* 46 (2): 1–27.
Piperno, Dolores R. 2011. The Origins of Plant Cultivation and Domestication in the New World Tropics: Patterns, Process, and New Developments. *Current Anthropology* 52 (S4): S453–S470.
Ponting, Clive. 2007. *A New Green History of the World: The Environment and the Collapse of Great Civilizations*. New York: Penguin.
Powell, Barry B. 2012. *Writing: Theory and History of the Technology of Civilization*. New York: John Wiley & Sons.
Price, T. Douglas, and Ofer Bar-Yosef. 2011. The Origins of Agriculture: New Data, New Ideas. *Current Anthropology* 52 (S4): S163–S174.
Price, T. Douglas, and Gary Feinman. 2010. *Images of the Past*. 6th ed. New York: McGraw Hill.
Renfrew, Colin. 1986. Introduction: Peer Polity Interaction and Socio-Political Change. In *Peer Polity Interaction and Socio-Political Change*, ed. C. Renfrew and J. F. Cherry, 1–18. New York: Cambridge University Press.
Rosaldo, Renato. 1989. *Culture and Truth: The Remaking of Social Analysis*. Boston: Beacon Press.
Sampson, Geoffrey. 1985. *Writing Systems: A Linguistic Introduction*. London: Hutchinson.
Scarre, Chris. 2018a. Holocene Europe. In *The Human Past: World Prehistory and the Development of Human Societies*, ed. C. Scarre, 4th ed., 388–428. London: Thames and Hudson.
Scarre, Chris. 2018b. The World Transformed: From Foragers and Farmers to States and Empires. In *The Human Past: World Prehistory & the Development of Human Societies*, ed. C. Scarre, 4th ed., 174–197. London: Thames & Hudson.
Schmandt-Besserat, Denise. 1996. *How Writing Came About*. Austin: University of Texas Press.
Scott, James C. 2017. *Against the Grain: A Deep History of the Earliest States*. New Haven, CT and London: Yale University Press.
Sherratt, Andrew. 1981. Plough and Pastoralism: Aspects of the Secondary Products Revolution. In *Pattern of the Past: Studies in Honour of David Clarke*, ed. I. Hodder, G. Isaac, and N. Hammond, 261–306. Cambridge, UK: Cambridge University Press.
Shipman, Pat. 2014. And the Last Shall Be First. In *Animal Secondary Products: Domestic Animal Exploitation in Prehistoric Europe, the Near East and the Far East*, ed. H. J. Greenfield, 40–54. Havertown, PA: Oxbow Books.
Smail, Daniel Lord, Mary C. Stiner, and Timothy Earle. 2011. Goods. In *Deep History: The Architecture of Past and Present*, ed. A. Shryock and D. Smail, 219–241. Berkeley: University of California Press.

Smith, Andrew E. 2005. *African Herders: Emergence of Pastoral Traditions*. Walnut Creek, CA: AltaMira Press.
Smith, David Livingstone. 2011. *Less Than Human: Why We Demean, Enslave, and Exterminate Others*. New York: St. Martin's Press.
Smith, Justin E. H. 2018. What Cave Art Means. *Art in America*. Penske Media Corporation. Retrieved June 22, 2023. https://www.artnews.com/art-in-america/features/cave-art-means-63549/.
Smith, Martin. 2017. *Mortal Wounds: The Human Skeleton as Evidence for Conflict in the Past*. Barnsley, South Yorkshire: Pen & Sword Military.
Stanish, Charles. 2017. *The Evolution of Human Cooperation: Ritual and Social Complexity in Stateless Societies*. New York: Cambridge University Press.
Stiner, Mary C., Timothy Earle, Daniel Lord Smail, and Andrew Shryock. 2011. Scale. In *Deep History: The Architecture of Past and Present*, ed. A. Shryock and D. Smail, 242–272. Berkeley: University of California Press.
Stiner, Mary C., and Gillian Feeley-Harnik. 2011. Energy and Ecosystems. In *Deep History: The Architecture of Past and Present*, ed. A. Shryock and D. Smail, 78–102. Berkeley: University of California Press.
Sutton, Mark Q., and E. N. Anderson. 2010. *Introduction to Cultural Ecology*. 2nd ed. New York: Rowman & Littlefield.
Trigger, Bruce. 2004. Writing Systems: A Case Study in Cultural Evolution. In *The First Writing: Script Invention as History and Process*, ed. S. D. Houston, 39–70. Cambridge: Cambridge University Press.
Turner, Jonathan. 2007. *Human Emotions: A Sociological Theory*. London: Routledge.
Tykot, Robert H. 2003. Determining the Source of Lithic Artifacts and Reconstructing Trade in the Ancient World. In *Written in Stone: The Multiple Dimensions of Lithic Analysis*, ed. P. N. Kardulias and R. W. Yerkes, 59–86. Lanham, MD: Lexington Books.
Volk, Anthony A., and Jeremy A. Atkinson. 2013. Infant and Child Death in the Human Environment of Evolutionary Adaptation. *Evolution and Human Behavior* 34 (3): 182–192.
Vollhardt, Johanna Ray, and Maggie Campbell-Obaid. 2016. The Social Psychology of Genocide and Mass Atrocities. In *The Social Psychology of Good and Evil*, ed. A. G. Miller, 2nd ed., 159–184. New York: The Guilford Press.
Von Petzinger, Genevieve. 2016. *The First Signs: Unlocking the Mysteries of the World's Oldest Symbols*. New York: Atria Books.
Voss, Barbara L. 2008. Sexuality Studies in Archaeology. *Annual Review of Anthropology* 37:317–336.
Watkins, Trevor. 2008. Supra-Regional Networks in the Neolithic of Southwest Asia. *Journal of World Prehistory* 21 (2): 139–171.
Watkins, Trevor. 2018. From Mobile Foragers to Complex Societies in Southwest Asia. In *The Human Past: World Prehistory and the Development of Human Societies*, ed. C. Scarre, 4th ed., 198–229. London: Thames and Hudson.
Wenke, Robert J., and Deborah J. Olszewski. 2007. *Patterns in Prehistory: Humankind's First Three Million Years*. 5th ed. New York: Oxford University Press.
Zeder, Melinda A. 2011. The Origins of Agriculture in the Near East. *Current Anthropology* 52 (S4): S221–S235.

Chapter 5
The Urban Revolution I: Polygroups, or Early Complex Societies, 6000–1000 BCE (8000–3000 BP)

Abstract A few thousand years after settlement and domestication started, human groups continued growing in demographic and geographic size in several world regions, becoming relatively complex. These societies, which I call "polygroups," developed permanent inequality, populations in the thousands, true cities, multiple levels of political leadership, increasing but inconsistent male supremacy, and monumental architecture. They were relatively unstable, and started to encroach on existing bands and villages. Their size forced village groups to grow in a violent and uneven process to match them. The development of complex societies was the beginning of the second major social transformation in human history.

Keywords Complex societies · Chiefdoms · Social inequality · Urbanization · Monumental architecture · Megaliths

Introduction

Domestication and settlement together comprised the first social transformation in the human past. The consequences of agriculture—more people, more possessions, more violence, worsening health, elaborated kinship, and others—were revolutionary. Sometime after 6000 BCE (8000 BP), another type of society developed around the world. Many historical accounts discuss the Agricultural Revolution as if it had led directly to states; the existence of an intermediate form of society between villages and states seems obvious, yet only anthropologists and archaeologists acknowledge it. I devote this chapter to analyzing that form, which eventually, and only sometimes, led to states. Here, I argue that the second revolution in human social organization led to what scholars call "complex societies." Following Childe ([1936] 1966), I call it the "Urban Revolution."

While this phrase stands for many social changes, not merely the development of cities, cities are a good place to start. Until this period, villages were small concentrations of people. Large ones such as Çatalhöyük were rare, and that was more of a super-village: almost all structures there were residential, with minor wealth differences. After 6000 BCE (8000 BP), larger demographic concentrations, which had

T. C. Burr, *Societies Past and Present*, Emerging Globalities and Civilizational Perspectives, https://doi.org/10.1007/978-3-031-91532-1_5

internal structures distinct from those of villages, appeared. First, cities had clear divisions of labor across them: different districts had different functions, such as religious ceremonies, residences, craft production, storage, and market trade (Hansen 2000). Second, as towns and cities concentrated power by concentrating people, skills, material resources, ideas, and information in a small space, permanent hierarchies of leadership developed; leaders used their power to rule people in outlying rural villages. Thus, large wealth and gender inequalities often developed, plus vertical rankings of groups. Finally, consistent warfare, occupational specialization, and monumental architecture also developed.

These changes describe a "chiefdom," a common way archaeologists talk about social complexity. However, many scholars (e.g., McIntosh 1999; Pauketat 2007; Grinin and Korotayev 2017a; Milner and Wills 2018; Graeber and Wengrow 2021) show that larger groups developed that were *not* organized as chiefdoms but were still large, complex social units. While chiefdoms centralized power, some complex societies were still based in villages (not cities) and were heterarchical, joined by complex kinship ties, sodalities across many villages, or alliances. An elite always existed, with the legitimacy or resources to get others to act (Feinman and Nicholas 2016; Stanish 2017), sometimes only seasonally (Graeber and Wengrow 2021). These societies were also more complex than groups in simple villages but in ways different from chiefdoms. Thus, I have adopted the word *polygroup* for small, early complex societies, either chiefdoms or their "analogues" (Grinin and Korotayev 2017b), to denote a group of a few leaders and thousands of followers. New terms are always awkward, so people understandably resist them, but the concept of "chiefdom" simply does not include the range of all early complex societies.

Again, human capacities made possible these increases in social complexity. Building on previous developments and again in response to new problems, informal leaders around the world took individual and collective action to weave together villages into larger, more complicated social units, working on people's grief and fear and on their deeply held supernatural beliefs. This required many centuries, with failures along the way, to accomplish, and many others helped, turning human inventiveness to new technologies, beliefs, and relationships. These changes and larger group sizes (nested and interlinked) made possible much larger interaction zones between these incipient supergroups.

Here, I examine the development of early complex societies in four steps. First, I trace interconnected chronological trends: additional domestication sites and geographic expansion, the development of polygroups, and remaining hunter-gatherers. Second, I review theories of polygroup formation. Third, I analyze social patterns of early complex societies. Finally, I examine broader patterns between such societies, and between them and bands and villages. I stop around 1000 BCE (3000 BP) because complex societies had appeared in the Americas by then and because a fourth type of society had developed in Afro-Eurasia, which will be analyzed in Chap. 6.

Telling the Tale: The Origins and Expansion of Complex Societies

From around 6000 BCE (8000 BP), two major trends existed. First was the development of additional farming regions and the geographic expansion of farming from both earlier and later domesticating regions. The other was a delayed consequence of agriculture: complex societies. We will now trace how both trends developed and interacted around the world.

Between 5000 BCE and 2000 BCE (7000 BP–4000 BP), more societies around the world started plant and animal management and domestication. Again, these processes took several millennia, and again, later domesticating regions seem to have done it independently (Larson et al. 2014:6140). While management and domestication developed in these regions, existing farming societies continued expanding demographically. In other places, some central power brought multiple villages together into one political unit. This required a formal political structure of leaders, sometimes overlapping with a permanent socioeconomic elite. Warfare and gender inequality developed—according to some, they developed together. Some villages grew into towns or small cities. Everywhere complexity appeared, monumental, ceremonial architecture was associated with an elite.

The first change occurred just after 6000 BCE (8000 BP) in Mesopotamia, between the Tigris and Euphrates Rivers. These run inconsistently parallel from southeastern Anatolia through what is now Syria and Iraq, joining 120 miles (200 kilometers) before draining into the Persian Gulf. In this era, people upriver seemed to continue infill (Lemke 2018) in this sparsely populated region, using irrigation canals from these rivers to water distant fields. Downstream, despite annual flooding and silting, people also started using irrigation. Temples became increasingly larger and probably more important over time and spread upriver. To the east, in the Zagros Mountains (now western Iran), people built temples (Matthews 2018:432–435). By 4500 BCE (6500 BP), there were towns of 1000 to 3000 people and clear inequalities in housing and goods. After 4200 BCE (6200 BP), communities grew even larger and more numerous. Irrigation farming produced surpluses, so trade within the Southwest Asian interaction sphere greatly increased. Temples were key institutions, feeding temple workers and lending capital to long-distance merchants (Matthews 2018). Evidence of warfare, including sling bullets and massacred skeletons, increased over time. Agropastoralism spread into Central Asia from Iran to east of the Caspian Sea by 6500 BCE (8500 BP) (Baumer 2012), and weaponry and defensive walls appeared in Anatolia and east of the Caspian after 3000 BCE (5000 BP) (Ferguson 2006).

Social complexity also developed in the Nile Valley. The introduction of farming helped the population increase. From 4500 BCE (6500 BP) onward, there were numerous villages where some larger population concentrations existed, wealth inequalities increased, craft specialization developed, the trade area enlarged, and weaponry aimed at humans (such as maces) appeared (Ferguson 2006; Köhler 2011; Connah 2018). As population centers grew in size and number, they traded with

Southwest Asia. By 3800 BCE (5800 BP), towns controlled villages along the narrow valley. By 3600 BCE (5600 BP), there was regular long-distance trade, and larger population centers each controlled many villages. Sometime around 3350 BCE (5350 BP), one political entity controlled much of Upper Egypt (Köhler 2011; Anđelković 2011).

Across Africa, most people remained mobile foragers; in the north, mobile pastoralism continued. Domestication of animals appeared in a few more places. Agropastoralism reached the middle Nile Valley by 4000 BCE (6000 BP). Herders kept cattle, sheep, and goats in what is now Mali after 3000 BCE (5000 BP), also cultivating pearl millet (Connah 2018). Permanent settlements of herders appeared after 3500 years in Mali, the Lake Chad basin, and "south of the great bend of the Niger River," and grain agriculture started (Smith 2005:92, 123). In Ethiopia, people herded local cattle by 1500 BCE (3500 BP). Domesticated cattle also reached southern Africa after 2000 BCE (4000 BP) (Connah 2018:364, map).

Farming continued moving northward in Europe, and farming societies slowly became more complex. Towns of hundreds of houses grew in Ukraine after 4000 BCE (6000 BP). In southern Spain, a fortified settlement appeared by 3000 BCE (5000 BP) with mass burials in large tombs showing economic inequalities and long-distance trade. By 2700 BCE (4700 BP), a large town grew in central Spain. To the north on the Atlantic, farmers sometimes lived near complex foragers for centuries; other times they replaced them suddenly, peacefully or violently (Scarre 2018b). They built large monuments, seemingly without much social hierarchy. Around 2500 BCE (4500 BP), Yamnaya immigrants came from the Caucasus into Central Europe. Within a thousand years, elites developed and controlled long-distance trade (Scarre 2018b).

In South Asia between 5500 and 4500 BCE (7500–6500 BP), the region around Mehrgarh had domesticated water buffalo; people now developed pottery and began grain storage. They produced agricultural surpluses after 4500 BCE (6500 BP) with specialized craft activities (Jain 2006). Rice cultivation started in central India after 4000 BCE (6000 BP), either imported from East Asia or developed locally. Animal domestication started in eastern and southern India after 2500 BCE (4500 BP). Pottery and jewelry production spread across northwestern South Asia, and villages shipped surpluses throughout the region. Fortifications became common by 3000 BCE (5000 BP) (Coningham 2018), as did formal planning of towns. Larger settlements dominating smaller ones became common by 2600 BCE (4600 BP) in the Indus Valley.

Rice culture spread throughout the middle Yangzi River valley from 4500 BCE (6000 BP) onward. Plowing for rice and irrigation appeared by 4000 BCE (6000 BP), and population increased greatly beyond this region after 3300 BCE (5300 BP) (Higham 2018a). By 2500 BCE (4500 BP), there were three sizes of settlements, the largest "probably measured in the tens of thousands" (Higham 2018a:247). As most people were cremated, cemeteries are unusually complete, showing economic inequality: most have an urn and pottery, but a few have rich goods. Warfare was a "way of life" in the region by 1800 BCE (3800 BP) (Ferguson 2006:486).

East Asian agriculture spread southward and eastward. After 3500 BCE (5500 BP), millet farmers moved into Korea; local hunter-gatherers mixed these practices with their own, and rice cultivation appeared there after 2000 BCE (4000 BP) and in Southeast Asia in the same period (Higham 2018a). Independently, people in New Guinea developed soil mounds for cultivation after 5000 BCE (7000 BP), and "rectilinear grid-like ditching appeared by 2000 BCE" (4000 BP) for taro and banana cultivation (Bellwood and Hiscock 2018:275).

In the Americas, similar changes occurred. Maize domestication started "by 5000 BCE" (7000 BP) in Mesoamerica, and there were "semi-permanent base camps" by 3000–2000 BCE (5000–4000 BP) (Browman et al. 2018:306), in which people mixed hunting, gathering, and agriculture. Sedentary villages became common after 1600 BCE (3600 BP), growing in size after 1400 BCE (3400 BP). By 1200 BCE (3200 BP) on the Gulf of Mexico, two complex, "Olmec" societies appeared with monumental architecture, hillside agricultural terracing, and hierarchical ranking (Webster and Evans 2018). Villages surrounded each town. Other polygroups developed in western Mexico between 1500 and 400 BCE (3500–2400 BP) (Webster and Evans 2018). People brought agriculture to North America by 2100 BCE (4100 BP) (Browman et al. 2018). Warfare appeared in the Appalachian Mountains by 2500 BCE (4500 BP) (Ferguson 2006), and irrigation canals by 1000 BCE (3000 BP). Domesticated gourds came from the Mesoamerican Gulf Coast to North America (Browman et al. 2018), yet people also domesticated local plants.

After 3000 BCE (5000 BP), the Andes had large ceremonial centers, including monumental architecture (Moseley and Heckenberger 2018). People traded obsidian over long distances and intensified horticulture. Ceremonial centers had few indications of residence. The coastal center of Caral in Peru had a central authority by 2400 BCE (4400 BP). A later, much larger site, after 1800 BCE, had some agriculture. There is pictographic evidence of warfare (Moseley and Heckenberger 2018). In Amazonia, mobile foragers became sedentary agriculturalists over a few millennia, started using pottery just after 6000 BCE (8000 BP), and started managing and then domesticating plants by 2000 BCE (4000 BP) (Browman et al. 2018) (Fig. 5.1).

In short, new centers of plant and animal domestication appeared, and farming continued to spread. Again, this may (or may not) have been a peaceful process. DNA analysis of skeletons shows that sometimes hunter-gatherers were absorbed into farming communities; other times foragers adopted farming—mostly piecemeal, sometimes entirely; in some places, farmers simply replaced foragers (Bellwood 2013), and in others, farmers continued processes of infill into completely unpopulated land (Graeber and Wengrow 2021).

Still, foragers populated most of the world. Some remained simply organized; others became more complex; and a few were quite complex. For instance, all Australian peoples remained foragers. However, through most of the Holocene, Australian peoples kept diversifying materially, managing "plant and animal resources by burning vegetation, constructing dams and drainage channels, detoxifying poisonous plants, and storing and replanting seeds" (Bellwood and Hiscock 2018). New Guinean domestication never moved to Australia, although the lands

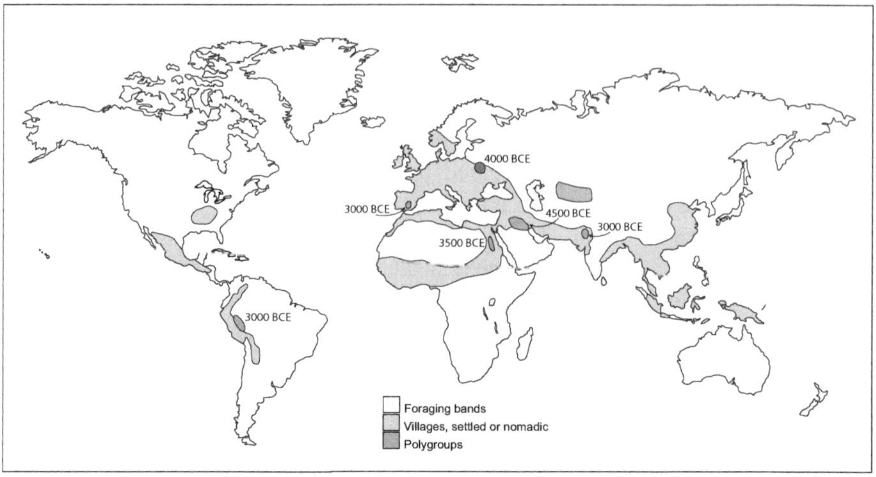

Fig. 5.1 Polygroups, Villagers, and Foragers, ~3000 BCE (5000 BP). Sources: Moore (1995); Black (2005); O'Brien (2010); Haywood (2011)

were still connected (Davidson 2014). After a cooling, drying trend around 2000 BCE (4000 BP), Australian peoples moved to smaller, more standardized tools.

While Siberian peoples remained foragers throughout the Holocene, they continued changing in other ways. Pottery spread across northern Asia (Hommel 2018). In the region around Lake Baikal (southern Siberia), foraging peoples purposely buried some of their dead and shifted between land animals and lake animals for protein (sometimes relatively suddenly), although different groups adopted different strategies at the same time (Weber et al. 2021).

While a few North American peoples domesticated plants, most remained foragers. By 3500 BCE (5500 BP), Louisiana foragers built mounds around plaza-like areas. Two thousand years later, 1700–1200 BCE (3700–3200 BP), peoples to the north at Poverty Point built larger monumental architecture, either as foragers or food managers (Browman et al. 2018). They created a large interaction zone from the Gulf of Mexico to the Great Lakes. Archaeologists disagree about whether it was a town with an elite or a ritual gathering site (Browman et al. 2018; Graeber and Wengrow 2021). After 4000 BCE (6000 BP), many peoples started management practices. Ohio River valley foragers were egalitarian but occasionally fought (Browman et al. 2018). On the East Coast, people started using pottery around 2500 BCE (4500 BP). In the Plains and Great Basin, people remained simple hunter-gatherers. On the West Coast, while people remained foragers, there was "nearly endemic warfare" in the northwest by 2500 BCE. In California, "hierarchical ranking, sedentism, and associated territorial competition and conflict" appeared after 1300 BCE (3300 BP) (Browman et al. 2018:324).

Between 6000 and 1000 BCE (8000–3000 BP), groups uniting multiple villages appeared around the world. Towns subordinated villages. Agriculture, social inequality, pottery use (i.e., storage), warfare, and monumental architecture almost

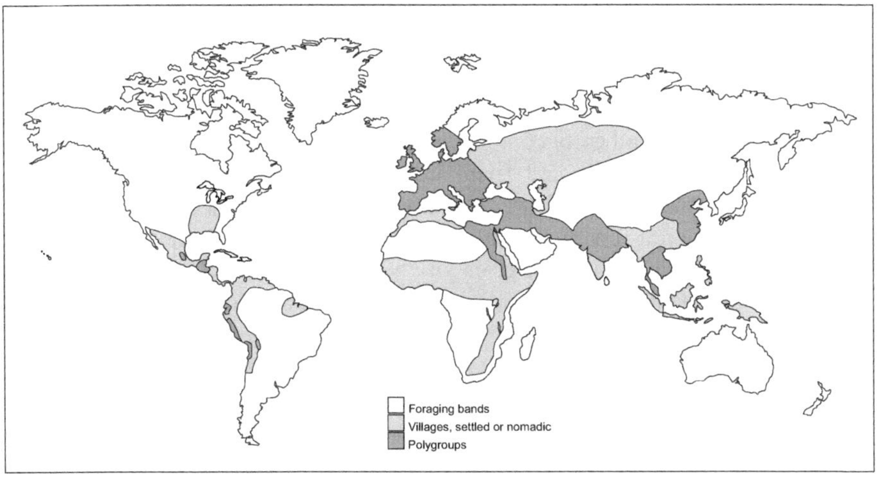

Fig. 5.2 Polygroups, Villagers, and Foragers, ~1000 BCE (3000 BP). Sources: Black (2005); Bahn ([2000] 2009); O'Brien (2010); Haywood (2011); Houston (2018)

always coincided with towns in Southwest Asia, Egypt, the Indus Valley, parts of China, two regions in Mesoamerica, parts of the Andes, southwestern North America, and Europe. In northwest North America, most of these trends appeared without agriculture. Elsewhere in North America, in much of South America, all of Australia, and much of Asia and Africa, hunter-gatherers continued their ways of life (Fig. 5.2).

Explaining the Development of Social Complexity: Debates

While many archaeologists resist using typologies in favor of analyzing processes over time (Peregrine 2012; Stevenson 2016; Earle 2017), any social entity can last for centuries despite significant changes. Multivillage groups involved relations "beyond … personal relationships in face-to-face interaction" so "new forms of relations, control, and leadership must have appeared within them" (Grinin and Korotayev 2017a:69), and the different processes led to varied social structures. Furthermore, many of these societies were unstable, often with a nearly constant process of disintegration and attempts at re-formation.

Many archaeologists agree that early complex societies took diverse forms (Blanton et al. 1996; Flannery and Marcus 2012; Feinman and Nicholas 2016; Grinin and Korotayev 2017a). Many scholars use "chiefdom" as shorthand for all complex societies (e.g., Carneiro 2017; Earle 2017), but "chiefly analogues" also developed (Grinin and Korotayev 2017a). A chiefdom was a multivillage entity ruled by a chief at the center. Chiefly analogues include "self-governed cities, *poleis*, temple communities, and large rural communities; decentralized chiefless tribes;

and various complex acephalous [leaderless] sociopolitical systems" (Grinin and Korotayev 2017a:70). Many African societies included groups "in which central authority, often of a ritual nature, is paired with a power structure that is diffuse, segmentary, and heterarchical" (McIntosh 1999:4). Given this diversity, I use the term "polygroup" for all early complex societies.

Why did polygroups appear? There are two general explanations: they formed for practical reasons, or chiefs were successful aggrandizers; these are the aforementioned "structural-functionalism" and "conflict theory" (Tainter 1988), though Carneiro (1970) calls them theories of "voluntaristic" and "coercive" action. Structural-functionalist theorists argue that societies developed social structures to aid functioning. One motive would be developing networks for rituals, trade, marriage, and friendships, all of which fostered group cohesion. Forager fission–fusion had accomplished this, as did meeting places such as Gobëkli Tepe, but polygroups scaled this up. Another practical problem would be population pressure (i.e., population growth within geographic limits resulting in pressure on resources), forcing innovations such as the growth of more food, which might involve collective projects. Conflict theorists argue instead that aggrandizers wanted to rule over others, so they developed institutional structures to mostly benefit themselves. There are two subparadigms. In one, an individual or small group tried to take over internally, as a variant of free-riding. In the other, groups conquered external others. Carneiro (1970) classically argued that population growth within social or environmental limits led to population pressure, in turn leading to warfare over land.

Tainter (1988) argued that both functional and conflict-driven changes led to social complexity. Aggrandizers regarded opportunities to solve problems as opportunities to gain power. Thus, political inequality started as a deal—a form of reciprocity: anyone solving a major problem argued that they needed the power to implement the solution. Upon success, they "deserved" continued power. If accurate, this synthetic approach still does not tell us the step-by-step process to social complexity. The debate on its origins seems messier than the one on the Agricultural Revolution. Comprehensive overviews on the complexity debate like those on domestication (e.g., Price and Bar-Yosef 2011; Larson et al. 2014) seem rare, perhaps because the greater diversity of forms suggests a wider range of explanations. Thus, the following discussion may or may not be comprehensive. It still relies heavily on ethnographic inference since some ethnographically identified processes were "archaeologically invisible" (Flannery and Marcus 2012:353). Other collections (Fargher and Heredia Espinosa 2016; Carneiro et al. 2017; Chacon and Mendoza 2017) rely mostly—but not exclusively—on archaeological examples.

As in domestication and settlement, people probably did not purposely create large, multivillage entities or independent cities. Instead, would-be leaders made it up as they went along. Most already commanded resources, coming from a wealthy lineage with a lot of land or well-practiced war skills (Bogucki 1999; Hayden 2018). Thus, the idea of "natural" aggrandizers has limitations. (See below.)

Carneiro wrote the classic explanation for chiefdom formation (1970); he developed its latest form (2017) in response to decades of questions and criticisms. He assumed, then and now, that population pressure set groups against each other in warfare over land, the source of food, and thus of survival. As discussed, settled villages often developed feuds, implemented by raids and massacres. If war became ongoing, eventual alliances between two villages were likely, then three or more allied. Since a group under attack had little time for a collective decision, "the warriors of the village alliance would operate under the command of one man" who would likely be a great warrior tested in battle. If war is "frequent, intense, and prolonged, the military leader would find his services called upon more and more often. And not surprisingly he would become increasingly reluctant to surrender his wartime powers once peace had returned," eventually retaining them, as everyone else was now used to subordinating themselves to him. A cohort of warriors beholden to him for war spoils would help him enforce his new power. With this, "the basic elements of a chiefdom had come into being" (all quotes Carneiro 2017:36).

Two changes followed. Internally, how would this war leader make his rule permanent? At first such a position was "the personal power structure of one man" (Carneiro 2017:41); chiefs could train their sons in warfare and leadership, but their succession was not guaranteed. If successful, however, the family's rule might become permanent over generations. Externally, under endemic war, this new, larger social unit threatened other villages which would either ally with other villages "to retain a measure of [internal] independence" or be conquered by the new chiefdom (Carneiro 2017:39). Indeed, archaeologists have found in many areas that one, "primary" chiefdom often led to other, "secondary" chiefdoms.

LeBlanc (2006) posits a variant of this process. In endemic warfare, people planted crops close to their villages, leaving unplanted "buffer zones" outside, which they patrolled to prevent raids. Ethnographers have found buffer zones in South America, New Guinea, Africa, and Southeast Asia; archaeologists found them in the U.S. Southwest and Europe (LeBlanc 2006).

Many changes ensued. If a multivillage alliance became a chiefdom through conquest, people could plant now-internalized buffer zones. This would quickly double harvests across the entire society (compare Figs. 5.3 and 5.4). It would take three to eight generations for people to fill in the land, allowing generations of leaders to make their new elite status permanent (LeBlanc 2006). They might have distributed empty land to colonists and "asked" for a portion of its surplus, creating permanent payments. If the region included different ecological zones, a variety of crops grew across it, so the new chief(s) could make sure people in those zones traded, thus increasing food diversity and survival for everyone. Finally, leaders would change their consumption behaviors to signify elite status, adopting special clothes, housing, and foods. LeBlanc does not mention another benefit. If some people received new land, they would leave their old land, so those staying behind could take their former neighbors' lands; thus, everyone in the conquering society gained land. The conquered, of course, lost their lives or their freedom.

Power and problem-solving could operate through kinship once complexity had developed. Segmentary societies appeared on all continents but were "not universal" (Willer et al. 2017:428). Eventually some segments could gain more wealth and perhaps higher status than others. Ethnographically, status in some lineages is

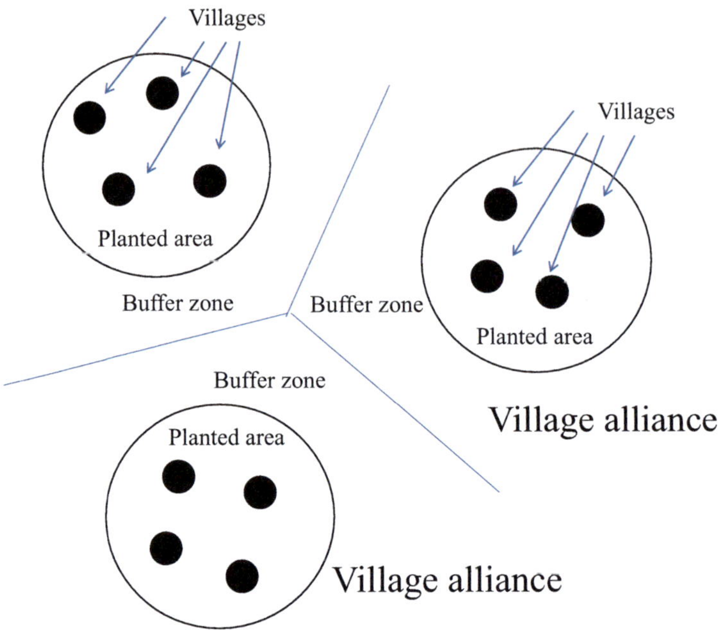

Fig. 5.3 Buffer zones in villages

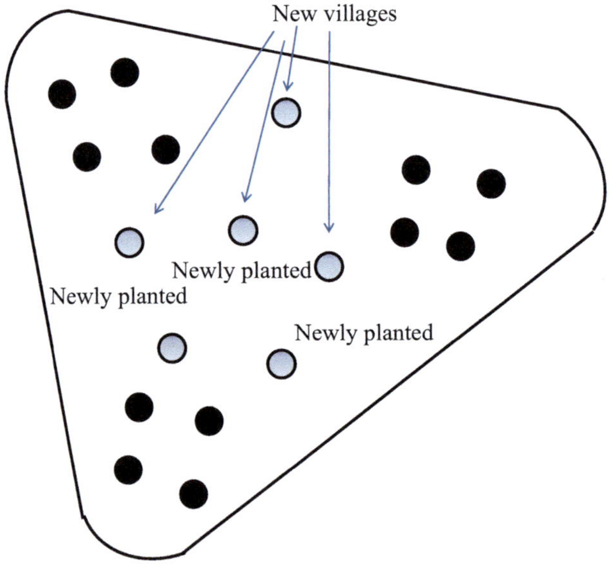

Fig. 5.4 Former buffer zones in a chiefdom

assigned by kin distance from the chief: the firstborn son is second highest in status, his first grandson third, his second grandson fourth, and so on. This clear structure helped solve collective problems: “Any higher status member of the chiefdom … can initiate collective action … toward common individual or collective goals” (Willer et al. 2017:430). This is an “influence-based” chiefdom. Over generations, everyone except the lead lineage experienced downward mobility, which created problems. Since war “is the normal condition of chiefdoms” (Willer et al. 2017:432), the populace faced being invaded; submission to the leading lineage established protection from that lineage. Thus, reversing Carneiro’s (2017) explanation, coercive chiefdoms could grow from influence-based ones.

Carneiro (2017) acknowledges that many scholars reject his account. For instance, Feinman (2017) argues that studies have “failed in the search for a unilineal pathway, a uniform proximate cause, or a universal set of unvarying properties that can account for early complex societies” (2017:462). Still, Carneiro (2017:463) argues that “sedentary populations numbering more than 2000–3000 [people] are almost always hierarchically organized.” To address this, Blanton et al. (1996) suggested that would-be leaders had two basic strategies available.

One strategy was “corporate,” providing in-group benefits, or “public goods.” This is a structural-functionalist strategy: improve society internally to gain power. Private goods have two characteristics: they are rival and excludable. If someone eats a piece of food, it is used up (rival), and they have stopped others from eating that food (excludable). On the other hand, public goods are nonrival and nonexcludable: one’s use does not degrade them for others, and everyone has access to them. Defense is a classic public good. Under social substitutability (Kelly 2000), attackers indiscriminately kill everyone among “the enemy,” so everyone must be defended. Since everyone benefits when the enemy is repulsed, defense is nonrival. Other public goods include public order, justice systems, roads,[1] sanitation, and collective ritual spaces.

The problem with public goods is that because they are nonexcludable, no single individual has an incentive to provide them. If one did, everyone would use it, free-riding on one person’s effort. However, if someone organizes the work, and everyone contributes a little, it ends free-riding, and everyone benefits from the public good. Therefore, a corporate strategy involves incipient leaders organizing collective projects to produce public goods, the populace paying the leaders in goods or labor, and the leaders taking a cut for their trouble, thus creating a situation of collective reciprocity (see Fig. 5.6 below).

The second strategy involved developing a “network” of resources external to one’s group. Network strategies were clearly conflict strategies: use external resources to dominate internally. Such resources include trade with or conquest of other groups or intermarriage with other elites. These increased leaders’ power but created fewer internal obligations, so leaders provided fewer public goods and

[1] While roads eventually wear out, one person using a road only barely degrades its condition for the next user.

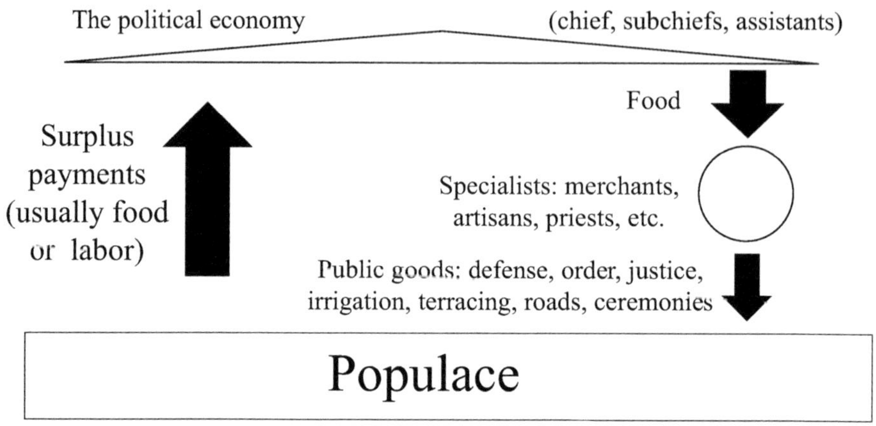

Fig. 5.6 Public goods in a corporate strategy

services (Blanton et al. 1996). Leaders following corporate strategies emphasized egalitarianism, downplaying their positions; network strategists displayed their own power lavishly to emphasize dominance. Leaders often mixed these strategies: they developed multiple power bases and simultaneously served and exploited their home groups.

Scholars have extended analysis of these two strategies. For instance, Peregrine (2012) posited three axes of actions creating complexity, not one. First, sources of power ranged from sacred to profane. Second, leaders constructed power through either corporate or network strategies. Finally, they implemented them through voluntaristic or terroristic methods. In conditions of resource uncertainty, aspiring leaders more likely followed network strategies, including warfare, but this might have intensified conflict (Peregrine and Ember 2016).

These strategies help us better understand the social complexity debate. One corporate strategy consisted of leaders holding feasts (Hayden 2003, 2014). Chiefdoms everywhere are associated with *megaliths*—large, heavy stones (Hayden 2003:227) including covered tombs or circles such as Stonehenge—or nonstone (earthen) structures. Burials in these structures usually have few bodies but expensive goods. The public purpose of megaliths is to improve society by pleasing the leaders' ancestors (Hayden 2003:234). Leaders hold feasts and "ask" feasters to build ancestral megaliths. To hold feasts (to increase consumption), the leader's group had to grow more crops and animals for meat (increase production). Competition between two lineages or clans forced a constantly intensifying production–consumption spiral. At some point, only one lineage—the leading lineage, which was now regularly receiving obligatory payments of food from everyone—was holding feasts (Hayden 2003, 2014).

Another path to power was to control trade. Leaders often displayed position through prestige goods, which were the dominant objects in long-distance trade in early social complexity (Hayden and Earle 2022:390–391, 397). Controlling "bottlenecks" or difficult points in long-distance commodity chains was one network strategy, including control over raw materials or secret manufacturing techniques, heavy investments in watercraft, or access to trade partners. Berea (2010) argues that merchants in Mesopotamia developed technologies and complex institutions to facilitate trade. These included developing tokens, the pottery wheel, and money; establishing coastal trade colonies to control trade routes; spotting comparative advantages between complementary, adjacent regions; and using temple granaries for famine relief as trading sites.

Some emphasize other foundations of power. Flannery and Marcus (2012) argue that there were three: supernatural beliefs, expertise, and bravery and toughness. Many leaders convinced their followers that they wielded or embodied supernatural forces; leaders were often regarded as semidivine, having "godly" powers to threaten the disobedient (Carneiro 2017:44). Expertise in social relations or toughness in warfare might legitimate supernatural claims. Some aspiring chiefs built temples "to the highest celestial spirits in their cosmos" (Flannery and Marcus 2012:260), while men's societies usually disappeared under complexity (Hayden 2018).

Still, secret societies were one path to social complexity. They were "widespread" in village-level societies (Hayden 2018:371), allowing aggrandizers to form ties across kin groups to create a small elite. They claimed to have secret ritual knowledge of the supernatural; engaged in terroristic practices (human sacrifice and cannibalism); concentrated material resources, often through the creation of obligations; protected their members from others; and often created regional networks of such organizations. They did not integrate societies but concentrated power in the aspiring elite (Hayden 2018). One tactic might have been to split groups into factions, setting them against each other to gain power. Human sacrifice was a common practice among early polygroups, existing "in early Germanic, Arab, Turkic, Inuit, American, Austronesian, African, Chinese and Japanese cultures" (Watts et al. 2016:228). The practice legitimated elites by sacrificing to the gods and also terrorized low-status groups into remaining submissive.

Others take a more demographic approach. Sheer numbers of people create thick webs of social interaction, which in turn have institution-building effects (Peterson and Drennan 2012). The co-occurrence of sedentism, domestication, and pottery always seems to have started groups on at least three paths to complexity. To examine this, scholars placed eleven archaeologically researched regions in the world on one artificial timescale; the common zero point was the first co-occurrence of sedentism, domestication, and pottery (Peterson and Drennan 2012:89). They then followed each region's chronological trajectory. The first path was the "amalgamation of smaller units through territorial expansion." A second strategy involved larger villages or towns, which "sucked [smaller ones] into" them by forced migration. Finally, "supra-local communities appear early and growth is largely internal to them" (all, Peterson and Drennan 2012:127). In all regions, the number of smaller settlements decreased, while the size of larger settlements increased. It took "about 1000 to 2000 years" from when sedentism, food production, and ceramics first coincided for large-scale societies "of 10,000 people or more" in most regions to appear (Peterson and Drennan 2012:118). Also, the "tax rate" was a maximum of one worker-day per year for public works, with other minimal taxes to fund an army and specialists.

Climate changes may have enabled social complexity. After the Holocene started, dramatic shifts in rainfall created "the framing *conditions* in which human societies mobilized to change their circumstances" (Brooke 2014:165, original emphasis). Social complexity seems to have occurred mostly in periods of population growth fed by moderate rainfall: in Mesopotamia and the Indus Valley after 4200 BCE, in the Andes after 3000 BCE, in Mesoamerica after 1500 BCE, and in North America around 1000 BCE (Brooke 2014).

Polygroups were notoriously unstable (Blanton et al. 1996; Bogucki 1999; Carneiro 2017). Archaeologically, they repeatedly formed and broke down, a pattern called cycling or oscillation: groups alternated between being larger and smaller social units over decades or centuries (Bogucki 1999). In Mesopotamia, "The whole history of early complexity … seems to be a messy boom-or-bust cycle, with only a very general long-term trend toward complexity" (Wenke and Olszewski 2007:339). Some argued that these cycles were about one generation in length (Blanton et al. 1996), and others suggest why: the original leader failed to install his son as successor (Carneiro 2017). Also, subleaders of smaller territories often rebelled (Bogucki 1999). Local climate change, incompetence, or bad luck may have stopped leadership processes or might have intervened after success, leading to the return of chronic warfare between villages. Finally, just as foragers greatly outnumbered individual aggrandizers in any band, the masses greatly outnumbered their rulers; leaders may have forgotten this to their regret (Bogucki 1999:269).

Much of this literature assumes chiefdoms; explanations only sometimes apply to other forms of complexity. One scholar argues that the "ethnographic record is replete with examples of chiefs or leaders … who … lack coercive powers" (Stanish 2017:67). Perhaps one path to complex inequality involved a leader trying to turn his purely ritual position into a coercive one. The existence of heterarchies, oligarchies, and confederations (Grinin and Korotayev 2017b), or individual leaders limited by other collective institutions, without any urbanization (McIntosh 1999) suggests collective, not individual, action to unite settled villages or herder camps. Other ethnographically found arrangements, such as powerful political leaders without elevated economic wealth or seasonal dictatorships alternating with seasonal egalitarianism, also imply that populaces sometimes limited leaders (Graeber and Wengrow 2021).

As scholars debate how and why social complexity developed, they suggest multiple factors: conquest, public goods provision, feasting, secret societies, human sacrifice, manipulating belief-systems, trade management, and demographic concentration. Most seem to be talking past each other, but the sheer number and variety of paths to, and resulting forms of, social complexity make this understandable. A comprehensive, worldwide overview of all archaeologically discovered cases of social complexity would be tremendously useful but nightmarishly difficult to compile. And processual archaeologists are right: time passes and these structures change, so what qualifies as a "case" of initial social complexity? For now, the range of theories available constitutes merely a beginning, however excellent, for understanding.

Social Patterns of Polygroups

Polygroups were complex for several reasons. First, they were simply larger, which increased the number of interconnections within and between societies. Second, there were more groups within any society. Third, the division of labor elaborated, with many types of specialists. Most people remained farmers, herders, or collectors. Here, I examine these social patterns.

Polygroup Politics

As discussed, multivillage groups appeared in many forms. "Chiefdom" is a term of political geography—a structure of chiefs held sway over multiple villages in a given region—so we must review and analyze the political structures of chiefdoms and analogues. Formation strategies seem to have been path dependent: different paths led to different social structures (Blanton et al. 1996; Peregrine 2012). Again, we must sometimes still use ethnographic analogy.

We start with chiefdoms. In a classic chiefdom, there is "some form of centralized authority … relatively permanent agencies of government, typically based on collection and redistribution of an economic surplus" (Lewellen 2003:32–33). It is not a heterarchy of multiple, equivalent leaders, but a hierarchy of one central leader operating through subleaders. Usually, one lineage or clan has more resources and power than any other does. Theorists have identified three types of chiefdoms: simple, complex, and paramount. In simple chiefdoms, the chief lives in a large town surrounded by villages, each often with its own headman. In "complex" chiefdoms, a chief in a large town appointed subchiefs in small towns, each of whom administered several villages within his territory. Finally, in a paramount chiefdom, the chief lived in a city, subchiefs lived in larger towns, lower chiefs lived in smaller towns, and a headman might be responsible for a village. (Fig. 5.5 shows a complex chiefdom.)

Actual governance required more. While chiefs usually appointed kin as subchiefs, each level needed assistants to carry out specific tasks: people to manage communications and transportation for intensified trade; warriors for warfare; religious assistants; and craftspeople to make weaponry, public works, and status symbols. Every level had such assistants (see Fig. 5.5).

Some criticize this "building block" model of chiefdoms in which the units simply stack on top of each other, and villages and smaller towns were unchanged (Pauketat 2007). In fact, as chiefs took over trade, defense, and justice, villages lost functions, becoming more dependent on and integrated with political centers (Johnson and Earle 2000), which eroded the importance of kinship.

Other kinds of polygroups, found ethnographically and inferred archaeologically, varied greatly. One form was the "monosettlement": a single, independent town or city, formed by demographic concentration, surrounded by a rural

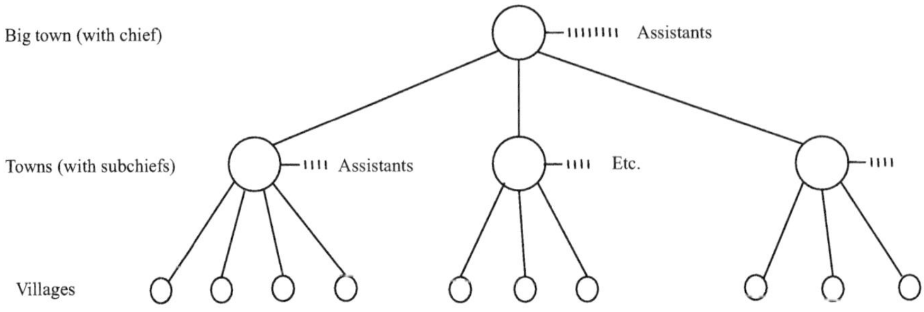

Fig. 5.5 Assistants in a complex chiefdom

hinterland which grew food for that center (Peterson and Drennan 2012). Priests, clans, representative councils, or a single leader ruled. "Horizontally integrated polysettlements" were sets of villages integrated by collective events in ritual centers, stratified clans, or secret societies. Confederations of multivillage entities ruled by a representative council also existed (Grinin and Korotayev 2017b; Hayden 2018; Graeber and Wengrow 2021). In the low-density, abundant-land environment in sub-Saharan Africa, mobilizing people at need (especially their intellectual resources) was more important than ongoing control over economic resources. Consequently, multivillage heterarchy was much more common (McIntosh 1999), taking the forms of age sets, secret societies or "title" societies (granting prestigious titles), councils of clan elders or of priests, or multiple, separate hierarchies. Kinship "segments" (clans or phratries) might live independently in decentralized settlements, uniting into a larger unit in the face of a need (e.g., defense) (Grinin and Korotayev 2017b). In many of these forms, there was no administrative structure of levels with assistants. Finally, at least ethnographically (and probably in prehistory) in highly seasonal environments, fission–fusion foraging continued. During one season, leaders ruled through authoritarian political structures, sometimes planting; in the other season, people moved back to egalitarian bands, or they rotated power between clans (Graeber and Wengrow 2021:98–111).

In polygroups with a "corporate" elite, rulers set up a redistribution system. This involved a center drawing in resources and distributing them out again (Polanyi 1957). What rulers took they gave to third parties, usually specialists, some of which became full-time, including carpenters, religious leaders, warriors, or construction workers; these provided public goods and services to solve collective problems: defense, conflict adjudication ("justice"), food storage, ceremonial (monumental) architecture, and irrigation and flood control. While it required a surplus, such "contributions" (proto-taxes) may have been minimal (Peterson and Drennan 2012). Figure 5.6 shows that redistribution was a form of indirect, collective reciprocity.

While chiefs thus controlled economic resources, they had to return most of them as services, but keeping merely a small portion could make them wealthy and powerful. Sometimes, however, chiefs had to impoverish themselves (Graeber and Wengrow 2021).

Not all polygroups relied on redistribution (Bogucki 1999:262–263). Network strategies (Fig. 5.7) rested on external resources. Independent towns or cities might need some public goods; others, as trade centers, relied on a flow-through of external goods. Confederations and segmentary societies might simply mobilize resources at need. Many large-scale societies were founded on collecting (Graeber and Wengrow 2021:Chaps. 4–5), in which political power was separated from economic resources.

Again, aspiring leaders followed both corporate and network strategies. Specific actions could fulfill either purpose. For instance, complex societies experienced much more warfare than autonomous villages (Bogucki 1999; Hayden 2003; Fagan 2004; Ferguson 2006). Warfare for conquest was a network strategy, allowing leaders access to slaves and strategic sites or fertile land; for defense, it was a corporate strategy, benefiting locals. Specific circumstances and mixed strategies led to the wide range of political arrangements outlined above.

Social complexity brought new interests to leaders. One reality had always existed: households, families, and commoner lineages, or the "subsistence economy," in which most people lived. With complexity, a conceptually new "level" appeared above it, called the "political economy" (Johnson and Earle 2000). This caused cohesion problems. Leaders using network strategies lowered or ignored cohesiveness between themselves and those living nearby. Corporate strategies explicitly increased cohesion. Furthermore, "the demographic scale [was] beyond … the one that can be organized by personal relationships in face-to-face interaction" (Grinin and Korotayev 2017a:69), so polygroups became abstract entities. Individuals could know hundreds of people across several villages—basically everyone; in polygroups of thousands, this was impossible. Still, leaders, whether individual or collective, could take a "bird's-eye view" of the society that the average person could not. Ceremonies involving thousands of people, activating deeply felt supernatural beliefs, enabled everyone to mentally encompass the entire society (perhaps in person in huge crowds) and make these "imagined communities" (Anderson 1991) outside of the ceremony, i.e., most of the time. This might increase

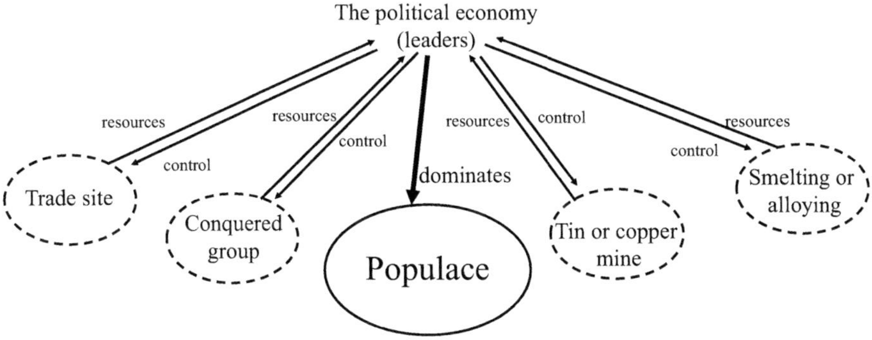

Fig. 5.7 External resources in a network strategy

cohesion. In contrast, network strategies were unstable from competition and from resentment since some people had more than others without any reciprocal benefits.

I suggest that new interests from creating a political economy might have had social–psychological consequences in polygroups. Villages often fought over land for survival, while polygroup leaders often fought over land and people to extract surplus. The larger the land they controlled, the more spoils they could access. Also, the sheer demographic size of polygroups (Grinin and Korotayev 2017a) made even minor emotional connection with every person impossible. Leaders could treat people as objects, without even a faked empathy to maintain their reputations. I think both changes facilitated aggrandizing through warfare. Treating people as objects would apply even more to outsiders, so while leaders could use dehumanizing ideologies as internal justifications for inequality, they could also use them to motivate troops to fight. Ideologies dehumanizing outsiders could have involved moral appeals to protect insiders: loved ones, homelands, gods, and others (Fiske and Rai 2015). For instance, many ethnographically studied villagers and polygroups in the Americas called themselves "human beings," implying that any outsiders were not (Smith 2011:58–59).

Complex societies appeared in multiple political forms. Some had hierarchies in which leaders had assistants; others formed informal alliances or confederacies based on clans or secret societies; and "monosettlements" were internally organized hierarchically. Many created a political economy, a group of leaders with motivations different from average people. Some leaders used assistants to provide public goods; others relied on unstable external resources. Warfare increased, which may have been aided by dehumanizing ideologies and impersonal relations.

Polygroup Economies

Polygroup economies were not dramatically different from those in villages, but they were more intensive and productive. They continued to rely on subsistence production; trade increased in size and distance; and the amount and quality of elite consumption goods increasingly displayed differences in rank.

The three basic forms of subsistence production—collecting, horticulture, and pastoralism—continued. Herds use a lot of land, usually grassland too dry to farm. There are three basic types of pastoralists. Nomadic pastoralists rely almost exclusively on their herds. They gather plants and are highly mobile. Second, seminomadic pastoralists have a seasonal round called transhumance, often farming along their route. Finally, agropastoralists "live in a permanent village and grow crops" (Sutton and Anderson 2010:228) while moving animals between pastures locally. Since a lactating cow will produce roughly four times as many calories in milk as she will in meat (McNeill and McNeill 2003), meat became a special-occasion food for pastoralists (Hayden 2003), while milk and (sometimes) blood were daily foods. Pastoralists obviously ate plant food, grown themselves or acquired through trade. Pastoralists slaughtered most bulls after two years, but let cows live much longer (Hayden 2003).

Secondary animal products developed more. Milking practices arrived in England after 5000 BCE (7000 BP), probably starting with sheep or goats (Greenfield and Arnold 2014). Traction of cattle-pulled weighted sledges seems to have originated in Anatolia around that time, which replaced the practice of having cattle step on grain to thresh it (separate the kernel from the stalk). Later, people used sheep as pack animals, a major increase in energy use. Mesopotamians may have also used animals before 4000 BCE (6000 BP) for pulling ards (early plows) (Brooke 2014; Scarre 2018a). In China, land unsuitable for farming became exploitable with herd animals being used primarily for wool (Li et al. 2014). Central Asian nomads domesticated horses by 4000 BCE (6000 BP), perhaps using them to manage herds 500 years later (Baumer 2012). From its invention around 3500 BCE (5500 BP), the wheel spread from Mesopotamia into Eastern Europe and Central Asia with cattle-pulling wagons (Baumer 2012).

Settled polygroups started using new agricultural technologies. After 5900 BCE (7900 BP), Mesopotamians developed irrigation systems (Matthews 2018); hillside terracing appeared in present-day Jordan by 6000 BCE (8000 BP) (Scarre 2018a:189). Village-level peoples in China used simple irrigation systems for rice paddies by 4000 BCE (6000 BP) (Barnes 2015:122). Small-scale polities in the Andes also practiced irrigation (Bogucki 1999:299). Ards (early wooden plows) made agriculture more productive (Scarre 2018a). Oxen-pulled ards effectively doubled the amount of land one farmer could plow, shifting labor in the household into more productive tasks (Bogucki 1999:228–229). The ard spread across Eurasia, into Egypt, and across northern and eastern Africa.

Peoples in the Americas and sub-Saharan Africa continued horticulture or pastoralism, often with supplementary foraging or collecting. While pastoral nomadic herds experience boom-and-bust demographic cycles from high birthrates and sudden epidemics (Hayden 2003), these disadvantages were preferable to using land too dry for farming. The invention of the wheeled wagon in Eurasia, which spread quickly (Baumer 2012:90), allowed Central Asia and East Africa to become major regions for pastoral nomads after 2000 BCE (4000 BP) (Connah 2018).

Regional divisions of labor expanded with social complexity. People around the world had always made their own tools, clothing, pottery, and artistic objects (Carballo 2012). Now full-time craft specialists (artisans) making metals, clothing, weaponry, and tools became attached to leaders, resulting in the expected increase in skill that specialization brings. They might have started by working for leaders but, over time, may have become independent, trading for a living. Either way they depended on the surplus extracted from the majority. Craft specialists appeared in Iran by 5000 BCE (7000 BP) (Matthews 2018), in the Indus Valley before 2600 BCE (4600 BP) (Coningham 2018), in China after 3000 BCE (5000 BP) (Bogucki 1999), and in Mesoamerica by 1000 BCE (3000 BP) (Pool 2012). Many produced elaborately crafted and painted ceramics; others produced textiles, figurines, jewelry, weaponry, and tools. Groups often acquired rare products through long-distance trade, while such specialists were often rare in sub-Saharan polygroups (McIntosh 1999).

Two forms of trade developed. Ethnographic inference suggests that much trade in and between villages, and perhaps polygroups, was still embedded in personal relationships as gift exchange. Most bartered objects were "staple" goods, basic foods such as wheat in Eurasia, rice in East and South Asia, and corn and beans in the Americas. They were bulky to transport and subject to rot. To handle these bulk goods, people started using some objects as general trade items, including shells, amber, beads, and cattle as "commodity monies" (Singh 2013; Davies 2016). Specialized craft production meant rarer objects began to exist and could circulate, leading to "wealth finance" based on special commodity monies (Bogucki 1999; Johnson and Earle 2000). In polygroups, metals increasingly operated this way, such as the use of copper to fashion items like "axes, spears, knives, swords, hoes, and spades" (Davies 2016:47). In Mesopotamia after 6000 BCE (8000 BP), some sites seem oriented to pottery production; obsidian made up much of what was traded. This seemed to intensify after 5000 BCE (7000 BP) (Matthews 2018:432, 434).

With elite expansion, consumption changed. As discussed, elites following corporate strategies downplayed lavish consumption; network-oriented elites publicly wore special clothes, ate special foods, had larger houses, and kept some objects to themselves—all to display rank (Bogucki 1999; Johnson and Earle 2000; Hayden 2003; LeBlanc 2006; Flannery and Marcus 2012). In graves, consumption goods varied greatly; in China, a few graves were "richly endowed," while most had a few goods (Higham 2018a:247). In western Mexico, graves had "[w]eapons, tools, ornaments, and elaborate pottery," plus figurines (Webster and Evans 2018:601). European grave goods also show such distinctions (Scarre 2018b:417–420). Elite consumption thus stimulated craft specialization and trade in prestige goods (Bogucki 1999:272).

With social complexity, economic processes and structures changed again. Humans gradually used animals for milk, blood, traction, and wool. Most societies still used digging-stick horticulture, but pulled ards began to transform farming. Pastoralists increasingly used secondary products from animals and traded with settled horticultural peoples or grew their own food. Craft producers specialized in production of elite display and practical tools. This stimulated trade within and between societies, enabling lavish elite consumption.

Kinship and Gender in Polygroups

Diverse kinship forms began to appear around the world after 6000 BCE (8000 BP). This included unilineal descent groups, both patrilineal and matrilineal; unilocal residence systems; and regular polygyny, almost exclusively among elites. Technological changes seem to have started altering gender roles; still, families had to remain flexible.

Evidence for unilineal descent groups is spotty. Corporate kinship groups seem to have existed in Mesoamerica after 2000 BCE (4000 BP) (Carballo 2012), but there is no indication of which type. However, matrilineality is a strong tendency in coastal East Asia and seems to have deep historical roots (Nelson 2013). Patrilineality seems to have appeared among pastoralists in East Africa at least by 1000 BCE (Wadley 2013). Both matrilineal and patrilineal descent groups seem to have developed at the core of the Hopewell Interaction Sphere (Claassen 2013). Many other complex peoples probably developed unilineal systems; perhaps many remained flexible.

Marriage in early polygroups is even more difficult to know. The high mortality rate and consequent importance of fertility probably made most marriages and relationships broadly heterosexual. Still, ethnographic data allows us to infer that there remained no necessary relationship between marriage and sexual expression, even in complex societies. In many places, heterosexual marriages may have existed, but kin affiliations and other relations were more important (Nelson 2013). For instance, there seem to have been homosexual relationships between Indo-European warriors (Neill 2009). Furthermore, romantic love has always been possible, but most marriages usually involved a purposeful intergroup alliance.

The gendered household division of labor seems to have remained flexible, culturally diverse, and locally dependent on circumstances in this period. In complex Mesoamerican societies, "[p]roduction was organized by gender complementarity, gender collaboration, or by the performance of identical tasks by male and female workers" (Brumfiel 2013:577) rather than by a clear male/female task binary. Early cross-cultural ethnographic research showed that pottery production was always predominantly female; more recent studies question this finding (Bolger 2013). While textile production also "strongly genders feminine," its gender division varies greatly; men and children were often part of it (Costin 2013:196). Finally, while "[w]omen warriors are unusual in most societies" (Hutson et al. 2013:61), many have existed, most famously among steppe peoples in Central Asia. Overall, in polygroups, tendencies to "gender" labor more seem to have existed, yet these depended closely on local circumstances.

Broad trends in kinship and gender in early complexity are difficult to infer. Unilineal descent systems seem to have spread, but available evidence is thin. Local circumstances seem more important in household divisions of labor, marriage, and sexuality.

Polygroup Inequalities

The lack of conclusions is frustrating because so many theories of social complexity infer lineages ethnographically, especially in inequality, which became permanent in social complexity. Ethnographically, one or a few lineages and assistants controlled more resources than average people. This unequal distribution of resources became institutionalized—taken for granted, expected—so some anthropologists refer to "ranked" societies (Flannery and Marcus 2012). A few lineages—and their individual members—were "above" all other lineages, at least in everyone's minds, and it was strongly linked to the further development of agriculture—for rice production (Cohen 2011) and for wheat, barley, corn, and other grain production.

Societies in Southwest Asia, Egypt, China, and Mesoamerica right after domestication showed villages containing dwellings of roughly the same size, and no other wealth disparities. Still, in villages, "there were wealth differences and private ownership of resources and produce" (Hayden 2003:173). Over time, especially in polygroups, a few skeletons were buried with rich, numerous grave goods, while all other burials had only a few goods. Buildings also became differentiated—both larger and smaller existing together (Bogucki 1999:233, 237–238; Higham 2018a). This evidence for elites has been found in China, Europe, and Mesoamerica.

Such wealth differences could have happened in different ways. Lineages might have made interlineage alliances, often shifting. Inequality might have increased as one lineage experienced bad luck or made bad decisions which reduced lineage resources compared to others—it would "sink" in comparison to them (Bogucki 1999). Conscious, purposeful interlineage spirals of competitive feasting were also possible (Hayden 2003), driven by the few existing aggrandizers; the pressure to produce more would have increased a lineage's or clan's wealth in the long run. Male secret societies often had entrance requirements. Over time, this might have produced the phenomenon, repeatedly observed ethnographically, of three informal ranks in a village: "rubbish men," average men, and Big Men (Flannery and Marcus 2012:112–116). The wealth of Big Men was at first in relationships (allies) which they transformed into economic resources. Continuing this process might have led to much greater economic inequalities, which the wealthy then sometimes transformed into political positions.

Many archaeologically researched polygroups developed economic elites of some sort. These include some in Mesopotamia by 4000 BCE (6000 BP) (Matthews 2018), in Egypt along the Nile at the same time (Connah 2018), in the Indus Valley after 3000 BCE (5000 BP) (Coningham 2018), by 2500 BCE (4500 BP) in China (Higham 2018b), in Mesoamerica (Webster and Evans 2018), and in the Andes (Moseley and Heckenberger 2018). There is evidence that in sub-Saharan Africa and North America, economic wealth was often decoupled from political power (McIntosh 1999; Graeber and Wengrow 2021), but even in these regions, economic wealth and political position seem to have become increasingly linked over time. Ultimately, people were born into high positions in societies in which "[a]ll lineages [were] ranked relative to one another" (Flannery and Marcus 2012:196). Sociologists call this "ascription": as individuals were born into preranked groups, each person's status was a function of group membership, regardless of individual desires or achievement. Ethnographically, the Tlingit and Haida foraging chiefdoms of the Northwest Coast in North America had three categories of persons translated into English as nobles, commoners, and slaves, with the oldest noble lineages or clans leading society. In simple chiefdoms, only one or two such lineages might exist; in complex societies, there might be several with only one lineage being the most important. However, they would be endogamous, e.g., a noble could only marry someone in another noble lineage. Commoner lineages were obviously vastly more numerous, following local rules of exogamy and endogamy. Slaves, obtained by conquest or trade, started to be relatively common in chiefdoms and might or might not have married at all (Flannery and Marcus 2012).

The social–psychological foundations of inequality are important. In bands and villages, the cognitive recognition of groups-within-groups (Gamble et al. 2014) was always based on a horizontal metaphor: one's group was within one's village, which was within a larger geographic area populated by other people. The groups were wider and wider. Dominance hierarchies had always been individual-to-individual; in social complexity, vertical thinking was applied to groups (such as lineages), which were mentally above or below each other. In many cases, elite individuals thought of themselves as better than everyone else "below" them. Anger, contempt, or disdain for enemies geographically separated now flowed to people close by—"inferiors" in one's own group, perhaps supported by new, dehumanizing ideologies (Haslam et al. 2013).

Gender inequality also became more prevalent in polygroups, but it was not automatic (Hutson et al. 2013). Ethnographically, there is merely a trend of increasing gender inequality in complex societies, as there are many gender-egalitarian examples (Bonvillain 2001). The Tlingit and Haida, mentioned above, were organized by matrilineal clans and practiced avunculocal (uncle) residence. Men were chiefs and heads of households. Feasts (potlatches) were particularly important, and both men and women could host them. There were slight cultural norms that women defer to men, yet men often relied on women's production for feasts, and women had their own property and independence from men. In contrast, the Mpondo of what is now South Africa were male dominant. While women grew food for households, men tended cattle for household and for trade. There were strong norms of female deference to males, a set of beliefs justifying them, and a double standard around sexual behavior. Pastoral nomads tended to be touchy and prone to aggressive male dominance (Hayden 2003; Cao et al. 2021).

The archaeological evidence in social complexity is more equivocal. Evidence of health disparities between women and men varies greatly across Mesoamerica, the Andes, and Eurasia; sometimes the sexes are equally healthy, and sometimes men are healthier (Hutson et al. 2013). Brumfiel (2013) finds some gender inequalities in Mesoamerica, but not always.

Archaeological and historical theorizing about gender inequality increased in the 1980s. Lerner (1986), and Coontz and Henderson (1986), argued variously that commodification starting in early complexity later led to deeper gender inequality; Lerner focused on male control of female sexuality, and Ehrenberg (1989) argued that women's economic power began reducing with the appearance of secondary products, which consequently reduced their social power. In contrast, Bonvillain (2001) argues that it is warfare that increases gender inequality. Johnson and Earle (2000) argue that extreme gender inequality appears in the "male supremacy complex":

> male monopoly over weapons, training of males for combat and bravery, female infanticide, training of females to be the passive rewards for masculine performance, patrilineal bias in property and descent, prevalence of polygyny, competitive male sports, intense male puberty rituals, ritual uncleanliness of women, bride price, and other male-centered institutions. (Johnson and Earle 2000:128)

Bonvillain (2001) argued that three factors affected gender inequality. First, when men fought close at home, they had more power; fighting at a distance, they were often gone, so women had more power (Bonvillain 2001). Second, matrilocality gave woman marital power, as kin were allies. Patrilocality meant that a woman often had only the allies she literally created—the birth of a son often provided her more status. Finally, when women produce much of the group's food, they tend toward equality; a peripheral role in food production tends toward gender inequality.

These increasing economic and gender inequalities suggest some speculations on aggrandizing. Archaeologists seem to stop analyzing aggrandizing once they have explained social complexity; as we will see, it showed up in other guises as complexity intensified. Yet, if aggrandizing personalities are randomly born everywhere, they must have kept appearing after social complexity developed, but elaborate social structures must have reshaped their behavior. I suggest that in ranked societies, aggrandizing began to take three forms. First, it continued as a rare and randomly distributed personality characteristic in all ranks. Lower-ranked aggrandizers would strive for position within their social world—a clan, a village, or a town. The few "natural" aggrandizers within any elite would attempt to push their peers toward more aggrandizing strategies or policies. Second, psychologists argue that most people have "loss aversion" (Müller and Greifeneder 2007): they value keeping what they have over risking what they might gain. Loss aversion might have led to elite ideologies that justified inequalities; most would defend their position without being natural aggrandizers (Hayden 2014:50). Finally, political scientists will recognize situational aggrandizing: a new external situation appears, forcing leaders to organize alliances to meet the threat. A new expansionist polity nearby offers any group only two options: conquer or be conquered. This is neither natural nor cultural, but it produces aggrandizing behavior anyway.

Permanent economic and gender inequalities began in polygroups, with the associated mental ranking relying on a vertical metaphor. Some societies had several elite lineages, commoners as the bulk of society, and slaves. Prehistorically, gender inequality was variable, depending on multiple factors, yet women's power reduced in at least a few complex societies.

Ideas in Polygroups

Ideas became more elaborate and multifaceted in polygroups. Rituals became larger and more public and were often led by elites, who also seem to have instituted new moralities. Information storage continued developing. Individual and collective technologies diversified.

As discussed, strategic deployment of belief-systems was sometimes part of polygroup formation. Ceremonies were meant to impress an audience and to "create a sense of common origins, purpose and destiny among strangers and to sanctify the class divisions of society" (Johnson and Earle 2000:250). Polygroups connected thousands of people who knew little of each other. Sometimes leaders needed to tie the identities of most people to themselves; other times collective projects (construction, war) were easier with higher group cohesion. Archaeologists have found evidence of ceremonies in many parts of the world, nearly always accompanied by monumental architecture; they seem to be translations of the vertical metaphor of relations between people into massive literal things reaching toward the sky. Elites restricted access to the tops or insides of pyramids, mounds, and megaliths (Hayden 2003). Temples appeared in Mesopotamia after 5900 BCE (7900 BP); over the centuries, people expanded them or built new ones above the old ones. Monuments before inequality appeared in Europe, but they were increasingly associated with elites (Scarre 2018b). Between 1800 and 400 BCE (3800–2400 BP), monumental architecture decorated in supernatural motifs developed in the Andes (Moseley and Heckenberger 2018) and in Mesoamerica after 1200 BCE (3200 BP) (Webster and Evans 2018).

Belief-systems in complex societies remained broadly animistic, but just as villagers modified them with ancestor veneration, now leaders emphasized their ancestors as influential spirits (Hayden 2003). Within such belief-systems, successful leaders claiming the approval of their ancestors was logical. Consequently, massive ceremonies and monumental architecture had multiple purposes: Increasing social cohesion; reassurance that someone was in charge; and instilling terror in populaces since human sacrifice increasingly supported complexity (Chirot 2012; Watts et al. 2016). Gods "demanding" sacrifices cowed people into submission.

In contrast, "little is known about the rituals and religions of common people" (Hayden 2003:240). For instance, many archaeological sites in Europe show that those buried with few grave goods, i.e., commoners, may have been concerned with fertility, as many phallic symbols, and other ritual and practical objects, were often buried. Scholars speculate that they were used to ask the spirits to intercede "to achieve fertility, wealth, respite from adverse conditions, or personal goals" (Hayden 2003:240). Everyday cults were likely practiced in home alongside elite-managed, large-scale, megalith-raising feasts; as elites increasingly had high-level shamans, villages likely retained local shamans for their inhabitants (Hayden 2003).

It remains difficult, perhaps impossible, to reconstruct morality systems among early polygroups, but ethnographic inference offers clues. As discussed, reciprocal obligations in villages had often narrowed to corporate kin groups. However, leaders had to move the mass of people away from "an intuitive sense of sociability, egalitarianism, and reciprocity, to sources *outside* the individual" (Friesen 2015:30, original italics) in order to legitimate their new, unequal positions. Thus, leaders' late ancestors (supposedly) became concerned with the moral behavior of the average person. This required combining the previously separate religion with morality (Friesen 2015). Leaders claimed to know the ancestors' wills and turned these into moral codes. Thus, moral systems were perhaps more complex: locally, within the village or kin group, and regionally, with moral obligations to leaders. Interestingly, elites in secret societies and perhaps beyond violated moral codes with impunity, engaging in sexual exploitation, murder, property destruction, physical attacks, and even cannibalism (Hayden 2018:344).

Technology changed in two ways. First, information storage began to become more elaborate. Elites used proto-writing differently in Mesopotamia from previous villager practices. Mesopotamian tokens were found in elite graves (Schmandt-Besserat 1996). Egyptians developed an early form of true writing sometime around 3000 BCE (5000 PB) (Connah 2018:367), and high-status ceramics with pictograms existed in China a little later (Barnes 2015:143, 146; Higham 2018b:554). This was part of the larger movement to elite goods. Second, some societies started using metals. People were smelting copper (superheating the ore, which separated the metal from the ore) at Çatalhöyük as well as lead, "a metal not native to Turkey" (Alcock and Cherry 2018:473). In what is now Bulgaria, people worked silver and gold from the fifth millennium BCE; use and display of copper, silver, and gold spread throughout Europe (Bogucki 1999). Eventually peoples across Eurasia, perhaps in Africa, and in the Americas learned to smelt copper. Complex American societies worked copper and gold, yet metals never became practically important; they continued relying on stone (Bogucki 1999).

Finally, both individual and collective technologies helped increase production. The ard pulled by oxen scraped the surface faster and deeper than any handheld digging stick could. Irrigation took two forms: storage and release of rain or floodwaters, or canals diverting river flows. They needed "constant cleaning and repair" (Scarre 2018a:189). Terracing—which appeared in rainier, hillier areas of Southwest Asia, Mesoamerica, the Andes, and Southeast Asia—is a method of cutting flat levels out of hillsides in "steps" and shoring them up with retaining walls, which produces high yields. Monumental architecture was a form of collective technology, and complex societies developed techniques to build them over time. People continued making rock art, figurines, jewelry, and housing decoration. However, pottery became increasingly elaborate. Metals made possible new jewelry and other decorations.

Even while important ideological and technical changes occurred, much remained similar. Belief-systems remained fundamentally supernatural. Elites increasingly emphasized their legitimacy through public religious ceremonies, sometimes appealing to their own ancestors as godlike beings. Morality seems to have become more complex. Information-storage systems, metals, and new agricultural systems developed, as did new forms of artistic expression.

The Demographics of Complexity

A similar story was true demographically: there were real continuities and major changes. Mortality and fertility remained high. Demographic concentration in towns and small cities was new, and both the size of groups and total global population increased greatly.

The health decline after the agricultural transition took thousands of years (Cohen and Crane-Kramer 2007), so as social complexity increased, mortality probably increased or stayed the same. Going to war created risks, but it also created opportunities for warriors; many people might see the risk as worth it. Likewise, fertility also likely went up slightly; every intensification of farming brought slight fertility increases and perhaps slightly better child survival (Brooke 2014). "High incidences of malaria" found in skeletons in Mesopotamia between 6000 and 4200 BCE (8000–6200 BP) suggest "devastating diseases were a recurring feature" of the period (Matthews 2018:432). Concentrating populations in large settlements spread diseases very quickly and perhaps originated "endemic childhood diseases" (Brooke 2014:241). Grief and fear probably remained high, as the increase in warfare raised the stakes and risks in people's lives.

Societies increased in size in three ways. First, multivillage groups are obviously larger than single villages. Different authors suggest different numbers, but they are always in the thousands of people (Peterson and Drennan 2012; Scarre 2018a). Second, some argue that population density and political structure correlate: the denser the society, the more formal political structure (Lewellen 2003). Their logic is simple: more people on the same plot of land produces more conflict and land overuse, and leaders can adjudicate disputes, set up formal rules of interaction, and help people extract more from the earth. Some disagree; sub-Saharan Africa polygroups seem to have experienced an *inverse* relationship between density and formalization (McIntosh 1999). Instead, management of sheer numbers and increasing interconnections produced political formalization. In general, though, polygroups were denser than villages, though that density varied. Pastoral nomads in large grassy areas might have densities of two persons per square mile, while (ethnographically) the Trobriand Islanders had densities of 100 people per square mile, and Kali Loro had upwards of 1850 people per square mile. Finally, total global population increased many times through this period (Brooke 2014; Livi-Bacci 2017). Global population around 9000 BCE (11,000 BP) was somewhere around seven million people, while six thousand years later (around 3000 BCE, or 5000 BP), it was about 38 million (Brooke 2014:259). This population was concentrated in originally domesticating societies, which, thousands of years later, also started on the path to hereditary inequality: Mesopotamia, China, the Nile Valley, India, Mesoamerica, and Peru.

Multiple cities, towns, and villages were usually organized in a hierarchy, following the political hierarchy. In chiefdoms, the chief lived in the largest settlement, and subchiefs lived in smaller settlements, although still larger than villages. Specialist assistants often lived near leaders, not in villages. Unlike Çatalhöyük, towns and cities had a spatial division of labor: in addition to large residential areas, there were "specialized areas for craft production, public meetings, markets, large temples [and] special residences for elite rulers" (Matthews 2018:438)—in other words, neighborhoods. Many cities were planned settlements, such as those in the Indus Valley (Coningham 2018:525–528). Likewise, towns and cities became places to which people migrated, whether pushed by warfare or pulled by opportunity. People now met strangers daily (Smith 2019), creating both opportunities and difficulties.

Pastoral nomads had multiple demographic patterns. The most mobile followed an older form, fission–fusion dynamics, coming together in large, temporary settlements and otherwise dispersing in low-density living. Some remained transhumant, moving in small nomadic camps with the seasons; others had permanent settlements but moved part of the year; and some were sedentary. Ethnographically, there is a direct but loose relationship between sedentism and settlement size; as one increased, the other generally did (Galaty and Johnson 1990). The maps (above) showing "villages, sedentary and nomadic" include such quasi-sedentary pastoralists, as well as horticulturalists and agropastoralists. Furthermore, pastoralists traded with settled horticulturalists for plant foods, while others grew their own, becoming semisedentary agropastoralists; gathering wild plants was still often an option as well.

Social complexity started true urban living. While mortality and fertility changed only slightly even though diseases became even more prevalent, the human population grew, and concentrations of humans provided new experiences.

Environment and Complexity

The environmental problems we saw debated earlier started to show up in polygroups. These included salinization, waterlogging, deforestation and desertification, and perhaps species extinction. Farmers replaced natural vegetation with single crops and often used only one species—sheep, goats, cattle, or pigs—in a practice called "monoculture." This often led to soil erosion and overgrazing. In addition, increased populations made permanent buildings, fortifications, tools, and fires out of wood, leading to deforestation (Headrick 2020). Between 6000 and 1000 BCE (8000–3000 BP), deforestation occurred in the Mediterranean basin, in western Europe, in Mesopotamia, and in China. Over time, as the human population grew, these patterns intensified. Also, had previous cycles of greenhouse gases and thus global warming and cooling continued, the earth would have entered a new ice age around 3000 BCE (5000 BP). One climate scholar argues that human farming created enough greenhouse gases to prevent this, so the "Holocene" climate is in fact a human creation (Brooke 2014:286–287; Headrick 2020:60–65).

The Larger Context of Social Complexity

Polygroups existed within larger cultural and political-economic geographic patterns, extending and changing village patterns. Regional cultural patterns became more elaborate, more entrenched, and wider geographically; trade and political interaction systems also expanded.

Social complexity accelerated regional cultural differentiation. Languages were still diverse around the world, but clear regional traditions in other areas of life started to develop. For instance, in Mesopotamia in the Ubaid period (5900–4200 BCE [7900–6200 BP]), temples became extremely important, overseeing "many aspects of daily life, including land and labor management, distribution of food, and … the correct procedures for religious rites" (Matthews 2018:434). Temples and other traditions, such as pottery styles, spread northwest up the Tigris and Euphrates valleys and southeast into the Persian Gulf. Rather than build new temples on old ones, Egyptian polygroups after 4000 BCE (6000 BP) saw "ephemeral building materials used," especially along the lower Nile (Connah 2018:366). Two millennia later, Andean polygroups developed without pottery or woven clothing, yet two kinds of distinctive monumental architecture appeared. Common were steep pyramids with flat tops; less common, but more distinctive, were "semi-subterranean ritual structures," i.e., "sunken courts" (Moseley and Heckenberger 2018:640–641), that seemed to embody beliefs about humans emerging from within Mother Earth. Finally, Mesoamericans developed an array of elements: courts for ball games, "prismatic" obsidian knives, and the massive carved stone heads of the Olmecs (Webster and Evans 2018:594–598). Each region also developed a staple crop: rice in China, corn (maize) in Mesoamerica, potatoes in the Andes, and barley and wheat in Mesopotamia and Egypt.

Polygroups rarely existed alone. Villages had always traded, and complex societies sometimes fostered rivals through "secondary" formation. Warfare consequently increased, but so did trade—in both amount and distance. Many goods moved sometimes hundreds of miles in a "down-the-line" fashion, that is, moving from group to group to group. In addition, long-distance trade consisted primarily of prestige goods that were usually easily transportable such as metals, rare animal products, "exotic" goods found only in one place, and so on (Hayden and Earle 2022); typically these were rare objects from far away. Some of this occurred along coastlines in small boats, and river transportation also became more important. Commodity chains for goods such as bronze, shells, obsidian, pottery, and textiles were the foundations for long-distance trade networks. These networks naturally consisted of two-way trade—one commodity going one way, another the other. Triangular trade, and more complex patterns, were obviously also possible.

Trade systems of hundreds and thousands of miles developed. (This section is taken from Smith 2009.) Some extended from Italy to Sweden; others linked Egypt, Mesopotamia, and the Indian Ocean. Trade in obsidian (volcanic glass) through Southwest Asia was especially intense but also occurred in central Europe (e.g., Poland and Hungary) and what is now Italy and Greece, hundreds of miles from their sources. Deposits of tin and copper, the elements of bronze, never occurred in the same place. Leaders thus took control of the copper and tin trades, and of bronze manufacture. When metal tools superseded obsidian, amber—sourced in Scandinavia—became an important prestige good, sent to Britain, Greece, and Turkey. There was interaction across the entirety of Eurasia by 2000 BCE (4000 BP), and perhaps earlier (Bogucki 1999:271). Chiefdoms straddling the desert between the Pacific and the Andes stimulated food trade between these zones, and trade extended beyond the eastern slopes of the Andes into the Amazon basin. Likewise, the Olmec people between 1500 and 500 BCE (3500–1500 BP) engaged in long-distance trade "over a wide part of Mesoamerica" (Bogucki 1999:273).

Some scholars argue that this trade expansion led to "world-systems," unequal trade systems with cores and peripheries rather than peer–polity interaction spheres. Supposedly, the interaction sphere in Southwest Asia discussed previously was a world-system, and by 4000 BCE (6000 BP), it had links into the Mediterranean, through the Persian Gulf to the Indus Valley, and into Central Asia. The development of social complexity occurred within it (Grinin and Korotayev 2017b). Early on, ideas flowed more than trade items. Still, other regions started to experience larger trade and interaction systems, especially in China and Mesoamerica. Whether these amounted to world-systems or merely peer–polity interactions spheres is still debated.

Conclusion

The Urban Revolution was a true social transformation from life in villages to life in, or subject to, towns and cities. It usually but not always created a distinction between "rural" and "urban," and all that followed made societies more complex. Researchers suggest multiple theories of complexity based on those changes: populations grew and concentrated spatially, warfare increased, technologies became more complex, permanent inequalities developed, trade increased, corporate kin groups became common, and women's positions in relation to men started to erode in a few regions. All these were possible because of human capabilities of concepts and beliefs, imagination of past and future, moral codes, technological innovation, and nested and interwoven groups. Still, mortality and fertility remained high. Most people sustained a supernatural attitude toward the world and generally remained farmers or foragers.

The character of groups changed with social complexity. While based on the human group-within-a-group capacity, we might term multivillage collectivities *societies* rather than *groups*, given the sheer number of people included and the complex relations uniting them. But how cohesive or united were they? Stanish (2017) downplays the power of leadership; Graeber and Wengrow (2021) argue that people in some places reversed harsh inequalities and retained personal autonomy. Others (Flannery and Marcus 2012; Carneiro 2017) think the inequalities were consistently sharp and harsh, reducing autonomy. In many places, polygroups took centuries of ongoing violence to establish permanently, suggesting resistance and an ongoing tension between cohesion and conflict based on arrogance among the few and resentment among the many. Leaders often deployed (manipulated?) preexisting supernatural beliefs that explained the world, and often reduced the fear and grief of constant death, to foster cohesion and to soothe the sense of unfairness that inequality generated. Still, monumental architecture turned the metaphoric placement of people "higher" and "lower" into something literal, a visual reminder of inequalities. Furthermore, while most inequalities were geographically or monumentally marked, wherever gender inequality developed, it existed within every household or kin group (De Beauvoir [1949] 1952). Yet, it did not automatically appear in every complex society.

Indeed, a global survey of human groups five millennia ago—impossible at the time, but retrospectively possible now—reveals increasing diversity. Many still lived as mobile foragers in most of the Americas, much of Africa and northern Asia, and Australia. In much of Central, South, and East Asia, in much of Europe, in Southeast Asia and New Guinea, and a few places in the Americas, more peoples lived as horticulturalists, pastoralists, or foragers in villages. Only in a few places—Southwest Asia, Egypt, parts of East Asia, Mesoamerica, and the Andes—did people live in increasingly complex societies; sometimes these were linked as segments or confederations rather than as hierarchies. Instead of a "natural" progression everywhere of increasing size and complexity, only a few complex societies appeared in only a few places—only under unusual circumstances. Still, in a few thousand more years, social complexity intensified.

References

Alcock, Susan E., and John F. Cherry. 2018. The Mediterranean World. In *The Human Past: World Prehistory and the Development of Human Societies*, ed. C. Scarre, 4th ed., 469–514. London: Thames and Hudson.

Anđelković, Branislav. 2011. Political Organization of Egypt in the Predynastic Period. In *Before the Pyramids: The Origins of Egyptian Civilization*, ed. E. Teeter, 25–32. Chicago: The Oriental Institute of the University of Chicago.

Anderson, Benedict Richard O' Gorman. 1991. *Imagined Communities: Reflections on the Origin and Spread of Nationalism*. New York: Verso.

Bahn, Paul, ed. [2000] 2009. *Atlas of World Archaeology*. London: Brown Reference.

Barnes, Gina L. 2015. *The Archaeology of East Asia: The Rise of Civilization in China, Korea, and Japan*. Havertown, PA: Oxbow Books.

Baumer, Christopher. 2012. *A History of Central Asia: The Age of the Steppe Warriors*. Vol. I. New York: I. B. Tauris.

Bellwood, Peter. 2013. *First Migrants: Ancient Migration in Global Perspective*. Malden, MA: Wiley Blackwell.

Bellwood, Peter, and Peter Hiscock. 2018. Australia and the Indo-Pacific Islands During the Holocene. In *The Human Past: World Prehistory and the Development of Human Societies*, ed. C. Scarre, 4th ed., 261–302. London: Thames and Hudson.

Berea, Anamaria. 2010. Trade as a Premise for Social Complexity. *Journal of the Washington Academy of Sciences* 102 (2): 17–38.

Black, Jeremy, ed. 2005. *World History Atlas*. 2nd ed. New York: Dorling Kindersley.

Blanton, Richard E., Gary M. Feinman, Stephen A. Kowalewski, and Peter N. Peregrine. 1996. A Dual-Processual Theory for the Evolution of Mesoamerican Civilization. *Current Anthropology* 37 (1): 1–14.

Bogucki, Peter. 1999. *The Origins of Human Society*. Malden, MA: Blackwell.

Bolger, Diane. 2013. Gender, Labor, and Pottery Production in Prehistory. In *A Companion to Gender Prehistory*, ed. D. Bolger, 161–179. Malden, MA: Wiley-Blackwell.

Bonvillain, Nancy M. 2001. *Women and Men: Cultural Constructs of Gender*. Upper Saddle River, NJ: Prentice Hall.

Brooke, John L. 2014. *Climate Change and the Course of Global History: A Rough Journey*. New York: Cambridge University Press.

Browman, David L., Gayle J. Fritz, and Brie Anna S. Langlie. 2018. Origins of Food Producing Economies in the Americas. In *The Human Past: World Prehistory and the Development of Human Societies*, ed. C. Scarre, 4th ed., 303–343. London: Thames and Hudson.

Brumfiel, Elizabeth M. 2013. The Archaeology of Gender in Mesoamerica: Moving Beyond Gender Complementarity. In *A Companion to Gender Prehistory*, ed. D. Bolger, 564–584. Malden, MA: John Wiley & Sons, Inc.

Cao, Yiming, Benjamin Enke, Armin Falk, Paola Giuliano, and Nathan Nunn. 2021. Herding, Warfare, and a Culture of Honor: Global Evidence. Working Paper No. 29250, National Bureau of Economic Research, Cambridge, MA.

Carballo, David. 2012. Households in Ancient Mesoamerica: Domestic Social Organization, Status, Economies, and Rituals. In *The Oxford Handbook of Mesoamerican Archaeology*, ed. D. L. Nichols and C. A. Pool, 684–696. New York: Oxford University Press.

Carneiro, Robert. 1970. A Theory of the Origin of the State. *Science* 169 (3947): 733–738.

Carneiro, Robert L. 2017. The Chiefdom in Evolutionary Perspective. In *Chiefdoms: Yesterday and Today*, ed. R. L. Carneiro, L. E. Grinin, and A. V. Korotayev, 15–60. Clinton Corners, NY: Eliot Werner.

Carneiro, Robert L., Leonid E. Grinin, and Andrey V. Korotayev, eds. 2017. *Chiefdoms: Yesterday and Today*. Clinton Corners, NY: Eliot Werner.

Chacon, Richard J., and Ruben G. Mendoza, eds. 2017. *Feast, Famine, or Fighting? Multiple Pathways to Social Complexity*. New York: Springer.

Childe, V. Gordon. (1936) 1966. *Man Makes Himself*. 4th ed. Bungay, Suffolk: Fontana.

Chirot, Daniel. 2012. *How Societies Change*. 2nd ed. Thousand Oaks, CA: SAGE/Pine Forge Press.

Claassen, Cheryl. 2013. Gender in the Prehistory of the Eastern United States. In *A Companion to Gender Prehistory*, ed. D. Bolger, 527–543. Malden, MA: John Wiley & Sons.

Cohen, David Joel. 2011. The Beginnings of Agriculture in China: A Multiregional View. *Current Anthropology* 52 (S4): S273–S293.

Cohen, Mark Nathan, and Gillian M. M. Crane-Kramer. 2007. Editors' Summation. In *Ancient Health: Skeletal Indicators of Agricultural and Economic Intensification*, ed. M. N. Cohen and G. M. M. Crane-Kramer, 320–343. Gainesville: University Press of Florida.

Coningham, Robin. 2018. South Asia: From Early Villages to Buddhism. In *The Human Past: World Prehistory and the Development of Human Societies*, ed. C. Scarre, 4th ed., 515–546. London: Thames and Hudson.

Connah, Graham. 2018. Holocene Africa. In *The Human Past: World Prehistory and the Development of Human Societies*, ed. C. Scarre, 4th ed., 344–387. London: Thames and Hudson.

Coontz, Stephanie, and Peta Henderson, eds. 1986. *Women's Work, Men's Property: The Origins of Gender and Class*. London: Verso.

Costin, Cathy Lynne. 2013. Gender and Textile Production in Prehistory. In *A Companion to Gender Prehistory*, ed. D. Bolger, 180–202. Malden, MA: John Wiley & Sons.

Davidson, Ian. 2014. Hunter-Gatherers in Australia: Deep Histories of Continuity and Change. In *The Oxford Handbook of the Archaeology and Anthropology of Hunter-Gatherers*, ed. V. Cummings, P. Jordan, and M. Zvelebil, 368–404. New York: Oxford University Press.

Davies, Glyn. 2016. *A History of Money*. 4th ed. Revised and updated by Duncan Connors. Cardiff: University of Wales Press.

De Beauvoir, Simone. (1949) 1952. *The Second Sex*. New York: Bantam Books.

Earle, Timothy K. 2017. Chiefs, Chieftaincies, Chiefdoms, and Chiefly Confederacies: Power in the Evolution of Political Systems. In *Chiefdoms: Yesterday and Today*, ed. R. L. Carneiro, L. E. Grinin, and A. V. Korotayev, 233–256. Clinton Corners, NY: Eliot Werner.

Ehrenberg, Margaret. 1989. *Women in Prehistory*. Norman and London: University of Oklahoma Press.

Fagan, Brian M. 2004. *People of the Earth: An Introduction to World Prehistory*. 11th ed. Upper Saddle River, NJ: Pearson-Prentice Hall.

Fargher, Lane F., and Verenice Y. Heredia Espinosa. 2016. *Alternative Pathways to Complexity: A Collection of Essays on Architecture, Economics, Power, and Cross-Cultural Analysis in Honor of Richard E. Blanton*. Boulder: University Press of Colorado.

Feinman, Gary M. 2017. Multiple Pathways to Large-Scale Human Cooperative Networks: A Reframing. In *Feast, Famine, or Fighting? Multiple Pathways to Social Complexity*, ed. R. J. Chacon and R. G. Mendoza, 459–478. New York: Springer.

Feinman, Gary M., and Linda M. Nicholas. 2016. Framing the Rise and Variability of Past Complex Societies. In *Alternative Pathways to Complexity: A Collection of Essays on Architecture, Economics, Power, and Cross-Cultural Analysis in Honor of Richard E. Blanton*, ed. L. Fargher and V. Y. Heredia Espinosa, 271–290. Boulder: University Press of Colorado.

Ferguson, R. Brian. 2006. Archaeology, Cultural Anthropology, and the Origins and Intensification of War. In *The Archaeology of Warfare: Prehistories of Raiding and Conquest*, ed. E. N. Arkush and M. W. Allen, 469–523. Gainsville: University Press of Florida.

Fiske, Alan Page, and Tage Shakti Rai. 2015. *Virtuous Violence: Hurting and Killing to Create, Sustain, End, and Honor Social Relationships*. New York: Cambridge University Press.

Flannery, Kent, and Joyce Marcus. 2012. *The Creation of Inequality: How Our Prehistoric Ancestors Set the Stage for Monarchy, Slavery, and Empire*. Cambridge, MA: Harvard University Press.

Friesen, Bruce K. 2015. *Moral Systems and the Evolution of Human Rights*. New York: Springer.

Galaty, John G., and Douglas L. Johnson. 1990. Introduction: Pastoral Systems in Global Perspective. In *The World of Pastoralism: Herding Systems in Comparative Perspective*, ed. J. G. Galaty and D. L. Johnson, 1–31. New York: Guildford Press.

Gamble, Clive, John Gowlett, and Robin Dunbar. 2014. *Thinking Big: How the Evolution of Social Life Shaped the Mind*. London: Thames & Hudson.

Graeber, David, and David Wengrow. 2021. *The Dawn of Everything: A New History of Humanity*. New York: Farrar, Straus, and Giroux.

Greenfield, Haskell J., and Elizabeth R. Arnold. 2014. The Secondary Products Revolution, Mortality Profiles, and Practice of Zooarchaeology. In *Animal Secondary Products: Domestic Animal Exploitation in Prehistoric Europe, the Near East and the Far East*, ed. H. J. Greenfield, 186–205. Havertown, PA: Oxbow Books.

Grinin, Leonid E., and Andrey V. Korotayev. 2017a. Chiefdoms and Their Analogues: Alternatives of Social Evolution at the Societal Level of Medium Complexity. In *Chiefdoms: Yesterday and Today*, ed. R. L. Carneiro, L. E. Grinin, and A. V. Korotayev, 63–108. Clinton Corners, NY: Eliot Werner.

Grinin, Leonid E., and Andrey V. Korotayev. 2017b. The Afro-Eurasian World-System: Genesis, Transformation, Characteristics. In *Routledge Handbook of World-Systems Analysis*, ed. S. J. Babones and C. Chase-Dunn, 30–38. New York: Routledge.

Hansen, Mogens Herman. 2000. Preface. In *A Comparative Study of Thirty City-State Cultures: An Investigation*, ed. M. H. Hansen, 9–10. Copenhagen: Kongelige Danske Videnskabernes Selskab.

Haslam, Nick, Steven Loughnan, and Elise Holland. 2013. The Psychology of Humanness. In *Objectification and (De)Humanization. 60th Nebraska Symposium on Motivation*, ed. S. J. Gervais, 25–51. New York: Springer.

Hayden, Brian. 2003. *Shamans, Sorcerers and Saints: A Prehistory of Religion*. Washington, DC: Smithsonian.

Hayden, Brian. 2014. *The Power of Feasts: From Prehistory to the Present*. New York: Cambridge University Press.

Hayden, Brian. 2018. *The Power of Ritual in Prehistory: Secret Societies and Origins of Social Complexity*. New York: Cambridge University Press.

Hayden, Brian, and Timothy Earle. 2022. Political Economy Perspectives in Trade Before and Beyond Civilizations. In *Trade Before Civilization*, ed. J. Ling, R. J. Chacon, and K. Kristiansen, 385–415. New York: Cambridge University Press.

Haywood, John, ed. 2011. *The New Atlas of World History: Global Events at a Glance*. Princeton, NJ and Oxford: Princeton University Press.

Headrick, Daniel R. 2020. *Humans Versus Nature: A Global Environmental History*. New York: Oxford University Press.

Higham, Charles. 2018a. East Asian Agriculture and Its Impact. In *The Human Past: World Prehistory and the Development of Human Societies*, ed. C. Scarre, 230–260. London: Thames and Hudson.

Higham, Charles. 2018b. Complex Societies of East and Southeast Asia. In *The Human Past: World Prehistory and the Development of Human Societies*, ed. C. Scarre, 4th ed., 547–589. London: Thames and Hudson.

Hommel, Peter. 2018. What's the Point? Globalization and the Emergence of Ceramic-Using Hunter-Gatherers in Northern Eurasia. In *Globalization in Prehistory: Contact, Exchange, and the "People Without History"*, ed. N. Boivin and M. D. Frachetti, 15–42. New York: Cambridge University Press.

Houston, Rob, ed. 2018. *History of the World, Map by Map*. New York: Dorling Kindersley.

Hutson, Scott R., Bryan K. Hanks, and K. Anne Pyburn. 2013. Gender, Complexity, and Power in Prehistory. In *A Companion to Gender Prehistory*, ed. D. Bolger, 45–67. Malden, MA: John Wiley & Sons.

Jain, V. K. 2006. *Prehistory and Protohistory of India: An Appraisal: Palaeolithic—Non-Harappan—Chalcolithic Cultures*. New Delhi: D.K. Printworld.

Johnson, Allen W., and Timothy Earle. 2000. *The Evolution of Human Societies: From Foraging Group to Agrarian State*. 2nd ed. Stanford, CA: Stanford University Press.

Kelly, Raymond C. 2000. *Warless Societies and the Origin of War*. Ann Arbor: The University of Michigan Press.

Köhler, E. Christiana. 2011. The Rise of the Egyptian State. In *Before the Pyramids: The Origins of Egyptian Civilization*, ed. E. Teeter, 123–126. Chicago: The Oriental Institute of the University of Chicago.

Larson, Greger, Dolores R. Piperno, Robin G. Allaby, Michael D. Purugganan, Leif Andersson, Manuel Arroyo-Kalin, Loukas Barton, et al. 2014. Current Perspectives and the Future of Domestication Studies. *Proceedings of the National Academy of Sciences* 111 (17): 6139–6146.

LeBlanc, Steven A. 2006. Warfare and the Development of Social Complexity: Some Demographic and Environmental Factors. In *The Archaeology of Warfare: Prehistories of Raiding and Conquest*, ed. E. N. Arkush and M. W. Allen, 437–468. Gainsville: University Press of Florida.

Lemke, Ashley K. 2018. Hunter-Gatherers and Archaeology. In *Foraging in the Past: Archaeological Studies of Hunter-Gatherer Diversity*, ed. A. K. Lemke, 3–18. Louisville: University Press of Colorado.

Lerner, Gerda. 1986. *The Creation of Patriarchy*. New York: Oxford University Press.

Lewellen, Ted C. 2003. *Political Anthropology: An Introduction*. 3rd ed. Westport, CT: Praeger.

Li, Zhipeng, Roderick B. Campbell, Katherine R. Brunson, Jie Yang, and Yang Tao. 2014. The Exploitation of Domestic Animal Products from the Late Neolithic Age to the Early Bronze Age in the Heartland of Ancient China. In *Animal Secondary Products: Domestic Animal Exploitation in Prehistoric Europe, the Near East and the Far East*, ed. H. J. Greenfield, 56–79. Havertown, PA: Oxbow Books.

Livi-Bacci, Massimo. 2017. *A Concise History of World Population*. 6th ed. Chichester, UK: Wiley-Blackwell.

Matthews, Roger. 2018. Peoples and Complex Societies of Ancient Southwest Asia. In *The Human Past: World Prehistory and the Development of Human Societies*, ed. C. Scarre, 4th ed., 429–468. London: Thames and Hudson.

McIntosh, Susan Keech. 1999. Pathways to Complexity: An African Perspective. In *Beyond Chiefdoms: Pathways to Complexity in Africa*, ed. S. K. McIntosh, 1–30. New York: Cambridge University Press.

McNeill, J. R., and William H. McNeill. 2003. *The Human Web: A Bird's-Eye View of World History*. New York: W. W. Norton.

Milner, George R., and W. H. Wills. 2018. Complex Societies of North America. In *The Human Past: World Prehistory and the Development of Human Societies*, ed. C. Scarre, 4th ed., 670–702. London: Thames and Hudson.

Moore, R. I., ed. 1995. *Atlas of World History*. Rand McNally.

Moseley, Michael E., and Michael J. Heckenberger. 2018. From Village to Empire in South America. In *The Human Past: World Prehistory and the Development of Human Societies*, ed. C. Scarre, 4th ed., 636–669. London: Thames and Hudson.

Müller, Patrick A., and Rainer Greifeneder. 2007. Loss Aversion. In *Encyclopedia of Social Psychology*, ed. R. Baumeister and K. Vohs, 535–536. Thousand Oaks, CA: SAGE.

Neill, James. 2009. *The Origins and Role of Same-Sex Relations in Human Societies*. Jefferson, NC and London: McFarland & Company.

Nelson, Sarah Milledge. 2013. Gender and Archaeology in Coastal East Asia. In *A Companion to Gender Prehistory*, ed. D. Bolger, 333–350. Malden, MA: John Wiley & Sons.

O'Brien, Patrick, ed. 2010. *Atlas of World History*. 2nd ed. New York: Oxford University Press.

Pauketat, Timothy R. 2007. *Chiefdoms and Other Archaeological Delusions*. Lanham, MD: AltaMira Press.

Peregrine, Peter. 2012. Power and Legitimation: Political Strategies, Typology, and Cultural Evolution. In *The Comparative Archaeology of Complex Societies*, ed. M. E. Smith, 165–191. New York: Cambridge University Press.

Peregrine, Peter N., and Carol R. Ember. 2016. Network Strategy and War. In *Alternative Pathways to Complexity: A Collection of Essays on Architecture, Economics, Power, and Cross-Cultural Analysis in Honor of Richard E. Blanton*, ed. L. Fargher and V. Y. Heredia Espinosa, 259–270. Boulder: University Press of Colorado.

Peterson, Christian E., and Robert D. Drennan. 2012. Patterned Variation in Regional Trajectories of Community Growth. In *The Comparative Archaeology of Complex Societies*, ed. M. E. Smith, 88–137. New York: Cambridge University Press.

Polanyi, Karl. 1957. In *Trade and Market in the Early Empires: Economies in History and Theory*, ed. Karl Polanyi, Conrad M. Arensberg, and Harry W. Pearson. Glencoe, IL: Free Press.

Pool, Christopher A. 2012. The Formation of Complex Societies in Mesoamerica. In *The Oxford Handbook of Mesoamerican Archaeology*, ed. D. L. Nichols and C. A. Pool, 169–187. New York: Oxford University Press.

Price, T. Douglas, and Ofer Bar-Yosef. 2011. The Origins of Agriculture: New Data, New Ideas. *Current Anthropology* 52 (S4): S163–S174.

Scarre, Chris. 2018a. The World Transformed: From Foragers and Farmers to States and Empires. In *The Human Past: World Prehistory and the Development of Human Societies*, ed. C. Scarre, 4th ed., 174–197. London: Thames and Hudson.

Scarre, Chris. 2018b. Holocene Europe. In *The Human Past: World Prehistory and the Development of Human Societies*, ed. C. Scarre, 4th ed., 388–428. London: Thames and Hudson.

Schmandt-Besserat, Denise. 1996. *How Writing Came About*. Austin: University of Texas Press.

Singh, Supriya. 2013. *Globalization and Money: A Global South Perspective*. Lanham, MD: Rowman & Littlefield.

Smith, Andrew E. 2005. *African Herders: Emergence of Pastoral Traditions*. Walnut Creek, CA: AltaMira Press.

Smith, David Livingstone. 2011. *Less Than Human: Why We Demean, Enslave, and Exterminate Others*. New York: St. Martin's Press.

Smith, Monica L. 2019. *Cities: The First 6,000 Years*. New York: Viking.

Smith, Richard L. 2009. *Premodern Trade in World History*. London and New York: Routledge.

Stanish, Charles. 2017. *The Evolution of Human Co-operation: Ritual and Social Complexity in Stateless Societies*. New York: Cambridge University Press.

Stevenson, Alice. 2016. The Egyptian Predynastic and State Formation. *Journal of Archaeological Research* 24:421–468.

Sutton, Mark Q., and E. N. Anderson. 2010. *Introduction to Cultural Ecology*. 2nd ed. New York: Rowman & Littlefield.

Tainter, Joseph A. 1988. *The Collapse of Complex Societies*. Cambridge, UK: Cambridge University Press.

Wadley, Lyn. 2013. Gender in the Prehistory of Sub-Saharan Africa. In *A Companion to Gender Prehistory*, ed. D. Bolger, 313–332. Malden, MA: John Wiley & Sons.

Watts, Joseph, Oliver Sheehan, Quentin D. Atkinson, Joseph Bulbulia, and Russell D. Gray. 2016. Ritual Human Sacrifice Promoted and Sustained the Evolution of Stratified Societies. *Nature* 532 (7598): 228–231.

Weber, Andrzej W., Christopher Bronk Ramsey, Rick J. Schulting, Vladimir I. Bazaliiskii, and Olga I. Goriunova. 2021. Middle Holocene Hunter-Gatherers of Cis-Baikal, Eastern Siberia: Chronology and Dietary Trends. *Archaeological Research in Asia* 25 (March):

Webster, David, and Susan Toby Evans. 2018. Mesoamerican Civilization. In *The Human Past: World Prehistory and the Development of Human Societies*, ed. C. Scarre, 4th ed., 590–635. London: Thames and Hudson.

Wenke, Robert J., and Deborah J. Olszewski. 2007. *Patterns in Prehistory: Humankind's First Three Million Years*. 5th ed. New York: Oxford University Press.

Willer, David, Pamela Emanuelson, Yamilette Chacon, and Richard J. Chacon. 2017. How Chiefdom and Early State Structures Resolve Collective Action Problems. In *Feast, Famine, or Fighting? Multiple Pathways to Social Complexity*, ed. R. J. Chacon and R. G. Mendoza, 417–452. New York: Springer.

Chapter 6
The Urban Revolution II: Consolidation and Expansion of Complexity in Agrarian States, Empires, and Civilizations, 3500 BCE–1450 CE

Abstract States were the fourth type of human society, evolving through extensions of polygroup formation. States repeatedly expanded and contracted, but scholars disagree on whether they overtaxed their populations, grew demographically beyond state structures, or overused their physical environments. Over five millennia, states and empires grew geographically and demographically, increasing the size of state-conditioned cultural regions (civilizations), state systems, and trade networks. States relied on varied forms of intensive agriculture; massively increased markets and trade; and the consumption of diverse goods. State bureaucracies were complex to administer and often at war. Formal book religions developed, and inequality worsened, especially female subordination to males in all states. Life was difficult, with even higher mortality and fertility rates, more people living in cities, and an increase in people moving. States also seemed to have increasingly degraded their physical environments.

Keywords Agrarian states · State systems · Civilizations · Agriculture · World-systems · World religions · Rise and fall

Introduction

After complex societies developed, foraging bands continued to exist in some areas; foraging and horticultural villagers lived in other places; pastoral nomads in others; and foraging, horticultural, and pastoral complex societies in yet others. Over millennia, the global territory of foraging bands shrank; villager territories expanded, then started to shrink; and polygroup territories kept growing. Around 5000 years ago, a yet more complex kind of society developed: the state. In this chapter, I argue two points. First, state making was an intensification of complexity processes, so it was a second phase of the Urban Revolution. Second, this change resulted in a far more complex type of society and another new way of life. Early complex societies (polygroups) seem rarely to have expanded from their original areas (e.g., Bogucki 1999; Moore 2014). However, states often became large empires. This tendency

T. C. Burr, *Societies Past and Present*, Emerging Globalities and Civilizational Perspectives, https://doi.org/10.1007/978-3-031-91532-1_6

accelerated; by 1450 CE, states dominated the planet demographically and controlled much of the planet geographically.

States continued and strengthened trends of social complexity. Farming intensified. Scholars call states relying on agropastoralism "agrarian" states, though horticultural and pastoral states existed. States expanded trade and industry, while consumption became stratified. Economic and gender inequalities greatly increased, and mortality and fertility increased again. Informal networks of polygroup leaders gradually became formal, hierarchical bureaucracies. State leaders developed pantheons of elite spirits, i.e., gods; later, "universalizing" religions developed. Arts remained both part of daily life and a specialized practice, while technology became more sophisticated. Global population increased, people continued infill processes, and cities grew in size and number and were networked. They affected their physical environments more than any previous society did. The "oscillation" of chiefdoms became the "rise and fall" of states, on longer timescales. Interaction systems between states grew geographically.

Again, human capacities made this new phenomenon possible. Historians have long moved past the Great Man theory of history, yet unusual individuals—Sargon, Alexander, Genghis Khan—did exist and took unusual, and unusually effective, action. Still, only long-standing patterns of leaders-and-followers enabled them to sway thousands to action and therefore (violently) include millions in their new societies. This still rested on the groups-within-groups and multiple-groups capacities. Humans elaborated on the organizational and technological innovations of their ancestors, such as the "levels" of political and subsistence economies, new kinship forms, trade forms, gender roles, and symbolic communication (writing). Emotional ties remained crucial interpersonally, yet the expanded political economy allowed more dehumanization. All this made possible an increase in violence and oppression, yet it also provided opportunities for artistic elaboration and even greater cultural diversity.

I first define "the state," then trace the development and expansion of agrarian states in a 5000-year period. I then review theories of state making, continuing the complexity debate. Next, I analyze explanations of the "rise and fall" of states—i.e., repeated expansion and decline—plus the growth of civilizations, state systems, and world-systems, and relations between states and smaller peoples. Finally, I analyze seven aspects of states. They evolved before all regions developed primary polygroups, so we move back roughly 2500 years. Also, the evidence changes: prehistorical evidence still supports earlier periods and nearly all knowledge of the Americas. States usually developed formal writing, so documents analyzed by historians mostly support the understanding of Afro-Eurasian states. Ethnographic inference is rarely necessary.

Telling the Tale: The Development of the State Around the World

The question of what states were helps us understand how they formed. Classically, a "state" consists of territory, people, an autonomous ("sovereign") organization managing the land and people—usually consisting of a four-level settlement hierarchy (Flannery 1998) centered on cities—and a claimed monopoly of legitimate force. This classic definition applies to territorial states (Hansen 2000). Some define "empires" as states that conquered other states (Flannery and Marcus 2012); others define them as "direct administration of different communities from an imperial centre" (Watson 1992:16; also, see Bang 2021:12–20). Some distinguish between "independence, hegemony, dominion, and empire" (Watson 1992:14). Independent states make both external and internal decisions. Hegemony means one state influences external decisions of others in a system. States with dominion over others somewhat influence their internal decisions.

These definitions ignore, though also overlap with, what some call the "city-state" (Nichols and Charlton 1997; Hansen 2000; Yoffee 2005), an independent city surrounded by a rural hinterland; as mentioned, some (Grinin and Korotayev 2017) define self-governing cities as chiefly analogues. Others argue that "the evolution of ancient states was unlikely to have passed through a stage like chiefdoms" (Yoffee 2005:44), though they acknowledge that "there are early states that are not city-states" (45). Yet others say that city-states were usually remnants of collapsed territorial states (Marcus 1998:92–93). City-state scholars assert that they developed in regions together, creating a "city-state culture" (Hansen 2000) or "civilization" (Yoffee 2005). Territorial states were always multiethnic; some argue that city-state regions were also (Yoffee 2005:49), while others disagree (Nichols and Charlton 1997:9).[1] And city-states often became territorial states; for instance, the three city-states of the Triple Alliance became, through external conquest, the Aztec empire (Smith 2012).

The first, "primary," states appeared where early complex societies did: in chronological order, Mesopotamia, the Nile Valley, the Indus River Valley, the Yellow River Valley, the Andes, and Mesoamerica. These always declined, with new, larger, "secondary" states subsequently developing; these too collapsed, with even larger, later replacements. Expansion often involved conquest of smaller complex societies, people in villages, and even hunter-gatherers. Later, states developed in Europe, Southeast Asia, Central Asia, and western and southern Africa. These underwent the same expansion, prosperity, decline, and collapse cycles. In North America, northern Amazonia, other parts of Africa, and many areas in Asia, complex societies appeared but did not become states. By 1500 CE, states controlled much of the

[1] The word "ethnic" is a modern concept (Bonilla-Silva 1999), but archaeologists use it as we do now (e.g., Brumfiel and Robin 2012) for culturally constructed group identities. Unless I learn of a better term, I will use it this way.

world's people and landmass, but people in low-density bands and villages still lived in large areas of the planet.

A (very abstract) chronology can provide a timeline for reference. States first appeared in Southwest Asia and northeast Africa. Some claim that the first true city with a division of labor across it was the city-state of Uruk in Lower Mesopotamia, developing around 3500 BCE; others argue that Uruk was part of a region of city-states (Yoffee 2005). In present-day southwestern Iran, the city-state of Susa developed, often copying styles from Uruk. The latter collapsed around 3100 BCE (Matthews 2018). Simultaneously, one chief unified the Nile chiefdoms, a kingdom that lasted until 2575 BCE (Connah 2018). From 2900 to 2350 BCE, "a large number of independent city-states" developed in Lower Mesopotamia (Matthews 2018:442) (Fig. 6.1).

After 2600 BCE, the number of states increased. A state-like society developed in the Indus Valley with five major cities, but evidence for inequalities is lacking. East Africans founded Kerma on the Nile around 2500 BCE, trading with Egypt to the north (Connah 2018), and entered a period of prosperity that lasted until 2150 BCE. After 2334 BCE, the city-state of Akkad conquered other city-states in Mesopotamia. This first empire collapsed in 2200 BCE (Flannery and Marcus 2012; Matthews 2018). The state of Ur controlled Lower Mesopotamia and western Iran from 2112 to 2004 BCE. Between 2125 and 1975 BCE, Egypt experienced political collapse and chaos (Connah 2018). Elam in western Iran conquered Ur. Complexity in the Indus Valley declined after 1900 BCE, yet trade networks expanded (Coningham 2018).

After 2000 BCE, state areas expanded. Napata developed south of Kerma on the Nile. The city-state of Minos formed on the Mediterranean island of Crete (Alcock and Cherry 2018). Egypt entered another period of prosperity (Connah 2018).

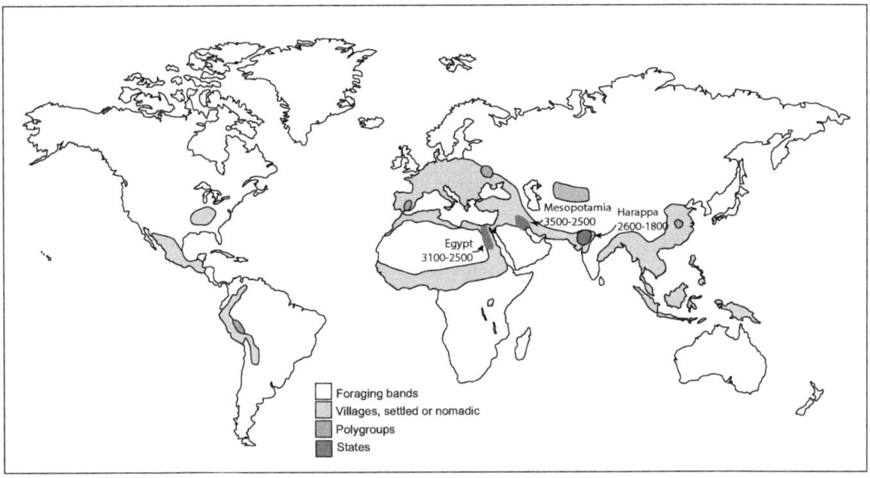

Fig. 6.1 First states and kin-ordered peoples, 2500 BCE. Sources: Black (2005); O'Brien (2010); Houston (2018)

Amorites and Hurrians moved into Mesopotamia, establishing city-states that competed across the region from 2000 to 1700 BCE (Matthews 2018). Babylon conquered much of Lower Mesopotamia and Elam (1792–1750 BCE), while Hittites conquered most of Anatolia by 1700 BCE. Hittites, Babylonians, Egyptians, and others in Southwest Asia exchanged ambassadors and traded extensively with each other.

States also developed in East Asia. Trade brought bronze there (Higham 2018), and the Xia dynasty conquered this area after 1900 BCE, spurring development of rival cities. The even larger Shang dynasty expanded after 1500 BCE and lasted around 400 years, many small states surrounding it. In the Yangzi Valley to the south, rival states appeared (Higham 2018). Migrants from China had settled Taiwan before 2000 BCE; afterward, Austronesian farming peoples from Taiwan started a seaborne diaspora southward to the Philippines (Bellwood and Hiscock 2018).

A new state system formed between the Mediterranean and Persian Gulf after 1500 BCE. Egypt broke apart again, then reconstituted and flourished. Mycenae in Greece replaced Minos as a Mediterranean power after 1450 BCE. Kassites (replacing Babylonia), Assyrians in the north, Hittites farther north in Anatolia, Egyptians, and the Mycenae all traded ambassadors and goods for several hundred years (Alcock and Cherry 2018; Connah 2018; Matthews 2018). This system ended in the "Bronze Age collapse" after 1200 BCE. Aramaeans, Israelites, and Sea Peoples moved through Southwest Asia, and Phrygians migrated into southwest Europe (Alcock and Cherry 2018; Matthews 2018). Multiple South Asian states developed after 1200 BCE. Austronesian peoples reached present-day Vietnam and Indonesia, especially Sumatra, by 1000 BCE. Between 1250 and 750 BCE, they also fanned out across the southwestern Pacific Ocean into islands such as Vanuatu, Tonga, and Samoa (Bellwood and Hiscock 2018) (Fig. 6.2).

After 1000 BCE, the eastern Mediterranean revived, iron replaced bronze, states expanded in size and number, and they developed in the Americas. Ironmaking may have developed in Southwest Asia and in sub-Saharan Africa independently (Connah 2018; Matthews 2018); it had major effects only after 900 BCE (Matthews 2018:460). The first state in the Andes, Sechín Alto, developed in the Casma Valley, collapsing between 900 and 800 BCE (Moseley and Heckenberger 2018). After 1000 BCE, Phoenician city-states in present-day Lebanon explored the western Mediterranean and established colonies (Alcock and Cherry 2018). A city-state civilization developed on the Greek peninsula between 700 and 480 BCE, colonizing northwestern Mediterranean coasts (Alcock and Cherry 2018). The Assyrian empire expanded again in southern Anatolia, peaking in the 600s BCE (Matthews 2018).

After 500 BCE, social complexity became common globally. In North America, Northwest Coast hunter-gatherer chiefdoms developed, traded, and fought (Milner and Wills 2018). Ohio Valley Adena peoples built ceremonial mounds, while nearby complex Hopewell peoples traded with others in the Great Lakes, Yellowstone, the Atlantic, and the Gulf of Mexico (Milner and Wills 2018). Egypt revived again after 715 BCE (Connah 2018). Phrygia developed in Anatolia by 800 BCE, while the Iranian Achaemenid empire conquered Mesopotamia, Anatolia, and the Indus

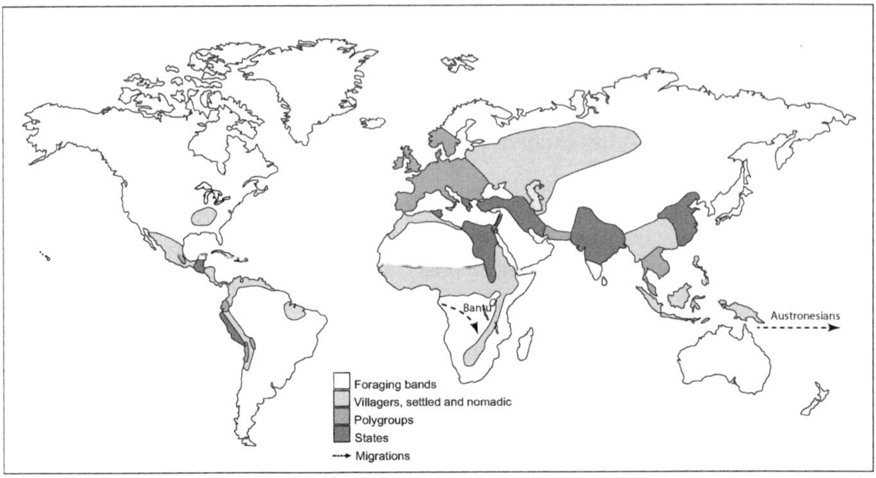

Fig. 6.2 States, kin-ordered peoples, and migrations, 1250–750 BCE. Sources: Moore (1995); Black (2005); Bahn ([2000] 2009); O'Brien (2010); Haywood (2011); Bellwood and Hiscock (2018); Connah (2018)

Valley by 500 BCE (Matthews 2018; Coningham 2018). Several kingdoms developed across northern and eastern India (Coningham 2018). Between 770 and 481 BCE, the East Asian state of Zhou broke into "Warring States" (Higham 2018). Austronesians from Borneo crossed the Indian Ocean to Madagascar, off the southern African coast, around 500 BCE (Bellwood and Hiscock 2018). Massive Bantu migrations started in present-day Cameroon around 1000 BCE, moving eastward and southward for a millennium (Connah 2018).

Global change then accelerated. After the Olmec declined, populations in Mesoamerica grew rapidly, and soon larger "polities … became more common," including Monte Albán in Oaxaca, Mayan city-states on the Yucatán peninsula, and Teotihuacán, which by 400 BCE emerged triumphant from interstate wars in the Valley of Mexico (Webster and Evans 2018:601). Southward, Chavín de Huántar influenced "northern and central Peru between 400 and 200 BCE" (Moseley and Heckenberger 2018:646). The Pukara state developed around 400 BCE near Lake Titicaca; these states collapsed, and the Moche and Nazca arose. Northern Amazon basin and Caribbean peoples developed chiefdoms after 1 CE, trading and fighting. In Eurasia, Alexander "the Great" of Macedonia conquered the entire eastern Mediterranean, most of Southwest Asia, and the Indus Valley (Matthews 2018). At his death (323 BCE), his empire split three ways. The Mauryan empire conquered most of South Asia, breaking into small states in the second century BCE (Coningham 2018), while three small kingdoms ruled southern India for centuries (Trautmann 2016). A state developed in present-day Thailand by the third century BCE (Higham 2018). The short-lived Qin empire conquered all Warring States in China (481–221 BCE); the Han dynasty then took over (Higham 2018). Contemporaneously, the city-state of Rome expanded throughout the western Mediterranean (Alcock and Cherry 2018).

A vast and stable Eurasian interaction zone existed for a few centuries (Fig. 6.3). The Roman empire conquered the entire Mediterranean, plus Britain (Alcock and Cherry 2018; Scarre 2018b). The East African state of Meroë replaced Napata and traded with Roman Egypt. The Satavahana dynasty ruled much of southern India for the next few centuries. The Kushan empire spread south from Central Asia into northern India, while a state ruled Sri Lanka (Coningham 2018). By 1 CE, states developed in Central and Southeast Asia, which facilitated trade across the Eurasian "Silk Road" in Central Asia and across the Indian Ocean.

Afro-Eurasia stability ended after 200 CE, while American polygroups expanded. After 400 CE, Germanic invaders conquered the western Roman empire, creating several small states. The eastern empire survived as Byzantium. Meroë disintegrated in the 300s CE, replaced by Aksum on the Ethiopian Plateau (Connah 2018). In the West African inland Niger Delta, cities formed early in the first millennium, trading with the Saharan oasis state of Garamantes (Mattingly 2017; Connah 2018). The Gupta empire conquered much of India; Central Asian nomads broke it up around 500 CE (Trautmann 2016). The Han empire broke into three states; the southernmost traded with new Vietnamese and Thai states (400–900 CE) and into the Indian Ocean (Higham 2018). Korean kingdoms broke away from Han dominance in 57 BCE, the Great Silla later uniting all three (668–918 CE). Japanese chiefdoms formed Yamato state after 200 CE, lasting to the early 700s (Higham 2018). The Moche confederacy of city-states controlled some of the Andes (100–700 CE) (Webster and Evans 2018). In North America, pottery spread through the Colorado Plateau, while Plains foragers began farming maize (Milner and Wills 2018).

After 600 CE, Eurasian states revived; American and African states expanded. A new political–religious movement, Islam, conquered the region from Spain to the

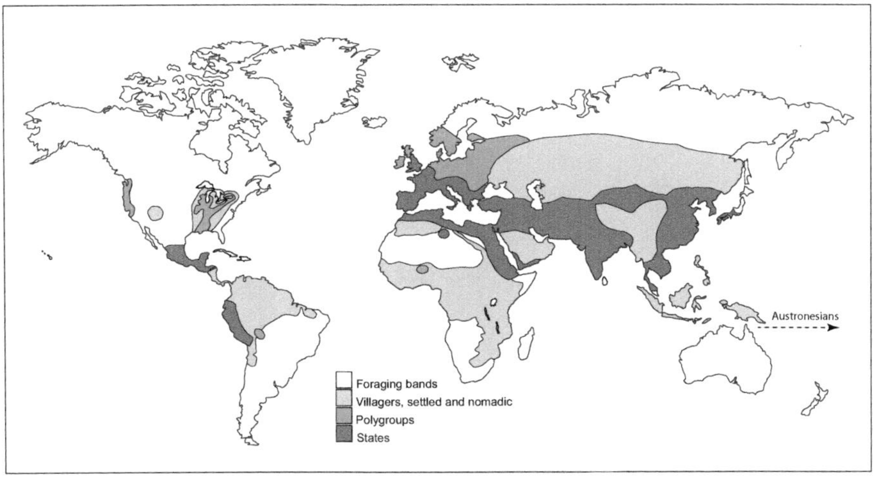

Fig. 6.3 States and kin-ordered peoples, ~1 CE. Sources: Black (2005); Bahn ([2000] 2009); O'Brien (2010); Haywood (2011); Moore (2014); Houston (2018)

Indus River by 750 CE, ruling it for 300 years (Hodgson 1974). In the late 700s CE, Charles "the Great" (Charlemagne) consolidated a new empire in western-central Europe. On his death, it broke in two and then into more, smaller states. Viking invaders from Scandinavia raided coastal Western Europe, establishing new kingdoms, and Magyars invaded Central Europe (Rietbergen 2021). South Asia broke up into five state regions for several centuries; the states often fought with each other (Trautmann 2016:87 [map], 88). The Tang dynasty reunited China (Holcombe 2017) (Fig. 6.4).

After 900 CE, states became larger and more numerous, and more polygroups developed. In West Africa, the Wagadu state formed before 1000 CE; after it collapsed, the larger Mali, to the east, incorporated its lands (1200s–1300s); and after Mali declined, the empire of Songhay controlled an even larger region. Numerous states formed on the Bight of Benin, including Akan, Oyo, and Edo (Connah 2018). Small principalities and city-states multiplied across Europe. Western European states repeatedly invaded the eastern Mediterranean in the 1000s, 1100s, and 1200s in "Crusades," establishing small states (Rietbergen 2021). Kiev developed in Eastern Europe in the late 800s CE, expanding across the Russian plain and fighting and trading with Scandinavian and Central European states and Byzantium (Riasanovsky and Steinberg 2019), then declining. The Song dynasty succeeded the Tang for several centuries (Holcombe 2017).

Afro-Eurasian states grew more after 1200 CE. Zimbabwe developed in southern Africa, trading with Swahili city-states on the East African coast (Connah 2018). Central Asians invaded the Ganges Valley around 1000 CE; two centuries later, Central Asian Turks invaded again, establishing the Delhi sultanate (Trautmann 2016). Pastoral Mongols conquered all of Central Asia, then invaded Eastern Europe—destroying the declining Kievan state—took over most of Southwest Asia,

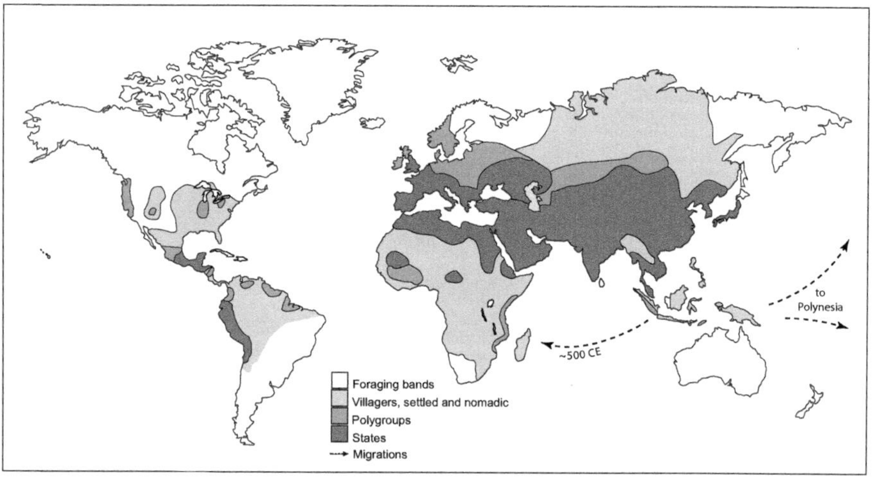

Fig. 6.4 States, kin-ordered peoples, and migrations, ~700–800 CE. Sources: Black (2005); Bahn ([2000] 2009); O'Brien (2010); Haywood (2011); Moore (2014); Connah (2018); Houston (2018)

and conquered nearly all of China and Korea. This Mongol empire then broke up into four smaller Eurasian ones in the 1300s (Holcombe 2017; Riasanovsky and Steinberg 2019; Rietbergen 2021). A later successor, the Central Asian Timur ("Tamerlane"), reconquered part of the empire (Khalid 2021). The Ming dynasty took over China from the Mongols (Holcombe 2017); the Ottoman empire conquered the eastern and southern Mediterranean; and Muscovy succeeded Kiev, uniting much of eastern Europe (Riasanovsky and Steinberg 2019). The Angkor state controlled much of Southeast Asia from 800 to after 1400 CE (Higham 2018).

Maize spread in North America, supporting multiple chiefdoms. Cahokia, possibly a city-state, developed in the Mississippi River Valley between 800 and 1100 CE (Milner and Wills 2018:680–681). A ritual center in Chaco Canyon in present-day New Mexico had influence for hundreds of miles (Milner and Wills 2018). Social complexity then shifted northward to Mesa Verde, but "by 1300 CE farming groups had abandoned the Colorado Plateau" in response to incoming hunter-gatherer peoples, warfare, and drought (Milner and Wills 2018:691).

Mesoamerican and Andean states expanded and contracted. After Teotihuacán declined, the Tula ("Toltecs") expanded from 900 to 1200 CE, then "violently collapsed" (Webster and Evans 2018:627). Mayan Puuc states of northern Yucatán developed after the 700s CE; from 900 to sometime after 1200 CE, Chichén Itzá dominated the region. After it collapsed, the area depopulated. The Mayapan state ruled Yucatán (1260s–1440s), then devolved into several chiefdoms (Webster and Evans 2018). The Nahua conquered or allied with many city-states in the Basin of Mexico in the 1300s, then conquered independent polities across Mesoamerica, creating the Aztec empire (Webster and Evans 2018). In the Andes, the Chimú empire conquered its rival, Lambayeque, in 1370 CE. The Inca empire to the south conquered Chimú in the 1400s (Moseley and Heckenberger 2018), spreading along most of the Pacific Coast (Fig. 6.5).

After 1 CE, people colonized the rest of the Pacific Ocean islands. They sailed to and lived on Rapa Nui (Easter Island), the Cook Islands, Hawaii, New Zealand, and others. A few centuries after 1000 BCE, a powerful chiefdom controlled each island in the Hawaii chain, while in what is now New Zealand, the northern Maori were villagers and southern Maori foragers (Bellwood and Hiscock 2018). People in bands and villages continued in the northern regions of North America and in the Caribbean; in much of Amazonia and in the Southern Cone; in parts of central and southern Africa; in far northern Europe; in Siberia; and in Oceania and Australia.

The accompanying maps (Figs. 6.1, 6.2, 6.3, 6.4, and 6.5) greatly oversimplify the situation. State rise and fall combined with expansive state claims means that the maps show areas that states controlled only temporarily. Still, this story of social change between 3500 BCE and 1500 CE reveals five patterns. First, states developed independently of each other in both hemispheres. Second, the first state in a region seemed to stimulate secondary state formation. Third, the first state's development and decline usually initiated repeated cycles of rise and fall. Fourth, states interacted in three ways: state rulers traded ambassadors; state armies fought each other; and states traded with each other. Finally, despite long reversals such as the Bronze Age collapse, the Han/Silk Road/Roman collapse, and the Black Death, the total area of state control grew.

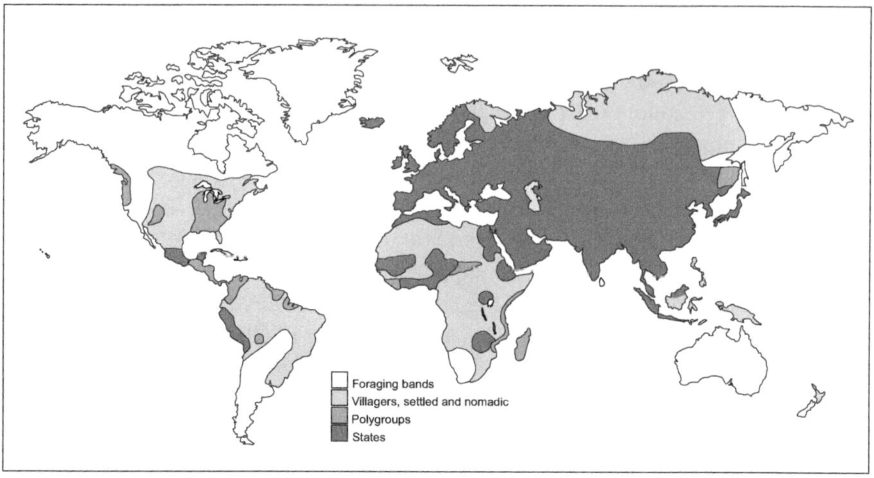

Fig. 6.5 States and kin-ordered peoples, ~1300–1400. Sources: Black (2005); Bahn ([2000] 2009); O'Brien (2010); Haywood (2011); Moore (2014); Bellwood and Hiscock (2018); Houston (2018)

The next two sections address these themes. The first examines explanations of state formation and collapse. Scholars usually discuss these separately, yet they raise overlapping issues, so I discuss them sequentially. The second addresses interactions between states and smaller societies; civilizations; and world-systems.

Explaining the Rise and Fall of States: Debates

Social scientists, especially archaeologists, analyze state formation in a debate echoing the complexity discussion. Most address primary state formation—the first state ever in a region. Secondary state formation seems to parallel secondary polygroup formation: once one state developed, surrounding polygroups faced the defensive grow-or-succumb dilemma. Some question the primary–secondary dichotomy (Blanton and Fargher 2008; Earle 2016; Feinman and Nicholas 2016), suggesting that each society's path was unique. Others argue that chiefdoms were not state precursors (Yoffee 2005). The arguments still generally cluster into structural–functionalist (corporate/public goods) or conflict (network/domination) explanations.

Two caveats apply. Most polygroups never became states, but "chiefly competition continued indefinitely" (Flannery and Marcus 2012:362; Carneiro 2017); primary states appeared only in a few places. Also, "state formation processes [were] more uneven, disjointed, and geographically variable than standard models." Aspiring state leaders developed "multiple sources of power," and new states were fragile, sometimes failing (Stevenson 2016:427).

As with polygroups, the "corporate" strategy of providing public goods is a common structural–functionalist explanation of state formation (Blanton et al. 1996). One asserted public good was irrigation (Wittfogel 1957), though irrigation sometimes appeared after states did, not before. Studying 30 primary and secondary states, Blanton and Fargher (2008) concluded that since populations can free-ride or resist if their taxes provide them no benefits, rulers must "provide public goods, control the agency of officials (bureaucratization), and relinquish some aspects of their power" to legitimate their rule (Blanton and Fargher 2008:252). Others argue that rulers had incentives to create public goods quickly because "most first-generation states … were underpinned heavily by … the agrarian production of the local population" (Feinman and Nicholas 2016:286). Finally, some think that a minimum population of 100,000 people supported state formation rather than large territory, inequalities, or public goods (Turchin et al. 2018:11, 13).

Hayden (2003, 2014) has long emphasized feasting as an internal domination strategy. Building monumental architecture had multiple purposes: by intensifying production to feed workers, it increased the surplus and the population's carrying capacity, spawned specialist bureaucracies, and inspired awe among the state's internal subjects to legitimate royal rule (Hayden 2003:357). Mesopotamian, Egyptian, Chinese, sub-Saharan African, Andean, Mesoamerican, and Austronesian states continued human sacrifice (Chirot 2012; Watts et al. 2016), perhaps an internal splitting strategy.

The "network" strategy of using external resources to dominate one's group (Blanton et al. 1996) took many forms. Warfare helps predict state making (Turchin et al. 2013). Some three-level chiefdoms conquered surrounding chiefdoms, making the latter into governorships, incorporating and formalizing their informal administrative structures. In Mesopotamia, the first states followed Peterson and Drennan's (2012) demographic concentration model: lands defensively and quickly emptied into a few growing cities. If chiefdoms had unplanted buffer zones around them (LeBlanc 2006), a conquering chiefdom (a new state) could plant more land, the surplus increasing much faster than the population did. The king could distribute new land to previously circumscribed subjects, gaining their loyalty, while enslaving conquered peoples. For others, slave raiding was "perhaps the most significant motive driving state policy" (Willer et al. 2017:438), though resources to capture slaves were limited. State creators were sometimes "aggressive usurpers," lower-ranked members of a chiefly elite; "it often took several generations of aggressive rulers to create a kingdom" (Flannery and Marcus 2012:385). Confederations or heterarchical segmentary societies, such as the Mongols, sometimes fought together to create states (Bronson 1988). Constant warfare between pastoralist confederacies might also have created states, but ones with "a federal or feudal character" (Gibson 2017:192). Other network-oriented scholars question warfare as an explanation (Ferguson 2006). Köhler (2011) suggests that trade was important for Egypt, while there was little monumental architecture in the first Indus Valley civilization, no indication of rulers, and few remains of warfare (McIntosh 2008). Likewise, in peer–polity interaction spheres, states more likely grew from managing trade, and they were more authoritarian (Feinman and Nicholas 2016).

Again, some argue that the physical environment influenced human action (Brooke 2014). Carneiro's (2017) theory of environmental circumscription, operating most stringently in Egypt and least so in China, applied to both chiefdoms and states. Primary state formation was usually a response to "a discrete natural catastrophe, droughts in Egypt and Mesopotamia, floods in China and Peru, that led to *population collapse and cultural crisis*" (Brooke 2014:189, original italics). A climate crisis around 3200–2900 BCE, in which the first states formed in Mesopotamia and Egypt, "saw a significant intensification of human economies" (Brooke 2014:183).

Some reject the distinction between primary and secondary states; others use it. Since "any state [primary or secondary] was a fluctuating mix of power strategies" (Earle 2016:302), financing conquest from previous conquests and fostering trade to tax were both useful. Others say that once "first-generation kingdoms had begun to form, neighboring societies hastened to nucleate in order to avoid being taken over" (Flannery and Marcus 2012:389): a leader's new realization that he *could* construct a larger realm added force to the other new realization that he *had better* do it. Still, many agree that state leaders made up their path as they went along: "The participants in this enormous process seem not to have known what they were creating" (Harris 1995:227). Most leaders were "people of great ability and charisma" (Fagan 2004:350), already having elite status, possessing military ability, and being personally aggressive (Flannery and Marcus 2012). Since they had a "relatively tenuous hold on power," they "used all the means possible to reinforce" it, including feasting, conquest, ideological claims, and other strategies (Hayden 2003:356). Indeed, some think states never functioned smoothly; they were probably "quite ramshackle contraptions" (Cowgill 1988:253; Graeber and Wengrow 2021:Chapter 10).

Overall, then, scholars suggest that, as in polygroups, leaders followed "corporate" or "network" strategies to form the first states (Blanton et al. 1996). In one, they gained power by providing defense against invasion, public order and adjudication of disputes, infrastructure, urban planning, public ceremonies, science, and food storage. In the other, they conquered other peoples and their land, ritually sacrificed or enslaved people, and managed external trade. Many scholars argue that leaders combined these strategies to take power.

However successfully leaders made states, their descendants ultimately could not keep them. Scholars have analyzed imperial decline for centuries. Chinese historians long ago noted their dynastic cycle, usually attributed to leadership decay and factional disputes (Fairbank et al. 1989:70–74). North African scholar Ibn Khaldûn argued that decline stemmed from imperial overextension and the corruptions of luxury ([1377] 1958). Gibbon ([1776–1789] 1960) also emphasized moral failures in Rome's long collapse.

Instead, scholars now emphasize structural factors, often those mentioned for state formation. We must first define "collapse." Many systems of complexity, including human social systems, are hierarchically constructed in units, so they are "decomposable" into those units (Simon 1962). Some use decomposability (Renfrew 1986) to explain imperial collapses as sudden events, usually involving overspecialization and overintegration. Others have long objected to the word "collapse" (Yoffee 1988; Yoffee 2005; McAnany and Yoffee 2010). Often only the political structure ended: "Collapse ... ensues when the center is no longer able to secure resources from the periphery." Life went on in towns and villages, not necessarily "incurring ... severe social troubles" (Yoffee 1988:13, 15; Yoffee 2010). People could even flee cities to repopulate the countryside (Yoffee 2005), reversing demographic concentration (Peterson and Drennan 2012). Others argue that internal political instability predicted dramatic population collapses. In China for at least 13 centuries, when instability increased, millions died and many others avoided having children, greatly reducing population size (Turchin 2003:165).

The "Dynamic Model" of state formation (Marcus 1998) opposes sudden collapse (Renfrew 1986); instead, a cycle of rapid expansion, slow decline, and new, quick expansion occurred. Primary states formed quickly through one chiefdom conquering others, making them provinces. Eventually, surplus overextraction led provinces to break away until the state collapsed. A breakaway province formed a secondary state through a second conquest, again declining slowly until succeeded by tertiary state conquests and collapses, and so on. As rise-and-fall dynamics continued, empires increasingly succeeded states. Marcus (1998) argues that her Dynamic Model applied to Egypt, the Andes, the Maya, the Valley of Mexico, Mesopotamia, and the Aegean. It likely applied also to China, West Africa, and Europe. States in these regions experienced expansions and collapses on scales of hundreds of years; subsequent states were usually larger than previous ones.

Scholars argue that some empires formed peacefully, including ancient Mediterranean city-states that coalesced into federations (Scheidel 2021), Teotihuacán, and the Incan empire (Hornborg 2018; Webster and Evans 2018). Still, I think that after primary state formation, conquest became the most common empire-formation strategy. This chapter's Appendix lists 39 empires that grew partly or only from conquest, yet it is incomplete. Furthermore, as Bang (2021) points out, following Hui (2005), often interstate rivalry ended up with one state conquering the others. This implies that "it is difficult to" distinguish "between defense and aggression among states locked in mutual rivalry within the same system"; in this model, defensive aggression drives conquest, which "eliminates rivals by absorption" (Bang 2021:48–49). Since the reversal of this process also occurred often—large empires decaying into many states in a state system—interstate rivalry became globally more common over time.

While the Dynamic Model postulates some decomposition into and re-formation from constituent units, there were more thorough collapses (Scheidel 2017; Yoffee 2005). Elites always suffered in state collapse; sometimes subject populations also suffered, while other times they threw off elite oppression. The eastern Mediterranean Bronze Age collapse of 1200–1000 BCE (in Egypt, the "Third Intermediate") devastated everyone, but most severely "the rich and powerful" (Scheidel 2017:273). Much later, the collapse of the western Roman empire shrank elite landholdings, while archaeological evidence shows improvements in commoners' health measures (Brooke 2014; Scheidel 2017 is unsure). The burning of Teotihuacán around 700 CE and the Chinese Tang dynasty collapse two centuries later involved, respectively, a populace-wide disaster and a deadly elite purge (Scheidel 2017:262–263, 278). Total ending of long traditions occurred rarely. The Achaemenid empire replaced Mesopotamian institutions over a few centuries (Yoffee 1988, 2005); the end of Egyptian Pharaonic traditions under Alexander and the Romans seems similar.

Overall, then, scholars identify three types of collapse. One is political collapse or "state failure" (Yoffee 1988; Scheidel 2017); the capital loses control over provinces, which remain intact (Marcus 1998). In another, "population as a whole contracts, settlements shrink or are abandoned, and economic practices regress to less sophisticated levels" (Scheidel 2017:258). Third, millennia-long traditions occasionally end completely (Yoffee 2005).

Scholars proffer multiple explanations for collapse. Despite other definitional differences, two scholars separately think that "maintaining large-scale inegalitarian structures for long periods of time" was simply difficult (Marcus 1998:94) since the "earliest states" had "a good deal of bungling and [generated] considerable conflict" (Yoffee 2005:132). Some note how pastoral nomads often conquered states in North Africa, South Asia, and East Asia (Ibn Khaldûn [1377] 1958; Bronson 1988; Brooke 2014), but this still leaves most collapses unexplained.

One theory emphasizes long-term costs of public goods provision (Tainter 1988). Food provides energy, so states extracted energy (agricultural surplus) from peasantries to feed bureaucrats who provided public goods. This society-wide deal of indirect reciprocity (diagrammed in Chap. 5) through public goods legitimated the state: peasants accepted limited extractions in return for benefits. Order and prosperity increased birthrates, which also increased population density and the need for more public goods, as well as elaborated state bureaucracies. However, each additional bureaucrat provided diminishing returns. At some point, government extracted more than it provided, but long before that, populaces became disaffected and local officials held back taxes for personal benefit. Rebellions, out-migrations, invasions, and collapse ensued (Tainter 1988).

Other explanations analyze tensions in state structures. Bureaucracies always contained administrative, religious, and military hierarchies. Centers and peripheries, with different aristocracies and ethnicities, were always at odds. Bureaucratic infighting reduced elite cohesion, sparking fruitless struggles wasting resources (Kaufman 1988). As all territorial states were multiethnic entities (Flannery and Marcus 2012; Bang 2021), peripheries often resented outside rule, needing only an excuse to rebel. Central overextraction due to infighting led peripheral groups to break away or refuse to pay (Eisenstadt 1963; Yoffee 2005:133–140).

Another explanation for state breakdown is a "demographic-structural" model. Goldstone ([1991] 2016) developed this theory for agrarian states after 1500 CE. He treated population growth as "exogenous," or independent of (generated outside of) social structures. Basically, population grows faster than agrarian state social structures grow, they are overwhelmed, and the state breaks down. Specifically, population growth directly and indirectly affects three social structures: state finances, positions within the elite, and the masses of peasants and urban workers. Population growth directly affects all three by making necessary increases in public goods, forcing increased state expenditure; by increasing the population of elites (bureaucratic, military, and religious) faster than the number of elite positions available (no matter how many the state adds); and by forcing diminishing returns on peasants' land, increasing the number of young people (who more likely rebel when disaffected) as a proportion of total population and increasing urban populations past the number of jobs available. Indirectly, population growth increases real prices since population grows past what food the land can produce. Price increases worsen already strained state finances, antagonize already angry unemployed elites (who use consumption goods as a positional strategy), and impose ever-greater peasant and worker misery. All three movements may occur separately or in various ways, but when all three worsen at once, they multiply each other's effects so that state breakdown becomes likely. Peasants and workers rebel, and disaffected elites lead them; civil wars led by different elite factions can also break out.

Turchin (2003) argued that modifying this theory explained state breakdown through the entire agrarian era (3500 BCE–1500 CE). Since state breakdown itself causes population decline, Turchin treats population growth as endogenous, modifying Goldstone's model. This change suggests that "secular cycles" will occur, so "state breakdown should occur at somewhat irregular intervals 2–3 centuries apart" (Turchin 2003:148). He documents this process for China (2003:165), but it applies generally, with variability by region and time period. Political instability (rebellions and civil wars) varies inversely with population size. Population collapse results from mass deaths from rebellions and civil wars and from foregone marriages and births since "people choose to marry later and to have fewer children" in chaotic periods (Turchin 2003:120, 179, 184). Once population reduced, the elite group shrank, peasants had much more available land per capita, the scarcity of labor increased wages, and the cycle started over.

Some historians argue for environmental causes of social collapse since humans have always overused their environments (Chew 2001; Diamond 2005; Ponting 2007), either causing (Chew 2001) or contributing to (Ponting 2007) the problem. They say deforestation leading to soil erosion, the salinization of fields, overextraction of soil nutrients, and irrigation overwatering ("waterlogging") reduced agricultural yields. Some disagree: "ancient and medieval societies in *reasonably robust ecological situations* were sustained for considerable periods of time" (Brooke 2014:266, original italics). Instead, exogenous environmental change—climate shifts, volcanic activity, and epidemics—brought down states (Brooke 2014). For instance, the Bronze Age collapse seems to have resulted from a new cold, dry climate and disastrous earthquakes. Global cooling and drought forced Central Asians westward into Europe and southeast into China, ending the western Roman Empire and the Han dynasty. The Chinese dynastic cycle may have resulted from cycles of monsoons from the Pacific; reduced rainfall set off crises. The Andean Moche state and Teotihuacán collapsed from droughts (Brooke 2014:Chapter 8). Others agree: changes in Ethiopian rainfall affected Nile River flow and the Egyptian dynastic cycle but also argue that human overuse was a factor (Ponting 2007).

A popular book (Diamond 2005) has derailed the debate somewhat. Many scholars angrily object to its characterization of elites "choosing" to "fail" (McAnany and

Yoffee 2010), emphasizing instead the resilience of those who survived. I agree with the objections; I also think that resilience matters only during hardship, which occurred when millions died through rebellions, conquests, epidemics, or famines. Many factors seem to have led to state collapse, including public-goods creep, squabbling elites, population growth, or environmental overuse. New states usually recomposed themselves in the regions, increasingly through conquest. Understanding these factors certainly requires more research.

The Larger Context: Frontiers, Civilizations, and World-Systems

Repeated cycles of rise and fall led to three larger social patterns. First, state expansion affected small-scale societies far beyond formal state limits. Second, long-term regional state dynamics deepened institutional patterns, producing "civilizations." Finally, long-distance interstate relations developed geographically vast, intercivilizational trade and warfare systems.

States influenced bands, villages, and polygroups on their imperial edges. There, different peoples met in a situation of inequality. Some scholars distinguish three areas: boundaries, or formal lines; borders, zones past boundaries; and frontiers, zone beyond the borders (Feuer 2016). Others call the frontier region and beyond the *tribal zone*, an "area continuously affected by the proximity of a state, but not under state administration" (Ferguson and Whitehead 1992:3). Here, state influence "ripples out far beyond the region of direct contact" (Hall et al. 2011:263). New technologies, migrants, officials, and traders spread beyond state borders. Trade and war generated new social forms: tribes, "secondary states, open-ended alliance networks, autonomous villages, clan segments and extended families, [and] specialized bandit groups" (Ferguson and Whitehead 1992:13). They also stimulated competition between new groups. When expanding, states and empires often conquered nearby polygroups that their own influence had previously created. Villages or forager bands farther out also began adopting metals, textiles, permanent leaders, and complexity. A related concept, the "shatter zone," highlights the long-distance disruption, instability, and reorganization that polygroups and states forced on other peoples (Raffield 2021).

Three examples illustrate how the tribal zone worked. After 2500 BCE, "European societies ... did not adopt writing and urbanization" from eastern Mediterranean societies, but they did adopt copper and tin, making bronze, and trading "gold, silver, amber, and probably also slaves and woolen textiles," which tended "toward [making them] more complex societies" (Kristiansen 2018b:105; Scarre 2018b). Much later, as the Roman empire expanded, it often exploited frontier lands economically, then conquered them (Feuer 2016). The Central Asian steppe, too dry to support agriculture, was beyond the Chinese border. Steppe pastoral nomads interacted with successive Chinese dynasties for two millennia. This interaction tended to unify steppe peoples over time. Sometimes they raided and extracted tribute; later, they invaded and conquered (e.g., the Mongols). Always they traded with successive dynasties (Feuer 2016).

Overview studies of tribal and shatter zones before 1500 CE are rare (Hall et al. 2011), so I infer that similar processes occurred just outside of states in Mesoamerica, the Andes, Africa, and elsewhere in Asia. Since states changed nearby smaller-scale peoples so much, I will now sometimes group together what were bands, villages, and smaller polygroups as "kin-ordered societies" (Wolf 1982). Ethnographers have long resisted using social-evolution "types"; constant interactions between such groups, and between them and states, suggest why.

Second, conquest-driven expansion formed multiethnic states and empires (Flannery and Marcus 2012; Bang 2021), creating collective action problems. Subject populations constructed their identities and loyalties culturally; they were kin-based and local, overlaid by religious beliefs and perhaps linguistic similarity. Thus, governing them was difficult (Turchin 2003). The conquered resented taxation, and coercion or buying them off with public goods was expensive. And how could states recruit conquered peoples with suspect loyalties into bureaucracies? Conquest started the internalization of informal frontier processes into formal "assimilation," which helped states overcome some of these problems. The likelihood of warfare increased across cultural divides, creating strong pressures toward assimilation behind the line (Turchin 2003), but assimilation often took several centuries. When it was forced, the decision was easy, though it generated resentment.[2] If given a choice, ethnic solidarity and continuing resentment under harsh repression were good reasons not to convert. Sometimes, however, cultural patterns provided incentives; for instance, Islamic law superseded local ethnic norms, standardizing behaviors across a wide area, so state leaders and merchants often became Muslims (Oka 2018).

Interestingly, only some empires assimilated populations; others tolerated local cultures. Early Mesopotamian empires enforced conversion to buttress the ruler's claims to legitimacy, but later rulers, such as Sargon of Akkad (*circa* 2300 BCE), tolerated and honored defeated gods to avoid religious objections to his legitimacy (Roberts 2011:18). Almost four millennia later, the Ottomans also followed toleration over assimilation (Fisher 1979). Still, many other states enforced cultural assimilation on conquered peoples. The Tang dynasty did so (Abramson 2008), the Mauryan emperor Asoka encouraged Buddhism (Coningham 2018), and Kiev's ruler adopted Christianity around 1000 CE and enforced it on his people (Riasanovsky and Steinberg 2019).

Ongoing assimilation in a region through repeated cycles of rise and fall fostered administrative, religious, and scientific traditions resulting in regional elite cultures, or "civilizations." *Pace* some scholars (e.g., Trigger 2003), I do not equate states with civilizations. States were and are primarily political entities—collective

[2] Today, many might call assimilation "cultural genocide." Though this is anachronistic, it is not necessarily wrong.

actors—coercing most members into them. Civilizations were and are regional cultural traditions formed by sequences of successive states; they were less coercive, relying more on enculturation. Furthermore, they transcended the life of any specific state or empire,[3] developing only after *repeated* imperial expansions and collapses in a region. Again, after humans populated the planet, cultural diversity developed around the world. Civilizations were a special type of cultural region: they offered the cultural–institutional beliefs, values and morals, and practices that states could use to govern populations.

This definition only partly overlaps with others. Despite exceptions (Wilkinson 1995), most scholars define civilizations culturally. Some define them as cultural regions (Renfrew 1986; see also Braudel 1995). Others define them as regional cultures of city-states (Yoffee 1988, 2005; Hansen 2000). A few define civilizations synonymously with states (e.g., Trigger 2003). Some focus on a common elite culture in a region, acknowledging that local cultures exist within (Redfield 1956; Melko 1995; and see Bang 2021:39–40). Recently, scholars have extended this anti-essentialist understanding (Hall and Jackson 2007), defining civilizations as "internally highly differentiated and culturally loosely integrated" (2010:6). Thus, states never imposed full uniformity, so cultural diversity within civilizations was normal. Only one scholar I found rejects any state-based definition (Fernández-Armesto 2001).

The etymology of the word "civilization" leads to this definition. The original Latin word was *civis*, member of a community, which led to civitas, the community itself (i.e., Rome). *Civis* also led to *civilis*, making someone a member of a community, from which we get *civilize*: to "be more advanced … to enlighten, refine, and educate … to make more sophisticated" (*Oxford English Dictionary* 2024). States are centered in cities (civitas), so it was the culture of cities into which people were "civilized." City people often sneered at rural people for being undeveloped, crude, and unsophisticated (e.g., "the idiocy of rural life" in *The Communist Manifesto*), and they often saw "uncivilized" rural peoples as living in the past: "*out there* is equated with the *back then*" (Friedman 2018:543, original italics). In contrast to this elitist definition, I define "civilization" neutrally. In the singular, civilization consists of elite traditions of any urbanized region developed over multiple imperial cycles. In the plural, multiple rise-and-fall cycles created civilizations in specific places: e.g., Europe, Mesoamerica, and East Asia. As we will see, the definition of "civilization" as a common culture among city-states (Yoffee 1988, 2005) is not necessarily much different from one focused on territorial states; city-state regions could also experience repeated imperial cycles.

The "starting" dates in Table 6.1 indicate only midpoints in long periods of development. Given the continuation of the repeated rise-and-fall processes that created them, civilizations always changed over time, and some civilizations ended millennia ago.

[3] Thanks to Ino Rossi for this language.

Table 6.1 Civilizations in world history

Region	Start date	Ending date[a]
Mesopotamian (Tigris/ Euphrates Valleys)	3500 BCE	300 BCE
Pharaonic (Nile Valley)	3100 BCE	300 BCE
Indian (South Asia)	2500 BCE	
Chinese (Huang He/Yangtze Valleys)	2000 BCE	
Greco-Roman (Mediterranean)	500 BCE	500 CE
Japanese (Japanese islands)	400 BCE	
Andean (western South America)	400 BCE	
Mesoamerican (Valley of Mexico)	400 BCE	
Mayan (Yucatan peninsula)	300 BCE	
Indochinese (Southeast Asia)	1 CE	
Orthodox/Slavic (Eastern Europe)	500 CE	
Ethiopian (East African Highlands)	300 CE	
Latinate (Western Europe)	500 CE	
Islamic (Southwest Asia/North Africa)	700 CE	
Swahili (East African coast)	1000 CE	

Sources: Melko (1995); Yoffee (1988, 2005); DeCorse (2012); Horton (2012); McIntosh (2012); Moseley and Heckenberger (2018); Webster and Evans (2018); Finneran (2007)
[a]Dates are approximate. See Chap. 8 for more

Three examples illustrate this definition. The classic one is China. The perhaps mythical Xia dynasty (2100–1600 BCE) gave way to the very real Shang dynasty, which the Zhou kingdom later replaced. The Zhou fragmented into around 170 ministates (722–481 BCE), which over two centuries consolidated into seven major "Warring States" (403–221 BCE). By this time, rice culture, writing through characters, and traditions of statecraft were firmly established. During the Warring States, Kong Fuzi ("Confucius") developed his philosophy of interpersonal obligations. The short-lived Qin dynasty united the Warring States, creating a formal bureaucracy, and adopted the harsh code of "Legalism," which the subsequent Han dynasty combined with gentler Confucian ideas. The Qin also standardized writing, weights and measures, and currencies, which the Han also accepted. The Tang dynasty retained all this and split the Han bureaucracy into six ministries. It also developed a bureaucratic recruitment system based on examinations of Confucian writings. The subsequent Song dynasty adopted all this, spreading Confucian values across the region and firmly established the scholar-gentry—a much larger group than bureaucratic officials—as the elite. Even when pastoral groups invaded and took over (e.g., the Mongols), they adopted these Chinese institutions (Fairbank 1992).

A similar though shorter process played out in the eastern highlands of Africa—Ethiopia. Wheat and barley were in use by 1000 BCE, as were sheep, goats, and cattle. A few centuries later, people cultivated a local grain, teff, and social complexity and a local elite seem to have developed, evidenced by monumental architecture (Finneran 2007; Phillipson 2012). Early in the first millennium CE, the city of Aksum formed a state and "flourished during the first seven centuries AD" [sic] (Phillipson 2012:47). It expanded into southern Arabia and had trade links with the Mediterranean. Aksum developed writing for its language, Ge'ez, and slowly developed a formal bureaucracy to replace tributary relationships. Its rulers also adopted Christianity by 350 CE; it spread throughout the kingdom in the next few centuries (Phillipson 2012). By 650, the state and its population declined, perhaps from land overuse (Finneran 2007:210). While Aksum had been the capital, after its decline, smaller kingdoms in the next few centuries moved their capitals seasonally through their realms as, effectively, tent cities. These kingdoms established power southeastward for the next few centuries, as monasteries spread and entrenched Christianity despite Muslim proselytizing from across the Red Sea. A regional non-Christian queen supposedly defeated the local Christian king after 950 CE (Derat 2020), but this is debated (Finneran 2007). Sometime after 1000 CE, the Christian Zagwe dynasty in Lalibela, south of Aksum, established central control over the highlands; it claimed descent from Aksumite kings, retained the mobile capital, wrote all documents in Ge'ez despite its spoken version having died off (in an imperfect parallel to Latin in Europe), and supported further Christian monastic expansion as a legitimization strategy (Finneran 2007; Phillipson 2012; Derat 2020). In 1270, the "Solomonic" dynasty, also claiming Aksumite descent, overthrew the Zagwe (Phillipson 2012:228) and conquered northward. This dynasty continued writing in Ge'ez for all documents, supported more Christian monasteries, and encouraged Christian assimilation through church building (Ayenachew 2020:70–71). It also continued the mobile capital tradition; the tent city seasonally exceeded a population of 30,000 (Ayenachew 2020:71).

In Mesoamerica, the process varied but led to a similar outcome. Politically, city-states were the foundation of imperial organization. The Olmec polity around 1000 BCE was either a complex society or a full state (Evans 2012; Feinman 2012). Across the region, subsequent states were always centered on cities; as the centuries passed, some dominated others. Scholars debate whether Monte Albán, Teotihuacán, Tula, and Chichén Itzá were capitals of empires, linchpins of city-state confederations, or merely regional hegemons (Evans 2012; Feinman 2012). Even the Aztec empire consisted of a "Triple Alliance" of three city-states and several tributary polities they collectively controlled (though Tenochtitlán became primary); they bordered similarly organized empires in the highlands (Feinman 2012; Smith and Sergheraert 2012). Culturally, writing, ballcourts, and common religious themes developed throughout the region (Evans 2012). The Olmecs and neighboring polities seem to have developed writing systems based on logo-phonic glyphs (Justeson 2012; Grube 2012; Urcid 2012), and pantheons of the sun as god of war, earth as a crocodile, shark, or turtle, and gods of rain and lightning existed throughout the region (Taube 2012; Reilly 2012). The urban center Monte Albán to the west of

Oaxaca and Mayan city-states to the east and north also adopted these pantheons, glyphs, and ballcourts and modified them, as did Teotihuacán to the northwest. The classic Mayan city-states, the Aztecs, and their local enemies all relied on these cultural elements to rule (Evans 2012; Feinman 2012; Urcid 2012; Diel 2012).

The three examples illustrate the definition of "civilizations" as regional cultures developed through sequences of state expansions and declines, generally taking more than a millennium to reach full fruition. Earlier states established an institutional pattern for rule; later states adopted these, modifying, diffusing, and entrenching them through assimilation; and ultimately this built up a "civilization" in the region. The Chinese and Ethiopians relied more on diffusion of cultural elements from a center; Mesoamerican states already possessed or shared them. For instance, polities adapted logo-phonetic glyphs to many different languages, but their underlying patterns were similar. Although specialists of the three regions discussed will recognize how incomplete these accounts are, they still illustrate the basic idea. One could narrate a similar, though regionally unique, story for Mesopotamia after 3500 BCE, for South Asia after 2600 BCE, for the Andes after 1000 BCE, for West Africa after ~750 CE, and for other areas where states developed.

The final long-term trend involved the growth of social systems even larger than civilizations, still rooted in cities. Some were peer–polity interaction spheres of the special type called state systems. Some argue that Sumer was a system of city-states with one city hegemonic over others, and classical Greece was a hegemonic duopoly of two dominant city–states (Watson 1992). Scholars disagree about power relations between Mayan city–states (Marcus 1998; Yoffee 2005). Territorial states also formed in systems; examples include the eastern Mediterranean between 2000 and 1100 BCE (Watson 1992; Matthews 2018); the Chinese "Warring States" period (481–221 BCE) (Higham 2018); ongoing conflict between Mesopotamian and Greek states before 200 BCE (Alcock and Cherry 2018); Indian states between the Gupta empire and the Delhi sultanate (500–1200 CE) (Trautmann 2016); and states in the Valley of Mexico. Many state systems existed within civilizations; others developed between them. When city-based states did not fight, they traded, leading to an even larger social form, world-systems. As discussed, these encompassed limited "social worlds," not the whole planet. Wilkinson (1995) argues that interstate wars and interurban trade define civilizations, not culture, so he equates civilizations with world-systems; others define them culturally, as I do (Kristiansen 2018a). Thus, another definition of world-systems could be patterns of interaction between civilizations.

As mentioned, some characterize the trade system centered on Southwest Asia before 5000 BCE as a world-system. It expanded after states formed (Grinin and Korotayev 2017). A short history of it would be useful; it constitutes a partial retelling of the chronology above from a larger perspective. At first, the most important trade was in copper and tin, sources of which were far from Southwest Asia (Kristiansen 2018a, b). This provided a platform of sorts for all other trade, which increased greatly after state formation. In polygroups, precious goods or "wealth finance" constituted the main traded items; once states developed, trade in basic goods that most people needed, or "staple finance," became dominant (Hayden and

Earle 2022), though trade in prestige objects was still important. By 3000 BCE, Mesopotamian and Egyptian kings traded representatives with each other. These merchants carried textiles, leather, jewelry, weapons, and aromatic oils; they negotiated and increased regular, consistent trade (Smith 2009). Soon merchants operated independently, developing local and long-distance markets. By 2000 BCE, Mesopotamian and Egyptian states traded copper, gold, ivory, and jewelry with the Indus Valley, which sent Tibetan jewelry and Southeast Asian cloves westward (Smith 2009). Mesopotamian states also imported copper from the Alps and Wales, tin from northern Bohemia and Cornwall, and amber from the Baltic (Scarre 2018b). Minos, on Crete, exported timber, then olives and wine. Mycenae exported olive oil and wine to Scandinavia and Sicily. By 2000 BCE, a new trade system had formed in the Yangzi River Valley around the Erlitou and Shang states. China and Mesopotamia traded through Central Asia (Smith 2009) (Fig. 6.6).

This system shrank drastically during the Bronze Age collapse, partly because iron was more easily produced locally. It grew again with growth in specialized iron products and the development of the Mediterranean Phoenician trade network (Smith 2009; Kristiansen 2018a). The system expanded even more with the Achaemenid and Alexandrine empires. Some argue that there were two systems, Mediterranean-Mesopotamia and East-South Asia, at this point. Trade in iron, precious stones, and bulk products developed greatly after 500 BCE, and the establishment of Roman and Han empires created one massive trade system from the Atlantic Ocean to the Pacific (Beaujard 2018). Central Asian pastoral groups became the "nomadic periphery" of this system (Grinin and Korotayev 2017:26, 31), having commercial links with both the Han and with Rome. The Han empire expanded westward, building roads that started the Silk Road; the Romans built roads, ports,

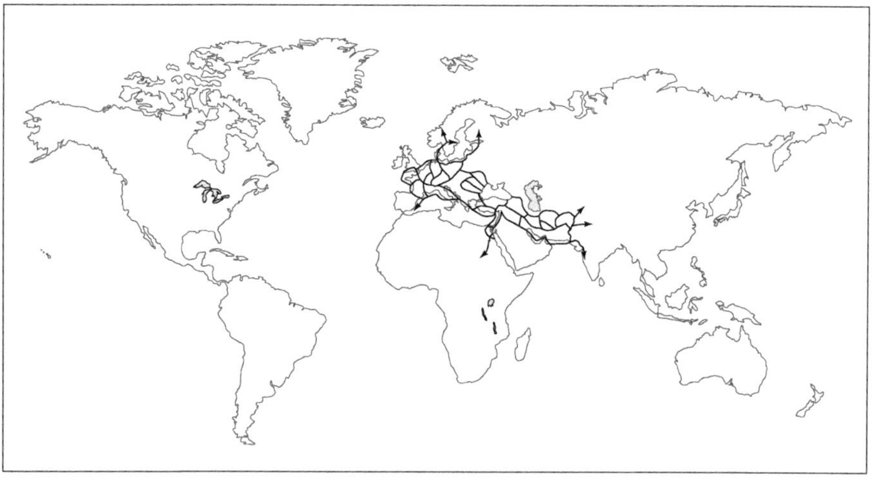

Fig. 6.6 Continental trade systems, 2500–1800 BCE. Sources: Moore (1995); Black (2005); Bahn ([2000] 2009); O'Brien (2010); Houston (2018)

and canals around the Mediterranean. Both efforts spurred a tremendous increase in trade (Smith 2009). Once merchants learned Indian Ocean seasonal storm patterns, Malays, Indians, and eastern Africans navigated around it once a year. Chinese silks, perfumes, resins, cosmetics, and spices went westward, while Roman copper, tin, iron, silver, and gold went eastward. This system had multiple cores and peripheries, consisting of a series of regional economies closely linked to each other (Abu-Lughod 1989; Oka 2018). Political and intellectual elites distrusted the merchants that made the system work (Oka 2018), but politicians tolerated them because they generated wealth (Fig. 6.7).

With Roman and Han decline, the system shrank again (Grinin and Korotayev 2017). Trade increased again with Islamic conquests and the resurgent Chinese Tang dynasty. By the tenth century CE, West African states exported gold northward and eastward through the system; the Crusades opened trade between the Levant and western Europe (Smith 2009). A large trade in Central Asian and East African slaves developed in Southwest Asia and the Indian Ocean (Pomeranz and Topik 2018:map, 168–169; Kent 2021). Rice and papermaking, originating in East Asia, spread as far as Spain, while cotton from India spread to Egypt and Spain. Soldiers in Palestine "wore chain mail from the Caucasus and carried steel swords smelted in India from iron mind in present-day Tunisia" (Pomeranz and Topik 2018:21). After their destructiveness, Mongol conquests fostered more trade. In the early 1400s, establishment of the Ottoman empire reduced European trade (Smith 2009; Grinin and Korotayev 2017).

A large-scale trade system existed in the Americas; it may or may not have amounted to a world-system. Two trade systems in Mesoamerica, the Mayan lowlands and the Mexican highlands (the Valley of Mexico), traded with each other

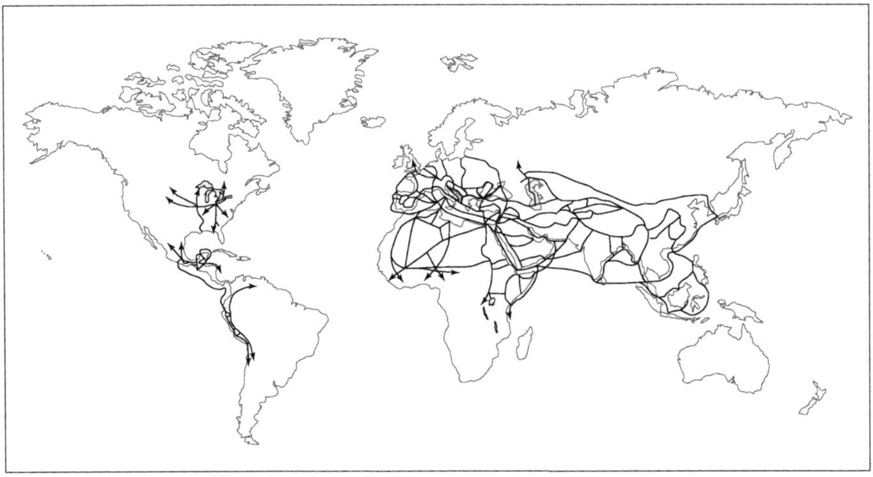

Fig. 6.7 Continental trade systems, ~1 CE. Sources: Black (2005); Bahn ([2000] 2009); O'Brien (2010); Ballesteros Danel and Arango (2023)

(Hirth and Pillsbury 2013). Chaco Canyon and Casas Grandes had "deep and enduring" trade ties with Mesoamerica (Peregrine and Lekson 2012:66), "mediated by peoples of the desert region of northwest Mexico" (Carballo 2012:169). Cahokia, near present-day St. Louis, controlled a large North American trade system which seems only tenuously linked to Chaco and Mesoamerica. While some argue that "Andean civilization did not depend significantly on long-distance commodity trade" (Earle 2018:466), goods still moved throughout the Andes on road systems built up over centuries (Carballo 2012). Finally, Mesoamerica and the Andes traded regularly. Similar ships, similar objects (hairless dogs, pottery styles, figurines, and clothing), and similar practices embodied in objects (shaft tombs, diving techniques, specialized paints, metallurgy, and axe-monies) appear in both areas far inland, showing integration of these foreign elements into local trade systems (Ballesteros Danel and Arango 2023).

These systems operated on several basic principles. First, any such system was a complex network of connections, and commodity chains constituted their individual strands. Trade in each commodity—Roman and then West African metals, Chinese silks, ceramics, and spices—occurred along a "chain" of markets, each with multiple sellers and buyers. The Silk Road, made up of Central Asian and Indian Ocean routes (Haksöz and Uşar 2012), generally transported luxury goods through such urban markets. Second, state institutions supported trade in their areas: militaries protected markets, multilingual translators supported those who wrote contracts and bills of lading, and financiers loaned money (Haksöz and Uşar 2012). Furthermore, ideologies such as Islam during the caliphate and Buddhism in the Mongol empire integrated the system, uniting various merchant communities (Chase-Dunn and Hall 1997). Indeed, ethnic diasporas seemed to be the foundation for much of the trade in the Afro-Eurasian world-system; for instance, a Jewish firm in Cairo had branches in India, Iran, and Tunisia (Pomeranz and Topik 2018). Third, the diversity of the system promoted the diffusion of innovations throughout, and this accelerated systemic changes. That is, changes in one place stimulated changes in others, although they seemed locally generated (Grinin and Korotayev 2017). This happened through cities (Chase-Dunn and Hall 1997), which were the nodes in the network, sites of markets, and foundations for states. Thus, the system was a network of urban nodes in states; in turn, the system influenced those states. Finally, despite all this trade, most economic output was grown, traded, and consumed locally. Only a little entered long-distance trade networks as luxury goods, often on land, while bulk goods generally went by sea at shorter distances (Chase-Dunn and Hall 1997; Oka 2018) (Fig. 6.8).

Overall, then, over five millennia, states transformed the lives of people living within them and changed groups outside them. As they expanded geographically, they also developed multistate trade systems, integrating most of Afro-Eurasia and, to a lesser degree, the Americas.

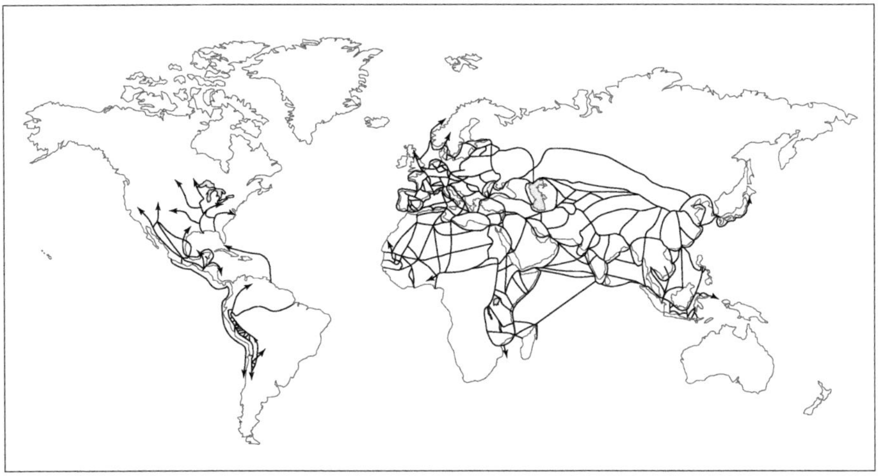

Fig. 6.8 Continental trade systems, ~1400 CE. Sources: Wolf (1982); Moore (1995); Santon and McKay (2005); Bahn ([2000] 2009); O'Brien (2010); Carballo (2012); Houston (2018); Ballesteros Danel and Arango (2023)

Social Patterns of States and Empires

In states and empires, people developed a broadly similar way of life. They relied on intensive farming, developed money, expanded trade, and invented new status goods. States became war and tax machines that also provided public goods. Fertility rates increased again and populations expanded, though mortality rates also increased. People invented new technologies, including writing; formalized idea systems; and developed scientific systems. Economic inequality intensified, and men usually subordinated women. Finally, states tended to degrade their physical environments.

State Economies

Plows in Afro-Eurasia and raised fields in the Americas intensified agriculture. Manufacturing consisted mostly of handcraft production. Internal and external trade increased greatly, stimulating finance, law, and accounting, and new forms of consumption.

Plows became common in Eurasia and differ from horticulture (stick-digging) in two ways. A plow digs more deeply into the soil than hand tools do. It also lifts and overturns soil, keeping weeds down. The ard was a digging stick people pulled with ropes and guided with handles, scraping the land deeper than hoes could. Farmers drew ards through fields in rows, then planted seeds in the rows (Sinclair and

Sinclair 2010). Eventually animals, not people, pulled the ard. The ard spread to Egypt and other parts of Afro-Eurasia—becoming the plow in some places—but not to sub-Saharan Africa (Mazoyer and Roudart 2006; Scarre 2018a).

Agriculture developed differently in the Americas. In Mesoamerica, some farmers grew crops on half their fields, letting the other half lay fallow for 2–3 years, then reversing the fields. Hill terracing increased the available land. Raised-field agriculture, developed in South American tropical wetlands, spread to the Andes and appeared differently in Mesoamerica. In South America, it involved "mounding up a raised bed of earth … flanked by furrows," then planting crops on top of the raised beds, running water in the furrows (Moore 2014:161). In Mesoamerica, farmers built raised square islands in lakes, called *chinampas*. They defined the islands by planting trees in four corners, then built up the islands within the corners by dredging muck from the lake. The lake always kept the islands wet, enabling high fertility and sustaining high Mesoamerican populations. The Maya used orchards integrated into forests; small, terraced gardens; chinampas; and swidden (Sutton and Anderson 2010:296). The land sloping from the Andes Mountains down to the Pacific changes every few miles, creating microzones with different foods; people in zones traded food, increasing their food diversity.

The division of labor across states became more complex. Farmers grew surplus food, some of which states and property owners took as taxes or rents, respectively. They used the surplus to employ specialists, often handcraft artisans, to make clothes, jewelry, tools, cookware, weaponry, shoes, barrels, special foods, buildings, and furniture. Handcraft specialists appeared in the Indus Valley society 5000 years ago, China 4000 years ago, Europe and Mesoamerica at least 3000 years ago, and Peru around 1500 years ago (Shimada 2007; Higham 2018). Craftsmen and craftswomen worked independently for rulers or elite families or worked from home, organizing themselves into associations. Eventually workshops became larger; by the eleventh century CE, Chinese blast furnaces—larger than in any other region in the world at that time—produced vast quantities of iron (Hartwell 1966).

States also improved energy sources. Human muscle was most important before use of secondary animal products; only gradually did humans use horses, cattle, camels, llamas, reindeer, yaks, and elephants to plow, pull wheeled wagons, lift water from wells, and carry loads over distances (Smil 2017:60–61). Good roads helped both pack animals and wagon pulling. They also developed inanimate energies. Sails made fast water travel possible, obviously increasing trade. People invented the waterwheel in Eurasia by the first century BCE; it lifted and moved water in irrigation and milled grains (Smil 2017:62–63). After 500 CE, Southwest Asians developed windmills; this technology eventually spread to Europe and East Asia.

Larger regional divisions of labor stimulated trade. Except in the Andes, urban markets converted the surplus from agricultural product to money—from staple to wealth finance (Johnson and Earle 2000)—through sales increasingly relying on metal monies. Graeber (2011) argues that credit and debt (personal obligations) always developed first, and items representing these debts (money) developed later. The first monies measured debts to the state and between state agencies and were denominated in dates, olive oil, barley, and other foods. For long-distance trade, precious metals, stored in rods, bags, or bars, were more useful (Roberts 2011). As these superseded barter and commodity monies, Chinese dynasties started issuing metal coins in standardized weights, and Mediterranean states also used coins (Davies 2016). Later, financial markets appeared in which one could borrow or lend money at interest. Loans, contracts, and insurance developed. Rich merchants often served as, and sometimes developed into, bankers.

As discussed above, there were multiple ways to sell goods. In ancient Sumer, "Long-distance trade generally took the form of gift exchanges between rulers" conducted through state representatives (Roberts 2011:14). These agents soon engaged in private trade, becoming merchants. Some merchants became "tax-farmers," tax-collectors sharing a percentage of their take with the state. As centuries passed, most states relied on markets, places where multiple sellers sold to multiple buyers (Rosenbaum 2000). All empires, except in the Andes, developed multiple levels of markets. Smaller village markets sold to bigger town markets, which sold to larger city markets (Nolan and Lenski 2011:154). Markets provided towns and cities with perishable food from local hinterlands and less perishable food from longer distances. Large cities, especially port cities, were the nodes in the Afro-Eurasian and American world-systems.

Two examples illustrate both levels of markets and their operations. Chinese village markets operated on a weekly schedule. In the smallest markets, families traded locally; larger villages transshipped agricultural goods to towns and cities. Average people increasingly engaged in trade in addition to producing their own food (Skinner 1964). In the Aztec empire, markets existed in nearly every town and city; the capital's massive market sold nearly anything produced in the empire (Berdan 2005). Aztec merchants were mostly state officials, so trade "was largely in tribute goods" (Pomeranz and Topik 2018:32). In both, other state officials provided services such as keeping order, adjudicating disputes, and enforcing weights and measures (Skinner 1964; Berdan 2005). This third-party enforcement was a public good. Reputations throughout ethnic diasporas increased trust between transaction partners across long distances. However, strangers could meet locally in markets, and states enforcing market rules made transactions more trustworthy (McMillan 2002:Chapter 5; Blanton and Fargher 2008).

Inequalities in consumption intensified. "Consumerism" slowly developed in states in which "people formulate their goals in life partly through acquiring goods that they clearly do not need for subsistence or for traditional display" (Stearns 2006:vii). Militaristic elites early in imperial conquest valued simple lifestyles; their descendants often enjoyed luxurious food, dress, furniture, and housing. Monetization and long-distance trade increased elite consumerism globally before 1500 CE. Elites displayed their high status with special goods: swords for European nobility, silks for Chinese officials, and private transportation everywhere. Elite consumption often stimulated craft production, especially among artists and musicians. While most people were poor by today's standards (Stearns 2006:3–4), over the millennia, average people started buying elite goods. In response, elites passed "sumptuary" laws punishing commoners who used elite objects (Hollander 1984; Trigger 2003:143–144).

Over time, states grew richer. There are two types of economic growth, "extensive" and "intensive" (Jones 1988) or "Smithian" and "Schumpeterian" growth (Brooke 2014). *Extensive* growth is the geographic and demographic expansion of a society, in two forms. First, people can spread out geographically. Each new household is about as wealthy as the previously existing ones, but the total amount of wealth grows. Wealth per person can also grow, but slowly. Second, if a king conquers a larger area, taxing more people, he is richer, but the people are not. In *intensive* growth, innovations—new technologies or forms of organization—help each plot of land or each artisan's workshop produce more than it used to, so each farmer or artisan is richer. Repeated over a large area, this increases the average amount of wealth per person, and the same tax rate can generate more taxes. There were often periods of intensive economic growth, "efflorescences," which often occurred during reconstructions after collapses (Goldstone 2002:357). They were not merely recoveries, but real increases in wealth per person from innovations. People often regarded them later as golden ages since they tended to end.

In states, agricultural production increased greatly, producing surpluses that people used for handcraft industries, for trade in markets, and for new consumption goods.

State Politics

As states transformed economic activity, they also changed political structures. Governments both provided public goods and services and helped elites exploit populations, and each required taxation; in turn, public goods, exploitation, and taxation all required bureaucracies. While state leaders recruited nonkin people into bureaucracies and developed militaries, this took centuries, perhaps millennia, to accomplish fully. Some, perhaps many, states used dehumanizing ideologies in conquest.

Basic public goods were similar around the world. States always provided defense (after conquest, of course), judicial services, and public order (Blanton and Fargher 2008:133). Monumental architecture was obviously public. Provision of transportation infrastructure within and between urban areas was also common, as were public water supplies, though cities did not always provide sanitation. Finally, monetary systems, well-regulated (i.e., trustworthy) markets, and public order always facilitated trade (Blanton and Fargher 2008).

State taxation paid for these services and helped elites get rich, but taxation was always difficult, for three reasons (Ardant 1975). First, productivity was low: any harvest yielded only a "small multiple" of the number of seeds planted (Ardant 1975:174), and up to half of any farmer's crop went to taxes or to landlords' rents (Nolan and Lenski 2011). Second, most market systems were not developed enough to enable collecting taxes in money, or metal money was scarce. Thus, most states operated primarily on staple finance: collecting food as taxes (Willer et al. 2017:443). Finally, calculating tax payments was difficult in semiliterate societies. Everyone had reason to hide product, making tax collection contentious. Only in the Song dynasty in China after 1000 CE did a state rely mostly on taxes in money, and it still proved unstable (Liu 2015).

Providing public works also required labor. Chiefly "assistants" gradually became state bureaucrats in formal, hierarchical organizations: leaders commanded subordinates who commanded lower subordinates, and so on in a "pyramidal chain" (Willer et al. 2017:442). Polygroups involved face-to-face control; a leader's representative could show up anywhere in person, limiting the chiefdom's size. In states, bureaucracies divided labor enough so that each position had narrow powers, reducing insurrections, and long-distance bureaucratic communications enabled rule over vast territories. However, the great diversity of polygroups (McIntosh 1999; Grinin and Korotayev 2017) means that early states were also diverse, so the foundations of the earliest kingdoms varied. Some emphasized the personal power of the monarch, which made delegation to others difficult (Graeber and Wengrow 2021).

Thus, pyramidal bureaucracies probably took centuries or millennia to develop. Some scholars say that kings usually wanted to involve themselves ever more deeply in their realms, so they developed principles of "*impersonal* equivalence" or "fictions of legal equality" (Graeber and Wengrow 2021:425, original emphasis). However, as most empires formed through conquest, co-optation of or alliance with local elites was always necessary: "Community leaders became de facto state agents" (Graeber and Wengrow 2021:424) who "would take care of most of the actual government on the ground" (Bang 2021:32). For instance, the Aztec Triple Alliance ruled other city-states in the Valley of Mexico directly, through appointed officials and a bureaucracy, but ruled conquests farther out indirectly, by conquering then backing the local ruler, requiring tributary payments (Smith 2012:51, 56). In general, the center threatened local elites enough militarily to keep their loyalty but supported them with that same military to allow them local control. Furthermore, "In the long run, although seemingly cyclical, the history of the aristocratic empires is one of slow consolidation and accumulating complexity" (Bang 2021:34). This process seemed to proceed most completely in China: "each new dynasty gradually [widened] its hold on government and elite society" (Bang 2021:34; also, see Pirie 2021:247). As bureaucracies were established in the capital, and increasingly in regions, they consisted of at least three hierarchies: the tax-collection system, public works (roads, irrigation systems), and a justice system to adjudicate disputes. Other bureaucracies supporting education or science sometimes developed.

Nearly all states had military institutions, which expanded as states did. However, institutional patterns varied around the world. For instance, in pre-Han China and post-Roman Europe, feudal systems provided land and serfs—farmers tied to the land—to war leaders in return for military service (Anderson 1978; Fairbank and Reischauer 1978). In the Islamic caliphate just before 1000 CE, governors of provinces were also generals, and judges were also religious leaders because they implemented Islamic law locally (Sourdel 1983). In the Gupta empire, the king maintained a standing army, while provincial governors were not military, and the priestly class did not administer law. Chinese generals were mostly subordinate to civilians (Hodgson 1974; Sourdel 1983; Hucker 1973).

Recruitment into these bureaucracies also varied. Unlike chiefdoms, states had too many positions to fill with kin (Nolan and Lenski 2011). To become a Chinese government official, the state required aspirants to pass three levels of examinations on classic Confucian texts, which took years of study; fewer than 10% passed at each level. Early on, rich men financed their sons' studies; later, merchants and peasants received classical educations to become officials (Fairbank and Reischauer 1978). In contrast, in Islamic Southwest Asia around 1000 CE, generals received plots of land for their salaries, which eventually became hereditary; they left recruitment of lower officials to local tradition (Ashtor 1976; Hodgson 1974). Medieval English recruitment was yet different; only rich landowners entered government service (Blanton 2016).

In early states, large bureaucracies may not have been necessary. City-states had public works, but only on a small scale. Some city-states had assemblies for all citizens, such as in the Greek *poleis* around 500 BCE. Others were ruled by "competing elites and factions"; in sub-Saharan Africa, controlling labor sometimes mattered more than controlling land (Nichols and Charlton 1997; McIntosh 1999). In territorial states, relations between conquerors and conquered took several forms besides formal incorporation (Watson 1992). Some kings paid annual tribute to the conqueror, otherwise staying independent. Some empires exercised dominion over client states, such as the Romans over Judea in Herod's reign, with occasional internal interference.

Empires increased in size and complexity over time (Marcus 1998). By 500 BCE, states controlled not quite 10% of the earth's landmass. By 1000 CE, 1500 years later, states controlled at least 20% of its landmass. Five hundred years later, the figure approached 40% (Myrdal 2012:41). States also increased their control over people, but Table 6.2 shows no specific trend in increasing state control over populations, at least when considering the two largest empires at any time.

Rice culture always supported very large populations, so South Asia and East Asia have been the world's most populated areas for over three millennia. Demographically, the largest empire was usually a Chinese dynasty, and the second largest usually controlled the north Indian plain. Together, they consistently controlled 30–40% of the world's population (Myrdal 2012). Chinese and Indian states

Table 6.2 Two largest empires' shares of world population (only if larger than 15%)

	Largest empire	Second largest empire
400 BCE	15	<15
200	26	18
1 CE	29	21
200	32	21
400	18	16
600	23	<15
800	20	<15
1000	24	<15
1200	21	<15
1400	21	<15

Source: Myrdal (2012:43)

and, we can infer, others around the world managed ever-larger and more complex populations, suggesting improvement in statecraft over time.

Overall, then, each agrarian state's bureaucracies, considered together, collected taxes, usually in food; provided public goods; made elites rich; and changed the basis of recruitment. Sometimes states remained subordinate but quasi-independent of others.

States and Ideas

Major cultural developments in regional civilizations included writing and writing-based religions, formal moral systems, science, artistic development, and new technologies.

Assimilation into a religion or a philosophy reduced several collective-action problems. One was interpretive heterogeneity, especially for local elites: a broadly shared belief-system provided a shared vocabulary and common metaphors. Shared beliefs also justified inequalities, i.e., provided legitimacy for Catholic and Orthodox priests, Muslim scholar-judges, and Confucian-educated officials. Shared beliefs rendered co-religionists "one of us" at least sometimes, so they reduced defection, and probably free-riding, at all levels.

Religion in early states seems to have extended animism. People did not distinguish "between what we perceive as the natural, supernatural, and social realms" but believed in "powerful forces that possessed human-like personalities" (Trigger 2003:411, 412). Pantheons of gods developed in diverse societies such as Rome and Mayan city-states (Mark 2012; Rüpke 2018). People tried to persuade gods—anthropomorphized spirits—to intervene on their behalf. Deities regulated aspects of the cosmos and were male, female, bi-sexed, or alternating. Different deities did different things: some addressed natural phenomena (rain, sun, wind, and so on), while others managed humans, illness and health, natural landforms, and fertility.

People in early states believed in a cosmic power suffusing the entire universe (Trigger 2003). While some scholars have argued that these societies were consequently monotheistic, others argue that creator gods did not always rule over other gods (Trigger 2003). True monotheism developed in Eurasia in a supposed "Axial Age," from around 900 BCE to 300 BCE (Jaspers 1953), rooted in the expanding Afro-Eurasian world-system. In this period, religion became universalizing and transcendental. Chinese philosophers such as Laotzi and Kong Fuzi; the writers of the Hindu Upanishads and Siddhartha Gautama, the Buddha; Zarathustra, the founder of Zoroastrianism; the Hebrew prophets in Judea; and ancient Greek philosophers all abandoned pantheons of local gods for universal philosophical issues, with and without monotheism (Armstrong 2006), appealing to "each person in the world" (Loftin 2000:137). This pattern fostered the first monotheistic religions in central Afro-Eurasia: Zoroastrianism, Hinduism, Buddhism, Judaism, Christianity, and Islam. Hinduism both retained pantheons of gods and emphasized the one underlying Reality. Confucianism is a philosophy but operated in China as a quasi-religion. Religions in the Americas remained rooted in pantheons of gods. Some universalizing religions developed de facto pantheons, such as Christian saints.

Scholars strenuously debate the Axial Age, either asserting its reality and influence (Armstrong 2006) or denying its reality (Provan 2013). In any case, universalizing religions spread throughout most of Eurasia. Zoroastrianism remained in Iran. Rome put down a Judean rebellion and exiled the Jews throughout the empire. Christianity followed, then moving to northern Europe, the Caucasus, and East Africa. Hinduism remained dominant across South Asia. Buddhism spread into East and Southeast Asia. Islam spread westward across North Africa and into the Iberian peninsula, and eastward to the Indus Valley. After Afghan conquerors established the Delhi sultanate in India, Islam spread across India, though thinly, and through trade, it expanded into sub-Saharan Africa and maritime Southeast Asia (SarDesai 2013).

These religions all followed a similar institutional path: one or a few people founded them; followers wrote down the founders' ideas; and the growing priesthood developed the ideas over centuries (Smith 1991). Religious institutions spread as key components of civilizations; each set of ideas elaborated and adapted to local conditions. Some sort of priestly group developed and recruited people (usually men), collected monies, distributed charity, held public ceremonies, counseled locals, and usually advised loyalty to the state (Smith 1991).

Religions spread much more easily because all states developed writing systems. There were three main traditions. There is debate about whether Mesopotamia or Egypt developed the Afro-Asiatic tradition (Powell 2009; Fischer 2018). Either way, it spread from one to the other, then eastward to the Indus Valley and westward to the Mediterranean, each geographical movement requiring adaptation: "The greatest changes in writing systems seem to occur with speakers of other languages who borrow and adapt an ill-fitting system" (Fischer 2018:119). Greeks developed a purely phonetic system: each symbol represented a sound. The East Asian system was based on glyphs (characters), possibly under inspiration from West Asia (Fischer 2018), possibly independently (Powell 2009). The Mesoamerican glyph system originated independently of these around 2500 years ago (Powell 2009; Fischer 2018). Writing only sometimes derived directly from semasiographic systems in kin-ordered societies (Trigger 2004). For instance, while Mesopotamian tokens seem not to have influenced later writing systems (Schmandt-Besserat 1996), some Chinese pictograms seem to have become characters later (Barnes 2015; Higham 2018). The Incan empire developed *quipu*, a system of colored strings of knots storing information (Urton 2017).

Writing eventually became important to statecraft. Most state leaders were literate, but literacy in any civilization's populace varied greatly. For instance, perhaps 10–15% of the Roman male population was functionally literate, the female population less so (Everett 2009). However, literacy in the Roman army was much higher. Furthermore, literacy was a variable; some could write their names and read simple documents, while others could read and write more complex text. Furthermore, writing was important at all levels, and the less literate could often hire a writer (Kolb 2016). Literacy increased in Christian Europe but still remained low. In

Achaemenid Iran, literacy was actually higher in the provinces than in the center (Kolb 2016), while it seems to have remained mostly, but not exclusively, an elite phenomenon in Mesoamerica, though there were public inscriptions (Jiménez and Smith 2008). In contrast, schools teaching literacy in China started over 2500 years ago. The Song dynasty attempted to have all children in schools. By 1270 CE, every fifty households had to set up a school for their children (Wang et al. 2009). Still, estimates of effective literacy by percentage of the population seem rare; furthermore, "[w]omen generally were denied access to education in ancient China" (Wang et al. 2009:401), so "children" generally meant "boys."

Religion changed in states in two other ways. First, monumental architecture became even larger. Egyptian, Mayan, and Aztec pyramids and European cathedrals took centuries to build, while Chinese emperor burials were also elaborate. Second, non-state cults for commoners developed alongside state religions (Hayden 2003). For instance, in Roman cities, people formed urban mystery cults, often composed of people from the same region. They offered rural-to-urban migrants friendships, marriages, and spiritual meaning. Ceremonies provided powerful emotional experiences to bond with other members.[4]

The state, religion, and morality came together in legal systems: "Laws provide rules for moral conduct" (Pirie 2021:12), and legal systems were usually based in religions. The earliest laws came from Ur in 2112 BCE; four centuries later, Hammurabi developed a comprehensive legal code regulating agriculture and commerce (Pirie 2021), one that provided a model for Mesopotamian regimes for the next two millennia. While states in the Americas and Africa operated on customary rules, plus monarchical and bureaucratic edicts—not on legal systems based on "general rules" (Pirie 2021:11)—Texcoco, one of the three city-states of the Aztec Triple Alliance, had a well-developed court system of many levels, and its principles somewhat resembled Legalism in China (Offner 1983); there seems to have been a legal code, though it was not comprehensive. In contrast, some Eurasian states "created objective standards, explicit sets of rules" that "outlast[ed] the authority of the lawmakers" (Pirie 2021:11, 12). While laws were always tools of statecraft, they gave at least some power to ordinary people: "People have … quoted laws to challenge government decisions, resist abuses of power, and seek justice" (Pirie 2021:11). Indeed, some aspects of Mesopotamian law seem designed to mitigate the growing economic inequalities in states (Pirie 2021:42–44).

Legal systems rested on quite varied civilizational belief–systems. Mesopotamian law rested on "casuistic" foundations—that is, laws were written in an "if–then" mode: if someone broke a law, then this was the punishment (Pirie 2021:Chapter 1). This case-based method allowed Mesopotamian systems to be adapted to varied religious contexts. In contrast, "In the Hindu tradition there was never any clear divide between law and religion" (Pirie 2021:45). They were rooted in the sacred texts (Vedas), written around 1000 BCE by Indo-Aryan invaders from Central Asia. Subsequent power struggles between the incipient priestly caste (Brahmins) and

---

[4] I found nothing on private religious associations in other empires, though they must have existed.

warrior caste (Kshatriyas) over the next millennium led to Brahmanic legal systems. Their "overwhelming emphasis ... is on the correct behaviour of individuals according to their caste, family, and life stage," emphasizing duties, not rights (Pirie 2021:54). Chinese law started with the Zhou dynasty (around 1000 BCE) whose "kings took it for granted that imposing punishments was the way to maintain order" (Pirie 2021:73). The philosopher Kong Fuzi ("Confucius") instead emphasized individuals' obligations to others, especially in families. The short-lived Qin state returned to the older philosophy of punishments. The successor Han dynasty relied on both harsh Legalism and Confucian ideals (Ruskola 2012; Pirie 2021:89). The recruitment examinations started in the Tang dynasty required deep knowledge of Confucian texts, which permeated the legal codes that subsequent dynasties issued at their beginnings, although they always expanded previous codes and maintained a punitive Legalist bent. Yet as a multiethnic empire, the Chinese state "allowed multiple forms of law" on its peripheries (Ruskola 2012:266). Roman law originated in class conflict (Pirie 2021:99–101): commoner plebeians demanded a code of rules from elite patricians, and the Twelve Tables, descending from Mesopotamian traditions, laid out its principles. During imperial expansion, Rome established a legislature and a court system. Roman law continued partly under Gothic invaders, partly through Charlemagne's rule and partly from subsequent expansion of Christianity, especially to Britain and Germany. This sparked a revival of Byzantine emperor Justinian's *Digest* of Roman law, which became the foundation of most European law, except in England (Pirie 2021). Islamic law originated supernaturally, in the word of God (the *Qur'an*) as dictated to the Prophet Muhammad, though most law was adaptation of Qur'anic ideas; thus, implementation of God's Way (*Shari'a*) was always imperfect. This adaptation required "sophisticated conceptual frameworks" for application to daily life (El Fadl 2012:297), partly "inspired" by Greek and Jewish legal traditions (Pirie 2021:123). Islamic scholar-judges gradually developed a massive body of law, one that rarely influenced public policy, though qadis judged caliphs and other rulers, a position that often created "troubled relations" between the two groups (Pirie 2021:139).

Most states encouraged science (McClellan and Dorn 1999). While it was often separate from technology, much science arose from confronting practical problems: Constructing monumental architecture or irrigation systems. States developed staffs of engineers who learned practical mathematics; a quest for "pure" knowledge came later. Tax collection and budgets also encouraged mathematical thinking, as did astronomy to understand agricultural cycles, the basis of state power. Astronomy-based calendars developed in every area of primary states. In every case, religion and calendars were closely intertwined (McClellan and Dorn 1999).

States also changed the arts. Artists were craft artisans, and art took increasingly varied styles around the world. Monarchs and elites in the provinces were often artists' customers, commissioning specific projects for personal or public service. In Iranian empires for five millennia, much public art and architecture demonstrated the power of the king, and court manners and artistic styles often diffused from the center to peripheries (Franke et al. 2021). The Chinese imperial court was "the most generous and consistent patron of religious art," yet scholar-officials, elite women, merchants, and ordinary people also sponsored funerary and other art (Lee and Xu 2023:1). Monarchs and wealthy patrons sponsored art in Europe (Stokstad 2008). Royal and elite sponsorship of art also seems to have occurred in the Americas. Of course, elites could afford professional artists; most people could not and made their own visual arts, music, and theater.

Technologies slowly, continuously, and cumulatively improved around the world, with effects on agricultural productivity, trade, warfare, and urbanization. Civilizational technologies developed in three long periods—during state formation, through 500 CE, and between 500 and 1400 CE (Headrick 2009, the source of the following discussion). All primary states developed irrigation systems with terraces, dikes, sluices, or canals. People gradually improved personal technologies such as clothing and pottery. Central Asians domesticated horses by 4000 BCE; Mesopotamians developed the wheel after 3000 BCE; and these were combined after 1700 BCE to make war chariots (Headrick 2009; Baumer 2012), which increased warfare across Eurasia. Iron weaponry after 1000 CE helped increased it more. Roman aqueducts carried water hundreds of miles to major cities. Successive Chinese dynasties constructed canals between rivers for trade and water control. Rome, China, and the Incas built major road systems that supported the movement of goods, armies, and information. Ships and ports also developed. Special collars made horses more useful, improving agriculture and transportation. The Islamic empire diffused Chinese wood-based paper, the compass, and other inventions. The Mongol empire fostered diffusion of gunpowder and movable type from China to Europe.

In states, religion started changing from pantheons of multiple gods to monotheistic systems. States increasingly imposed cultural assimilation on subject populations, always an incomplete process. States developed writing systems, and technology and art elaborated.

State Inequalities

The semiformal inequality of polygroups became institutionalized in states as "stratification," a geological metaphor mentally arraying people into permanent "levels." People in the upper "strata" commanded greater material, political, or intellectual resources. Broadly, people who avoided physical labor had more power than those who engaged in it.

According to some, early state stratification systems were similar (Trigger 2003; Nolan and Lenski 2011). In theory, the king commanded all societal resources; in practice, this was limited, but kings and emperors were still the richest of all. Elites were "no more than a few percent of the population" (Trigger 2003:147). They were mostly hereditary landowners, but sometimes high state officials, including generals (Trigger 2003:147–155). Specialists below them included bureaucrats, soldiers and police, merchants, artisans, and servants. Merchants ranged from poor to rich, having more power in city-states. Most people were farmers, "probably amounting to

80 percent or more of the total population" (Trigger 2003:155). They were sometimes tied to land and sometimes free. Wageworkers and slaves were at the bottom. Shang China, the Aztecs and Incas, and early Mesopotamians had small populations of slaves, while Egyptians and Yoruba states "enslaved large numbers of foreign captives" (Trigger 2003:158).

Agrarian inequality is difficult to measure because quantitative data only sometimes survived. Examples may help. Most people in the Roman empire lived at a low subsistence level; the hierarchy was not elaborate (Milanovic 2011). The top 1% consisted of fewer than half a million people in a population of 50–55 million (Milanovic 2011:46–52); their incomes were about 500 times the average income. Only a small middle class existed. A Han "estate economy" formed (von Glahn 2016:134–142). Groups including "senior officials, imperial kinsmen, and the consort families, the kinfolk of empresses" (von Glahn 2016:135) amassed large estates, often with hundreds of slaves or free households working the land. Some sources speak of "middling families" consisted of "perhaps half the total population" (von Glahn 2016:136), so the rest were poorer. A millennium later, around 2–3% of the English population was wealthy (Dyer 1989:32–33); 85% were rural. In the Aztec empire, there was "a continuum of social differentiation in ancient Mesoamerica ... [not] discrete classes" (Brumfiel and Robin 2012:675), but less than 10% of the population was wealthy.

Qualitative, cultural definitions supplement quantitative measures. Early states did not have groups classified primarily by occupation, while later states did (Trigger 2003:144). For instance, the India caste system relies on two concepts, *varna* and *jati*. In sacred texts, the four varnas were the Brahmins (priests), the Kshatriyas (warriors), the Vaishyas (merchants and artisans), and the Shudras (farmers). In reality, jatis organized stratification; these closed occupational groups of intermarrying lineages varied locally across South Asia. Some suggest that they resulted from state conquest of kin-ordered peoples (SarDesai 2008). Elite jatis followed elaborate rules to avoid "polluting" contact with lower castes. As centuries passed, rules of endogamy rigidified, and new jatis formed through wars and invasions. In contrast, medieval Europeans lived in a feudal-manorial system evolved from Roman patterns of landownership. In response to Viking invasions (700–1000 CE), farmers put themselves under protection of large landowners, becoming "serfs": neither slaves, since they had rights, nor free, since they could not leave. The king granted lands with serfs to elite subordinates who lived off the income and served the leader in war. These elite positions eventually became hereditary. Enterprising European serfs had a few, difficult paths to freedom and upward mobility (Anderson 1978). The Aztec noble class was divided into three ranks. The lowest nobles, the *pillis* or *pipiltin*, made up a varying percentage of the populace depending on the region of the empire: 19% in one region, 4% in another (Davies 1987:121). They were sometimes poor, working as artisans, while commoners, the *macehuales*, were occasionally rich (Davies 1987:121–123).

Despite this institutional diversity, some argue that in all empires, landowning elites developed values of duty, honor, and glory, and contempt for work and money (Kautsky 1982). Honor—one's public reputation and thus self-esteem—consisted of doing one's duty to one's king and country, especially in warfare. Surrender tarnished one's honor, but death in battle was glorious. (Triumph in battle was, of course, better.) Working for money was shameful. Expensive, quasi-militaristic leisure pursuits such as dueling, fencing, jousting, and hunting displayed one's freedom from labor. Merchants threatened elites because they were rich but worked for a living. Disdain for merchants was, therefore, especially important (Kautsky 1982).

Elite ideologies socialized people into beliefs about superiority and inferiority of persons (Kautsky 1982:197–205). Vertical linguistic metaphors were common in states; elite ideologies ascribed inferior characteristics to peasants, merchants, and others, so elite dehumanization of others (Haslam et al. 2013) may have intensified. In early states, the unequal supernatural order justified the unequal social order (Trigger 2003:672). Certainly, people in ancient Greece, Rome, and Islamic areas used dehumanizing language about slaves, which was translated into legal codes about slave treatment (Smith 2011:110–111). However, inequality created a sense of unfairness among the populace against elites (Turner 2007), though much ideology suppressed it. State breakdown allowed violent expression of these feelings, especially when elites broke the society-wide deal of public goods for taxes, which always happened eventually.

In states and empires, rank became stratification. Only a few were fabulously wealthy, while most people were poor. Cultural systems of stratification overlaid economic ones, and similar values seem to have possessed elites around the world.

State Demographics

States had old and new demographic patterns. Mortality and fertility increased, global population grew more quickly, cities increased in size and number, and migrations increased.

Mortality and fertility increased slightly from polygroups to states. Scholars estimate life expectancy at roughly 20 years around 10,000 BCE. As discussed, average lifespan subsequently increased. Around 1 CE, global average lifespan was 22 years; a millennium later, it was 27 years (Livi-Bacci 2017)—yet it seems that half still died before age 20 (Volk and Atkinson 2013). Annual attritional mortality continued, while mortality catastrophes such as floods, droughts, wars, and epidemics continued (Bocquet-Appel and Bacro 2008).

States offer textual evidence of grief's impact. King David of Israel publicly mourned his son, Absalom, despite that son rebelling against him (2 Samuel 18: 33, 19: 1–7). The Roman philosopher Seneca discussed the normality of grief, recommending the acceptance of the feeling but not overdoing it publicly (1925, III:129–149). In India, the Buddha advised a woman driven mad by grief for her dead son to go door-to-door to find a family who had never suffered death, but she could not (Kornfield 1993:48–52); this helped her accept his death.

After mortality catastrophes, population bounced back quickly because fertility remained high (Turchin 2003:165). The total fertility rate (TFR) in states was between five and eight children (Boone 2002; Livi-Bacci 2017). In good times, fertility increased; in mortality catastrophes, many women died, reducing fertility. High mortality and fertility combined to produce a now-familiar age structure: many children, some adults, and a few elderly. This was the general pattern for populations in states, but various catastrophes, such as waves of plague or wars, could distort this pattern (e.g., Herlihy and Klapisch-Zuber 1978:189–190, 192–193).

The increased fertility rate, and gradual improvements in food production, allowed the global population to grow through the millennia. Good estimates of global population growth from 3500 BCE to 1500 CE are difficult to find (McNeill and McNeill 2003:29). Still, some think that by 1500 CE, global population amounted to 477 million people (Brooke 2014:259). Other estimates are similar, as shown in Table 6.3.

The table illustrates variations in mortality: Roman-Han collapse (200–400 CE) and the Black Death (1300s CE) reduced populations drastically, yet they recovered and increased again.

Cities also grew. In 2250 BCE, the largest city was around 30,000 people, and there may have been 240,000 people living in cities around the world. By 100 CE, there were five million people in cities; the largest city had 450,000 people. By 1500 CE, over seven million people globally lived in cities, and the largest held 672,000 souls. Often cities grew haphazardly, with buildings simply added to their edges; others were well planned, often on grids. Cities were densely populated, often with filthy conditions: excrement, dead bodies, and garbage were sometimes left on streets (Smith 2019). Some cities built sewage-management systems, but these were often incomplete (Lees 2015). Rural-urban migration—brought on by slavery, war, and economic opportunity—was part of an acceleration of movement. Rulers encouraged skilled groups to migrate to cities. Also in this period, seafaring

Table 6.3 World population growth, 3000 BCE–1500 CE

Date	Approximate world population, millions
3000 BCE	14
2500 BCE	19
2000 BCE	27
1500 BCE	38
1000 BCE	62
600 BCE	110
400 BCE	153
200 BCE	165
1 CE	252
200 CE	257
600 CE	208
1000 CE	253
1200 CE	400
1400 CE	375
1500 CE	461

Source: Sanderson (1995:103)

explorers settled Polynesia (New Zealand, Hawaii, Tahiti, etc.) (Christian et al. 2014:208–209).

As empires increased in size and number, what we would now call multicultural empires became normal. The Roman empire contained Germanic speakers in the north; Jews, Egyptians, and North Africans; peoples in Anatolia and the Fertile Crescent; and the dominant Romans spread thinly throughout. Centuries later, the Islamic empire included Arabic speakers, plus South Asians, Persians, North Africans, and Spaniards. The Mongol empire encompassed Slavs in central Europe, Central Asians, Arabic- and Persian-speaking societies, and most of China. The Aztec empire contained a welter of peoples across Mesoamerica.

In states and empires, mortality and fertility increased yet again. Global population grew over time with "temporary" decreases in populations due to epidemics and collapse. Urbanization led to more and larger cities, and all empires were multicultural.

Kinship and Gender in States

Overall, states consistently subordinated women to men, but the forms and levels of subordination varied greatly around the world. Families remained economically and emotionally crucial, but states took over functions from families, including defense, justice, trade, credit, religion, and education (Nolan and Lenski 2011).

In bands and villages, women were effectively equal with men; in some poly-groups, their power had started to decline, a trend that accelerated in states. Some argue that this subordination was consistent (Bonvillain 2001; Nolan and Lenski 2011); others agree broadly but identify crucial exceptions for some times and places, while noting that subordination took extremely varied forms (e.g., Kent 2021). Others argue that female subordination was limited to a broad band of land across Afro-Eurasia (Hughes and Hughes 2005), yet women in Mesoamerican and sub-Saharan African states also consistently had less power than men, though they were possibly more valued than elsewhere (Kellogg 1995; Sheldon 2017).

Examples illustrate why the debate exists. In the Indus Valley society (2600–1800 BCE), women chose their husbands and performed public rituals; elite women could read. Millennia later, husbands—not wives—made family decisions (at least publicly), and sons—not daughters—inherited wealth. Women gained status from having sons, not daughters (Hughes and Hughes 2005; SarDesai 2008). The later Brahminic codes also "regarded women with considerable suspicion" (Pirie 2021:66). Women in early Israel had social power, which eroded "as Jewish tribes moved from confederacy to statehood" (Hughes and Hughes 2005:24). Elite Japanese women also had power in the early state, especially in the matrilineal system, but over time, families became more patrilineal (Hughes and Hughes 2005). The Mongol conqueror Genghis Khan trusted his daughters more than his sons to administer his empire; after his death, his sons systematically destroyed their sisters' power (Kent 2021). Women's power in sub-Saharan Africa also seems to have eroded as political centralization proceeded (Sheldon 2017).

Elsewhere, female subordination was minimal, or it was severe but eased over time. Women under Egyptian Pharaohs enjoyed many rights. Queens had appointment power and later ruled on their own. Ordinary women were legally equal to men; they bought and sold property and were important in cloth and oil manufacture (Hughes and Hughes 2005). In early Arab states, fathers could bequeath nothing to their daughters; the Prophet Muhammad required fathers to leave half as much to their daughters as they left their sons, a major improvement (Smith 1991). Confucian ideology denigrated Chinese women, who still had considerable property ownership and business involvement (Hughes and Hughes 2005). In the Aztec empire, gender parallelism structured daily activities; there were both male and female administrators at markets, both kinds of gods, male and female priests, and separate property systems (Kellogg 1995). In Greek city-states, Athenian women were severely controlled, while Spartan women had much more freedom (Kent 2021).

State leadership was predominantly male. Men populated armies, bureaucracies, and ruling councils, though queens ruled many states in Afro-Eurasia and in the Americas (Socolow 2015; Kent 2021). Among the Aztecs and Incas, only men went to war and managed the state (Kellogg 1995:Chapter 3). Kinship structures often reinforced this. Since Indian and Chinese families were both patrilineal and patrilocal, "the family" consisted of men, their sons, their grandsons, and their wives. Daughters married out, giving families little incentive to invest in them (Sanderson 2014). Girls were often married young; this created a maturity, and thus a power, differential between husband and wife, reinforcing patriarchy. As states developed in sub-Saharan Africa, many societies shifted from matriliny to patriliny (Sheldon 2017).

Moral codes consistently restricted female more than male sexuality. In practice, most people were monogamous—but most people were farmers. Elite women were often influential, but most married one man and remained sexually faithful, while elite men often had multiple wives or mistresses. In most states, women usually had considerable informal power. This varied greatly; in patrilocal systems, new daughters-in-law had little power, while matriarchs had much, especially those who bore sons. In plow-based agricultural systems, men took over farming, while women worked inside more often (Ehrenberg 1989).

The concept of "female subordination" is useful but simplistic. It accurately describes a long-term trend but also obscures how women got around or creatively used male-favorable laws, wielded much informal power, and made crucial economic contributions to their societies. As the examples indicate, it also ignores tremendous diversity in gender systems. Still, this change from the rough gender equality of early human societies helps us understand those more complex realities (Hughes and Hughes 2005; Kellogg 2005; Sheldon 2017; Kent 2021).

Same-sex sexual relationships continued in states. Mesopotamian statuettes show male–male sex, kings had male lovers, and there were often goddess cults that included male homosexual attendants, with whom non-elite men had sex during rituals. Many Mesoamerican and Andean temples also had these. Homosexuality also seems to have been accepted in ancient Egypt. Mesopotamian rulers used eunuchs sexually as boys and later employed them as officials. Central Asian or "Indo-European" warriors commonly took lovers among each other, and same-sex love was common in China throughout its history. Homosexuality was common in Greek city-states among both men and women, especially in initiation rituals. Being the receiving partner was the only shame in Rome, not homosexual acts themselves. Ascetic Christian sects started condemning same-sex acts, an attitude that spread throughout the church. Islamic attitudes toward same-sex relations have long been ambivalent: condemned in theory but accepted in practice, especially due to the commonly strong segregation of the sexes (Neill 2009).

As states increased demographically and geographically, more women became subordinate to men; however, the forms this took varied greatly.

State Environments

States also increasingly degraded their environments at levels much higher than those in kin-ordered societies (Ponting 2007), which, as mentioned, may have influenced state collapse.

There were three main problems, two related to irrigation: waterlogging and salinization. Over-irrigation added water to a region's underground water table, carrying it away too slowly. Eventually the area became too moist to farm. The constant inflow of new water also brought salts up from water tables, eventually reducing fertility. Finally, trees and their ground cover hold soil in place; deforestation made land useful for farming or pasturage, but prevailing wind and rain patterns in an area eroded the topsoil, creating a desert. Erosion silted rivers, which ruined downstream areas, especially ocean ports at river mouths (Chew 2001; Ponting 2007).

Historians argue that states overused land to generate surplus, which degraded their land and led to imperial decline (Chew 2001; Ponting 2007). For instance, after 3000 BCE, lower Mesopotamia experienced waterlogging and salinization. After 2400 BCE, crop yields dropped 40% and dropped another two-thirds by 1700 BCE. The Harappan state that developed around 2300 BCE managed the flood-prone Indus River, but deforestation led to erosion and collapse. These trends took "hundreds of generations" to play out completely, especially deforestation (Ponting 2007:73), which occurred all around the Mediterranean Sea. Soil erosion from deforestation appeared in Greece by 650 BCE and started in Italy and Sicily after 300 BCE. Deforestation continued in Spain after Rome's collapse. Waterlogging, salinization, deforestation, and erosion also occurred in the Americas. Deforestation caused silting of chinampa waterways, lowering horticultural productivity. Skeletons from just before 800 CE show higher infant and female mortality and more deficiency diseases, then population collapse.

Conclusion

The development of states extended the Urban Revolution, the second great social transformation that started in polygroups. Agrarian states were the fourth form of human society to develop, the most complex to that time. In the subsequent five millennia, social evolution "slowed down" (Sanderson 1995). There were no new forms: states simply developed and grew. Their governments recruited groups beyond leaders' kin, usually local elites, as literate bureaucrats. They relied on intensive farming, encouraged trade, and used the surplus to generate public goods and to fuel more conquests. They supported and relied on formally organized religions which often legitimated them. Technological life improved very slowly. The mortality rate remained high, the average person being subject to invasions, floods, droughts, and epidemics. Even in good years, people had many children but lost around half of them before adulthood, and few people lived to old age. States centered on large cities and included millions of people. Most men dominated women, who carved out their own power in multiple ways. In turn, most men were subordinated to wealthy landowners and government officials.

Within this transformation, deep continuities existed. The global human population kept growing. Technologies improved, production intensified, and trade systems grew geographically. Political systems continued expanding and formalizing. Human creativity elaborated, taking form in complex and written idea systems. Inequalities continued increasing. Mortality and fertility rates remained high. Humans accelerated their impact on the physical environment.

Both the transformations and the continuities relied on basic human capacities: nested and multiple groups rooted in emotional bonds; ongoing cultural cumulation; and imagination. Calling agrarian states "super-groups," as some do, misses a few issues. Agrarian states and empires were too large to discuss as sets of individuals, so "societies" became abstractions to their leaders, as regions on maps. Most people likely did not think of entire "societies"; their knowledge was mostly personal and local. The physical size and necessary abstraction among leadership of their realms intensified state leaders' dehumanization of populaces they ruled.

I use the word "civilizations" to denote the long-term cultural effects of the repeated rise and fall of states or state systems in given regions. States and civilizations are not the same. As always, people in any region passed on assumptions (concepts and belief-systems), feeling-patterns, and institutions intergenerationally. States expanded and declined within these regions, slowly changing what everyone passed on. Rulers used cities to concentrate people and resources to influence the regions around them. Civilizations existed within geographically larger world-systems: Trade networks also based in cities.

Furthermore, even skeptics (e.g., Graeber and Wengrow 2021) recognize that, over several millennia, dominance and subordination became permanent for most people in civilizations. The constant rise and fall, usually taking many generations, suggests a long-term political instability—the arrogance-versus-resentment problem in polygroups, now multiplied many times—while assimilation increased long-term cultural stability in civilizations. Thus, complex institutions helped unify demographically diverse and geographically large states. Still, as historical sociologists and processual archaeologists (Sztompka 1994; Earle 2016) remind us, unity was always a moving target, and ultimately always failed. When this happened, sometimes local peoples were glad their overlords were gone; other times, they died with them.

Despite this rise and fall, the overall trend throughout humanity was increasing complexity and ramifying diversity. This chapter shows that generic statements about "how life was" in agrarian states must give way to knowledge of trends that occurred haltingly and fitfully, with periods of warfare and collapse, and periods of rebuilding and enrichment. Likewise, foraging bands; foraging, horticultural, and pastoral villages; foraging, horticultural, and pastoral polygroups; and horticultural, pastoral, and agropastoral states all coexisted on this planet by 1500 CE. Within each of these kinds of groups, many possible cultural patterns existed. Few peoples were isolated; instead, within the basic patterns of trade-systems, frontiers, religious proselytization, and conquest, there was a tremendous variety of ways for these societies to interact. Over the next five centuries, the pace of these changes accelerated.

Appendix

List of empires partly or completely formed by conquest. (Empires without an ending date continue into the period addressed in the next chapter.)

Akkadian empire (Anatolia to eastern Iran), 2350–2154 BCE
Third Dynasty of Ur (Mesopotamia), *circa* 2100–2000 BCE
Babylonian empire (Mesopotamia), 1792–1595 BCE
Mittani empire (Mesopotamia), 1550–1340 BCE
Kassite state (Mesopotamia), 1600–1155 BCE
Elam/Ansan empires (Iran), 1500–1100 BCE
Assyrian empire (Mesopotamia), 800–750 BCE
New Kingdom (Egypt), 1550–1069 BCE
New Babylonian empire (Mesopotamia), 626–539 BCE
Achaemenid Persian empire (eastern Mediterranean to the Indus River), 550–331 BCE
Carthaginian empire (western Mediterranean), 550s–146 BCE
Athenian empire (Greece/Mediterranean), 470s–332 BCE
Alexandrine empire (Greece to the Indus River), 336–323 BCE[a]
Mauryan empire (South Asia), 321–185 BCE
Qin (China), 221–202 BCE
Han (China), 202 BCE–220 CE
Roman (entire Mediterranean), 167 BCE–476 CE[a]
Parthian empire (Iran), 238 BCE–224 CE
Sasanian empire (Iran), 224–642 CE
Kushan empire (West and South Aasia), *circa* 50–250 CE
Monte Albán (Mesoamerica), *circa* 200 BCE–500 CE
Byzantine empire (eastern Mediterranean), 641–1453 CE
Wari and Tiwanaku (Andes), 650s–1000s CE
Tang dynasty (China), 650–907 CE

Srivijayan empire (SE Asia), 670s–1350s CE
Abbasid caliphate (Spain to the Indus Valley), 650s–930s CE
Carolingian empire (Europe), 771–843 CE[a]
Khmer empire (Southeast Asia), 802–1431 CE
Song dynasty (China), 960–1279 CE
Tula (Mesoamerica), 900–1150 CE
Chimor and Lambayeque (Andes), 900–1470 CE
Venetian empire (Adriatic and Mediterranean), 1202– CE[b]
Mongol empire (Eastern Europe, Central Asia, and East Asia), 1206–1259 CE[a]
Delhi sultanate (South Asia), 1210s– CE[b]
Mali empire (West Africa), 1235– CE[b]
Purépecha empire (Mesoamerica), *circa* 1300– CE[b]
Ottoman empire (Anatolia), *circa* 1300– CE[b]
Aztec empire (Mesoamerica), 1430– CE[b]
Songhay empire (West Africa), 1464– CE[b]

[a]These empires broke up into smaller, successor empires at the latter date.
[b]These empires continued beyond the end of the period discussed in this chapter.

Sources
Multiple articles in Scarre, ed., *The Human Past* (2018).
All articles in Bang et al., eds., *The Oxford World History of Empires* (2021).

References

Abramson, Marc S. 2008. *Ethnic Identity in Tang China*. Philadelphia, PA: University of Pennsylvania Press.

Abu-Lughod, Janet L. 1989. *Before European Hegemony: The World System A.D. 1250–1350*. New York: Oxford University Press.

Alcock, Susan E., and John F. Cherry. 2018. The Mediterranean World. In *The Human Past: World Prehistory and the Development of Human Societies*, ed. C. Scarre, 4th ed., 469–514. London: Thames and Hudson.

Anderson, Perry. 1978. *From Antiquity to Feudalism*. London: Verso.

Ardant, Gabriel. 1975. Financial Policy and Economic Infrastructure of Modern States and Nations. In *The Formation of National States in Western Europe*, ed. C. Tilly, 164–242. Princeton, NJ: Princeton University Press.

Armstrong, Karen. 2006. *The Great Transformation: The Beginning of Our Religious Traditions*. New York: Alfred A. Knopf.

Ashtor, Eliyahu. 1976. *A Social and Economic History of the Near East in the Middle Ages*. London: William Collins Sons.

Ayenachew, Deresse. 2020. Territorial Expansion and Administrative Evolution under the 'Solomonic' Dynasty. In *A Companion to Medieval Ethiopia and Eritrea*, ed. S. Kelly, 57–85. Boston, MA: Brill.

Bahn, Paul, ed. [2000] 2009. *Atlas of World Archaeology*. London: Brown Reference.

Ballesteros Danel, Andrea, and Antonio Jaramillo Arango. 2023. Mesoamerican–South American Pre-Columbian Pacific Contacts: Evidence, Objects, and Traditions, 1500 BCE–1532 CE. In *The Cambridge History of the Pacific Ocean*, ed. A. P. Hattori and J. Samson, 613–632. New York: Cambridge University Press.

Bang, Peter Fibiger. 2021. Empire—A World History. In *The Oxford World History of Empire*, ed. P. F. Bang, C. A. Bayly, and W. Scheidel, 1–90. New York: Oxford University Press.
Barnes, Gina L. 2015. *The Archaeology of East Asia: The Rise of Civilization in China, Korea, and Japan*. Havertown, PA: Oxbow Books.
Baumer, Christopher. 2012. *A History of Central Asia: The Age of the Steppe Warriors*. Vol. I. New York: I. B. Tauris.
Beaujard, Philippe. 2018. The Birth of a Single Afro-Eurasian World-System (Second Century BC-Sixth Century CE). In *Trade and Civilization: Economic Networks and Cultural Ties, from Prehistory to the Early Modern Era*, ed. K. Kristiansen, T. Lindkvist, and J. Myrdal, 242–250. New York: Cambridge University Press.
Bellwood, Peter, and Peter Hiscock. 2018. Australia and the Indo-Pacific Islands During the Holocene. In *The Human Past: World Prehistory and the Development of Human Societies*, ed. C. Scarre, 4th ed., 261–302. London: Thames and Hudson.
Berdan, Frances. 2005. *The Aztecs of Central Mexico: An Imperial Society*. In *Case studies in Cultural Anthropology*, ed. G. Spindler, 2nd ed. Belmont, CA: Thomson/Wadsworth.
Black, Jeremy, ed. 2005. *The Atlas of World History*. London: Dorling Kindersley Limited.
Blanton, Richard, and Lane Fargher. 2008. *Collective Action in the Formation of Pre-Modern States*. New York: Springer.
Blanton, Richard E., with Lane F. Fargher. 2016. *How Humans Cooperate: Confronting the Challenges of Collective Action*. Boulder, CO: University Press of Colorado.
Blanton, Richard E., Gary M. Feinman, Stephen A. Kowalewski, and Peter N. Peregrine. 1996. A Dual-Processual Theory for the Evolution of Mesoamerican Civilization. *Current Anthropology* 37 (1): 1–14.
Bocquet-Appel, Jean-Pierre, and Jean-Noël Bacro. 2008. Estimation of an Age Distribution with Its Confidence Intervals Using an Iterative Bayesian Procedure and a Bootstrap Sampling Approach. In *Recent Advances in Palaeodemography: Data, Techniques, Patterns*, ed. J.-P. Bocquet-Appel, 63–82. Dordrecht, The Netherlands: Springer.
Bogucki, Peter. 1999. *The Origins of Human Society*. Malden, MA: Blackwell.
Bonilla-Silva, Eduardo. 1999. The Essential Social Fact of Race. *American Sociological Review* 64 (6): 899–906.
Bonvillain, Nancy M. 2001. *Women and Men: Cultural Constructs of Gender*. Upper Saddle River, NJ: Prentice Hall.
Boone, James L. 2002. Subsistence Strategies and Early Human Population History: An Evolutionary Ecological Perspective. *World Archaeology* 34 (1): 6–25.
Braudel, Fernand. 1995. *A History of Civilizations*. Richard Mayne, translator. New York: A. Lane.
Bronson, Bennet. 1988. The Role of Barbarians in the Fall of States. In *The Collapse of Ancient States and Civilizations*, ed. N. Yoffee and G. L. Cowgill, 196–218. Tucson, AZ: University of Arizona Press.
Brooke, John L. 2014. *Climate Change and the Course of Global History: A Rough Journey*. New York: Cambridge University Press.
Brumfiel, Elizabeth M., and Cynthia Robin. 2012. Class and Ethnicity in Ancient Mesoamerica. In *The Oxford Handbook of Mesoamerican Archaeology*, ed. D. L. Nichols and C. A. Pool, 673–683. New York: Oxford University Press.
Carballo, David M. 2012. Trade Routes in the Americas Before Columbus. In *The Great Trade Routes: A History of Cargos and Commerce Over Land and Sea*, ed. P. Parker, 166–170. London: Conway Publishing.
Carneiro, Robert L. 2017. The Chiefdom in Evolutionary Perspective. In *Chiefdoms: Yesterday and Today*, ed. R. L. Carneiro, L. E. Grinin, and A. V. Korotayev, 15–60. Clinton Corners, NY: Eliot Werner.
Chase-Dunn, Christopher, and Thomas D. Hall. 1997. *Rise and Demise: Comparing World-Systems*. Boulder, CO: Westview Press.
Chew, Sing C. 2001. *World Ecological Degradation: Accumulation, Urbanization, and Deforestation, 3000 B.C.–A.D. 2000*. Walnut Creek, CA: AltaMira Press.

Chirot, Daniel. 2012. *How Societies Change*. 2nd ed. Thousand Oaks, CA: SAGE.
Christian, David, Cynthia Stokes Brown, and Craig Benjamin. 2014. *Big History: Between Nothing and Everything*. New York: McGraw-Hill Education.
Coningham, Robin. 2018. South Asia: From Early Villages to Buddhism. In *The Human Past: World Prehistory & the Development of Human Societies*, ed. C. Scarre, 4th ed., 515–546. London: Thames & Hudson.
Connah, Graham. 2018. Holocene Africa. In *The Human Past: World Prehistory and the Development of Human Societies*, ed. C. Scarre, 4th ed., 344–387. London: Thames and Hudson.
Cowgill, George L. 1988. Onward and Upward with Collapse. In *The Collapse of Ancient States and Civilizations*, ed. N. Yoffee and G. L. Cowgill, 244–275. Tucson, AZ: University of Arizona Press.
Davies, Nigel. 1987. *The Aztec Empire: The Toltec Resurgence*. Norman, OK: University of Oklahoma Press.
Davies, Glyn. 2016. *A History of Money*, Revised and updated by Duncan Connors. 4th ed. Cardiff: University of Wales Press.
DeCorse, Christopher R. 2012. West African Forest Kingdoms. In *The Oxford Companion to Archaeology*, ed. N. A. Silberman, vol. III, 365–368. New York: Oxford University Press.
Derat, Marie-Laure. 2020. Before the Solomonids: Crisis, Renaissance and the Emergence of the Zagwe Dynasty, (Seventh-Thirteenth Centuries). In *A Companion to Medieval Ethiopia and Eritrea*, ed. S. Kelly, 31–56. Boston, MA: Brill.
Diamond, Jared. 2005. *Collapse: How Societies Choose to Fail or Succeed*. New York: Viking Press.
Diel, Lori Boornazian. 2012. Nahua and Mixtec Pictorial Books: Religion and History Through Visual Text. In *The Oxford Handbook of Mesoamerican Archaeology*, ed. D. L. Nichols and C. A. Pool, 869–878. New York: Oxford University Press.
Dyer, Christopher. 1989. *Standards of Living in the Later Middle Ages: Social Change in England c. 1200–1520*. New York: Cambridge University Press.
Earle, Tim. 2016. Pathways to Power: Corporate and Network Strategies, Staple and Wealth Finance, and Primary and Secondary States. In *Alternative Pathways to Complexity: A Collection of Essays on Architecture, Economics, Power, and Cross-Cultural Analysis in Honor of Richard E. Blanton*, ed. L. Fargher and V. Y. Heredia Espinosa, 291–308. Boulder, CO: University Press of Colorado.
Earle, Timothy. 2018. Trade-Light: The Political Economy of Polynesian and Andean Civilizations. In *Trade and Civilization: Economic Networks and Cultural Ties, from Prehistory to the Early Modern Era*, ed. K. Kristiansen, T. Lindkvist, and J. Myrdal, 441–470. New York: Cambridge University Press.
Ehrenberg, Margaret. 1989. *Women in Prehistory*. London: British Museum Publications.
Eisenstadt, Shmuel N. 1963. *The Political Systems of Empires*. New York: Free Press of Glencoe.
El Fadl, Khaled Abou. 2012. The Islamic Legal Tradition. In *The Cambridge Companion to Comparative Law*, ed. M. Bussani and U. Mattei, 295–312. New York: Cambridge University Press.
Evans, Susan Toby. 2012. Time and Space Boundaries: Chronologies and Regions in Mesoamerica. In *The Oxford Handbook of Mesoamerican Archaeology*, ed. D. L. Nichols and C. A. Pool, 114–126. New York: Oxford University Press.
Everett, Nicholas. 2009. Literacy from Late Antiquity to the Early Middle Ages. In *The Cambridge Handbook of Literacy*, ed. D. R. Olson and N. Torrance, 362–385. New York: Cambridge University Press.
Fagan, Brian M. 2004. *People of the Earth: An Introduction to World Prehistory*. 11th ed. Upper Saddle River, NJ: Pearson-Prentice Hall.
Fairbank, John King. 1992. *China: A New History*. Cambridge MA: Belknap Press.
Fairbank, John King, and Edwin Reischauer. 1978. *China: Tradition and Transformation*. Boston, MA: Houghton Mifflin.
Fairbank, John K., Edwin O. Reischauer, and Albert M. Craig. 1989. *East Asia: Tradition & Transformation*, Revised ed. Boston, MA: Houghton Mifflin.

Feinman, Gary M. 2012. Mesoamerican States and Empires. In *The Oxford Handbook of Mesoamerican Archaeology*, ed. D. L. Nichols and C. A. Pool, 726–738. New York: Oxford University Press.

Feinman, Gary M., and Linda M. Nicholas. 2016. Framing the Rise and Variability of Past Complex Societies. In *Alternative Pathways to Complexity: A Collection of Essays on Architecture, Economics, Power, and Cross-Cultural Analysis in Honor of Richard E. Blanton*, ed. L. Fargher and V. Y. Heredia Espinosa, 271–290. Boulder, CO: University Press of Colorado.

Ferguson, R. Brian. 2006. Archaeology, Cultural Anthropology, and the Origins and Intensification of War. In *The Archaeology of Warfare: Prehistories of Raiding and Conquest*, ed. E. N. Arkush and M. W. Allen, 469–523. Gainsville, FL: University Press of Florida.

Ferguson, R. Brian, and Neil L. Whitehead, eds. 1992. *War in the Tribal Zone: Expanding States and Indigenous Warfare*. Santa Fe, NM: School of American Research Press.

Fernández-Armesto, Felipe. 2001. *Civilizations: Culture, Ambition, and the Transformation of Nature*. New York: Touchstone.

Feuer, Bryan. 2016. *Boundaries, Borders and Frontiers in Archaeology: A Study of Spatial Relationships*. Jefferson, NC: McFarland.

Finneran, Niall. 2007. *The Archaeology of Ethiopia*. New York: Routledge.

Fischer, Steven R. 2018. *A History of Language*. London: Reaktion Books.

Fisher, Sidney. 1979. *The Middle East: A History*. 3rd ed. New York: Knopf.

Flannery, Kent. 1998. The Ground Plan of Archaic States. In *Archaic States*, ed. G. M. Feinman and J. Marcus, 15–58. Santa Fe, NM: School of American Research Press.

Flannery, Kent, and Joyce Marcus. 2012. *The Creation of Inequality: How Our Prehistoric Ancestors Set the Stage for Monarchy, Slavery, and Empire*. Cambridge, MA: Harvard University Press.

Franke, Ute, Ina Sarikhani Sandmann, and Stefan Weber, eds. 2021. *Iran: Five Millennia of Art and Culture*. Munich: Hirmer Publishers.

Friedman, Jonathan. 2018. Postscript: Getting the Goods for Civilization. In *Trade and Civilization: Economic Networks and Cultural Ties, from Prehistory to the Early Modern Era*, ed. K. Kristiansen, T. Lindkvist, and J. Myrdal, 534–546. New York: Cambridge University Press.

Gibbon, Edward. [1776–1789] 1960. *The Decline and Fall of the Roman Empire*. Abridged by D. M. Low. New York: Harcourt, Brace.

Gibson, D. Blair. 2017. Chiefdom Confederacies and State Origins. In *Chiefdoms: Yesterday and Today*, ed. R. L. Carneiro, L. E. Grinin, and A. V. Korotayev, 179–194. Clinton Corners, NY: Eliot Werner.

Goldstone, Jack A. 2002. Efflorescences and Economic Growth in World History: Rethinking the 'Rise of the West' and the Industrial Revolution. *Journal of World History* 13 (2): 323–389.

Goldstone, Jack A. [1991] 2016. *Revolution and Rebellion in the Early Modern World*. Twenty-fifth anniversary edition. New York: Routledge.

Graeber, David. 2011. *Debt: The First 5,000 Years*. Brooklyn, NY: Melville House.

Graeber, David, and David Wengrow. 2021. *The Dawn of Everything: A New History of Humanity*. New York: Farrar, Straus and Giroux.

Grinin, Leonid E., and Andrey V. Korotayev. 2017. The Afro-Eurasian World-System: Genesis, Transformation, Characteristics. In *Routledge Handbook of World-Systems Analysis*, ed. S. J. Babones and C. Chase-Dunn, 30–38. New York: Routledge.

Grube, Nickolai. 2012. Maya Writing. In *The Oxford Handbook of Mesoamerican Archaeology*, ed. D. L. Nichols and C. A. Pool, 845–854. New York: Oxford University Press.

Hakstöz, Çagri, and Damla Durak Uşar. 2012. Silk Road Supply Chains: A Historical Perspective. In *Managing Supply Chains on the Silk Road: Strategy, Performance, and Risk*, ed. Ç. Haksöz, S. Seshadri, and A. V. Iyer, 3–26. Boca Raton, FL: CRC Press.

Hall, Martin, and Patrick Thaddeus Jackson, eds. 2007. *Civilizational Identity: The Production and Reproduction of "Civilizations" in International Relations*. New York: Palgrave Macmillan.

Hall, Thomas D., P. Nick Kardulias, and Christopher Chase-Dunn. 2011. World-Systems Analysis and Archaeology: Continuing the Dialogue. *Journal of Archaeological Research* 19 (3): 233–279.

Hansen, Mogens Herman. 2000. *A Comparative Study of Thirty City-State Cultures: An Investigation Conducted by the Copenhagen Polis Centre*. Copenhagen: Det Kongelige Danske Videnskabernes Selskab.

Harris, Marvin. 1995. The Emergence of Pristine States. In *Sociological Worlds: Comparative and Historical Readings on Society*, ed. S. K. Sanderson, 221–228. New York: Oxford University Press.

Hartwell, Robert. 1966. Markets, Technology, and the Structure of Enterprise in the Development of the Eleventh-Century Chinese Iron and Steel Industry. *The Journal of Economic History* 26 (1): 29–58.

Haslam, Nick, Steven Loughnan, and Elise Holland. 2013. The Psychology of Humanness. In *Objectification and (De)Humanization. 60th Nebraska Symposium on Motivation*, ed. S. J. Gervais, 25–51. New York: Springer.

Hayden, Brian. 2003. *Shamans, Sorcerers and Saints: A Prehistory of Religion*. Washington, DC: Smithsonian.

Hayden, Brian. 2014. *The Power of Feasts: From Prehistory to the Present*. New York: Cambridge University Press.

Hayden, Brian, and Timothy Earle. 2022. Political Economy Perspectives in Trade Before and Beyond Civilizations. In *Trade Before Civilization*, ed. J. Ling, R. J. Chacon, and K. Kristiansen, 385–415. New York: Cambridge University Press.

Haywood, John, ed. 2011. *The New Atlas of World History: Global Events at a Glance*. Princeton, NJ: Princeton University Press.

Headrick, Daniel. 2009. *Technology: A World History*. Oxford: Oxford University Press.

Herlihy, David, and Christiane Klapisch-Zuber. 1978. *Tuscans and Their Families: A Study of the Florentine Catasto of 1427*. New Haven, CT: Yale University Press.

Higham, Charles. 2018. Complex Societies of East and Southeast Asia. In *The Human Past: World Prehistory & the Development of Human Societies*, ed. C. Scarre, 4th ed., 547–589. London: Thames & Hudson.

Hirth, Kenneth G., and Joanne Pillsbury. 2013. *Merchants, Markets, and Exchange in the PreColumbian World*. Washington, DC: Dumbarton Oaks Research Library and Collection.

Hodgson, Marshall. 1974. *The Venture of Islam*. Vol. I–III. Chicago, IL: University of Chicago Press.

Holcombe, Charles. 2017. *A History of East Asia: From the Origins of Civilization to the Twenty-First Century*. 2nd ed. New York: Cambridge University Press.

Hollander, Stanley. 1984. Sumptuary Legislation: Demarketing by Edict. *Journal of Macromarketing* 4 (1): 4–16.

Hornborg, Alf. 2018. Long-Distance Exchange and Ritual Technologies of Power in the Pre-Hispanic Andes. In *Trade and Civilization: Economic Networks and Cultural Ties, from Prehistory to the Early Modern Era*, ed. K. Kristiansen, T. Lindkvist, and J. Myrdal, 471–493. New York: Cambridge University Press.

Horton, Mark. 2012. East Africa: Coastal Towns of East Africa. In *The Oxford Companion to Archaeology*, ed. N. A. Silberman, vol. I, 438–441. New York: Oxford University Press.

Houston, Rob, ed. 2018. *History of the World, Map by Map*. New York: Dorling Kindersley.

Hucker, Charles. 1973. Political Institutions. In *An Introduction to Chinese Civilization*, ed. J. Meskill with J. M. Gentzler, 551–586. New York: Columbia University Press.

Hughes, Sarah Shaver, and Brady Hughes. 2005. Women in Ancient Civilizations. In *Women's History in Global Perspective*, ed. B. G. Smith, vol. II, 9–46. Urbana, IL: University of Illinois Press.

Hui, Victoria Tin-bor. 2005. *War and State Formation in Ancient China and Early Modern Europe*. New York: Cambridge University Press.

Ibn Khaldûn. [1377] 1958. *The Muqaddimah: An Introduction to History*. Translated by Franz Rosenthal. New York: Pantheon.

Jaspers, Karl. 1953. *The Origin and Goal of History*. Translated by Michael Bullock. London: Routledge & K. Paul.

Jiménez, Robert T., and Patrick H. Smith. 2008. Mesoamerican Literacies: Indigenous Writing Systems and Contemporary Possibilities. *Reading Research Quarterly* 43 (1): 28–46.

Johnson, Allen W., and Timothy Earle. 2000. *The Evolution of Human Societies: From Foraging Group to Agrarian State*. 2nd ed. Stanford, CA: Stanford University Press.

Jones, Eric. 1988. *Growth Recurring: Economic Change in World History*. 2nd ed. New York: Oxford University Press.

Justeson, John. 2012. Early Mesoamerican Writing Systems. In *The Oxford Handbook of Mesoamerican Archaeology*, ed. D. L. Nichols and C. A. Pool, 830–844. New York: Oxford University Press.

Kaufman, Herbert. 1988. The Collapse of Ancient States and Civilizations as an Organizational Problem. In *The Collapse of Ancient States and Civilizations*, ed. N. Yoffee and G. L. Cowgill, 219–235. Tucson, AZ: University of Arizona Press.

Kautsky, John H. 1982. *The Politics of Aristocratic Empires*. Chapel Hill, NC: The University of North Carolina Press.

Kellogg, Susan. 1995. *Law and the Transformation of Aztec Culture, 1500–1700*. Norman, OK: University of Oklahoma Press.

Kellogg, Susan. 2005. *Weaving the Past: A History of Latin America's Indigenous Women from the Prehispanic Period to the Present*. New York: Oxford University Press.

Kent, Susan Kingsley. 2021. *Gender in World History*. New York: Oxford.

Khalid, Adeeb. 2021. *Central Asia: A New History from the Imperial Conquests to the Present*. Princeton, NJ: Princeton University Press.

Köhler, E. Christiana. 2011. The Rise of the Egyptian State. In *Before the Pyramids*, ed. E. Teeter, 122–125. Chicago, IL: The Oriental Institute of the University of Chicago.

Kolb, Anne. 2016. Literacy in Everyday Life—Problems and Results. In *Literacy in Ancient Everyday Life*, ed. A. Kolb, 1–12. Berlin: Walter de Gruyter.

Kornfield, Jack. 1993. *Teaching of the Buddha*. With Gil Fronsdal. Boston, MA: Shambhala.

Kristiansen, Kristian. 2018a. Theorizing Trade and Civilization. In *Trade and Civilization: Economic Networks and Cultural Ties, from Prehistory to the Early Modern Era*, ed. K. Kristiansen, T. Lindkvist, and J. Myrdal, 1–24. New York: Cambridge University Press.

Kristiansen, Kristian. 2018b. The Rise of Bronze Age Peripheries and the Expansion of International Trade, 1950-1100 BC. In *Trade and Civilization: Economic Networks and Cultural Ties, from Prehistory to the Early Modern Era*, ed. K. Kristiansen, T. Lindkvist, and J. Myrdal, 87–112. New York: Cambridge University Press.

LeBlanc, Steven A. 2006. Warfare and the Development of Social Complexity: Some Demographic and Environmental Factors. In *The Archaeology of Warfare: Prehistories of Raiding and Conquest*, ed. E. N. Arkush and M. W. Allen, 437–468. Gainsville, FL: University Press of Florida.

Lee, De-nin D., and Jin Xu. 2023. China: Patronage. *Grove Art Online*. https://doi.org/10.1093/oao/9781884446054.013.90000371757.

Lees, Andrew. 2015. *The City: A World History*. New York: Oxford University Press.

Liu, William Guanglin. 2015. The Making of a Fiscal State in Song China, 960–1279. *The Economic History Review* 68 (1): 48–78.

Livi-Bacci, Massimo. 2017. *A Concise History of World Population*. 6th ed. Malden, MA: Blackwell.

Loftin, John. 2000. *The Big Picture: A Short History of World Religions*. Jefferson, NC: McFarland.

Marcus, Joyce. 1998. The Peaks and Valleys of Ancient States: An Extension of the Dynamic Model. In *Archaic States*, ed. G. M. Feinman and J. Marcus, 59–94. Santa Fe, NM: School of American Research Press.

Mark, Joshua J. 2012. The Mayan Pantheon: The Many Gods of the Maya. *World History Encyclopedia*. https://www.worldhistory.org/article/415/the-mayan-pantheon-the-many-gods-of-the-maya/. Retrieved 2 Mar 2024.

Matthews, Roger. 2018. Peoples and Complex Societies of Ancient Southwest Asia. In *The Human Past: World Prehistory and the Development of Human Societies*, ed. C. Scarre, 4th ed., 429–468. London: Thames and Hudson.

Mattingly, David. 2017. The Garamantes and the Origins of Saharan Trade. In *Trade in the Ancient Sahara and Beyond*, ed. D. J. Mattingly, V. Leitch, C. N. Duckworth, A. Cuénod, M. Sterry, and F. Cole, 1–52. New York: Cambridge University Press.

Mazoyer, Marcel, and Laurence Roudart. 2006. *A History of World Agriculture: From the Neolithic Revolution to the Current Crisis*, Translated by James H. Membrez. New York: Monthly Review Press.
McAnany, Patricia Ann, and Norman Yoffee, eds. 2010. *Questioning Collapse: Human Resilience, Ecological Vulnerability, and the Aftermath of Empire*. New York: Cambridge University Press.
McClellan, James E., III, and Harold Dorn. 1999. *Science and Technology in World History: An Introduction*. Baltimore, MD: The Johns Hopkins University Press.
McIntosh, Susan Keech. 1999. Pathways to Complexity: An African Perspective. In *Beyond Chiefdoms: Pathways to Complexity in Africa*, ed. S. K. McIntosh, 1–30. New York: Cambridge University Press.
McIntosh, Jane. 2008. *The Ancient Indus Valley: New Perspectives*. Santa Barbara, CA: ABC-CLIO.
McIntosh, Susan Keech. 2012. West African Savannah Kingdoms. In *The Oxford Companion to Archaeology*, ed. N. A. Silberman, vol. III, 369–372. New York: Oxford University Press.
McMillan, John. 2002. *Reinventing the Bazaar: A Natural History of Markets*. New York: W.W. Norton.
McNeill, J. R. R., and William H. McNeill. 2003. *The Human Web: A Bird's-Eye View of World History*. New York: Norton.
Melko, Matthew. 1995. The Nature of Civilizations. In *Civilizations and World Systems: Studying World-Historical Change*, ed. S. K. Sanderson, 25–45. Walnut Creek, CA: AltaMira.
Milanovic, Branko. 2011. *The Haves and the Have-Nots: A Brief and Idiosyncratic History of Global Inequality*. New York: Basic Books.
Milner, George R., and W. H. Wills. 2018. Complex Societies of North America. In *The Human Past: World Prehistory and the Development of Human Societies*, ed. C. Scarre, 4th ed., 670–702. London: Thames and Hudson.
Moore, R. I., ed. 1995. *Atlas of World History*. Rand McNally.
Moore, Jerry D. 2014. *A Prehistory of South America: Ancient Cultural Diversity on the Least Known Continent*. Boulder, CO: University of Colorado Press.
Moseley, Michael E., and Michael J. Heckenberger. 2018. From Village to Empire in South America. In *The Human Past: World Prehistory and the Development of Human Societies*, ed. C. Scarre, 4th ed., 636–669. London: Thames and Hudson.
Myrdal, Janken. 2012. Empire: The Comparative Study of Imperialism. In *Ecology and Power: Struggles Over Land and Material Resources in the Past, Present, and Future*, ed. A. Hornborg, B. Clark, and K. Hermele, 37–51. London: Routledge.
Neill, James. 2009. *The Origins and Role of Same-Sex Relations in Human Societies*. Jefferson, NC: McFarland & Company.
Nichols, Deborah L., and Thomas H. Charlton. 1997. *The Archaeology of City-States: Cross-Cultural Approaches*. Washington, DC: Smithsonian Institution Press.
Nolan, Patrick, and Gerhard Lenski. 2011. *Human Societies: An Introduction to Macrosociology*. Boulder, CO: Paradigm Press.
O'Brien, Patrick, ed. 2010. *Atlas of World History*. 2nd ed. New York: Oxford University Press.
Offner, Jerome A. 1983. *Law and Politics in Aztec Texcoco*. New York: Cambridge University Press.
Oka, Rahul. 2018. Trade, Traders, and Trading Systems: Macromodeling of Trade, Commerce, and Civilization in the Indian Ocean. In *Trade and Civilization: Economic Networks and Cultural Ties, from Prehistory to the Early Modern Era*, ed. K. Kristiansen, T. Lindkvist, and J. Myrdal, 279–319. New York: Cambridge University Press.
Peregrine, Peter N., and Stephen H. Lekson. 2012. The North American *oikoumene*. In *The Oxford Handbook of North American Archaeology*, ed. T. R. Pauketat, 64–72. New York: Oxford University Press.
Peterson, Christian E., and Robert D. Drennan. 2012. Patterned Variation in Regional Trajectories of Community Growth. In *The Comparative Archaeology of Complex Societies*, ed. M. E. Smith, 88–137. New York: Cambridge University Press.
Phillipson, D. W. 2012. *Foundations of an African Civilization: Aksum & the Northern Horn, 1000 BC–AD 1300*. Rochester NY: James Currey.

Pirie, Fernanda. 2021. *The Rule of Laws: A 4,000-Year Quest to Order the World*. New York: Basic Books.
Pomeranz, Kenneth, and Steven Topik. 2018. *The World That Trade Created: Society, Culture, and the World Economy, 1400 to the Present*. 4th ed. New York: Routledge.
Ponting, Clive. 2007. *A New Green History of the World: The Environment and the Collapse of Great Civilizations*. New York: Penguin.
Powell, Barry B. 2009. *Writing: Theory and History of the Technology of Civilization*. Malden, MA: Wiley-Blackwell.
Provan, Iain. 2013. *Convenient Myths: The Axial Age, Dark Green Religion, and the World That Never Was*. Waco, TX: Baylor University Press.
Raffield, Ben. 2021. Broken Worlds: Towards an Archaeology of the Shatter Zone. *Journal of Archaeological Method and Theory* 28:871–910.
Redfield, Robert. 1956. *Peasant Society and Culture: An Anthropological Approach to Civilization*. Chicago, IL: University of Chicago Press.
Reilly, F. Kent, III. 2012. Mesoamerican Religious Beliefs: The Practices and Practitioners. In *The Oxford Handbook of Mesoamerican Archaeology*, ed. D. L. Nichols and C. A. Pool, 764–775. New York: Oxford University Press.
Renfrew, Colin. 1986. Introduction: Peer Polity Interaction and Socio-Political Change. In *Peer Polity Interaction and Socio-Political Change*, ed. C. Renfrew and J. F. Cherry, 1–18. New York: Cambridge University Press.
Riasanovsky, Nicholas V., and Mark D. Steinberg. 2019. *A History of Russia*. New York: Oxford University Press.
Rietbergen, P. J. A. N. 2021. *Europe: A Cultural History*. New York: Routledge.
Roberts, Keith. 2011. *The Origins of Business, Money, and Markets*. New York: Columbia Business School Publishing.
Rosenbaum, Eckehard F. 2000. What Is a Market? On the Methodology of a Contested Concept. *Review of Social Economy* 58 (4): 456–482.
Rüpke, Jörg. 2018. *Pantheon: A New History of Roman Religion*. Princeton, NJ: Princeton University Press.
Ruskola, Teemu. 2012. The East Asian Legal Tradition. In *The Cambridge Companion to Comparative Law*, ed. M. Bussani and U. Mattei, 257–277. New York: Cambridge University Press.
Sanderson, Stephen K. 1995. *Social Transformations: A General Theory of Historical Development*. Cambridge, MA: Blackwell.
Sanderson, Stephen K. 2014. Family Systems in Comparative Perspective. In *Concise Encyclopedia of Comparative Sociology*, ed. M. Sasaki, J. Goldstone, E. Zimmermann, and S. K. Sanderson, 190–198. Boston, MA: Brill.
Santon, Kate, and Liz McKay. 2005. *Atlas of World History*. Bath, UK: Parragon Publisher.
SarDesai, D. R. 2008. *India: The Definitive History*. Boulder, CO: Westview Press.
SarDesai, D. R. 2013. *Southeast Asia: Past and Present*. 7th ed. Boulder, CO: Westview Press.
Scarre, Chris. 2018a. The World Transformed: From Foragers and Farmers to States and Empires. In *The Human Past: World Prehistory and the Development of Human Societies*, ed. C. Scarre, 4th ed., 174–197. London: Thames and Hudson.
Scarre, Chris. 2018b. Holocene Europe. In *The Human Past: World Prehistory and the Development of Human Societies*, ed. C. Scarre, 4th ed., 388–428. London: Thames and Hudson.
Scheidel, Walter. 2017. *The Great Leveler: Violence and the History of Inequality from the Stone Age to the Twenty-First Century*. Princeton, NJ: Princeton University Press.
Scheidel, Walter. 2021. Ancient Mediterranean City-State Empires: Athens, Carthage, Early Rome. In *The Oxford World History of Empire*, ed. P. F. Bang, C. A. Bayly, and W. Scheidel. New York: Oxford University Press.
Schmandt-Besserat, Denise. 1996. *How Writing Came About*. Austin, TX: University of Texas Press.

Seneca. 1925. *Ad Lucilium Epistulae Morales*, Translated by Richard M. Gummere. Vol. III. New York: Putnam.
Sheldon, Kathleen. 2017. *African Women: Early History to the 21st Century*. Bloomington, IN: Indiana University Press.
Shimada, Izumi, ed. 2007. *Craft Production in Complex Societies: Multicraft and Producer Perspectives*. Salt Lake City, UT: The University of Utah Press.
Simon, Herbert A. 1962. The Architecture of Complexity. *Proceedings of the American Philosophical Society* 106 (6): 467–482.
Sinclair, Thomas R., and Carol Janas Sinclair. 2010. *Bread, Beer & the Seeds of Change: Agriculture's Imprint on World History*. New York: CAB International.
Skinner, G. William. 1964. Marketing and Social Structure in Rural China: Part I. *The Journal of Asian Studies* 24 (1): Nov:3–43.
Smil, Vaclav. 2017. *Energy Transitions: Global and National Perspectives*. 2nd ed. Santa Barbara, CA: Praeger.
Smith, Huston. 1991. *The World's Religions: A Completely Revised and Updated Version of the Religions of Man*. San Francisco, CA: Harper Collins.
Smith, Richard L. 2009. *Premodern Trade in World History*. London: Routledge.
Smith, David Livingstone. 2011. *Less Than Human: Why We Demean, Enslave, and Exterminate Others*. New York: St. Martin's Press.
Smith, Michael E. 2012. *The Aztecs*. 3rd ed. Malden, MA: Wiley-Blackwell.
Smith, Monica L. 2019. *Cities: The First 6,000 Years*. New York: Viking.
Smith, Michael E., and Maëlle Sergheraert. 2012. The Aztec Empire. In *The Oxford Handbook of Mesoamerican Archaeology*, ed. D. L. Nichols and C. A. Pool, 449–458. New York: Oxford University Press.
Socolow, Susan Migden. 2015. *The Women of Colonial Latin America*. 2nd ed. New York: Cambridge University Press.
Sourdel, Dominique. 1983. *Medieval Islam*, Translated by J. Montgomery Watt. Boston, MA: Routledge & Kegan Paul.
Stearns, Peter. 2006. *Consumerism in World History: The Global Transformation of Desire*. 2nd ed. New York: Routledge.
Stevenson, Alice. 2016. The Egyptian Predynastic and State Formation. *Journal of Archaeological Research* 24:421–468.
Stokstad, Marilyn. 2008. *Art History: A View of the West*. Vol. I. Upper Saddle River, NJ: Pearson-Prentice Hall.
Sutton, Mark Q., and E. N. Anderson. 2010. *Introduction to Cultural Ecology*. 2nd ed. New York: Rowman & Littlefield.
Sztompka, Piotr. 1994. *The Sociology of Social Change*. Cambridge, MA: Blackwell.
Tainter, Joseph A. 1988. *The Collapse of Complex Societies*. Cambridge, UK: Cambridge University Press.
Taube, Karl A. 2012. Creation and Cosmology: Gods and Mythic Origins in Ancient Mesoamerica. In *The Oxford Handbook of Mesoamerican Archaeology*, ed. D. L. Nichols and C. A. Pool, 741–751. New York: Oxford University Press.
Trautmann, Thomas R. 2016. *India: Brief History of a Civilization*. 2nd ed. New York: Oxford University Press.
Trigger, Bruce G. 2003. *Understanding Early Civilizations: A Comparative Study*. Cambridge, UK: Cambridge University Press.
Trigger, Bruce G. 2004. Writing Systems: A Case Study in Cultural Evolution. In *The First Writing: Script Invention as History and Process*, ed. S. D. Houston, 39–70. Cambridge: Cambridge University Press.
Turchin, Peter. 2003. *Historical Dynamics: Why States Rise and Fall*. Princeton, NJ: Princeton University Press.
Turchin, Peter, Thomas E. Currie, Edward A. L. Turner, and Sergey Gavrilets. 2013. War, Space, and the Evolution of Old World Complex Societies. *Proceedings of the National Academy of Sciences*. https://doi.org/10.1073/pnas.1308825110. Retrieved 27 Sept 2013.

Turchin, Peter, Harvey Whitehouse, Andrey Korotayev, Pieter François, Daniel Hoyer, Peter Peregrine, Gary Feinman, Charles Spencer, Nikolay Kradin, and Thomas E. Currie. 2018. Evolutionary Pathways to Statehood: Old Theories and New Data. https://www.researchgate.net/publication/329722533_Evolutionary_Pathways_to_Statehood_Old_Theories_and_New_Data

Turner, Jonathan. 2007. *Human Emotions: A Sociological Theory*. London: Routledge.

Urcid, Javier. 2012. Scribal Traditions from Highland Mesoamerica (300–1000 AD). In *The Oxford Handbook of Mesoamerican Archaeology*, ed. D. L. Nichols and C. A. Pool, 855–868. New York: Oxford University Press.

Urton, Gary. 2017. *Inka History in Knots: Reading Khipus as Primary Sources*. Austin, TX: University of Texas Press.

Volk, Anthony A., and Jeremy A. Atkinson. 2013. Infant and Child Death in the Human Environment of Evolutionary Adaptation. *Evolution and Human Behavior* 34 (3): 182–192.

Von Glahn, Richard. 2016. *The Economic History of China: From Antiquity to the Nineteenth Century*. Cambridge, UK: Cambridge University Press.

Wang, Feng, Yaching Tsai, and William S.-Y. Wang. 2009. Chinese Literacy. In *The Cambridge Handbook of Literacy*, ed. D. R. Olson and N. Torrance, 386–417. New York: Cambridge University Press.

Watson, Adam. 1992. *The Evolution of International Society: A Comparative Historical Analysis*. London: Routledge.

Watts, Joseph, Oliver Sheehan, Quentin D. Atkinson, Joseph Bulbulia, and Russell D. Gray. 2016. Ritual Human Sacrifice Promoted and Sustained the Evolution of Stratified Societies. *Nature* 532:228–234.

Webster, David, and Susan Toby Evans. 2018. Mesoamerican Civilization. In *The Human Past: World Prehistory and the Development of Human Societies*, ed. C. Scarre, 4th ed., 590–635. London: Thames and Hudson.

Wilkinson, David. 1995. Central Civilization. In *Civilizations and World Systems: Studying World-Historical Change*, ed. S. K. Sanderson, 46–74. Walnut Creek, CA: AltaMira.

Willer, David, Pamela Emanuelson, Yamilette Chacon, and Richard J. Chacon. 2017. How Chiefdom and Early State Structures Resolve Collective Action Problems. In *Feast, Famine, or Fighting? Multiple Pathways to Social Complexity*, ed. R. J. Chacon and R. G. Mendoza, 417–452. New York: Springer.

Wittfogel, Karl. 1957. *Oriental Despotism: A Comparative Study of Total Power*. New Haven, CT: Yale University Press.

Wolf, Eric R. 1982. *Europe and the People Without History*. Berkeley, CA: University of California Press.

Yoffee, Norman. 1988. The Collapse of Ancient Mesopotamian States and Civilization. In *The Collapse of Ancient States and Civilizations*, ed. N. Yoffee and G. L. Cowgill, 448–468. Tucson, AZ. University of Arizona Press.

Yoffee, Norman. 2005. *Myths of the Archaic State: Evolution of the Earliest Cities, States and Civilizations*. New York: Cambridge University Press.

Yoffee, Norman. 2010. Collapse in Ancient Mesopotamia: What Happened, What Didn't. In *Questioning Collapse: Human Resilience, Ecological Vulnerability, and the Aftermath of Empire*, ed. P. A. McAnany and N. Yoffee, 176–206. New York: Cambridge University Press.

Yoffee, Norman, and George L. Cowgill, eds. 1988. *The Collapse of Ancient States and Civilizations*. Tucson, AZ: University of Arizona Press.

Chapter 7
The Second Globalization: Industrial Revolution and Global Conquest, 1450s–1910s CE

Abstract From the 1450s to just past 1900, Europeans conquered the world, created a global order, and developed industrial capitalist states, a new type of society. These three changes constitute the third social transformation in human history, leading to what some call "modernity." In the 1450s, a new European printing culture spawned multiple cultural changes. Europe also became increasingly commercialized and wealthier. Iberians conquered parts of the Americas and established a global system of bases; other European powers replaced them and expanded European presence in the Americas, India, and African coasts. After 1750, change accelerated so that science-driven industrialization tremendously expanded output, and therefore surplus. Control of profit-seeking organizations replaced landownership as the main source of wealth. Science and education became institutionalized, and both mortality and fertility fell dramatically, driving population growth.

Keywords European colonialism · Globalization · Industrial revolution · Modernity · Global governance · Global economy

Introduction

The Agricultural and Urban Revolutions played out over millennia and were spatially incomplete. Outside of state tribal zones, foragers and villagers remained beyond the influence of social complexity in parts of North and South America, in Australia and North Asia, in parts of Africa, and in other scattered areas. In this chapter, I argue that from the 1490s to the 1910s, a third transformation in human groups occurred, consisting of two major changes: Europeans developed a new type of society, and they conquered most humans on every inhabited landmass.

Much scholarship has focused on the first part of this transformation. Childe ([1936] 1966) took his term for it from the English Industrial Revolution (1760–1830, dates debated). Europeans did change their industries from handcrafts to steam-driven, large factories. Yet this occurred within other changes, together meriting the term "social transformation." Industry combined with an already-developing capitalism made Europeans rich. This resulted from, and also drove, scientific change,

T. C. Burr, *Societies Past and Present*, Emerging Globalities and Civilizational Perspectives, https://doi.org/10.1007/978-3-031-91532-1_7

which also transformed technology. New ideologies developed, challenging humanity's long-standing beliefs in the supernatural. New forms of political rule developed; societies urbanized in a way never seen before; states spent money on and set up agencies to foster education and literacy; and a gendered separation of work and home grew.

Scholars have grouped this complex of changes under the term "modernity." There is an enormous and long-standing debate in sociology and history about what it means to be modern and how modernity operates, when it started, and what its problems are (see Bauman 2001, Briggle 2008, and Shilliam 2010 for discussions and bibliographies). Sociologists often periodized history into "traditional" and "modern" eras. Modern reason and rationality could disrupt and replace tradition. Enlightenment-inspired rationalizing tendencies created new personal experiences, which many considered "progress." They were so disruptive, however, that they generated resistance (Bauman 2001). Anthropologists, economists, and political scientists also discuss aspects of modernity (Shilliam 2010).

Tradition was a general enough category that it allowed sociologists to recognize bands, villages, and polygroups, along with the "important" traditional societies, agrarian states. Much more often, they recognized only agrarian societies, ignoring the others. Yet, kin-oriented peoples existed around the world anyway. This highlights the second aspect of this transformation, global conquest, which is why I prefer to call the third transformation in human history the "second globalization." The first globalization was not transformative; foraging humans simply migrated to every continent, and societies started changing only as this process ended. The second globalization *was* transformative, being the "culmination of processes that have unfolded over millennia" (Boivin and Franchetti 2018:2), such as the expansion of peer–polity systems, civilizations, and world-systems in each hemisphere. These two hemispheric systems would probably have encountered each other eventually, but that could have happened in many ways.

It happened in a specific way. I have argued that agrarian empires formed primarily through subjugation of others. They defeated enemies, consolidated control, extracted wealth, and provided public goods (more extraction than provision—or what was the point?). Europeans extended this pattern, conquering nearly the entire planet through accidental plagues during capitalist agriculture, purposeful extermination, intercontinental enslavement, and industrial capitalist extraction. Europeans also combined ethnocentrism and dehumanization into a new form, racism, and they imitated predecessors in culturally assimilating the conquered, mostly by force. Yet, conquest seems so closely interrelated with the internal transformation of Europe that produced modernity that I consider the phrase the "global Industrial Revolution" a synonym for the "second globalization." In this chapter, we will see global conquest, internal transformation, and very uneven global assimilation. In the next, we will see how assimilation intensified, and I will trace the debate about how globalization and social transformation were related.

As the formation of states had done, this global effort rested on human capacities, stretching them beyond previous limits. Using legacies from other regions, Western European leaders' aggrandizing initiatives to improve their position within one preexisting hemispheric system led to unintended consequences for both hemispheres. Europeans thus developed new and even more oppressive forms of large-scale organization (groups within groups within …) over vast distances, always modified by the long cultural legacies and active resistance (and occasionally cooperation) of the people they encountered. Technological and social innovations accumulated enough to help people fashion a complex global social system.

This chapter outlines these changes in two parallel sections, addressing the periods 1450s–1760s and 1760s–1910s, respectively. Each has parallel subsections: they trace the chronological story; sketch global patterns in the period; and analyze European social change. I argue that from the 1490s to the 1760s, Europeans *changed* their society and conquered some others; from the 1760s to the 1910s, Europeans *transformed* their society and conquered most of the world. While European change from the 1490s to the 1760s enabled the later transformation, one did not guarantee the other. Instead, again, Europeans made it up as they went along.

European Expansion and Global Change, 1450s–1750s

In 1450, Europe was an expanding cluster of polities in northwestern Afro-Eurasia. Through 1760, Europeans explored most of the world's coastlines, conquering or trading. By that date, they had subjugated large areas in the Americas, some areas in India, and some African and Asian coasts. Beyond the coasts, agrarian states continued their own paths. European states commercialized, fought each other, and changed intellectually, but remained agrarian.

Telling the Tale: European Global Expansion, 1450s–1750s

In the 1450s, northwestern Europe was an expanding but peripheral part of the Afro-Eurasia interaction zone. By the 1760s, Europeans had conquered both American civilizations and expanded beyond them; taken over portions of India; and established bases around Africa, the Indian Ocean, and maritime Southeast Asia. Most agrarian states, polygroups, villagers, and bands remained out of reach, though Europeans increasingly influenced many indirectly.

In the 1300s and 1400s, European states became increasingly aware of their peripheral position in the Afro-Eurasian trade system. From the 1100s to 1300, European scholars read Arabic philosophers in Latin translation (Hasse 2021). A few explorers had brought back information about India and China in this period. Europeans also knew about and benefited from Chinese inventions such as the compass, gunpowder, and printing (Sherman and Salisbury 2008). In the late 1300s, Central Asian Turks expanded into Anatolia, conquering the capital of Byzantium in 1453 and creating the Ottoman empire. Northern Italian cities, the main terminus for Eurasian trade, claimed that this raised prices on goods from the east (SarDesai 2008:208).

For these reasons and others, Western Europeans sought to change their position in the Afro-Eurasian system. In the early 1400s, the Chinese admiral Zheng He had commanded massive fleets around the Indian Ocean, mostly in visits rather than for trade. Zheng He's voyages ended in the 1430s. In this period, Portugal, on the Iberian peninsula abutting the Atlantic, made use of new inventions such as the quadrant, the astrolabe, and the compass (from China) to explore the African coast southward to bypass Italian trade and Ottoman control over West African gold (Sherman and Salisbury 2008; Pomeranz and Topik 2018). Portuguese ships reached the Indian Ocean by 1498. The other Iberian state, Castile-Aragon (later, Spain), sponsored a voyage westward to reach "the Indies." Europeans realized that the continents this voyage encountered were mostly unknown to Afro-Eurasians. Subsequent European expeditions explored North America, while others encountered what is now Brazil. Exploration continued into the 1500s; one voyage circled the planet (Fig. 7.1).

The Iberian kingdoms conquered these areas new to them. This was a network strategy, to obtain external resources (Blanton et al. 1996). With the Christian pope's approval in 1494, the Iberian empires each claimed dominance over one planetary hemisphere. While gunpowder weapons had diffused across Eurasia before 1400, only the Iberians had ship-mounted cannon (Andrade 2016). Over the 1500s, the Portuguese used these ships to conquer a global network of fortified bases along the Brazilian, African, and Asian coasts (Boxer 1969; de Oliveira Marques 1976); the

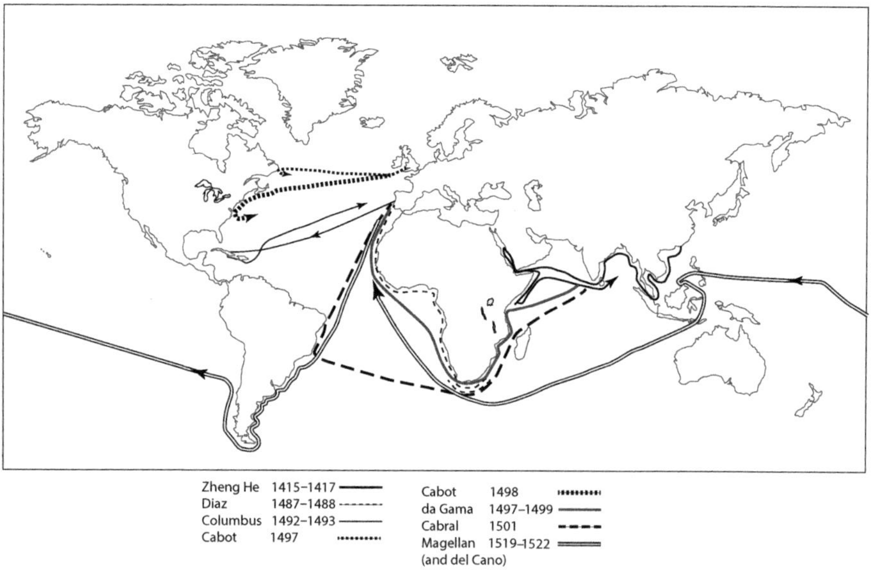

Fig. 7.1 Global Maritime Exploration, 1415–1522. Sources: Santon & McKay (2005); O'Brien (2010); Houston (2018)

Indian Ocean wing centered on Goa (SarDesai 2008). The Spanish conquered the Mesoamerican Aztec empire and the Andean Inca empire, aided by the Aztecs' resentful subject peoples; by an Incan succession crisis (Chasteen 2006); and accidentally, by Afro-Eurasian diseases to which the Americans had no immunity.

These conquests produced very different empires. The Portuguese stayed on the coasts because Afro-Eurasian agrarian empires easily stopped them. Still, the Portuguese had cannon mounted on ships, so they conquered ports and established sea bases. They forced Indian Ocean ship captains to buy a license (*cartaz*), impounding ships without one. Licensing "sales" helped them skim trade profits and allowed them to ship spices back to Europe (Boxer 1969; de Oliveira Marques 1976). Trade through Venice and Genoa slowly dwindled. The Spanish did conquer an archipelago in maritime Southwest Asia and named it the Philippines, after Spanish king Philip (SarDesai 2013). The Spanish restructured American economies for commercial agriculture, and the ensuing diseases caused a demographic collapse called "the Great Dying." Its scale is heavily debated (Mann 2013), but most American peoples died within a century in waves of epidemics, aided by economic exploitation and more violent conquest (Chasteen 2006). The Europeans were horrified—not from pity, but because they lost their labor force (Strayer 2011).

They still accomplished commercial restructuring. To acquire new labor, Europeans bought slaves from African coastal states, providing two supposed advantages. The African disease environment was the harshest in Afro-Eurasia, so the introduced diseases should have had little effect on Africans, but overwork weakened slaves, and they died of diseases anyway (Kent 2021). Africans also looked physically different from native Americans, so fleeing seemed pointless, but slaves often fled anyway from the tremendously harsh overwork. Consequently, African-descended populations in Brazil, the Caribbean, and areas in Mesoamerica grew slowly. In the 1500s, Russians moved eastward beyond the Ural Mountains into Siberia. They avoided pastoral states on the Central Asian grassy plains, conquering and sometimes enslaving native peoples organized in bands and villages (Lincoln [1993] 2007; Kivelson and Suny 2017).

Internal European political competition soon affected this global expansion. Before 1400, Europe consisted of a mix of territorial and city-based states, and larger states started to incorporate smaller ones. The influx of precious metals, foods, and information started increasing commercialization in Europe. By 1600, the Netherlands (the "Dutch") increasingly challenged the Iberians around the world. They dislodged neither the Spanish nor the Portuguese from the Americas, but in a series of wars, they replaced the Portuguese in the Indian Ocean and maritime Southeast Asia. The Dutch, the English, and the French also started exploring North America and trading with India (SarDesai 2008, 2013). All three governments granted chartered companies which had monopolies on trade and the power to make war. The Dutch dominated the world's oceans from roughly 1600 through 1700 (Boxer 1965; Neal and Cameron 2016) and increasingly controlled trade through maritime Southeast Asia (SarDesai 2013). By 1649, the Russians reached the Pacific Ocean, first establishing a string of trading posts to ship furs back to Moscow, then expanding this empire (Lincoln [1993] 2007). Just before and after

1700, the English and French fought several global wars, all of which the British won, replacing the Dutch globally.

By 1700, Europeans mostly held areas in the Americas and Siberia and remained on African and Asian coasts. The Portuguese retained Brazil and a few bases in Asia; the Spanish expanded in Mexico and the Andes. The Dutch stayed in maritime Southeast Asia. In the next fifty years, the British and French expanded into and competed over eastern North America, the Caribbean, and parts of India, the former ejecting the latter from all three regions in the Seven Years' War (1756–1763), while the Portuguese began to explore Amazonia (Chasteen 2006).

Most African and Asian agrarian empires continued their previous trajectories, and Europeans had little effect on many kin-oriented peoples. While the slave trade extended its influence inland, spurring secondary state formation and depressing population growth, African kingdoms largely kept Europeans out through their own power and through African diseases (Hopkins 1973). The Ottoman empire expanded into Southwest Asia, North Africa, and southern Europe through the 1500s. The Safavid empire also expanded in this period into what is now Iran, the two empires fighting and establishing a border in 1514 (Gelvin 2011). In the early 1500s, Central Asian descendants of Mongols conquered most of South Asia, establishing an empire for two centuries (Trautmann 2016). Several mainland Southeast Asian states arose, such as Pegu, Ayutthaya, Vietnam, and Cambodia; all traded with China and India. The Chinese Ming dynasty slowly collapsed over the period, and Qing invaders from the north conquered China and set up a new dynasty in 1644 (Holcombe 2017). Bands, villages, and a few polygroups in some areas of North America, the innermost areas of South America, the New Guinea highlands, and all of Oceania, including Australia, remained largely or completely unaffected by Europeans.

From after 1450 into the 1760s, Western Europeans expanded into the Americas, maritime Southeast Asia, and Siberia and based themselves on African and Asian coasts. Where they established themselves permanently, they effected major change in the local peoples, and they influenced those beyond, but large parts of the world remained beyond their reach.

Social Patterns: The Global Context, 1450s–1750s

European conquests and trade unevenly changed these societies. European activities created a thin but truly global economy; started an incipient cyclic pattern of global political dominance; helped accelerate global population growth; developed a new form of group identity, racism; and caused deep social change in the Americas and growing change in Afro-Eurasia.

Europeans united the previously separate hemispheric trade systems, creating a global economy. They initiated cross-Atlantic trade, which increased greatly (de Zwardt and van Zanden 2018). Europeans sold finished Asian and European goods to African coastal states, both of which provided slaves for the Americas, which then produced silver, sugar, tobacco, and other agricultural products for Europe and Africa. Silver went to Europe and, after 1650, much of it continued into South and East Asia. Finished European products also went to Spanish America (Neal and Cameron 2016; de Zwardt and van Zanden 2018). While in "the sixteenth and seventeenth centuries … Europeans did not make a significant dent in Indian Ocean commerce" (Oka 2018:302), American silver did begin to change Asian trade. Silver's value had varied between Afro-Eurasian regions before the American opening; after, massive American production began to equalize its value globally (de Zwart and van Zanden 2018). The Portuguese base-and-license strategy "created an alternative system of power" in the Indian Ocean (Oka 2018:302), the Dutch being even more powerful than the Portuguese. From 1500 to 1800, Ottoman, Safavid, and Mughal manufacturing increasingly oriented toward African and European markets, and the Chinese also began this reorientation. By 1700, "the revenues [in these areas] accruing from transregional and regional commerce were greater than agricultural revenues within each region" (Oka 2018:303). Most trade remained in the Indian Ocean, but it increasingly started to shift to Europe. In fact, "the total amount and value of [Afro-Eurasian] trade and commerce *between* any two core areas gradually increased and then dwarfed the amount and value of trade *within* each region by the eighteenth century" (Oka 2018:303). Most global trade remained in luxury goods (Lindqvist and Myrdal 2018), but Indian Ocean trade began to orient to raw materials; some of this was Europe–Asian trade, but much of it was African–South Asian or within Asia (Lindqvist and Myrdal 2018; Oka 2018).

Following the flow of products helps us understand the silver-based global economy (Flynn and Giraldez 1995). The Spanish mined silver across the Americas, sending most of it to Europe. Much of it went to Russia, Southwest Asia, and India; later, most of it ended up in China (Marichal 2006a), paying for goods, such as silks, gold, and spices, that went back to Europe (Strayer 2011). Asians considered European goods inferior, only wanting the silver (Haksöz and Usar 2012), which soon became the basic Chinese currency (Flynn and Giraldez 1995).

After silver, the second most important commodity was sugar, grown primarily in Brazil and the Caribbean by slave labor (Chasteen 2006). Domesticated plants and animals also flowed between the two hemispheres, a process called the "Columbian Exchange" (Crosby [1973] 2003). Potatoes, tomatoes, and corn went from the Americas to Afro-Eurasia, while Europeans brought pigs, cattle, horses, wheat, rice, sugarcane, and fruits and vegetables to the Americas. Spanish America also produced dyes, indigo, and cochineal (McCreery 2006; Marichal 2006b), that Europeans used to dye Chinese silks, so the silk and cochineal commodity chains intersected (Marichal 2006b). Since Europeans used African slaves to produce sugar, the slave and sugar commodity chains also intersected. Much of the Asia–Europe trade consisted of pepper, then other spices. China also started exporting tea (de Zwardt and van Zanden 2018). Multiple products flowed through the Silk Road/Indian Ocean system. The Ottoman empire imported fabrics, raw cotton, spices, indigo, and iron from India, and exported horses, herbs, buffalo skins, dates, rice, and other products. Persians also produced and exported silk (Haksöz and Usar 2012). Russians expanded into Siberia to increase the existing fur trade, as did the

Dutch in North America. There, as beaver were hunted out in one region, trappers and traders moved farther inland, constantly pushing forward a tribal or shatter zone that disrupted native peoples' social structures (Wolf 1982; Raffield 2021). Tobacco and sugar demand also grew in Europe, increasing production in Caribbean islands (Nater 2006). Production and trade of all commodities increased in the eighteenth century. The slave trade, with its horrific consequences for those sold—a 15 percent mortality rate on the passage across the Atlantic (Strayer 2011:455), and on arrival, a short lifetime of extremely hard work—peaked around 1750 and started to decline. Yemen had been the major transshipment point for coffee produced around the Indian Ocean in the 1500s. By 1750, Amsterdam nearly matched its volume; by 1789, American colonies produced over 80 percent of the world's coffee (Topik and Samper 2006). The British began to produce opium in South Asia to sell mostly to China (Holcombe 2017).

Commodity chain management changed over this period. In the sixteenth century, the Portuguese crown engaged in joint ventures with merchants who put up most of the financing. Later, the crown granted concessions to merchants (Haksöz and Usar 2012). Logistics and support industries also changed in the Indian Ocean. In the sixteenth century, unmanaged down-the-line trade continued. In the seventeenth, one commercial agent increasingly managed the entire process, though goods continued to be transshipped (Haksöz and Usar 2012). The Netherlands, England, and France created East India Companies to manage Indian Ocean trade and West Indies Companies for Atlantic trade. Their charters granted them the right to raise and field armies, which they used to create commodity monopolies. Unlike partnerships which expired, corporations were chartered indefinitely and were capitalized by stock markets (Haksöz and Usar 2012; Pomeranz and Topik 2018; de Zwardt and van Zanden 2018).

In a new form of imperial rise and fall, different European states politically dominated this incipient global system in sequence. Some argue that global power concentrated and diffused cyclically, in four phases of two to three decades each. In this account, the global system started in a state of anarchy; global war between two large alliances established the hegemony of the most powerful state. Through imitation and innovation, other states caught up and passed the hegemon, leading to another anarchic global system, leading to another global war, and so on (Modelski and Thompson 1966; Kennedy 1987 has a semi-similar account). To these scholars, the Portuguese were globally dominant through the 1500s; the Dutch dominant in the 1600s; and the English in the 1700s (Modelski and Thompson 1966). Others argue that hegemony was more fleeting, based on being the richest and most productive state. Profits came from being the world's entrepôt and controlling "the 'invisibles'—transport, communications, and insurance," plus banking and investment. Only the Dutch were briefly hegemonic in this period (Wallerstein 1980:38). Yet, others argue that the Genoese were hegemonic, though regionally, before the Dutch (Arrighi 2010). All these scholars imply that this form of rise and fall was different from agrarian empires. Agrarian empires rose and fell either in isolation or in regional state systems, while after 1500, hegemonic rise and fall was global. Hegemony rested not on land power but on sea power backed by ship-mounted guns. Finally, rather than disappearing after collapse, the hegemon continued its existence but lost its far-flung holdings to challenger states, and also lost its system-ruling power.

Europeans organized their empires in the Americas similarly; European patterns in Asia varied. The Spanish first created two viceroyalties (the *viceroy* was basically "vice-king") in Mexico and Peru. At first, claims to land were mostly on paper, but the Spanish moved armies and officials inland. As the empire slowly expanded, creating frontier shatter zones, the Spanish created two more viceroyalties, one in New Granada in northern South America and the other in the Río de la Plata region (now Argentina). Portugal also eventually established two viceroyalties (Chasteen 2006). Bureaucracies extended into smaller cities and towns to organize labor tribute, which only Indigenous peoples paid, and to engage in land confiscation during the high mortality of epidemics (Horna 2011). The Dutch in maritime Southeast Asia avoided administering any territory, preferring trade treaties with local rulers (SarDesai 2013).

The system's global scope and European changes in the Americas obscure how little political influence Europeans had in most of the world. Together, the Ottoman, Safavid, Mughal, and Ming empires included well over half of all humans; three of the four were Muslim, though most Mughal subjects were Hindu. Following and enforcing Islamic law eased trade through the Indian Ocean (Oka 2018). While increasing amounts of trade went back to Europe, for most of the period, Europeans within it were still only one set of players among many. Their actual influence in Afro-Eurasia was mostly coastal; only in West Africa and Bengal did European influence create frontier zones beyond their formal holdings (Trautmann 2016; Iliffe 2017). Even within new European empires, much of their control for decades was only on maps rather than actual dominance; Figures 7.2 and 7.3 show claims that took time to become real.

Global demographic patterns changed. Global population almost doubled, though it was unevenly distributed. European and Asian populations did double. Some growth resulted from the Columbian Exchange, providing corn (maize), potatoes, and cassava (manioc) to Afro-Eurasia (Brooke 2014); it also derived from a European reduction in mortality sometime after 1600 due to new public health measures, such as quarantines, and to agricultural improvements (Girard 2005; Mazoyer and Roudart 2006; Brooke 2014; Livi-Bacci 2017). Due to European conquest, enslavement, and diseases, at least two-thirds of the Mesoamerican and Andean populations died in the first century of Spanish occupation (Chasteen 2006; Livi-Bacci 2017), and more in the next century, the same seeming to have occurred in North America. More survived in those regions than in coastal Brazil where populations "were virtually wiped out" (Chasteen 2006:53); the slave trade then made Brazil and the Caribbean increasingly African. African populations forcibly transported to the Americas had varying fates; in Brazil and even more so in the Caribbean, fertility was low and mortality high, while in North America, Africans' fertility was high (with a total fertility rate around 8) and mortality only slightly higher than that of Europeans there. African lost millions of people during the slave trade, so the population stagnated (Brooke 2014; Livi-Bacci 2017:25, 53). This was

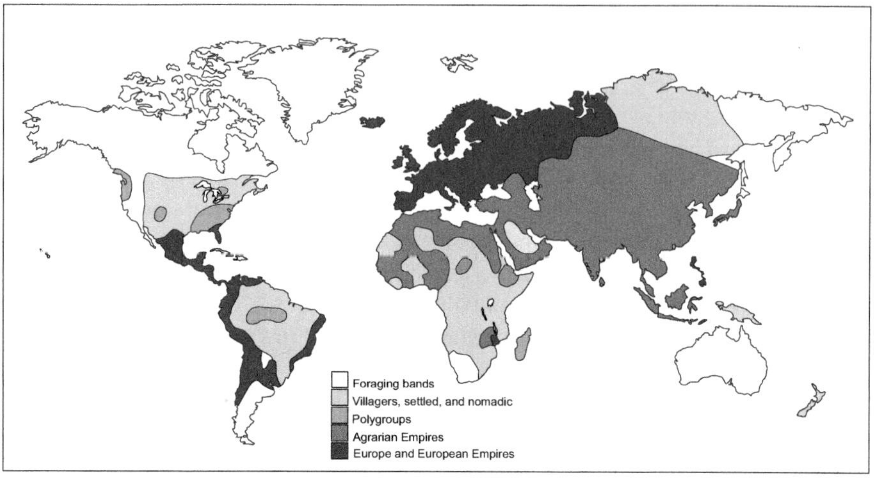

Fig. 7.2 Political Systems, ~1600. Sources: Moore (1995); Black (2005); Haywood (2011); Houston (2018)

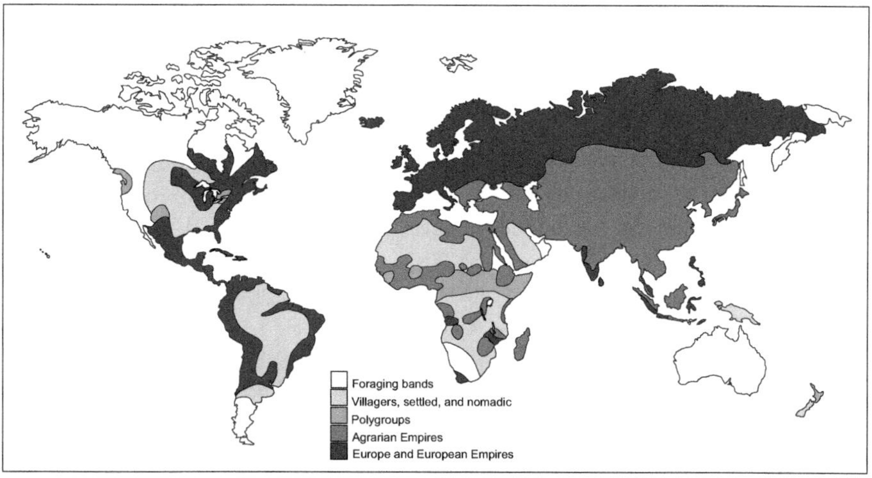

Fig. 7.3 Political Systems, ~1700. Sources: Black (2005); Haywood (2011); Houston (2018)

a result of the quickening of global migration. Cumulatively, upwards of two million Europeans settled in the Americas in this three-century period (Pomeranz and Topik 2018). Some Europeans moved to African and Asian coasts as merchants and missionaries. Many West Africans migrated inland to avoid the coastal slave trade, Chinese settlers moved westward, and Indigenous Americans moved away from the encroaching Europeans (Manning and Trimmer 2013:Chapter 7). Finally, the populations of North Africa and Southwest Asia had stagnated for five centuries, a pattern that continued for four more centuries in this period (Brooke 2014).

A seemingly new form of dehumanizing ideology developed in this period, eventually with global consequences. Some argue that before 1500, physical appearance did not seem to determine group memberships consistently: “Greeks and Romans did not establish color as an obstacle to integration in society” (Snowden 1983:63). Just before 1500, European treatment was inconsistent: some areas almost celebrated dark-skinned peoples, while other regions had prejudices against them (Fredrickson 2009; Howe 2002). Others disagree, arguing that Egyptians, Romans, Arabs, and Chinese distinguished other peoples by physical features and claimed that some of them were inferior on that basis (Smith 2011:112–113). Still, it does not seem that racism, a system of discrimination based on physical appearance, was built into these societies as deeply as it was after 1500. This new ideology seems to have originated in Iberia in a complex process (Fredrickson 2009); it made enslavement of Americans and Africans seem moral (Smith 2011; Fiske and Rai 2015), but one might question the underlying motives—as Bartolomé de las Casas, a Spanish cleric, did. After witnessing repeated, violent Spanish attacks on Indigenous Americans, he mounted a public campaign against the ideology that they were soulless beings. While the pope agreed, this had no effect on Spanish policy (Smith 2011:74–80).

In the Americas, new groups developed based on the mixing of “races.” Early, Spanish men married Mesoamerican and Andean elite women, creating new *mestizo* elites. Spaniards sent out from Spain (*peninsulares*) took the top spots, *criollos* (Spaniards born in the Americas) had middle administrative positions, and *mestizos* the lower positions. Most Indigenous peoples in Spanish America and Africans in Portuguese America mostly did manual labor. Intermarriage between the commoner natives and Spaniards slowly increased the mestizo population (Strayer 2011). Europeans and Africans, and Africans and Indigenous peoples, also had children together, creating two new, mixed groups. However, the empires imposed and supported notions of white superiority (Chasteen 2006). Dehumanizing ideologies commonly accompanied and supported slavery before 1500 (Smith 2011:106–111). They also supported African and American slavery, though American demographic collapse meant that it became more important for the Africans forced westward. European slave traders and owners justified their practices by saying that Africans had no souls or were animals (Smith 2011:114–119). Only with such beliefs could Europeans see slavery as “moral” (Fiske and Rai 2015), and even then, several Europeans disagreed, arguing that Africans were fully human (Smith 2011:114–115, 118).

New empires also began cultural assimilation practices; old ones sometimes continued them. Iberian conquerors brought Christianity with their new empires. Monastic orders aided colonial administration, and the Catholic Church eventually imposed its diocese–parish structure on every region (Super and Turley 2006). The Church also worked to convert native peoples; by 1800, most had accepted Christianity in some form, though many mixed Indigenous and African elements with Christian ideas, especially in Brazil (Chasteen 2006). The Spanish heavily Christianized the Philippines, again accepting adaptation to local traditions (SarDesai 2013). English colonists actively supported the Church of England and other Reformation sects along the North American Atlantic coast (Super and Turley 2006). The Ottoman, Safavid, and Mughal empires institutionalized Islam deeply in their regions, though South Asia still remained mostly Hindu (Oka 2018). Aggressive Portuguese proselytization of Christianity in maritime Southeast Asia proved to be counterproductive, so Islam kept spreading there (SarDesai 2013). Islam also spread in sub-Saharan Africa through trade and into the Balkans through Ottoman conquest. Because Moscow taxed Indigenous Siberians in furs and Russians in money, fur-trade expansion relied on nonassimilation of Siberian peoples, though Orthodox Christianity spread into Siberia anyway as Russian speakers migrated eastward (Kivelson and Suny 2017:69–70).

Europeans often increased gender inequalities. Early on, Spanish America and the English colony of Virginia were predominantly male; more women came later from Europe. The Iberian empires "were more rigidly patriarchal than many indigenous American and African societies" (Chasteen 2006:68). Men staffed all imperial positions, and white elite men prized and "defended" white women's sexual exclusivity, while Indigenous women often suffered sexual violence (Kellogg 2005). Spanish men pushed Aztec women out of market administration, reduced the gender-parallel religious structure for male-staffed Christian institutions, limited Indigenous women's property rights, and increasingly treated women as "legal minors" in the new Spanish court system (Kellogg 1995:112). Authorities punished women who resisted, though some women managed to live outside the system (Chasteen 2006). They were often leaders of revolts and maintained a reduced parallelism in their societies (Kellogg 2005).

Elsewhere, women had more mixed experiences with Europeans. Muslim, Portuguese, and Dutch traders in maritime Southeast Asia found that women were often the main merchants in the region; they were "appalled" at the women's wealth, independence, and commercial acumen and "hoped to make such women less independent and more obedient" (Pomeranz and Topik 2018:37). This started with intermarriage and putting up with independent, multilingual Asian and Eurasian wives for several centuries, which did aid in wealth creation. Still, European husbands sometimes manipulated European and Asian law to take over their family fortunes, though they often died young from local diseases, the women then taking other European husbands. The Dutch do seem to have inculcated European sexual mores into their families over time, but with only some success (Pomeranz and Topik 2018:38). Women were also traders in many West African societies, so they "played a visible role in mediating trade between African and European merchants" (Sheldon 2017:39). European administrators forbade liaisons with local women; their agents ignored this restriction, creating a group of wealthy and powerful mixed-race merchants and politicians of both sexes on West African coasts (Sheldon 2017).

European expansion may have benefited from environmental circumstances, and it had damaging environmental consequences. Droughts in the Americas before 1500 may have made populations more vulnerable to European diseases (Brooke 2014). Europeans intensified extraction of precious metals and timber in Europe and started such processes in the Americas. Declining output in the mid-1500s led to new policies that increased deforestation and damming. The Spanish developed a new agroecological regime relying on European crops to feed mine laborers (Moni 2007; Moore 2007). In North America and the Caribbean, tobacco, sugar, timber, and iron ore production all led to deforestation and soil erosion (Penna 2015). The Spanish introduction of herd animals led to deforestation and overgrazing (Finley 2007). Demographic collapse in the Americas led to reforestation, which captured significant amounts of greenhouse gases, especially carbon, and lowered global temperatures (Brooke 2014).

European influence also started reorganizing societies, but only in certain areas. In eastern North America and Siberia, the fur trade commercialized peoples in shatter zones beyond colonies or outposts, partly from the introduction of diseases and alcohol (Wolf 1982; Lincoln [1993] 2007). After 1763, the eastern coast of North America was exclusively British, and their colonists pushed the native peoples westward. The large chiefdoms that had been wiped out by diseases reorganized themselves as village societies in confederations or as bands (Wolf 1982). Historians debate how much the trade changed Africa. In this period, African slaves came mostly from people that African states captured in battle (Chasteen 2006). Their merchants sold these captives at European bases. West African polities increasingly took payment for slaves in firearms, which allowed African chiefdoms to become states and states to become commercialized empires, especially the Asante and Dahomey empires (Ajayi and Crowder 1985; Iliffe 2017). In three centuries, upwards of 22 million Africans were captured; over 10 million went to the Americas (9 million survived the passage), millions more died, some were domestically enslaved, and the rest entered the East African–Indian Ocean slave trade (Pomeranz and Topik 2018). 80 percent of those forced to cross the Atlantic went to Brazil or the Caribbean, 5 percent to North America, and the rest to Spanish dominions (Reader 1998:Chapter 38). Overall, the slave trade established a more violent political culture.

Asian agrarian empires changed, mostly from internal factors. The Ottomans granted trading rights to Europeans, which encouraged encroachment; after two centuries of flourishing, the empire began a typical agrarian decline (Gelvin 2011). Europeans increasingly involved themselves in Indian politics and expanded into maritime Southeast Asia. The Portuguese and Dutch remained mostly coastal, controlling trade, but when the Mughal empire collapsed in 1707, the British and French East India Companies expanded inland in Bengal, establishing alliances with small states. The British defeated the French at the Battle of Plassey in 1757. This established their supremacy in South Asia, supporting continued expansion (Trautmann 2016). The pastoral Central Asian Manchu took over the Chinese empire as the Qing dynasty in 1644; perhaps China was least affected by European influence. Both Korea and Japan became somewhat isolationist, but they retained links to states around them (Holcombe 2017).

After 1500, Europeans put Afro-Eurasia and the Americas in contact with each other, creating the thinnest beginnings of a global social system. In the Americas, they disrupted old ways and created new ones, often far more oppressive than previous patterns, and began constructing something new in Mesoamerica and the Andes.

They stimulated secondary state formation and commercialization in West Africa. Asian agrarian empires continued their own rise-and-fall trajectories, treating Europeans as minor players.

Social Patterns: European Change, 1450s–1750s

While Europeans caused change elsewhere, Europe itself also changed. A European state system developed by 1750; Europe became richer, improving agriculture and commercializing; a new marriage pattern developed; and revolutionary technology sparked cultural change.

A formal state system developed in Europe. Imperialism eased European circumscription (Stearns 1969), but European states fought each other, as in all other state systems. In each century, one state attempted and failed to conquer all of Europe. In response, European political leaders slowly developed the idea of a "balance of power" between states with constantly shifting alliances; in contrast to some previous systems, this prevented one state from conquering all the others. Still, larger states incorporated smaller ones through intermarriage, conquest, and treaties. Elite intermarriage generated some wars through conflicts about patrilineal claims to rule (Shennan 1974; Towns 2010:58), while Europe-wide Christian institutions and beliefs eased border redrawing (Tilly 1975:18–19). States also institutionalized a system of relations with each other, consisting of bureaucracies devoted to interstate relations, the exchanging of ambassadors, norms about embassy treatment, and international law including formal treaties and informal conventions. States also spied on each other (Hocking and Lee 2011).

European states also changed methods of internal rule. Despite rediscovery of models for popular participation (see below), northwest European monarchs increasingly relied on assemblies of regional leaders to raise money and eventually to make laws (Congleton 2011), such as the English Parliament in the 1200s. Similar national assemblies developed in the 1300s in Sweden and in the Netherlands and France in the 1500s. Nobles, merchants, and clergy usually represented their regions. The balance of power between monarchs and legislatures constantly shifted. Also, states developed more powerful bureaucracies, court systems, and militaries (Congleton 2011). Central and Southern European monarchs increasingly aspired to "absolute" power over their societies (Anderson 1974; Towns 2010). Males still dominated politics, though they tolerated queens as rulers (Towns 2010:62–64; Brooke 2014).

Europe also innovated economically. Dutch and British farmers learned to farm the entirety of their field rather than fallow some by rotating different crops that required different nutrients. This greatly increased agricultural productivity, supporting the acceleration of population growth (Mazoyer and Roudart 2006:Chapter 8). Transport innovations increased trade across Europe (Hoffman 2004), also increasing productivity. This increase and overseas trade helped commercialize Europe. Vast amounts of silver, tobacco, cotton, sugar, and fur flowed from the

Americas. Europe became richer, and markets developed more. Even before the 1400s, per-person (or "per capita") European income was higher than anywhere else globally, and it continued rising after. Commercialization and nearly constant warfare also encouraged the development of European financial institutions, including banking, merchant houses, government-chartered companies, stock markets, shipping, and insurance (Calomiris 2003; Maddison 2006). By 1750, England, France, and the Netherlands, and what later became Belgium, were more literate, more urbanized, and more commercial than the rest of Europe, with higher incomes, more complex distribution and retail systems, and a wider range of available consumer goods (Wiesner-Hanks 2006). This also changed retail institutions and consumer practices (Stearns 2006a:Chapter 2), including advertising, shop window displays, the use of salespeople to learn consumer desires, consumer credit with retailers, middle-class imitation of elite fashions, and increased sales of new foods, personal items, household goods, dining, and entertainment (McKendrick et al. 1982; Stearns 2006a).

Commercialization also changed European governance. As merchants increased their wealth, they began to challenge hereditary landowners' political power. This process went farthest in England. In the 1640s, Parliament executed the king; the resulting civil war established a short-lived republic. The restored king after 1660 slowly ceded power to merchants. In 1688, Parliament overthrew (without killing) another king, which some argue improved the commercial climate for innovation (Acemoglu et al. 2005).

European gender and family structures changed. What economic historians call the "European marriage pattern" developed after 1300. Elites remained patrilineal, but most marriages were neolocal and kinship was bilateral (Therborn 2004). The average woman married around age 25, much higher than in most societies, as neolocality required saving money for a new household (de Vries 2008). Around 10 percent of women never married. Both patterns restricted fertility in Europe, making the population increase remarkable. Men still had most economic resources and enjoyed more sexual freedom than women did (Smith 1989).

New technology sparked five intellectual movements in this period. The low rate of literacy before 1450 around the world was partly due to all books being written by hand. Few new books were written; few copies of any book existed; copying always introduced errors; and many books were lost over time (Eisenstein 1983). Information about movable type from China entered Europe in the 1400s. Europeans combined this with the screw wine press, creating a printing press. In the next three centuries, small, commercial printers published millions of books, pamphlets, handbills, and posters. Publishing many copies preserved books better, and copyediting (checking the typeset text) became standard procedure and reduced errors. The literate percentage of the population increased tremendously. Printers and publishers became meeting places for different authors, and elites built up large personal libraries (Eisenstein 1983). These supplemented traditional handwritten letters and public talks (Briggs and Burke 2009).

Philosophy changed greatly in this period. Arabic texts commenting on Aristotle had made their way into Europe through Iberian reconquest and through Sicily by

1300, heavily influencing scholarly thought (Hasse 2021). Italian scholars in the 1300s increasingly searched for classical Greek and Roman manuscripts in libraries; training in Greek from scholars in Constantinople aided the process (Brown 1988). Print preserved and diffused these books, supporting the "rebirth" (French: *renaissance*) of European learning (Lauer 1999). The Renaissance subtly undercut the Church's position: ancient thought was secular, so religious justification for government seemed hollow, and ideas about republics reappeared in Europe, leading people to question monarchies (Brown 1988).

The new printing culture supported long-overdue reform of the corrupt Catholic Church. It diffused calls for "re-formation," fragmenting European Christianity. Injunctions to read sacred Christian texts directly, bypassing priests (Cameron 1991), increased demand for printed Bibles, increased literacy rates, and combined with Renaissance notions of self-governance. Struggles during the Reformation intensified warfare in the state system (Wiesner-Hanks 2006).

Printing also opened a "public sphere" in Europe (Briggs and Burke 2009). Regular business publications, or newspapers, developed in the 1540s (McCusker 2005). The first general newspaper started in German lands in 1605 and spread throughout Europe in the next few decades so that they were, by mid-century, "the most widely distributed form of secular reading matter" (Weber 2006:399). Pamphlets, books, and newspapers in local languages, along with lending libraries and coffeehouses, enabled new debates about politics among previously uninvolved citizenries (Habermas 1989; Fortner et al. 2014), supported by traditional sermons, speeches, and gossip. This new public sphere[1] increased cultural cohesion in states: "imagined communities" previously based on ceremonies in monumental architecture were now based in print culture, helping people identify with a larger society beyond their region (Anderson 1991).

Printing also transformed science (Henry 2001). "Natural philosophers" increasingly used mathematics to understand the physical world; they emphasized observation, not authority, to learn; and they applied new information practically (Henry 2001). Since print circulated and stabilized scientific knowledge, it "enabled science to become fully cumulative for the first time" (Burns 2016:33). By 1700, regularly published journals, scientific conferences, royally funded research societies, university research, and overlapping scientific networks had developed (Burns 2016; McClellan and Dorn 1999). New information from around the world poured into Europe, changing cartography and expanding botany, zoology, geology, and pharmacology (Burns 2016).

Print culture, the Renaissance, scientific learning, and information from around the world flooding into Europe combined to create an elite intellectual movement retrospectively termed an "Enlightenment" (metaphorically shining light on the world) (Graeber and Wengrow 2021). Through in-person meetings, letter writing, and publications, Enlightenment philosophers argued that people could ignore tradition and remake society along rational lines to improve it. They argued that

[1] While Briggs and Burke (2009:80) note that Habermas denies that any public sphere developed in Europe before the 1700s, they argue that a weak version did exist.

institutionalized hierarchies were social creations, and that all humans had basic rights (Hampson 1968). Late in the movement, some argued that natural rights logically implied equality between the sexes and the end of slavery (Offen 2010).

European states regularized their legal systems. Most combined elements of Justinian's *Digest*, Christian ideas, and Enlightenment thought. They used legal systems to apply the same laws across their lands (Pirie 2021). The English system was least organized, with a multitude of courts. Incomplete efforts at systematization started in the 1620s; "by the 1750s, [Parliament] had accepted the necessity of legislating for the day-to-day government of [the] nation" (Pirie 2021:331). Spain imposed its laws on its American colonies, proceeding syncretistically, with Indigenous influences (Yannakakis 2023); the North American colonies imitated English law.

By 1750, European states were still basically agrarian, but changes in the economy, politics, gender and family, population, and idea-life created a new version of agrarian states.

The Transformation of the World, 1760s–1910s

After 1760, European global expansion accelerated. By 1914, Europeans invaded and conquered most societies in Asia and Africa; their descendants continued their expansion in the Americas. Also, "internal change" in Europe became "transformation": Europeans created a third new type of society called "industrial capitalist democracy." By 1914, these included Western European states; Canada, the United States, Australia, and New Zealand; and Japan.

Telling the Tale: The Atlantic Revolutions, 1760s–1820s

In this sixty-odd-year period, Europeans developed new economic, cultural, and political forms. American colonies gained independence from Britain, France, and Spain and Portugal, creating the United States of America, Haiti, and "Latin" American states, respectively. The French revolutionized government. Europeans ended the Atlantic slave trade, developed new forms of manufacturing and distribution, and continued geographic expansion. Again, some dislike the word "revolution," as it implies sudden change (Neal and Cameron 2016). Still, as argued previously, the context matters. After 5000 years of steadily growing agrarian states, the global Industrial Revolution developed in just a few centuries (Childe [1936] 1966).

By 1800, Europeans completed mapping the globe. In the 1760s and 1770s, French and British expeditions sailed and mapped the fringes of the Pacific Ocean. English contact on Hawaii in 1778 spurred secondary state formation: one chief conquered the other islands, creating a new state. In the following decades, British and American whalers called regularly in Tahiti and Hawaii. In 1788, Britain started settling convicts in Australia (Parry 1971).

Five interrelated international political changes occurred. English colonists in North America rebelled (1775–1783) against British taxes, trade limitations, and restrictions on expansion. Aided by France and Spain, they defeated Britain, which gave up the colonies. France incurred massive debt in the alliance. Bad harvests reduced tax revenues (1787–1788), creating a crisis.[2] The French king called an old legislature, the Estates General, in 1789. Commoner representatives established a National Assembly, sparking a ten-year revolution and a European war. An army general, Bonaparte, conquered much of Europe, proclaimed himself French emperor, and carried out reforms. A British-led alliance defeated Bonaparte and restored the monarchy (Stearns 1969).

These enabled the third and fourth changes. In 1791, slaves rebelled against white planters in the French Caribbean colony of Saint-Domingue. Supported by the British, they drove off rich white planters and poor white workers, establishing the new state of Haiti (Strayer 2011). France's invasion of Spain in 1808 triggered temporary independence of the Spanish American colonies; after French defeat in 1815, Spanish efforts to reassert control failed, partly because the British and Americans supported the rebellions. After independence, some of the new countries broke apart in further wars. Portugal negotiated Brazilian independence and established the only post-1492 American monarchy (Fagg 1977).

Enlightenment philosophers long objected to the African slave trade. Their arguments spurred a movement of anti-slavery societies, which led to legislation abolishing the trade across European states: Denmark in 1803; Britain, 1807; the United States, 1808; the Netherlands, 1814; and France, 1818. British naval efforts reduced but did not end the now-illegal trade, and 28 percent of all Africans forced to the Americas went after this legislation (Reader 1998:426).

Agricultural improvements before 1760 supported intensive economic growth. "For the first time, agriculture … support[ed] the needs of a nonagricultural population … more numerous than the agricultural population itself" (Mazoyer and Roudart 2006:331). Using scientific research, British inventors developed a steam-driven engine over the 1700s (Deane 1965; Landes 1969). Early engines provided ten times the energy that windmills and waterwheels did (Smil 2017), and power then increased. Textile factories increasingly used steam engines to increase production and reduce costs. This Industrial Revolution also changed materials. New chemicals provided new clothing dyes. Iron production surpassed textile production, and prices dropped greatly (Deane 1965; Fremdling and Solar 2010), enabling steam-driven railroads and ships (Fremdling and Solar 2010). Previous commercialization (McKendrick et al. 1982) intensified to absorb increased production.

The map below shows states' claims around 1825, many of which were (again) aspirational rather than factual. Despite European claims, bands remained in much of North America, parts of South America, Siberia, southern Africa, Australia, and a few other places. Village societies still existed in the Amazon basin, in much of North America and Siberia, in Africa and Central Asia, and in New Guinea. The

[2] Goldstone ([1991] 2016) disagrees with this explanation.

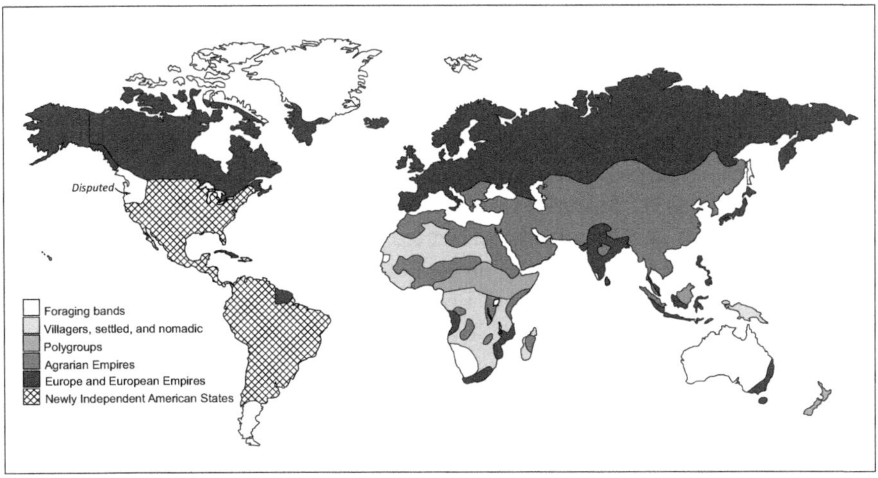

Fig. 7.4 Political Systems, ~1825. Sources: Black (2005); O'Brien (2010); Haywood (2011); Houston (2018)

British dominated the globe economically and militarily, continuing expansion in Canada, India, and the Caribbean and expanding trade with South America. All new American states expanded geographically, incorporating bands and villagers, and attempted to implement Enlightenment principles. West African coastal kingdoms declined, while Sahelian African states (e.g., Bornu and the Sokoto caliphate) expanded, as did the Qing dynasty and Russia in Siberia. The Ottoman and Safavid empires continued to decline (Fig. 7.4).

Telling the Tale: The Conquest of the World, 1810s–1910s

By 1815, Europeans had conquered some of the Americas and pieces of Afro-Eurasia. From 1815 to about 1870, European merchants and missionaries sought new markets and new souls. Military attempts to protect them often pulled European governments into empire in parts of Africa, Asia, and Oceania. From roughly 1870 to 1914, European governments more purposely conquered most of Africa and Oceania, and much of Asia. European-American "offshoots" also expanded geographically, violently incorporating kin-ordered peoples.

As of 1815, European movement into Africa and Asia was shallow except in India. The French annexed Algeria in 1830, took over Tahiti in 1842, and started political takeover in Southeast Asia in the 1860s. After 1757, the British expanded from Bengal into the rest of India. The British exported opium to China, drawing out Chinese silver reserves (Holcombe 2017), and defeated the Chinese in the Opium War of 1842. The British and French then imposed treaties "opening" Chinese markets. European whalers called at Hawaii, becoming involved in the new Hawaiian state. The Dutch continued expansion into the Spice Islands, and the

Portuguese expanded their possessions in southern Africa. After the slave trade formally ended, African states shifted to palm oil production for export to Europe. From Siberia, the Russians moved south into Central Asia (Betts 1978; Bowle 1974; Easton 1964; Fieldhouse 1966; Lloyd 1984).

Conquest accelerated after the 1860s. In the 1850s, scientists discovered that a daily dose of quinine provided resistance to malaria, a killer of Europeans in Africa. With each soldier's dose-a-day, plus steamships, telegraphs, and railroads, Europeans moved deeply inland and started to compete. At a Berlin conference (1884–1885), European diplomats parceled out a map of Africa, then sent troops in. In the ensuing "Scramble for Africa," Europeans conquered new African states such as the Zulu and the Sokoto caliphate. By 1905, the French possessed much of North and West Africa; the British had colonies all over Africa, including the strategic points of South Africa and Egypt (and the Suez Canal); and the Germans had territories in East Africa (Betts 1978; Bowle 1974; Easton 1964; Fieldhouse 1966; Lloyd 1984). Imperial borders bisected many African groups, linguistic and otherwise (Reader 1998:Chapters 46–47).

By 1914, Europeans had expanded into Asia. Russians conquered the Kazakhs and other Central Asian peoples. Britain had nearly full control over India and took over Australia and New Zealand; the French conquered most of Southeast Asia; and the United States wrested control of the Philippines from the Spanish and took over Hawaii, deposing its last queen. Afghanistan once again showed itself to be easily conquerable on the surface but impossible to govern in practice. Europeans never conquered China, Thailand, Iran, or the Ottoman empire but did tightly limit their ability to act (Betts 1978; Bowle 1974; Easton 1964; Fieldhouse 1966; Lloyd 1984).

European descendants' expansion in the Americas shows that European conquest was truly global. Before its civil war, the United States had moved westward into the land purchased from Napoleon in 1803. After 1865, the government sent its army to handle "the Indian problem," rounding up native peoples to clear the land for farmers and cattle ranchers. The United States bought Alaska from Russia in 1867, then sent colonists there after miners found gold in the Canadian Yukon (Sellers et al. 1985). After years of attempting "enlightened" relations with native peoples, Canadians unified in 1867 and moved violently against them. Argentinians conquered native peoples in the Pampas, creating massive ranches to sell beef to North America and Western Europe in refrigerated ships. Brazilians expanded into the Amazon rainforest, creating rubber plantations at the expense of kin-ordered peoples. Colombia, Venezuela, and Mexico also moved inland, conquering small, kin-ordered groups (Maybury-Lewis et al. 2009). Figure 7.5 shows European claims by 1914. At first, these were merely formal, but this also changed.

Social Patterns: The Global Context, 1810s–1910s

European activity fundamentally transformed the thin global system of the late eighteenth century into a comprehensive late nineteenth-century global system. Global trade and economic inequality increased many times, global political and

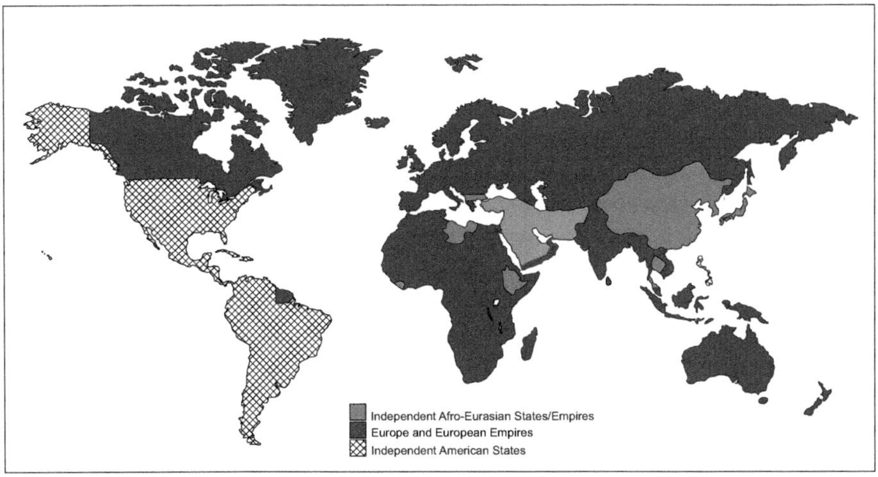

Fig. 7.5 Political Systems, 1900–1910. Sources: Moore (1995); Black (2005); Houston (2018)

Table 7.1 Carrying capacity of world shipping, 1780–1913 (in thousands of tons)

Year	Sail	Steam	Total (sail equivalent)
1780	3950	0	3950
1820	5800	20	5880
1850	11,400	800	14,600
1900	6500	22,400	96,100
1913	4200	41,700	171,000

Source: Maddison (2006, 97). The total is based on the equivalence of one steamship to four sailing ships, because of the larger size and increased reliability of steamships

cultural institutions developed, humans further degraded the physical environment, and more people moved around the world.

Between the 1760s and the 1910s, Europeans expanded and transformed the global economy that they had developed. Global trade expanded many times over the long nineteenth century (1760s–1914). Neal and Cameron (2016) argue that Europe mostly traded with itself; imports from the rest of the world were much lower in value than trade within Europe. Nevertheless, much of the increase was truly global. The following table of world shipping capacity provides a proxy for world trade to give an idea of the change (Table 7.1).

New energy sources partly drove this transformation. While use of human, animal, and wind energy increased after 1800, continuing innovation made steam engines dominant, in factories and transportation throughout Britain by 1840 and later in other countries. By 1650, coal had surpassed wood as the primary British fuel. Since steam engines ran on coal for heat, "Coal mining and steam engines reinforced each other's development" (Smil 2017:238). France, Japan, and the United States also shifted slowly from wood to coal as primary fuel (Smil 2017).

Other technological innovations allowed steam to generate electricity, and electrical systems transmitted it across wide areas (Smith 2017:262) to support industry, then residences. People had long used petroleum; new uses—lighting, automobiles, and flight—increased petroleum production in Russia, Canada, and the United States after mid-century (Smith 2017).

Industrial demand for raw materials restructured or created commodity chains. European companies extracted gold and diamonds from South Africa (Wolf 1982) and grew jute, cotton, and other crops in India, Egypt, and the U.S. South. Beef from Argentina and the United States; wool from Australia and New Zealand; rubber from Brazil and Malaya; more spices from the East Indies; sugar from Caribbean islands and Hawaii; tea from East and Southeast Asia; and palm oil, chocolate, and coffee from Africa were all part of restructuring. Merchants carried most of these raw materials to Europe and its "offshoots" (Canada, the United States, Australia, and New Zealand) for processing, sale, and consumption (Topik et al. 2006; Wolf 1982). Each commodity chain was structured as a series of global markets, centered on specific cities. European and North American corporations increasingly controlled commodity chains (Wolf 1982).

Financial flows also influenced global economic relations. After 1850, Latin American states increasingly borrowed investment money from Britain and France to increase exports (Chasteen 2006:200–202). The large sums invested supported the construction of Latin American roads, railroads, and ports to increase shipping. The United States also increasingly invested in Central America. Debt helped Europeans heavily influence Latin American states.

While most conquered states were somewhat commercialized, most kin-ordered peoples still relied on foraging, collecting, horticulture, and pastoralism, and on sharing, reciprocal gift-giving, and local barter trade. When they continued using commodity monies—cattle, cowrie shells, and others—they rarely cared about state-issued monies. However, commercialized agriculture spread in Europe itself and in Brazil, West Africa, and Southeast Asia after 1750 (Wolf 1982). Often, small local farmers started producing for cash sale to international markets.

Imperialism also diffused consumerism globally (Stearns 2006a). Department stores selling European fashions opened in Moscow. Highly commercialized China long sold its goods to Europe, so it resisted European consumerism until the 1890s. In contrast, many Japanese adapted European styles to local traditions after the 1850s. Latin American elites imported many European and U.S. goods. Urban African elites cautiously followed suit (Stearns 2006a). Middle-class consumerism barely began in Siberia, Central Asia, and most of Africa.

Global economic restructuring required new labor systems. The British anti-slavery regime meant that all workers were officially paid in (low) wages. Colonial governments solved the problem of subsistence producers rejecting state monies and labor "opportunities" by forcing colonial peoples to pay taxes in state-issued cash (Strayer 2011). This required them to seek wage employment, which conveniently provided labor for European economic activities. Imperial monies therefore became more common in European colonies over time.

Responding to changing imperial labor needs, global migration increased greatly, based on new railroads and steamships. This made "the period from 1850 to 1930 … the most intensive era of migration in human history" (Manning and Trimmer 2013:154). The African slave trade tailed off, while after 1840, other migrations increased. Around 56 million Europeans went to the Americas, mostly to North America. Thirty million Indians moved, mostly to Southeast Asia. 51 million Chinese migrated; 31 million of them moved northeastward, still within China, and 19 million into the Philippines. Much smaller, but cumulatively much larger, movements of peoples occurred over shorter distances in Africa, Latin America, and Asia (Manning and Trimmer 2013).

Global population grew immensely from the 1760s to the 1910s. By 1750, native American populations were growing again. Previously isolated peoples of Oceania—hunter-gatherers in Australia, villagers in New Zealand, and chiefdoms on Hawaii and Tahiti—experienced massive disease-driven die-offs, but their populations recovered and grew quickly toward 1900. Demographers estimate that the global population grew around 2½ times in this 150 year period, a tremendous acceleration over the past ten millennia: 770 million in 1750, 954 million in 1800, 1.24 billion in 1850, and 1.63 billion in 1900 (Livi-Bacci 2017:25). European populations nearly tripled, while Asian populations almost doubled. African populations grew very slowly, not even increasing by 50 percent. Due to migrations, populations in the Americas grew over *nine times* from 1750 to 1900 (Livi-Bacci 2017).

Population growth within increased commercialized resource extraction had severe environmental consequences. Europeans overhunted elephants in Africa and Asia, and "by 1920 at least six great whale populations had collapsed" in the Pacific Ocean (Simmons 2008:140). Over 220 million hectares of tropical forests were cut down for commercial agriculture and ranching, especially in Southeast Asia, Oceania, and Latin America (Simmons 2008). In Africa, colonial borders cut across some pastoral nomads' migratory routes, so they became commercial ranchers (Simmons 2008:122–124). Commercial agriculture for cash crops often replaced local plants and animals and indigenous land practices, which frequently led to deforestation and soil depletion (Finley 2007). Companies harvested guano (bird excrement) from Chilean islands and exported it to North America and Western Europe as fertilizer (Penna 2015). Coffee plantations spread through the tropics, requiring pesticides and chemical fertilizers; coffee bean shells soiled waterways and killed migratory birds (Penna 2015). Finally, this activity increased the amount of carbon in the atmosphere. Methane from Chinese rice agriculture had long been "the most significant single source of greenhouse gases until roughly the 1830s" (Brooke 2014:495). U.S. forest clearing, then tropical deforestation, and finally industrialization then superseded China's influence on greenhouse gases (Brooke 2014), which still had minimal climate effects.

The global economy reshaped patterns of global inequality through deindustrialization. Some argue that before 1860, the global "terms of trade"—the global price of exports from Latin America, Asia, or Africa relative to the global price of their imports—improved, so they increasingly exported raw materials to and imported manufactured goods from Europe (Williamson 2011). This restructured these

Table 7.2 Economic growth rates by global region, 1820–1913

Location	Percent per year 1820–1870	Percent per year 1870–1913
Western Europe	1.68%	2.11%
U.S. and Australia	4.31%	3.92%
Japan	0.41%	2.44%
Latin America	1.22%	3.48%
China	−0.37%	0.56%
India	0.38%	0.97%
Africa	0.75%	1.32%

Source: Maddison (2006:640)

economies to raw materials production, shifting resources from homegrown industries. In 1750, per capita levels of industrialization in these regions were similar to European levels; they fell by 40 percent by 1860, while European industrialization increased by 300 percent (Williamson 2011). After 1860, they started to reindustrialize slightly (Williamson 2011:196), but the raw materials emphasis slowed long-term economic growth in Asia, Africa, and Latin America. Others argue that deindustrialization in specific areas, such as British policy in India for a range of industries (coal, shipbuilding, iron, and textiles), was a consequence of purposeful policy, not merely responses to changing terms of trade (Parthasarathi 2011; Beckert 2015; Pomeranz and Topik 2018:282–283).[3] Still, European and British offshoots experienced much higher economic growth of industrialization throughout the period. Table 7.2 shows the trend, which is expressed in "Gross Domestic Product (GDP)," the total monetary value of goods and services produced in one year in one country.[4]

This was not primarily due to population growth. In 1820, the richest countries had incomes nearly three times as great as incomes of poor countries. In 1870, the ratio of richest to poorest was 4.8:1, and by 1913, the ratio was nearly 6:1. The chart below shows the details.[5] GDP per capita does not take into account internal inequalities, but these averages make clear that the new global economy had become quite unequal in the nineteenth century: Europe and European offshoots produced 73 percent of world output in 1913 (Carreras and Josephson 2010) (Table 7.3).

The practice of categorizing people by physical features grew with the transatlantic slave trade and slavery. Partly from the new global economic inequality, Europeans concluded that they were biologically superior to all others. Racism became a dominant ideology in Europe, North America, and Oceania, supported by

[3] For the massive debate on Indian deindustrialization, see Parthasarathi (2011), Ray (2011), and Williamson (2011).

[4] Gross Domestic Product has numerous problems (Argandoña 2008). Still, since GDP and GDP/cap correlate with measures of health, lifespan, quality of life, and state power, it can be a "quick and dirty" measure of wealth.

[5] These are estimates of total economic activity, including subsistence and commercial activities. Clark (2010) explained through example how one can estimate subsistence production in commercial terms.

Table 7.3 Estimated per capita GDP ($), 1820–1913: regional comparisons

Region	1820	1870	1913
European offshoots (USA, Canada, Australia, NZ)	1202	2419	5233
Western Europe	1204	1960	3458
Eastern Europe	683	937	1695
Latin America	692	681	1481
Japan	669	737	1387
India	553	533	673
China	600	530	552
Africa	420	500	637

Source: Maddison (2006:642)

the science of the time. Only a small minority dissented (Howe 2002:86–88). Scholars debate whether preexisting racism made possible harsh slavery or whether it resulted from legitimating a labor system profitable to Europeans (Howe 2002:67–68; Smith 2011:112–122, 143–144). "Scientific" racism was paradoxical and contradictory (Howe 2002:88–94), as its forms varied between Europe, the Americas, Australia, and the colonies (Chasteen 2006; Fredrickson 2009). Many argue that Britain outlawed the slave trade to protect its new and growing wage labor system domestically; the high fertility of slaves in the United States helped render the trade irrelevant. But in a larger sense, "[e]mpire depended on a belief in superiority, and on arguments vindicating that belief" (Howe 2002:83).

European states built global political institutions in addition to those empires. Britain established a "liberal" free-trade policy (see below), enforcing this policy on China by war, though closing India to outside trade (Pomeranz and Topik 2018:248). European diplomatic institutions incorporated the Americas and Oceania (Hocking and Lee 2011). Some involved the growing body of international law (Reuter 2011). Europeans also developed informal global organizations. In the "Concert" of Europe (in a metaphor of an orchestra) after the Napoleonic wars, foreign ministers of five European states met regularly to work out collective problems. The formal system had broken down by 1823, but these states followed Concert principles until the Crimean War of 1854 (Elrod 1976) and more informally late into the nineteenth century. Some argue that the Concert-established peace between North Atlantic societies aided their industrialization and global imperialism (Abbenhuis and Morrell 2020).

International law also developed from handling practical problems. These included international mail, time zone irregularities, shipping coordination problems (e.g., incompatible measurements), and international radio standards. Starting in the 1840s and increasingly over the century, diplomats held conferences, then usually proposed an organization to handle the problem (Murphy 1994; Pomeranz and Topik 2018; Pirie 2021). The resulting Public International Unions (PIUs), early examples of global public goods, greatly facilitated international trade. PIUs issued regulations to solve the practical problems; while they were less oriented toward moral relations, they did rest on European assumptions (Pirie 2021).

Trade fairs, art exhibitions, and technical expositions were long European traditions; in this era, Europeans globalized them. After the 1851 London Crystal Palace Exhibition at which European and East Asian states exhibited, many countries held global exhibitions: France (1855, 1867, 1878, 1889, and 1900) and the Netherlands, Australia, Spain, Guatemala, Belgium, Sweden, Vietnam, Ireland, Italy, and China (Findling and Pelle 2008).

Political administration of African and Asian empires took varied forms, slowly turning claims to land existing only on maps into actual rule of large areas. In Africa and Asia, Europeans used two forms of colonial rule, direct and indirect, each adapted to local situations (Heath 2005). They started with a typical hierarchy: a governor, district leaders, and local leaders. In direct rule, a local European military officer collected taxes, administered justice, and implemented European laws in person. Large areas and populations made this difficult, so Europeans increasingly shifted to what they called "indirect rule," which was the old agrarian tactic of using local elites to administer the empire (Bang 2021). The lowest officer in the European bureaucracy ruled local leaders who implemented his orders (Heath 2005).

Indirect rule had problems. Rulers already existed in chiefdoms and small states, but small-scale peoples had only multiple, informal, and sometimes short-term leaders. Europeans often appointed permanent local leaders where there had been none, whom local peoples often ignored. People often understood the difficult situation their leaders were in, yet resented them anyway since they became tools of the conquerors. If the leader failed, European administrators brought in the colonial army (Heath 2005). Since many Siberian and Central Asian peoples on Russia's frontiers were mobile herders, in the 1600s, Russia had established more of a protectorate system than a true empire, drawing tribute from and controlling, to some degree, the movements of these nomads. The relationship was between these nomadic peoples' leaders and the Tsars, not based on territorial occupation. After 1800, Russians shifted to more traditional control through territory-based military governorships (Kendirbai 2020), though these often relied on indirect rule in Central Asia (Kivelson and Suny 2017).

Europeans also imposed their legal systems around the world with various results. U.S. settlers westward justified their claims on native lands by the local peoples not "improving," i.e., cultivating, it, and U.S. courts agreed (Pirie 2021:346). Successive governors-general had British legal scholars compile "a complete digest of Hindu and Muslim law" in India (Pirie 2021:352), but they applied it rarely, establishing a mostly British legal system in India, though with some local influences. The French used "dubious arguments" to expropriate land from Algerians (Pirie 2021:357). Under the guise of "civilizing" conquered peoples, the French set up legal systems across their African possessions. British law in Africa was applied in an *ad hoc* manner for a while. Latin American states had adopted mostly Spanish and Portuguese legal institutions, while the Japanese based their Meiji constitution on the German model. In most areas, local elites backed European legal projects (Pirie 2021) to maintain some power in the new system.

Print culture started to globalize. In the 1850s, companies laid telegraph cables across the English Channel. Oceanic cables crossed the Atlantic and reached India in the 1860s, East Asia in the 1870s, South America by 1880, and they circled Africa by 1890. Finally, a cable was laid from North America to East Asia across the Pacific in 1902. This global telegraphic network enabled news to diffuse globally. Newspapers around the world relied heavily on cable and international press agencies (Winseck and Pike 2007). Both cable networks and news companies were "overwhelmingly privately owned" (Winseck and Pike 2007:4–5), though states increasingly involved themselves as media networks expanded. Reporters of news developed "journalism" as a profession (Chapman 2005). Wireless communication—radio—also began to develop.

Europeans attempted to assimilate conquered peoples into Christianity. The Catholic Church continued expanding in Latin American states. Protestant churches proselytized among North American native peoples; eventually, educational expansion, racism, and Christian missionary zeal came together in the Christian primary school, in which Native American, First Nations, and Aboriginal Australian youth were forced from their parents and educated in English and Christianity (Loftin 2000:248). In Africa and Asia, Christian missionaries also worked for conversion, often ruefully recognizing their imperialist role (Super and Turley 2006). The few mission schools in Africa seem to have been less punitive than North American and Australian ones, but they still supported European assimilation (Sheldon 2017). Russian assimilation policy was tolerant, more linguistic than religious; Moscow accepted the cultural patterns of Siberians and Central Asians as long as they remained loyal (Kivelson and Suny 2017).

Equally regional was European influence on kinship and gender systems, which is debated. Despite tremendous variation, broad patterns in sub-Saharan Africa included widespread polygyny, bridewealth paid to the bride's family, and matrilineality, giving women an unusual amount of power in African states, especially in West Africa (Hanson 2002; Therborn 2004; Kent 2021). Europeans reduced women's political and economic power in these areas, often by eliminating traditional female rulers and imposing taxes on male labor, keeping women out of labor markets, reducing their economic power relative to men (Sacks 1982; Towns 2010). However, in patrilineal southern Africa, African women sometimes used European missionary schools as refuges from local oppression, and some female West African merchants enriched themselves in trade with Europe (Sheldon 2017). Islamic families in North Africa and Southwest Asia were patrilineal and patrilocal. Paternal first cousin marriage was strongly encouraged, seclusion and veiling of women were common, girls typically married by puberty, and divorce was easy for husbands but difficult for wives (Therborn 2004). European influence started public debates about practices of sex segregation, which writers conducted with both European and Islamic ideas, but efforts to change them sometimes disrupted woman-only social networks (Keddie 2007). East Asian and South Asian family systems were heavily patrilineal and patrilocal; parents almost always arranged marriages, and both valued boys much more than girls. South Asian girls often married at age 13, and marriage was considered a duty, but Chinese women married around 18 years and could dissolve their marriages. Europeans fought elite female foot-binding in China and *sati* (or elite widow-burning) in India and tried to raise average marriage ages for girls (Therborn 2004).[6] Southeast Asian families were based on bilateral descent, with easy divorce for both sexes, bridewealth paid to the bride herself, and polygamy mostly among elites (Therborn 2004). The Creole-American family system in the Caribbean originated in the Iberian conquest and the slave trade—combining a shortage of white women (for white men) with the sexual subordination of local women—so it was often violent and always racist, often leading to informal, unstable unions, matrifocality, and many births for unmarried women (Therborn 2004). Elsewhere in Latin America, native women lost power in some places yet retained it in others (Kellogg 2005). In North America and British-conquered Oceania, Europeans attempted to impose their gender norms on all native peoples (Hall 2007).

European imperialism seems to have had two broad impacts on gender relations globally. Some (e.g., Therborn 2004) argue forcefully that European colonization helped reduce women's subordination in regions with long state traditions, such as South and Southwest Asia and North Africa. However, imposition of European gender roles seems to have generally reduced the power and autonomy of women in small-scale, kin-ordered groups in sub-Saharan Africa, the Americas, and some areas of Asia (Stearns 2006b; Sheldon 2018). More comprehensive comparative studies are necessary to determine whether this proposition is at all accurate.

The idea of two basic tendencies in European imperialism on global gender patterns is, as any historian would (and should) point out, a major oversimplification. The previous quick overview of global patterns obscures tremendous variation between regions for most issues, not merely for gender, yet there are some valid generalizations to make. Most importantly, in this period, the new global system started remaking cultural regions. For instance, an export boom of raw materials strengthened the influence of commercial agriculture and mining in Latin American states, increasing economic inequality. It also strengthened local patrons or "caudillos" and patron-client patterns, but some groups advocated for implementing post-Enlightenment liberal ideas from Europe, leading to political instability for decades (Chasteen 2006). Post-slave-trade African states increasingly relied on commercial export agriculture to Europe, a trend intensified by a "cash crop revolution" late in the century (early in the colonial period) (Austin 2016:325). States in the Islamic world—North Africa and Southwest and Central Asia—provided few public goods such as infrastructure; with European expansion, some states undertook reforms of tax and trade systems, which stopped short of separating religious and political leadership (Ghanem and Baten 2016). Land started running out in South Asia, manufacturing increased late in the century, and inequality increased slightly, while living standards remained unchanged during British occupation (Roy 2016) despite improvements in transportation and communication. China did not experience European colonialism, but its per capita income grew much more slowly than that of Europe for several centuries (Ma 2016). Loss of the Opium War (1842) shocked the Chinese elite, which implemented only surface reforms. Putting down the

[6] This may have been a net improvement for South Asian women, or this account may be based on an uncritical reading of British sources. See Hall (2007) for how the Raj used European gender norms as a tool of legitimation.

Taiping Rebellion (1851–1864) cost 20 million lives (Pomeranz and Topik 2018:74) and damaged the regime. Even so, mining and manufacturing increased after 1895, as did the rail system. A lost war with Japan in 1895 started more serious changes. The Qing dynasty finally fell in 1911. In many areas, villagers in European empires remained kin-ordered, engaged in subsistence horticulture or pastoralism; foragers kept foraging. Wage labor, use of state-based monies, formal political structures, and education increased only gradually. In other areas, especially closer to the coasts, these were imposed more thoroughly.

This was part of a larger process of European global imperialism starting to incorporate civilizations around the world into one social system. This began in Spanish and Portuguese America and in coastal West Africa, parts of India, and Maritime Southeast Asia in the 1500s; by 1914, the process was global. The process centered on cities, as civilizations always had; as the nodes in a now-global economic network, cities received European influences first and slowly diffused them into their hinterlands. Trade, conquest, commercial crop cultivation and mining, wage employment in state-issued monies, consumerism, migration, international and European law, Christian proselytization, and the attempted imposition of similar gender norms around the world all spread as assimilating influences in the ways that civilizations had previously operated: primarily among elites, with tremendous cultural diversity within this overall elite similarity (Hall and Jackson 2007).

By 1914, Europeans had conquered most the world, setting up a global system. They increased global trade tremendously, commercialized labor in many areas, stimulated new migration flows, increased global economic and cultural inequalities, set up global political institutions, and started altering local kinship, marriage, and gender systems.

Social Patterns: A New Type of Society, 1810s–1910s

While Europeans expanded globally, they also created a new kind of society. Capitalism, a money-based, market-organized economy, gradually replaced subsistence production and consumption. Economic inequality was based more in monetary classes. Legislatures gained power, male suffrage expanded, and governments provided more public goods. Intellectuals and activists developed new political ideologies. Science and mass media expanded greatly and took new forms. Mortality and fertility reduced and rural-to-urban migration increased. Finally, the European marriage pattern changed.

In 1800, Britain was still mostly an agrarian state but with a small, growing, innovative industrial sector and a larger, dynamic commercial sector. By 1900, Britain was a fully industrial capitalist state, and France, the United States, Canada, Belgium, Germany, Japan, and Russia, were commercializing and industrializing. But what is "capitalism"?

For nearly all of the human past, most people engaged mostly in subsistence production and consumption. As we have seen, and as many have pointed out (e.g.,

Pomeranz 2000), markets existed in states for millennia, yet few define agrarian economies as capitalist. Even late in the agrarian era in long-monetized China, most people were only partly involved in markets. Some discuss different kinds of capitalism for these eras (Hartwell and Engerman 2003) such as European "mercantile" capitalism from the 1450s to the 1750s in which merchants traded commodities in nonindustrial production methods, or "agrarian" capitalism in Britain which turned land increasingly toward cash crops (Scott 2014).

Instead, I think that the gradual shift from primarily subsistence production and consumption, supplemented by market purchases, toward primarily money-using, market-based production and consumption of nearly all goods and services for nearly all people is new in world history. Furthermore, only industries capable of high production rates through machines powered by inanimate energy could sustain commercialized production and consumption. Thus, I define "capitalism" as an industrialized, commercial economy in which elite-owned profit-seeking organizations produce and sell goods and services in markets; most people sell their labor for money in labor markets to those organizations in order to buy goods and services in consumer markets to meet their needs and wants, which increasingly include leisure goods; and states manage the money supply (for use in markets), protect private property, enforce contracts, and actively structure and support economic activity (Weber 1946; Stearns 2006a; Neal 2014).

When and how did capitalism develop? This requires a long answer. Increased productivity in British, Belgian, and Dutch agriculture by 1750 supported it. Many farmers combined subsistence and commercial agriculture (Dennison and Simpson 2010:152–153). Productivity improvements (technical innovations and the geographic expansion of markets) reduced the share of agriculture in all European economies (Broadberry et al. 2010). After 1870, new technologies of cheap steel production further increased agricultural production, as did more improvements in transportation and imported fertilizers (Mazoyer and Roudart 2006:Chapter 9).

Steam-driven "factories," large buildings housing machines connected to steam engines, made textiles early in the century. Workers managed machines in long, boring hours under tight management control. The steam engine provided vastly more energy than muscle, wind, or water could provide, so each worker's output grew many times. However, during the classic English Industrial Revolution period (1760–1830), economic growth was still quite gradual; after 1820, such growth accelerated, so "there was at least a fifty-year delay before the benefits began to be manifested in the general economy," in part due to rapid population growth (Brooke 2014:485). Industrial practices spread slowly throughout the economy. Sewing machines in the 1850s made possible production of ready-to-wear clothing, not just cloth. Britain produced over half of all textiles in the world by 1850, growing their own wool and importing cotton from and exporting product to other countries (Broadberry et al. 2010:175–178). Other European and North American countries then increased their textile output.

New developments in electricity, steel, and chemistry led to a Second Industrial Revolution in the 1870s. Scientific experiments in electricity early in the 1800s (see below) bore technological fruit late in the century: Factory electrification improved layout and increased production (Landes 1969). The electrified inventions of the telegraph and telephone improved communications; this allowed ownership and coordination of multiple factories. Industry produced many existing products, such as furniture, soap, prepared foods, dishes, and silverware, and new products, including sewing machines, bicycles, and cameras. Economic growth accelerated (Brooke 2014).

Businesses also developed new forms of organization. Some argue that the British court system protected property rights sufficiently to support increased commercialization (Dennison and Simpson 2010:159–160). Still, the main innovation was the corporation. Before 1800, most firms were sole proprietorships or partnerships (Hejeebu 2003; Lamoreaux 2003). Corporations were government-chartered organizations to fulfill a specific need. Over the nineteenth century, many states allowed private businesses to incorporate with limited liability, allowing bankruptcy, which economic historians argue encouraged innovative risk-taking (Hejeebu 2003; Lamoreaux 2003). Joint-stock corporations sold "shares" of ownership on markets to anyone, while other corporations remained privately held, preferring bank financing (Hall and Soskice 2001).

Labor organization also changed. Wage labor had existed for millennia, but it always constituted a small percentage of jobs. Now, wage work became common, and the old craft guilds either were outlawed or wasted away. Workers increasingly organized in "unions" to bargain collectively with employers over wages or grievances. By 1875, Britain recognized unions; the French state allowed combinations of both firms and workers in 1884. U.S. businesses consistently and violently fought unions. By 1914, however, union activity increased across Europe and North America (Wrigley 2003).

Ongoing commercialization of Europe provided markets in which corporations sold the vastly increased output (Hamilton et al. 2011). Just as steamships became important on oceans, steamboats first widened markets geographically on rivers, especially in Britain and North America (Stearns 2006a; Dennison and Simpson 2010; Pomeranz and Topik 2018). Later, railroads and telegraphs integrated local markets into national ones, which further tied them to international markets. Governments helped develop rail networks around the world (Grafe et al. 2010). New business methods also helped increase global shipping, including more careful (though more racialized) accounting by East India companies and the establishment of local agents and warehouses in colonial ports, fostering ongoing relations with local producers. All this greatly reduced port and transport times, cutting costs (Pomeranz and Topik 2018).

Despite its importance, manufacturing never reached half the value of economic output or of employment. Instead, market services expanded exponentially. Intermediaries between producers and consumers such as wholesalers (merchants) and retailers started operating nationally and internationally (Grafe et al. 2010; Stearns 2006a:Chapter 2). Retail outlets selling consumer goods proliferated, including traditional shops for one line of goods, new "department" stores selling a range of items, chains of small shops in the United States, and mail-order retailers. Advertising became much more visual in newspapers, magazines, and outdoor wall posters.

Consumption also changed. Increasingly relying on monetary wages to buy consumer goods, workers became gradually more consumerist (Stearns 2006a) as incomes rose, new kinds of goods proliferated, and entertainment forms—including sports events, dance halls and vaudeville, moving pictures ("movies"), newspapers and magazines, and tourism—elaborated. Greater proportions of most people's consumption became discretionary rather than survival-oriented, especially among the new "middle classes" (discussed below). These groups partly imitated elites' conspicuous consumption but also developed their own styles, with national variations, across North America and Western Europe (Tiersten 2001) in clothing, hairstyles, housing, furniture, foods, and leisure goods (Stearns 2006a). Workers could afford only a few of these goods, many being desperately poor even though wages increased toward 1900. Changes in household organization increasingly associated consumption with women (Stearns 2006a:Chapter 5).

These transformations helped states in Europe, North America, and Oceania become tremendously wealthy. By 1913, they produced 73 percent of the value of world output (Carrerras and Josephson 2010), which was distributed differently from previous societies. Instead of social status based on birth, commercial (monetary) success increasingly defined one's social position. The basis of economic elites slowly shifted from landownership to business ownership, though some landowners financed factories for merchants or owned factories themselves (Gullickson 2003). Some merchants and artisans moved into industry; other artisans fought industrialization. Middle classes became quite varied. The upper-middle classes consisted of doctors, lawyers, editors, business managers, and higher clerks. The middle classes consisted of clerks, retail salespeople, lower civil servants, schoolteachers, and secretaries. Many worked in multifactory firms, large-scale merchants, department stores, and multicity railroads. As these organizations grew, the middle classes did also, which made moving "up" in social hierarchies easier (Blumin 1989). The working classes included domestic servants and factory, construction, and transportation workers (George 1982; Wright 1995; Gullickson 2003).

Landowners, the traditional agrarian elite, seem to have been gradually ceding position to merchants from the 1450s to the 1750s. By 1700, agricultural land was about four-sevenths of national capital (total productive wealth) in Britain and five-sevenths in France. By 1810, British agricultural land constituted about half of all capital, so other domestic capital had increased, and it had shrunk in France also. Agricultural land in the North American English colonies made up about half of all capital in 1770 (Piketty 2014:116–117, 151).

Extending these measures illustrates how landowners lost power; other measures demonstrate overall economic inequalities. By 1910, agricultural land was only around 10–20 percent of national capital in North America and much of Western Europe (Piketty 2014), because housing, net foreign capital, and factories, machines, and stores had increased in value so dramatically. In 1810, in both Britain and France, the top 10 percent of population owned over 80 percent of all capital; the top 1 percent owned 45 percent of all capital in France and 55 percent in Britain. The same figures for the United States for that year are 60 percent and 25 percent (Piketty 2014:340, 344, 348). Some economists use national income rather than GDP as the foundation of measuring inequality (Piketty 2014; Alvaredo et al. 2018). Clearing U.S. land of its native inhabitants made land cheap for European settlers, and immigrants reduced the influence of inherited wealth (Piketty 2014): national wealth was only 300 percent of national income. Over time, U.S. national wealth and inequality increased greatly, and by 1910, total national wealth was almost 500 percent of national income; the top 10 percent owned 80 percent of all capital, similar to Europe, and the top 1 percent owned 45 percent (Piketty 2014).

Greatly increased wealth made possible new social insurance systems as new public goods. Previously, "voluntary associations, family networks, charity organizations, guilds, churches, or local authorities" had provided such support (Schmitt et al. 2015:505). States instituted five insurance programs in the nineteenth century: "protection against sickness, injury, and unemployment; old-age pensions; and provisions for family members" (2015:504), a trend that accelerated after the 1880s. Germany introduced social insurance programs for sickness, work injury, and old age explicitly to tamp down threats of working-class revolution (Gregory and Stuart 2004:207). Denmark, Argentina, and the Austro-Hungarian empire in Central Europe introduced such programs just after 1900 (Schmitt et al. 2015).

Governments changed in other ways. The U.S. and French revolutions had few short-term effects on governance (Congleton 2011), but they set in motion long-term pressures. Three developments changed elite representative assemblies into legislatures that represented the entire population: monarchs slowly ceded power to legislatures; legislatures increasingly relied on coalitions of interests and ideologies; and most men gained the right to vote for their legislative representatives. During the eighteenth and nineteenth centuries, legislatures met more regularly. British monarchs gradually appointed leaders of bureaucracies from the legislature (Congleton 2011:336), especially a leading or "prime" minister, a practice other European states followed. Political coalitions slowly formalized into organized "parties." A series of bargains over new tax sources shifted power to legislatures from the crown in Europe and Japan (Congleton 2011).

Legislative and voting systems changed and developed. Each English colony in North America had a royally appointed governor and an elite-elected legislature. After U.S. independence, an elected "President" separate from the legislature governed in place of a monarch; individual states adopted this model with a governor, creating a "presidential" model of governance. In the traditional "parliamentary" model, ministers were chosen from the legislature (Congleton 2011). Also in the United States, each geographic district has *one* representative who wins by majority vote. Such "winner-take-all" systems tend strongly to produce two-party systems, as coalitions must win over 50 percent of the votes. In proportional representation (PR) systems common in Europe, each geographic district has *multiple* representatives; parties get a representative based on their percentage of the vote in that district. Proportional representation tends to produce three or more political parties because parties need much less than 50 percent to accomplish this. Suffrage, the right to vote, also expanded because of constant activism. In France, universal male suffrage over age 25 was allowed in 1792; subsequent governments limited it, but France returned permanently to universal male suffrage in 1875. U.S. property requirements for voting ended in the early nineteenth century. British male suffrage extended in steps—in 1832, in 1867, in 1884, and in 1918 (Congleton 2011:343–358). The Swedish government extended suffrage over the late nineteenth and early twentieth centuries.

This expanding suffrage forced increasing responsiveness to popular opinion. Since all European colonies in the Americas were the result of European "network" strategies (Blanton et al. 1996), European empires provided few public goods to those colonies (O'Brien 2013). Over time, in response to public opinion expressed in voting, European and North American states shifted to a "corporate" strategy in which states increasingly taxed populations to provide more public goods, such as infrastructure, police forces for public order, mass education systems, and business law (Yun-Casalilla and O'Brien 2012; O'Brien 2013). One major public good was a coherent, comprehensive legal system (Pirie 2021).

State structures changed in three other ways. First, recruitment into bureaucracies changed. From patronage, purchasing offices, and volunteerism common by the 1700s (Raadschelders 1998; Hollyer 2011), meritocratic recruitment became the norm in the late 1800s. States adopted this basic reform at different times: Prussia earliest, then France, and late in the nineteenth century, the United States, Japan, and other European states. This was especially true after educational systems expanded (Hollyer 2011). Second, social movements also developed—sustained, organized campaigns making claims on authorities using associations, meetings, rallies, and publications. While these existed before 1500, modern social movements developed from 1750 to 1850 in Europe and North America (Tilly and Wood 2012). Social movements included those for suffrage in the 1820s and 1830s in Britain and the abolitionist movement in the United States. Third, states turned the government's bank into what economists call a "central bank," the primary function of which was to maintain economic stability by regulating the money supply and being the lender of last resort for other banks. The Bank of England became the British state's central bank in the 1870s, and by 1900, there were 18 banks that qualified as central banks (Capie 2003), including the Bank of Japan.

The Enlightenment sparked new political–economic ideologies. Liberalism became a political movement to seek individual rights (liberty). European governments had many restrictions on economic activity, often benefiting landowners. Manufacturers and merchants wanted these restrictions loosened, so liberalism became the ideology of business groups. This program went farther in the United States and Britain than in Continental Europe; the latter were generally stronger, while the U.S. state was weaker and always more business-friendly (Neill 1953; Donohue 2003).

In response, other thinkers used Enlightenment critiques of inequality to develop the ideology of socialism, Marx's critique of capitalism being the most famous (Marx [1867] 1915). He argued that capitalist businesses always kept some of the value of workers' labor, exploiting them. He advocated for overthrowing the new capitalist system and "socializing" all businesses to make ownership collective. He advocated for state ownership to do this, though theorists also suggested private, nonprofit, local cooperatives (MacKenzie 1966).

These ideologies were political–economic, while the ideology of nationalism was political-cultural. Nationalists advocated for the use of cultural similarities as the basis of states. While it acknowledged emotion and culture and was a rebellion against Enlightenment rationalism, nationalism became a tool of statecraft. Defining "nations" as culturally similar groups, intellectuals argued that cultural similarity was a stronger foundation for democracy and universal suffrage than monarchy, i.e., it increased state cohesion. Leaders founded movements dedicated to cultural groups gaining political independence from typical agrarian multiethnic states (Kramer 2011), claiming that political borders should match geographic distribution of ethnicity. This movement, combined with the expansion of male suffrage, led governments beyond the old agrarian strategy of co-optation and integration of local elites across a broad area to "intensively appealing to the sentiments of the mass of population" (Bang 2021:49). Two movements succeeded in this goal. German-speaking states such as Prussia, Bavaria, and others were not unified in one state; Prussia waged several wars over the 1860s to unite into a state called "Germany," while states on the Italian peninsula fought wars to create "Italy."

Mass education, a new public good also supported by Enlightenment philosophers, supported nationalism. Prussia started compulsory, free education in 1763 and had near universal enrollments by the 1830s. Each of the northern U.S. states followed suit by 1860, Southern states doing so after the Civil War. Scandinavian countries also passed compulsory education laws in the 1830s and 1840s, the Japanese and Italians in the 1870s, the British in 1880, and the French in 1882. Enrollments rose over the twentieth century both in percentage of children enrolled and in number of years enrolled (Sanderson 1995:304–310). Such systems often included compulsory language education, purposely snuffing out local languages and dialects.

Scientific research and technological development became increasingly linked, producing a "second Scientific Revolution" early in the century. Theoretical and experimental streams of science became united after 1800 in physics, chemistry, and the life sciences. Science found its new institutional home in universities, especially in Germany, which linked the state, businesses, and research universities. This model spread internationally, changing old universities in Britain and France (McClellan and Dorn 1999). Professional organizations and regularly published journals proliferated. Businesses increasingly applied scientific findings to technology, leading to the Second Industrial Revolution as discussed above. Doctors applied germ theory to medicine and public health, saving lives.

Scientists also operated ever more globally. They "started to regard the globe as an entity to be analysed in its own right" (Fara 2009:207). Through new transportation and communication technologies, scientists increasingly traveled globally to investigate phenomena where they occurred: "networks of laboratories came to span the globe, constantly monitoring patterns of weather, tides, and other variable phenomena" (Fara 2009:212). This made other discoveries, such as quinine, possible. Institutionally, scientists started to form international professional organizations around 1880 (Drori et al. 2003:84–85).

Technological changes transformed mass media and thus various national public spheres (Briggs and Burke 2009). When steam engines drove rotary printing presses starting in the 1830s, they made possible cheap newspapers, so the number of newspapers expanded in Europe and North America, as did the number of magazines published. Transportation and communication changes brought publications to much wider audiences, deeply institutionalizing public spheres in North America and Western Europe. Press professionals, or "journalists," became part of the public sphere (Briggs and Burke 2009:Chapter 6).

New technologies transformed the arts and entertainment. The novel became a standard art form, some appearing in serial form in newspapers, thus mixing entertainment with "serious" news. Photography developed in the 1830s and 1840s, and sound recording became viable in the late 1870s. Moving pictures, or "movies," were developed in the 1890s (Cullen 2002). Governments and philanthropic capitalists set up free public libraries, another public good, to make books, magazines, and newspapers available to their populations.

These media served growing populations. Mortality started to decline in Europe before 1750, but this accelerated after that date in England and later in other states. National markets and improved agriculture helped end famine and generally increased health. Improved infant care, vaccination, and sanitation reduced disease-generated mortality (Brooke 2014; Livi-Bacci 2017). Consequently, with mortality declines, periodic mortality crises decreased, so mortality became more predictable. Due to "more food, better clothing, better houses, and more medical care" (Livi-Bacci 2017:131), all of which correlated with greater wealth, the mortality decline accelerated over the nineteenth century. The reduction started at different times: in France in 1785, in Sweden in 1810, and in Germany and Italy in 1876. Still, in Britain and the United States, conditions worsened for urban workers mid-century, but workers' wages, health, heights, and lifespans all started to increase after 1870 (Brooke 2014).

Even as mortality fell, fertility remained high. Once people were sure that children were likely to survive, couples could limit fertility. Still, educational systems and child labor laws requiring later entry into the workforce made children more costly, which seems to have initiated fertility reductions, often starting decades later than mortality declines (Livi-Bacci 2017). Fertility began falling first in Britain and then across Western Europe after 1800 (Livi-Bacci 2017). The gap between low mortality and still-high fertility allowed the European population to more than double from 1800 to 1900 (Livi-Bacci 2017).

Increased food production had other demographic effects. Agricultural improvements after 1800 improved productivity and reduced labor needs, so many left for the cities (Knox and McCarthy 2012). This drew more people into the market-based cash economy, which itself was spreading into rural areas (Higgs 1971; Stearns 1969). While cities had primarily been sites for political leadership since social complexity started, industrially generated urbanization changed this (Lees and Lees 2013). Many cities became primarily industrial centers, places to process and transform raw materials into finished goods: "Industrialization and cities grew hand in hand" (Knox and McCarthy 2012:40). Cities continued to develop divisions of labor, with financial, industrial, shopping, government, and most of all residential neighborhoods. Industry changed regional divisions of labor: larger cities had many functions, while smaller cities often specialized more. As cities grew so quickly, they built more infrastructure: "networks of water, sewage, gas, electricity, telegraph, telephone, and streetcar and subway lines were fundamental to processes of urban change" (Knox and McCarthy 2012:65).

Cities also developed patterns of suburbanization: smaller residential cities formed around the larger one. European cities had always grown by siting industry and workers outside defensive city walls. Over time, cities built new walls farther out to protect them. This process continued across Europe, while North American cities were usually built without walls and acquired a concentric, three-ring pattern: a central business district, industrial and working-class neighborhoods, and middle-class suburbs. A few cities also expanded by annexation, such as New York, Philadelphia, and others. Suburbanization in the United States started in the late 1800s (Monkkonen 1988; Lees and Lees 2013).

Industrialization and urbanization increasingly damaged the environment. Urbanization required expansion of cities' hinterlands, which increasingly became croplands, dairy lands, and ranchlands (Simmons 2008). Until 1850, the main fuel in North America was wood. Population expansion and industrialization led to increased deforestation, which led to "[s]oil erosion, runoff and flooding, the silting of streams and estuaries, and the clogging of main waterways" (Penna 2015:192). Between 1700 and 1920, approximately 315 million hectares of forest and woodland disappeared in the temperate zones of the planet, mostly in North America and Russia (Simmons 2008:125). Before better infrastructure, European and North American cities were filthy. People and animals produced huge amounts of waste, polluting waterways. Residential fireplaces and industries produced tremendous amounts of smoke that killed many. Diseases spread more easily in cities, especially with under-functioning sewage systems. In the 1830s, cities started providing public goods such as sewage systems, animal waste and garbage pickup, and street washing to handle these problems, mostly late in the century (Brooke 2014:505–507; Penna 2015:205–224).

The European Marriage Pattern transformed into an idealized system of "separate spheres" among the middle classes. Delayed marriage, neolocal residence, and bilateral descent remained; industrial capitalism and urbanization added additional elements. Before 1800, in most families, men farmed and women managed the household, but this division of labor was flexible: men were deeply involved in home life, while women helped harvest and tended animals. Women in urban family business were closely involved. In urbanizing, industrializing societies, the geographic separation of paid work and home became much clearer (Davidoff and Hall 1992). Increasingly, men left the house daily to go to a money-paying job in a separate area of the city. Women stayed home, took care of children, and managed the household. Thus, monetary employment, civil society, and the political public sphere became coded as masculine. The home-based, domestic private sphere was coded as feminine. The ideology assumed that men could "naturally" handle the rough business world of industrial and commercial capitalism, while women provided men moral and spiritual refuge in the home (Hilton 1994; Reagin 2008).

Practice was more complex than ideology. While middle-class women rarely worked, they continued in family businesses, shopped in public, performed charity work, and sometimes engaged in politics, usually for family issues. Middle-class American women joined social movement organizations such as the Red Cross and the Women's Christian Temperance Union. Some started to object to the continuation of male dominance in this new system. A U.S. women's movement arose from anti-slavery and anti-alcohol movements arguing for women's equality and the right to vote. Still, the ideology of domesticity became increasingly institutionalized in the North Atlantic world (Hilton 1994; Reagin 2008).

In some ways, the position of women worsened. They had sometimes been partly included in state offices; this changed to rigid exclusion (Towns 2010). European racist discourses of "civilization" were partly constructed with ideas of male activity and female passivity, and "elevating woman" involved protecting her from politics, which would otherwise degrade her (Towns 2010:67–77). Many European and "offshoot" states legally narrowed, restricted, or eliminated women's already narrow political rights to attend political meetings and join associations. Almost as soon as these laws were in place, women worked to change them. The first national women's suffrage movements started across Europe and in North America in the 1840s; international suffrage movements and associations formed in the 1870s and 1880s. Organizational formation increased up to World War I (Paxton and Hughes 2017:203). By 1914, feminist movements existed across Europe, and in Southwest Asia, India, Japan, the Pacific, and the Americas—most adapting European ideas to local cultural beliefs (Offer 2010)—and Finland, New Zealand, and Australia granted suffrage to women (Towns 2010; Paxton and Hughes 2017).

North Atlantic working-class women had experiences very different from those of middle-class women. Women and children worked in factories alongside their husbands and brothers, in washing, sewing, or taking in lodgers, or as domestic servants (Hilton 1994). Laws and norms against extramarital childbirth weakened across Europe in the eighteenth and nineteenth centuries. Young men and women new to cities often lived together outside of marriage to save money, and around 1850, one-third to one-half of all children were born to unmarried women (Therborn 2004). As wages increased, working-class marriage rates increased, the number of extramarital births decreased, and workers began to adopt the middle-class separate spheres model of women staying home after marriage (Therborn 2004).

In the "long nineteenth century"[7] (1760s–1910s), states in Europe, North America, Oceania, and East Asia developed a new type of society, the industrial capitalist democracy. Agricultural improvements, scientific discoveries, new technologies, and commercialization all supported industrialization, which expanded markets, enriched populations, changed class and gender systems, lowered mortality and fertility, sparked new social and ideological movements, supported new cultural trends, and intensified human impact on the physical environment.

[7] Historians usually take this phrase to mean 1789 to 1914; I have obviously stretched it somewhat.

Conclusion

The third social transformation in world history originated in Europe sometime between 1450 and 1850. It constituted a new globalization, one vastly more complex than the first one, with two aspects. Internally, Europe industrialized and commercialized; grew greatly in population; started to urbanize; and developed a new gender system, new forms of governance, and a new intellectual apparatus of science, education, ideology, and media. Externally, Europe conquered most of the world, putting all humans into one thin global social system. While the initiating period is debated, indicators of increasing wealth, decreasing mortality, and other factors improved greatly sometime after 1750 in Europe and North America and later in East Asia. It brought a much better life than ever before for some humans and a worse life for others.

Dehumanization and ethnocentrism help us make sense of this story. A few scholars celebrate European cultural uniqueness and superiority, a stance I reject. However, the descendants of European subject peoples sometimes imply that Europeans have been uniquely bloodthirsty in world history. I disagree. Instead, temporal cosmopolitanism (Robbins 2016) puts European global conquest in a long chronological context: it was clearly an extension and intensification of agrarian imperial conquests. Indeed, from the 1450s to the 1750s, some aspects (not all) of European expansion look like the mass slaughter of the Mongol conquests of the 1200s. And after 1450, peoples physically very different from each other came into contact in a sustained and large-scale way, providing a ready-made method of group distinction and domination—racism, the latest and most geographically expansive form of dehumanization, based in ethnocentrism.

Thus, all civilizations were forced into a single system to create something new. In agrarian states that remained independent—China, Siam (Thailand), Persia (Iran), Ethiopia, and others—European influences were clearly present but attenuated; these became something like frontier zones for the entire global system. Japan is an unusual case, engaging in imitation but on its own terms, assimilating those European patterns consciously and selectively. The Mesoamerican and Andean civilizations were largely but not totally replaced; the Western and Eastern African civilizations were also incorporated, though not nearly as thoroughly by 1914; and Southwest Asia and North Africa were much less consistently incorporated.

Through both creative and repellent actions, Europeans created a fourth kind of society, the industrial capitalist democracy, using it to incorporate many kinds of societies into one global social system. They did this through a complex mix of cooperation, reciprocity, cumulative learning, and violent aggrandizing. Even so, the global system they constructed was still very thin in 1914, reaching most humans but leaving a large minority altered only insignificantly. Over the next century, this would change.

References

Abbenhuis, Maartje, and Gordon Morrell. 2020. *The First Age of Global Industrialization: An International History, 1815–1918*. New York: Bloomsbury Academic.
Acemoglu, Daron, Simon Johnson, and James Robinson. 2005. The Rise of Europe: Atlantic Trade, Institutional Change, and Economic Growth. *American Economic Review* 95 (3): 546–579.
Ajayi, J. F. A., and Michael Crowder, eds. 1985. *History of West Africa*. Vol. I. 3rd ed. New York: Longman, Inc.
Alvaredo, Facundo, Lucas Chancel, Thomas Piketty, Emmanuel Saez, and Gabriel Zucman. 2018. *World Inequality Report 2018*. Paris: World Inequality Lab.
Anderson, Benedict R. O'. G. 1991. *Imagined Communities: Reflections on the Origin and Spread of Nationalism*. Revised and extended ed. London: Verso.
Anderson, Perry. 1974. *Lineages of the Absolutist State*. London: NLB.
Andrade, Tonio. 2016. *The Gunpowder Age: China, Military Innovation, and the Rise of the West in World History*. Princeton, NJ and Oxford: Princeton University Press.
Argandoña, Antonio. 2008. Gross Domestic Product (GDP). In *Encyclopedia of Business Ethics and Society*, ed. R. W. Kolb. Thousand Oaks, CA: SAGE.
Arrighi, Giovanni. 2010. *The Long Twentieth Century: Money, Power, and the Origins of Our Times*. New York: Verso.
Austin, Gareth. 2016. "Sub-Saharan Africa." 2016. In *A History of the Global Economy: From 1500 to the Present*, ed. J. Baten, 316–352. Cambridge, UK: Cambridge University Press.
Bang, Peter Fibiger. 2021. Empire—A World History. In *The Oxford World History of Empire*, ed. P. F. Bang, C. A. Bayly, and W. Scheidel, 1–90. New York: Oxford University Press.
Bauman, Zygmunt. 2001. Modernity. In *The Oxford Companion to Politics of the World*, ed. J. Krieger, 2nd ed. New York: Oxford University Press. Oxford Reference: https://www.oxfordreference.com/.
Beckert, Sven. 2015. *Empire of Cotton: A Global History*. New York: Vintage Books.
Betts, Raymond. 1978. *Tricouleur! The French Overseas Empire*. New York: Gordon & Cremonesi.
Black, Jeremy, ed. 2005. *The Atlas of World History*. London: Dorling Kindersley Limited.
Blanton, Richard E., Gary M. Feinman, Stephen A. Kowalewski, and Peter N. Peregrine. 1996. A Dual-Processual Theory for the Evolution of Mesoamerican Civilization. *Current Anthropology* 37 (1): 1–14.
Blumin, Stuart M. 1989. *The Emergence of the Middle Class: Social Experience in the American City, 1760–1900*. New York: Cambridge University Press.
Boivin, Nicole, and Michael D. Franchetti. 2018. *Globalization in Prehistory: Contact, Exchange, and the 'People Without History.'* New York: Cambridge University Press.
Bowle, John. 1974. *The Imperial Achievement: The Rise and Transformation of the British Empire*. London: Secker & Warburg.
Boxer, Charles. 1965. *The Dutch Seaborne Empire*. New York: Alfred A. Knopf.
Boxer, Charles. 1969. *The Portuguese Seaborne Empire, 1415–1825*. New York: Alfred A. Knopf.
Briggle, Adam. 2008. Modernity. In *International Encyclopedia of the Social Sciences*, ed. W. Darrity, 2nd ed., 230–232. Farmington Hills, MI: Gale.
Briggs, Asa, and Peter Burke. 2009. *A Social History of the Media: From Gutenberg to the Internet*. 3rd ed. Malden, MA: Polity Press.
Broadberry, Stephen, Giovanni Federico, and Alexander Klein. 2010. Sectoral Developments, 1870-1914. In *The Cambridge Economic History of Modern Europe, Vol. 2: 1870 to the Present*, ed. S. Broadberry and K. H. O'Rourke, 59–83. New York: Cambridge University Press.
Brooke, John L. 2014. *Climate Change and the Course of Global History: A Rough Journey*. New York: Cambridge University Press.
Brown, Alison. 1988. *The Renaissance. Seminar Studies in History*. London: Longman Group UK Limited.
Burns, William E. 2016. *The Scientific Revolution in Global Perspective*. New York: Oxford.

Calomiris, Charles. 2003. Banking, Modern Period. In *The Oxford Encyclopedia of Economic History*, ed. J. Mokyr. New York: Oxford University Press. Oxford Reference: https://www.oxfordreference.com/.

Cameron, Euan. 1991. *The European Reformation*. New York: Oxford University Press.

Capie, Forrest. 2003. Central Banking. In *The Oxford Encyclopedia of Economic History*, ed. J. Mokyr. New York: Oxford University Press. Oxford Reference: https://www.oxfordreference.com/.

Carrerras, Albert, and Camilla Josephson. 2010. Aggregate Growth, 1870–1914: Growing at the Production Frontier. In *The Cambridge Economic History of Modern Europe, Vol. 2: 1870 to the Present*, ed. S. Broadberry and K. H. O'Rourke, 30–58. New York: Cambridge University Press.

Chapman, Jane. 2005. *Comparative Media History: An Introduction: 1789 to the Present*. Cambridge, UK: Polity.

Chasteen, John Charles. 2006. *Born in Blood & Fire: A Concise History of Latin America*. 2nd ed. New York: W.W. Norton.

Childe, V. Gordon. (1936)1966. *Man Makes Himself*. 4th ed. Bungay, Suffolk: Fontana.

Clark, Gregory. 2010. *Review of Contours of the World Economy 1-2030 AD: Essays in Macro-Economic History* by Angus Maddison. H-Soz-u-Kult, H-Net Reviews.

Congleton, Roger D. 2011. *Perfecting Parliament: Constitutional Reform, Liberalism, and the Rise of Western Democracy*. Cambridge, UK: Cambridge University Press.

Crosby, Alfred. (1973) 2003. *The Columbian Exchange: Biological and Cultural Consequences of 1492*. 30th anniversary ed. Westport, CT: Praeger.

Cullen, Jim. 2002. *The Art of Democracy: A Concise History of Popular Culture in the United States*. New York: Monthly Review Press.

Davidoff, Leonore, and Catherine Hall. 1992. *Family Fortunes: Men and Women of the English Middle Class, 1780–1850*. London: Routledge.

de Oliveira Marques, A. H. 1976. *History of Portugal*. Vol. 1. 2nd ed. New York: Columbia University Press.

de Zwardt, Pim, and Jan Luiten van Zanden. 2018. *The Origins of Globalization: World Trade in the Making of the Global Economy, 1500–1800*. New York: Cambridge University Press.

Deane, Phyllis. 1965. *The First Industrial Revolution*. Cambridge, UK: Cambridge University Press.

Dennison, Tracy, and James Simpson. 2010. Agriculture. In *The Cambridge Economic History of Modern Europe, Vol. 1: 1700–1870*, ed. S. Broadberry and K. H. O'Rourke, 147–163. New York: Cambridge University Press.

Donohue, Kathleen G. 2003. *Freedom from Want: American Liberalism and the Idea of the Consumer*. Baltimore: Johns Hopkins University Press.

Drori, Gili S., John W. Meyer, Francisco O. Ramirez, and Evan Schofer. 2003. *Science in the Modern World Polity: Institutionalization and Globalization*. Stanford, CA: Stanford University Press.

Easton, Stewart C. 1964. *The Rise and Fall of Western Colonialism: A Historical Survey from the Early Nineteenth Century to the Present*. New York: Frederick A. Praeger.

Eisenstein, Elizabeth L. 1983. *The Printing Revolution in Early Modern Europe*. Cambridge, UK: Cambridge University Press.

Elrod, Richard B. 1976. The Concert of Europe: A Fresh Look at an International System. *World Politics* 28 (2): 159–174.

Fara, Patricia. 2009. *Science: A Four Thousand Year History*. New York: Oxford University Press.

Fieldhouse, D. K. 1966. *The Colonial Empires: A Comparative Survey from the Eighteenth Century*. New York: Dell Publishing.

Findling, John E., and Kimberly D. Pelle, eds. 2008. *Encyclopedia of World's Fairs and Expositions*. Jefferson, NC: McFarland.

Finley, Laura. 2007. Colonialism. In *Encyclopedia of Environment and Society*, ed. P. Robbins, 301–304. Thousand Oaks, CA: SAGE.

Fiske, Alan Page, and Tage Shakti Rai. 2015. *Virtuous Violence: Hurting and Killing to Create, Sustain, End, and Honor Social Relationships*. New York: Cambridge University Press.

Flynn, Dennis, and Arturo Giraldez. 1995. Born with a 'Silver Spoon': The Origin of World Trade in 1571. *Journal of World History* 6 (2): 201–221.

Fortner, Robert S., Ann Snesareva, and Ksenia Tsitovich. 2014. Media, Civil Society, and the Public Sphere: History and Current Thinking. In *The Handbook of Media and Mass Communication Theory*, ed. R. S. Fortner and P. M. Fackler, vol. I, 314–332. Malden, MA: Wiley.

Fredrickson, George. 2009. *Racism: A Short History*. Princeton, NJ: Princeton University Press.

Fremdling, Rainer, and Peter Solar. 2010. Industry. In *The Cambridge Economic History of Modern Europe, Vol. 1: 1700–1870*, ed. S. Broadberry and K. H. O'Rourke, 164–186. New York: Cambridge University Press.

Gelvin, James. 2011. *The Modern Middle East: A History*. 3rd ed. New York: Oxford University Press.

George, Peter. 1982. *The Emergence of Industrial America: Strategic Factors in American Economic Growth Since 1870*. Albany: SUNY Press.

Ghanem, Rima, and Joerg Baten. 2016. Middle East, North Africa, and Central Asia. In *A History of the Global Economy: From 1500 to the Present*, ed. J. Baten, 208–248. Cambridge, UK: Cambridge University Press.

Girard, Philippe. 2005. Columbian Exchange. In *Encyclopedia of World Trade: From Ancient Times to the Present*, ed. C. Northrup, vol. II, 192–193. Armonk, NY: M.E. Sharpe.

Goldstone, Jack A. [1991] 2016. *Revolution and Rebellion in the Early Modern World*. Twenty-fifth anniversary edition. New York: Routledge.

Graeber, David, and David Wengrow. 2021. *The Dawn of Everything: A New History of Humanity*. New York: Farrar, Straus, and Giroux.

Grafe, Regina, Larry Neal, and Richard W. Unger. 2010. The Services Sector. In *The Cambridge Economic History of Modern Europe, Vol. 1: 1700–1870*, ed. S. Broadberry and K. H. O'Rourke, 187–214. New York: Cambridge University Press.

Gregory, Paul R., and Robert C. Stuart. 2004. *Comparing Economic Systems in the Twenty-First Century*. 7th ed. Mason, OH: South-Western/Cengage.

Gullickson, Gay L. 2003. Social Class System and the Economy. In *Industrial Revolution in Europe, 1750-1914 (World Eras. Vol. 9)*, ed. J. R. Farr, 123–164. Detroit: Gale.

Habermas, Jürgen. 1989. *The Structural Transformation of the Public Sphere: An Inquiry into a Category of Bourgeois Society*. Cambridge, MA: The MIT Press.

Haksöz, Çagri, and Damla Durak Usar. 2012. Silk Road Supply Chains: A Historical Perspective. In *Managing Supply Chains on the Silk Road: Strategy, Performance, and Risk*, ed. Ç. Haksöz, S. Seshadri, and A. V. Iyer, 3–26. Boca Raton, FL: CRC Press.

Hall, Catherine. 2007. Of Gender and Empire: Reflections on the Nineteenth Century. In *Gender and Empire*, ed. P. Levine, 46–76. New York: Oxford University Press.

Hall, Martin, and Patrick Thaddeus Jackson, eds. 2007. *Civilizational Identity: The Production and Reproduction of "Civilizations" in International Relations*. New York: Palgrave Macmillan.

Hall, Peter, and David Soskice. 2001. *Varieties of Capitalism: The Institutional Foundations of Comparative Advantage*. New York: Oxford University Press.

Hamilton, Gary G., Misha Petrovic, and Benjamin Senauer, eds. 2011. *The Market Makers: How Retailers are Reshaping the Global Economy*. New York: Oxford.

Hampson, Norman. 1968. *A Cultural History of the Enlightenment*. New York: Pantheon Books.

Hanson, Holly. 2002. Queen Mothers and Good Government in Buganda: The Loss of Women's Political Power in Nineteenth Century East Africa. In *Women in African Colonial Histories*, ed. J. Allman, S. Geiger, and N. Musisi, 219–236. Bloomington and Indianapolis: Indiana University Press.

Hartwell, R. M., and Stanley L. Engerman. 2003. Capitalism. In *The Oxford Encyclopedia of Economic History*, ed. J. Mokyr. New York: Oxford University Press. Oxford Reference: https://www.oxfordreference.com/.

Hasse, Dag Nikolaus. 2021. Influence of Arabic and Islamic Philosophy on the Latin West. *The Stanford Encyclopedia of Philosophy*, ed. E. N. Zalta. https://plato.stanford.edu/archives/fall2021/entries/arabic-islamic-influence.

Haywood, John, ed. 2011. *The New Atlas of World History: Global Events at a Glance*. Princeton and Oxford: Princeton University Press.

Heath, Elizabeth. 2005. Colonial Rule. In *Africana: The Encyclopedia of African and African American Experience*, ed. K. A. Appiah and H. L. Gates Jr., 170–175. New York: Oxford University Press.

Hejeebu, Santhi. 2003. Firm: The Firm Before 1800. In *The Oxford Encyclopedia of Economic History*, ed. J. Mokyr, vol. 2. New York: Oxford University Press. Oxford Reference: https://www.oxfordreference.com/.

Henry, John. 2001. *The Scientific Revolution and the Origins of Modern Science*. 2nd ed. Houndsmills, UK: Palgrave.

Higgs, Robert. 1971. *The Transformation of the American Economy, 1865–1914: An Essay in Interpretation*. New York: John Wiley & Sons.

Hilton, Kathleen C. 1994. Domesticity. In *Encyclopedia of Social History*, ed. P. N. Stearns, 274–275. New York: Garland.

Hocking, Brian, and Donna Lee. 2011. Diplomacy. In *International Encyclopedia of Political Science*, ed. B. Badie, D. Berg-Schlosser, and L. Morlino, 660–669. Thousand Oaks, CA: SAGE.

Hoffman, Philip T. 2004. *Europe 1450 to 1789: Encyclopedia of the Early Modern World*. New York: Scribner.

Holcombe, Charles. 2017. *A History of East Asia: From the Origins of Civilization to the Twenty-First Century*. New York: Cambridge University Press.

Hollyer, James R. 2011. *Merit Recruitment in 19th and Early 20th Century European Bureaucracies*. Retrieved June 15, 2018. https://ssrn.com/abstract=1861469 or https://doi.org/10.2139/ssrn.1861469.

Hopkins, A. G. 1973. *An Economic History of West Africa*. New York: Columbia University Press.

Horna, Hernán. 2011. *A History of Latin America*. Uppsala, Sweden: Acta Universitatis Upsaliensis.

Houston, Rob, ed. 2018. *History of the World, Map by Map*. New York: Dorling Kindersley.

Howe, Stephen. 2002. *Empire: A Very Short Introduction*. New York: Oxford University Press.

Iliffe, John. 2017. *Africans: The History of a Continent*. 3rd ed. New York: Cambridge University Press.

Keddie, Nikki R. 2007. *Women in the Middle East: Past and Present*. Princeton, NJ: Princeton University Press.

Kellogg, Susan. 1995. *Law and the Transformation of Aztec Culture, 1500–1700*. Norman and London: University of Oklahoma Press.

Kellogg, Susan. 2005. *Weaving the Past: A History of Latin America's Indigenous Women from the Prehispanic Period to the Present*. New York: Oxford University Press.

Kendirbai, Gulnar. 2020. *Russian Practices of Governance in Eurasia: Frontier Power Dynamics, Sixteenth Century to Nineteenth Century*. New York: Routledge.

Kennedy, Paul. 1987. *The Rise and Fall of the Great Powers: Economic Change and Military Conflict from 1500 to 2000*. New York: Random House.

Kent, Susan Kingsley. 2021. *Gender in World History*. New York: Oxford.

Kivelson, Valerie A., and Ronald Grigor Suny. 2017. *Russia's Empires*. New York: Oxford University Press.

Knox, Paul L., and Linda McCarthy. 2012. *Urbanization: An Introduction to Urban Geography*. Boston: Pearson.

Kramer, Lloyd S. 2011. *Nationalism in Europe and America: Politics, Cultures, and Identities Since 1775*. Chapel Hill: University of North Carolina Press.

Lamoreaux, Naomi R. 2003. Firm: The Firm After 1800. In *The Oxford Encyclopedia of Economic History*, ed. J. Mokyr, vol. 2. New York: Oxford University Press. Oxford Reference: https://www.oxfordreference.com/.

Landes, David S. 1969. *The Unbound Prometheus: Technological Change and Industrial Development in Western Europe from 1750 to the Present*. London: Cambridge University Press.

Lauer, Josh. 1999. The Birth of Print Culture: The Invention of the Printing Press in Western Europe. In *Science and Its Times: Understanding the Social Significance of Scientific Discovery*, ed. N. Schlager and J. Lauer, vol. III, 404–411. Detroit: Gale Group.
Lees, Andrew, and Lynn Hollen Lees. 2013. Europe: 1800–2000. In *The Oxford Handbook of Cities in World History*, ed. P. Clark, 464–482. New York: Oxford University Press.
Lincoln, Bruce. (1993) 2007. *The Conquest of a Continent: Siberia and the Russians*. Paperback ed. New York: Cornell Paperbacks.
Lindqvist, Thomas, and Janken Myrdal. 2018. World Trade in the Fifteenth and Early Sixteenth Centuries. In *Trade and Civilization: Economic Networks and Cultural Ties, from Prehistory to the Early Modern Era*, ed. K. Kristiansen, T. Lindkvist, and J. Myrdal, 515–533. New York: Cambridge University Press.
Livi-Bacci, Massimo. 2017. *A Concise History of World Population*. 6th ed. Malden, MA: Wiley Blackwell.
Lloyd, T. O. 1984. *The British Empire, 1558–1984*. New York: Oxford University Press.
Loftin, John. 2000. *The Big Picture: A Short History of World Religions*. Jefferson, NC: McFarland.
Ma, Debin. 2016. China. In *A History of the Global Economy: From 1500 to the Present*, ed. J. Baten, 188–207. Cambridge, UK: Cambridge University Press.
MacKenzie, Norman. 1966. *Socialism: A Short History*. Revised ed. London: Hutchinson University.
Maddison, Angus. 2006. The World Economy, 1–2001 AD. In *The World Economy: Volume 1: A Millennial Perspective and Volume 2: Historical Statistics*. Paris: OECD Publishing.
Mann, Charles C. 2013. 1491. *The Atlantic*. Retrieved October 14, 2013. http://www.theatlantic.com/magazine/archive/2002/03/1491/302445/.
Manning, Patrick, with Tiffany Trimmer. 2013. *Migration in World History*, 2nd ed. London and New York: Routledge.
Marichal, Carlos. 2006a. Export Commodity and Global Money of the Ancient Regime, 1550–1800. In *From Silver to Cocaine: Latin American Commodity Chains and the Building of the World Economy, 1500–2000*, ed. S. Topik, C. Marichal, and Z. Frank, 25–52. Durham, NC and London: Duke University Press.
Marichal, Carlos. 2006b. Mexican Cochineal and the European Demand for American Dyes, 1550–1850. In *From Silver to Cocaine: Latin American Commodity Chains and the Building of the World Economy, 1500–2000*, ed. S. Topik, C. Marichal, and Z. Frank, 76–92. Durham, NC and London: Duke University Press.
Marx, Karl. (1867) 1915. *Capital: A Critique of Political Economy. Vol. I: The Process of Capitalist Production*. Translated from the 3rd German Edition by S. Moore and E. Aveling. Chicago: Charles H. Kerr and Company.
Maybury-Lewis, David, Theodore Macdonald, and Biorn Maybury-Lewis. 2009. *Manifest Destinies and Indigenous Peoples*. Cambridge, MA: David Rockefeller Center for Latin American Studies, Harvard University.
Mazoyer, Marcel, and Laurence Roudart. 2006. *A History of World Agriculture: From the Neolithic to the Current Crisis*. Translated by J. H. Membrez. New York: Monthly Review Press.
McClellan, James E., III, and Harold Dorn. 1999. *Science and Technology in World History: An Introduction*. Baltimore and London: The Johns Hopkins University Press.
McCreery, David. 2006. Indigo Commodity Chains in the Spanish and British Empires, 1560–1860. In *From Silver to Cocaine: Latin American Commodity Chains and the Building of the World Economy, 1500–2000*, ed. S. Topik, C. Marichal, and Z. Frank, 53–75. Durham, NC and London: Duke University Press.
McCusker, John J. 2005. The Demise of Distance: The Business Press and the Origins of the Information Revolution in the Early Modern World. *American Historical Review* 110 (2): 295–321.
McKendrick, Neil, John Brewer, and J. H. Plumb. 1982. *The Birth of a Consumer Society: The Commercialization of Eighteenth-Century England*. Bloomington: Indiana University Press.

Modelski, George, and William R. Thompson. 1966. *Leading Sectors and World Powers: The Coevolution of Global Politics and Economics*. Columbia: University of South Carolina Press.

Moni, Monir Hassan. 2007. Environmental History. In *Encyclopedia of Environment and Society*, ed. P. Robbins, 872–876. Thousand Oaks, CA: SAGE.

Monkkonen, Eric. 1988. *America Becomes Urban: The Development of U.S. Cities & Towns, 1780–1980*. Berkeley: University of California Press.

Moore, R. I., ed. 1995. *Atlas of World History*. Rand McNally.

Moore, Jason W. 2007. Silver, Ecology, and the Origins of the Modern World. In *Rethinking Environmental History: World-System History and Global Environmental Change*, ed. A. Hornborg, J. R. McNeill, and J. Martinez-Alier, 123–142. Plymouth, UK: AltaMira Press.

Murphy, Craig. 1994. *International Organization and Industrial Change: Global Governance Since 1850*. New York: Oxford University Press.

Nater, Laura. 2006. Colonial Tobacco: Key Commodity of the Spanish Empire, 1500-1800. In *From Silver to Cocaine: Latin American Commodity Chains and the Building of the World Economy, 1500–2000*, ed. S. Topik, C. Marichal, and Z. Frank, 93–117. Durham, NC and London: Duke University Press.

Neal, Larry. 2014. Introduction. In *The Cambridge History of Capitalism. Volume I: The Rise of Capitalism: From Ancient Origins to 1848*, ed. L. Neal and J. G. Williamson, 1–23. New York: Cambridge University Press.

Neal, Larry, and Rondo Cameron. 2016. *A Concise History of the World: From Paleolithic Times to the Present*. 5th ed. New York: Oxford University Press.

Neill, Thomas Patrick. 1953. *The Rise and Decline of Liberalism*. Milwaukee: Bruce Publishing Company.

O'Brien, Patrick, ed. 2010. *Atlas of World History*. 2nd ed. New York: Oxford University Press.

O'Brien, Patrick. 2013. Fiscal, Financial and Monetary Foundations for the Formation of Nation States in the West Compared to Imperial States in the East c.1415–c.1839. *Journal of Chinese Economic and Business Studies* 11 (3): 161–168.

Offen, Karen, ed. 2010. *Globalizing Feminisms, 1789–1945*. London and New York: Routledge.

Oka, Rahul. 2018. Trade, Traders, and Trading Systems: Macromodeling of Trade, Commerce, and Civilization in the Indian Ocean. In *Trade and Civilization: Economic Networks and Cultural Ties, from Prehistory to the Early Modern Era*, ed. K. Kristiansen, T. Lindkvist, and J. Myrdal, 279–319. New York: Cambridge University Press.

Parry, J. H. 1971. *Trade and Dominion: The European Overseas Empires in the Eighteenth Century*. London: Wedenfeld and Nicolson.

Parthasarathi, Prasannan. 2011. *Why Europe Grew Rich and Asia Did Not: Global Economic Divergence, 1600–1850*. New York: Cambridge University Press.

Paxton, Pamela Marie, and Melanie M. Hughes. 2017. *Women, Politics, and Power: A Global Perspective*. Thousand Oaks, CA: SAGE.

Penna, Anthony J. 2015. *The Human Footprint: A Global Environmental History*. 2nd ed. Malden, MA: Wiley Blackwell.

Piketty, Thomas. 2014. *Capital in the Twenty-First Century*. Belknap: Cambridge, MA.

Pirie, Fernanda. 2021. *The Rule of Laws: A 4,000-Year-Old Quest to Order the World*. New York: Basic Books.

Pomeranz, Kenneth. 2000. *The Great Divergence: Europe, China, and the Making of the Modern World Economy*. Princeton, NJ: Princeton University Press.

Pomeranz, Kenneth, and Steven Topik. 2018. *The World That Trade Created: Society, Culture, and the World Economy, 1400 to the Present*. 4th ed. New York: Routledge.

Raadschelders, Jos C. N. 1998. *Handbook of Administrative History*. New Brunswick: Transaction.

Raffield, Ben. 2021. Broken Worlds: Towards an Archaeology of the Shatter Zone. *Journal of Archaeological Method and Theory* 28:871–910.

Ray, Indrajit. 2011. *Bengal Industries and the British Industrial Revolution (1757–1857)*. New York: Routledge.

Reader, John. 1998. *Africa: A Biography of the Continent*. New York: Knopf.

Reagin, Nancy. 2008. Domesticity and Separate Spheres. In *The Oxford Encyclopedia of Women in World History*, ed. B. G. Smith. New York: Oxford University Press. Oxford Reference: https://www.oxfordreference.com/.

Reuter, Tina Kempin. 2011. International Law. In *21st Century Political Science: A Reference Handbook*, ed. J. T. Ishiyama and M. Breuning, 431–439. Thousand Oaks, CA: SAGE.

Robbins, Bruce. 2016. Prolegomena to a Cosmopolitanism in Deep Time. *Interventions* 18 (2): 172–186.

Roy, Tirthankar. 2016. South Asia. In *A History of the Global Economy: From 1500 to the Present*, ed. J. Baten, 249–281. Cambridge, UK: Cambridge University Press.

Sacks, Karen. 1982. An Overview of Women and Power in Africa. In *Perspectives on Power: Women in Africa, Asia, and Latin America*, ed. J. F. O'Barr, 1–120. Durham, NC: Duke University Center for International Studies.

Sanderson, Stephen K. 1995. *Social Transformations: A General Theory of Historical Development*. Cambridge, MA: Blackwell.

Santon, Kate, and Liz McKay. 2005. *Atlas of World History*. Bath, UK: Parragon Publisher.

SarDesai, D. R. 2008. *India: The Definitive History*. Boulder, CO: Westview Press.

SarDesai, D. R. 2013. *Southeast Asia: Past and Present*. 7th ed. Boulder, CO: Westview Press.

Schmitt, Carina, Hanna Lierse, Herbert Obinger, and Laura Seelkopf. 2015. The Global Emergence of Social Protection: Explaining Social Security Legislation 1820–2013. *Politics and Society* 43 (4): 503–524.

Scott, John. 2014. Capitalism. In *A Dictionary of Sociology*, 4th ed., 57–59. New York: Oxford University Press.

Sellers, Charles Grier, Henry May, and Neil R. McMillen. 1985. *A Synopsis of American History*. 6th ed. Boston: Houghton Mifflin.

Sheldon, Kathleen. 2017. *African Women: Early History to the 21st Century*. Bloomington: Indiana University Press.

Sheldon, Kathleen. 2018. Women and Colonialism. *Oxford Bibliographies*. Retrieved October 12, 2024. https://www.oxfordbibliographies.com/display/document/obo-9780199846733/obo-9780199846733-0067.xml#obo-9780199846733-0067-div1-0001.

Shennan, J. H. 1974. *The Origins of the Modern European State, 1450–1725*. London: Hutchinson.

Sherman, Dennis, and Joyce E. Salisbury. 2008. *The West in the World*. 3rd ed. New York: McGraw-Hill.

Shilliam, Robbie. 2010. Modernity and Modernization. In *The International Studies Encyclopedia*, ed. R. A. Denemark and R. Marlin-Bennett. New York: Oxford University Press. Oxford Reference: https://www.oxfordreference.com/.

Simmons, I. G. 2008. *Global Environmental History*. Chicago and London: University of Chicago Press.

Smil, Vaclav. 2017. *Energy Transitions: History, Requirements, Prospects*. 2nd ed. Santa Barbara, CA: Praeger.

Smith, Bonnie G. 1989. *Changing Lives: Women in European History Since 1700*. Lexington, MA: D. C. Heath.

Smith, David Livingstone. 2011. *Less Than Human: Why We Demean, Enslave, and Exterminate Others*. New York: St. Martin's Press.

Snowden, Frank M. 1983. *Before Color Prejudice: The Ancient View of Blacks*. Cambridge, MA: Harvard University Press.

Stearns, Peter. 2006a. *Consumerism in World History: The Global Transformation of Desire*. 2nd ed. New York: Routledge.

Stearns, Peter N. 1969. *Modern Europe, 1789-1914*. Glenview, IL: Scott, Foresman.

Stearns, Peter N. 2006b. *Gender in World History*. 2nd ed. New York: Routledge.

Strayer, Robert. 2011. *Ways of the World: A Brief Global History*. Boston: Bedford/St. Martins.

Super, John C., and Briane K. Turley. 2006. *Religion in World History: The Persistence of Imperial Communion*. New York: Routledge.

Therborn, Göran. 2004. *Between Sex and Power: Family in the World, 1900–2000*. London and New York: Routledge.
Tiersten, Lisa. 2001. *Marianne in the Market: Envisioning Consumer Society in Fin-de-Siècle France*. Berkeley: University of California Press.
Tilly, Charles. 1975. *The Formation of National States in Western Europe*. Princeton, NJ: Princeton University Press.
Tilly, Charles, and Lesley J. Wood. 2012. *Social Movements, 1768–2012*. 3rd ed. Boulder, CO: Paradigm.
Topik, Steven, and Mario Samper. 2006. The Latin American Coffee Commodity Chain: Brazil and Costa Rica. In *From Silver to Cocaine: Latin American Commodity Chains and the Building of the World Economy, 1500–2000*, ed. S. Topik, C. Marichal, and Z. Frank, 118–146. Durham, NC and London: Duke University Press.
Topik, Steven, Carlos Marichal, and Zephyr Frank, eds. 2006. *From Silver to Cocaine: Latin American Commodity Chains and the Building of the World Economy, 1500–2000*. Durham, NC and London: Duke University Press.
Towns, Ann E. 2010. *Women and States: Norms and Hierarchies in International Society*. New York: Cambridge University Press.
Trautmann, Thomas R. 2016. *India: A Brief History of a Civilization*. New York: Oxford University Press.
Vries, De. 2008. *The Industrious Revolution: Consumer Behavior and the Household Economy, 1650 to the Present*. Cambridge, UK: Cambridge University Press.
Wallerstein, Immanuel. 1980. *The Modern World-System II: Mercantilism and the Consolidation of the European World-Economy, 1600-1750*. Berkeley: University of California Press.
Weber, Max. 1946. Class, Status, and Party. In *From Max Weber: Essays in Sociology*, translated and, ed. H. H. Gerth and C. W. Mills, 180–195. New York: Oxford University Press.
Weber, Johannes. 2006. Strasbourg, 1605. The Origins of the Newspaper in Europe. *German History* 24 (3): 387–412.
Wiesner-Hanks, Merry E. 2006. *Early Modern Europe, 1450–1789*. New York: Cambridge University Press.
Williamson, Jeffrey G. 2011. *Trade and Poverty: When the Third World Fell Behind*. Cambridge, MA: The MIT Press.
Winseck, Dwayne R., and Robert M. Pike. 2007. *Communication and Empire: Media, Markets, and Globalization, 1860–1930*. Durham, NC and London: Duke University Press.
Wolf, Eric R. 1982. *Europe and the People Without History*. Berkeley: University of California Press.
Wright, Gordon. 1995. *France in Modern Times: From the Enlightenment to the Present*. 5th ed. New York: W.W. Norton & Company.
Wrigley, Chris. 2003. Unions. In *The Oxford Encyclopedia of Economic History*, ed. J. Mokyr, vol. 2. New York: Oxford University Press. Oxford Reference: https://www.oxfordreference.com/.
Yannakakis, Yanna. 2023. *Since Time Immemorial: Native Custom and Law in Colonial Mexico*. Durham, NC: Duke University Press.
Yun-Casalilla, Bartolomé, and Patrick K. O'Brien. 2012. *The Rise of Fiscal States: A Global History, 1500–1914*. New York: Cambridge University Press.

Chapter 8
Building a Global Society? Accelerating Globalization, 1910s–2010s

Abstract From 1914 to 1945, global wars and an economic depression between them redrew the global political and economic map; global power shifted from Western Europe to North America. From 1945 to 1990, decolonization expanded the European state system throughout the world while different states experimented with capitalist, socialist, and mixed economies within a long, tense global standoff. U.S. hegemony helped proliferate international organizations, both political and economic. By 1990, this experimentation resulted in a global capitalist system. Global economic inequality increased through 2000, then started to decrease. English, education, science, and mass media began to become global realities, though modified by local traditions. Global population grew dramatically as mortality fell across the planet, and fertility began to decrease. Populations increasingly clustered in urban areas, and many from developing states moved northward or to other developing states. Women's movements increasingly fought for rights such as suffrage and political representation, female health and education neared parity with that of men, and women slowly but surely improved their economic situations. Ultimately, these intertwined processes began to produce a truly global society.

Keywords Globalization · World society · Socialism · Capitalist world-economy · Global hegemony · Rise of the West · Great divergence

Introduction

While most social scientists apply the term "globalization" to the last few decades, I have already defined it in a much longer-term way. In the 1490s, Europeans intentionally started exploring the world. By 1914, they had incorporated all societies in one global system, though many of them only barely, and had started transforming their own region. Here I argue that, into the 2010s, global incorporation started to become global assimilation, internal European transformation intensified, and both processes tended toward building a global social system. It likely will not fulfill the ideals of twentieth-century world federalists (Baratta 2004) since aggrandizers control much of the project. Still, their power is not total, and most humans have gained

T. C. Burr, *Societies Past and Present*, Emerging Globalities and Civilizational Perspectives, https://doi.org/10.1007/978-3-031-91532-1_8

huge material improvements—longer life most of all—however unevenly distributed.

As discussed, societies are both entities and processes (Sztompka 1994; Earle 2016). My idea of a global social system is a still-thin entity and a very incomplete process—so far. It may remind sociologists of world-system analysis (Wallerstein 1974, 2004; Chase-Dunn and Hall 1997; McMichael 2012; Babones and Chase-Dunn 2012) or the world polity perspective (e.g., Thomas 1987; Meyer et al. 1997; Drori et al. 2003; Lechner 2009). I draw only somewhat on either approach. I think capitalism is far more adaptable than its critics want it to be, and I hope this book demonstrates far more cultural diversity than world polity scholars acknowledge.

Yet, the latter are correct that diversity is narrowing. Cold War "modernization" efforts starting after 1945 accelerated decades later as neoliberal "globalization." This continues today, constituting worldwide assimilation. Thus, some can talk about a global civilization of modernity (Eisenstadt 2004). Since I have defined civilizations as purely regional entities, I think any global social entity is something new and different. *Ecumene* is a geographic term denoting the inhabited world (Mayhew 2023). A historian, an anthropologist, and a political scientist have, respectively, defined "the global ecumene" to include all people and the interactions between them (McNeill 1963), to describe global cultural flows creating hybrid forms (Hannerz 1992), or to treat it as the (shallow) incorporation of all civilizations (Katzenstein 2007). Thus, the phrase "global ecumene" denotes the cultural aspect of the system; worldwide flows of ideas, people, and objects slowly but surely tend toward providing increasingly similar experiences. Here, I document that process over the twentieth and twenty-first centuries; the global ecumene is at a very early stage, so the continued existence of civilizations and cultural areas is no mirage.

The global social system is composed of a thoroughly interconnected global economy, an international state system, and a global ecumene. It consists of a vast range of interconnections at many levels between individuals, groups, organizations, and states. Does this extremely complex set of connections amount to a global *society*? I argue that we are slowly tending toward it, but it is only a tendency, once again based in human capacities. Now, shared abstractions made possible through imagination have become crucial to make sense of anything "global." The groups-within-groups and multiple-groups capacities combine with historical trends to lead at least a few to identify with humanity as a whole. The great range of institutions globally allows people around the world to specialize in a vast array of activities. The pace of technological change and cultural accumulation has increased tremendously; first print, and now electronic media, helped new ideologies proliferate and constantly recombine.

To substantiate this argument, this chapter has three sections. I first narrate the political–economic history of the twentieth century, ending sometime after 2010. The next section analyzes the debate about why Europe transformed itself and the world, i.e., how globalization and social transformation were or were not interrelated. Finally, I analyze the nascent global society through the previous sevenfold analysis. Throughout, I will use the vocabulary of the "global North," the developed states of the world, and the "global South," all the rest (also see Chaps. 9 and 10).

Historians rarely discuss the recent past, but since my goal is understanding how history shaped humanity's situation *now*, I must proceed with recent and current history.[1] Also, since good evidence for the early twentieth century was often rare, I sometimes emphasize the last few decades.

Telling the Tale: The Tumultuous Twentieth Century, 1914–2015

Much of world history has been turbulent, and recent world history has been no exception. We can analyze the period 1914–1945 as one global struggle between the United States and Germany, and the period since as the peak and decline of U.S. power.

World War, 1914–1945

Global conquest was in part European interstate competition playing out externally. Within ten years of running out of conquerable land, the colonial powers warred with each other in Europe in two alliances (Stearns 1969:203–206[2]): from 1914 to 1918, Britain, France, and Russia fought Germany, the Austro-Hungarian Empire, and the Ottoman Empire. The latter alliance fought a two-front war, defeating Russia within three years. However, by 1917, millions had died, and both sides were merely hanging on. U.S. entry into the war on the side of Britain and France proved decisive. Around ten million combatants and millions more of civilians died in Europe. Millions of North American, African, and Asian troops fought and died in Europe, while Japan conquered Korea (Jansen and Osterhammel 2017).

The war had four main consequences. First, the Central European Austro-Hungarian Empire broke into several states, Germany became an unstable republic subject to international control, and the Ottoman Empire shrank down to the state of Turkey (Jansen and Osterhammel 2017). Second, European slaughter horrified local colonial elites, and colonial armies' contributions went unrewarded. Colonial peoples who had resisted European rule gained new adherents, starting nationalist movements for independence. Third, the British and French took control of Southwest Asia from the Ottomans (Jansen and Osterhammel 2017).

[1] The existence of the journal *Current History* seems to show that historians do analyze recent events, but flipping through recent issues shows that most of its contributors are historical sociologists or historical political scientists.

[2] This is an example of Carneiro's (1970) idea of circumscription, but on a global scale. Stearns noted this empirically for one region and period one year before Carneiro developed the idea theoretically.

The war's fourth impact changed global politics. The Russian defeat created massive unrest culminating in a revolution that overthrew the Tsar. Months later, socialist revolutionaries took over and fought a civil war. Upon winning, socialists established the Union of Soviet Socialist Republics (U.S.S.R.), implementing their version of Marxist theory and ruling through the Communist Party (Judge and Langdon 2018). Some Asian and African nationalist movements allied with the U.S.S.R. against their European colonizers (Jansen and Osterhammel 2017), adapting previously European forms of nationalism. Sub-Saharan Africa experienced economic growth within the colonial empires, which built roads that helped expand cash crop production. Exports grew by 15 times into the 1950s, while nationalist movements remained local (Iliffe 2017). Such movements developed in Southwest Asia before World War I, but they were less local and more state-oriented (Gelvin 2011). Pre-1914 Indian nationalist movements and parties gained new urgency after 1918 (Trautmann 2016). After its 1911 collapse, China descended into civil war, with "warlords" (quasi-chiefdoms) controlling different areas. Japanese industrialization accelerated, as did control over its Korean colony (Holcombe 2017).

The global economy changed over the 1920s. Global trade declined, British and German exports falling dramatically. Commercial agriculture shrank in some areas and expanded in others. Income from foreign investments also fell, and prices rose in some countries (Neal and Cameron 2016). New European countries attempted self-sufficiency, many states increased wartime tariffs or imposed retaliatory ones, and the Soviet Union now traded only through the state. Economic punishments reduced the German standard of living (Neal and Cameron 2016), while economic growth continued in the 1920s in the United States, Japan, Australia, and most of Europe (Maddison 2006). New industrial process techniques such as the assembly line spread throughout the economy, raising wages and lowering prices for goods. Corporations developed into large, multifactory organizations. New forms of advertising and new "super" markets and chain stores appeared. Corporations developed even newer products: automobiles, household appliances, radios, and lighting and central heating. Home sewing declined in favor of store-bought clothes, and new forms of entertainment, such as phonographs, telephones, radios, and moving pictures ("movies"), became popular (Neal and Cameron 2016).

Economic expansion ended in 1929, and economies started shrinking; global trade shrank even more. The initial U.S. collapse in October was financial, but international trade restrictions reduced foreign trade by two-thirds in two years, making this "Great Depression" worse (Neal and Cameron 2016:341–349). Many states developed or expanded social welfare programs to alleviate the suffering, while Britain and France restructured colonial economies to export more raw materials. In the economic misery, many electorates, now including women, voted for fascist governments in Germany, Italy, Japan, and Brazil. Global economic growth eventually started again (Maddison 2006), but the economic damage had worsened political tensions.

The U.S.S.R. avoided this disaster but created others. State socialism meant government ownership of all economic enterprises; each industry consisted of one government-owned organization. The state set industrial production targets in

Five-Year Plans; enterprises met those targets—or else. The Soviet dictator, Stalin, ensured obedience through political purges that killed several million people, and his agricultural reform led to millions more dead, mostly Ukrainians (Judge and Langdon 2018).

The Great Depression affected Latin America and African and Asia colonies differently. Latin American exports to the United States collapsed in the 1930s, and nationalist leaders gained power in many countries; Mexico nationalized the foreign-owned oil industry. Still, U.S. policy in Latin America became far friendlier (Chasteen 2006). Export prices in sub-Saharan Africa declined; cash crop farmers and urban residents suffered, and colony-wide nationalist movements developed. Indian nationalist activity intensified and Japan invaded parts of China, intervening in the Nationalist–Communist civil war (Trautmann 2016; Iliffe 2017; Holcombe 2017).

These led to another, more global war. Japan invaded northern China in 1931 and 1937; Italy invaded previously independent Ethiopia in 1936; Germany invaded Poland in 1939, and Britain and France declared war; Germany invaded France in 1940 and fought an air battle with Britain; and in 1941, Germany invaded Russia, and Japan attacked the United States (Judge and Langdon 2018). Only South America avoided involvement, and battles occurred almost exclusively in Afro-Eurasia. The capitalist United States, Britain, and France allied with the socialist Soviet Union, produced more war materiel than Germany did, and enjoyed advantages in numbers and energy sources. Germany, Italy, and Japan ultimately lost (Judge and Langdon 2018).

The atrocities of the war should not obscure its underlying dynamic. Japanese massacres in China, U.S. firebombing of German and Japanese cities, the ferocious German invasion of Eastern Europe, U.S. use of nuclear weapons in Japan, and the Nazi Holocaust were either strategies in or byproducts of interstate competition. By 1945, the United States was globally hegemonic.

Cold War, Decolonization, and Development, 1945–1990

The postwar global situation involved major changes. First, the United States set up global political organizations. Second, African and Asian colonies destroyed the European empires, globalizing the European state system. Third, the U.S.-U.S.S.R. relationship quickly went from alliance to "cold" war, one carried out mostly in proxy wars in the new states and by an arms race. Finally, economic development became a major Cold War concern.

U.S. and British policymakers and economists strove to avoid a recurrence of the 1930s tariff wars. These powers formed the United Nations, successor to the League of Nations, and established three global economic organizations in 1944: the International Monetary Fund (IMF) to help states with short-term money problems; the World Bank to support long-term development projects; and a planned

International Trade Organization, which became merely the General Agreement on Tariffs and Trade (GATT), to regulate international trade in goods (Cohn 2016).

Afro-Asian nationalist movements won colonial independence, mostly in bloody wars (Shipway 2008), from 1945 to 1975. Philippine independence from the United States (1946) was expected. Indian independence (1947) was not, but Indians forced the British hand, and other British colonies also became independent in ensuing years: Pakistan (1947), Burma (Myanmar) and Sri Lanka (1948), Malaysia (1957), and others. Indonesians fought themselves free of the Netherlands by 1949, and French defeat in Indochina (May 1954) demonstrated vulnerability, which Algerians exploited in an uprising (November 1954) for independence. A few African colonies became independent in the late 1950s; fifteen gained independence in one year, 1960. Algeria gained independence in 1962, and Portugal gave up its old African colonies of Angola and Mozambique in 1975.

The first wave of colonial breakaway from European states (1775–1825) was called "independence"; this second wave (1946–1975) was called "decolonization." The U.S.–U.S.S.R. competition was a "cold" war since both states could destroy each other with nuclear weapons, so they used each other's new-state allies as proxy battlegrounds. Both funded insurgent movements in Latin America, Asia, and Africa; sent troops to Korea, Vietnam, and Afghanistan; and occupied land or military bases in Europe (Shipway 2008; Judge and Langdon 2018). During its World War II invasion of Germany, the U.S.S.R. occupied Poland, Czechoslovakia, Hungary, Bulgaria, Romania, and part of Germany; after the war it kept them and imposed socialism (Judge and Langdon 2018). Some African, Asian, and Latin American states devolved into oppressive, U.S.-supported, anti-communist dictatorships; some Soviet allies became oppressive one-party dictatorships. Often, states used the Cold War as cover to accomplish local purges.

Economic growth, i.e., "development," became important. Industrialization was important to all: elites wanted its power, and average people wanted materially better lives. For a while, Soviet-style industrialization seemed as viable as the Western, capitalist model, and many new states avoided anything about their old colonial masters. Also, ideological differences influenced Cold War contention globally. Capitalist states supported their corporations by funding insurgent movements, overthrowing unsympathetic regimes, and occasionally invading in Latin American, African, and Asian states; socialist states did the same in the same regions. New states allied to one or the other superpower, and insurgent movements usually adopted ideological conformity with the capitalist or socialist sponsor. Other, more independent developing states, including India, Tanzania, and Iraq, adopted mixed socialist–capitalist approaches. Some European states (e.g., Britain and France) became partly socialist, nationalizing some businesses, though drastically less so than the Soviet Union. Thus, all states now implemented capitalism, socialism, or hybrids in economic experiments (Judge and Langdon 2018).

Two other developments occurred. Chinese Communists won the civil war with the Nationalists in 1949 and reconstituted the Chinese state as a socialist dictatorship (Judge and Langdon 2018). Sino–Soviet relations deteriorated, leading to a war in 1969. Although the United States and U.S.S.R. seemed well matched in nuclear weaponry, the United States deployed globally, while the Soviets did not. The U.S. empire consisted mostly of military bases with landholdings here and there, superficially similar to the Portuguese model, but the United States projected the kind of power that the Portuguese could never have dreamed of (Judge and Langdon 2018).

The United States had other Cold War advantages. World War II shattered the Soviet economy; tens of millions of its citizens died, while the United States lost a small fraction of that; and the U.S. economy grew tremendously in the next 30 years. Western European and East Asian states shared this economic growth so much that some French economists call the period 1945–1973 *les trentes glorieuses*, "the thirty glorious years"; Japan, South Korea, Singapore, and Taiwan experienced the highest growth rates globally (Neal and Cameron 2016). Workers' wages increased spectacularly, and social insurance expanded to help those untouched by the prosperity. Homes increased in size and filled with furniture, electronics, and clothes; automobile ownership became general; and the eight-hour workday became standard, providing so much leisure time that entertainment became a major industry (Neal and Cameron 2016).

The long expansion ended in the 1970s; economic growth stayed low until the 1990s. Despite some stagnation and social problems, capitalist democracies grew faster and produced more wealth than the Soviet system did (Kivelson and Suny 2017). While the U.S.S.R. also grew greatly in the same period, ideological rigidity and an oppressive police state suppressed Soviet innovation, so it never achieved the same material standard of living of capitalist states (Judge and Langdon 2018). In 1978, the Communist Chinese state adopted a hybrid economic model, "market socialism," in which agriculture was mostly privatized and business was partly so, and this sparked rapid economic growth (McCauley 2017). In contrast, the Soviet Union continued stagnating. The issue is debated, but problems with farming seem crucial. After unusually bad harvests around 1980, the Soviet state sold petroleum and natural gas abroad to buy food; falling oil prices in 1984 created a crisis. New Soviet leaders found economic "restructuring" (*perestroika*) difficult. In contrast to 1953, 1956, 1968, and (almost) in 1980 when the U.S.S.R. put down Eastern Europe rebellions, it did nothing when revolutions in Eastern Europe overthrew socialist dictatorships in November 1989 (McCauley 2017). In 1991, a socialist coup in Moscow failed and the Soviet Union collapsed, becoming Russia again. Central Asian and Eastern European states became the last decolonizing states (McCauley 2017).

Cold War and decolonization reshaped the global political and economic map. Modern economic growth occurred globally, with state economic experimentation intensifying competition; Soviet economic stagnation seems to have ended the Cold War. Once again, Fig. 8.1 reflects state claims rather than realities, though it still conveys some truth. Kin-ordered peoples still exist, but they live within aspiring or successful industrial capitalist states.

Fig. 8.1 International State System, ~2010. Source: Jill Thomas

Intensifying Global Interaction, 1990–2015

The end of state socialism led again to a global economy. The global capitalist economy was different from that before 1914, influenced by technological and ideological changes.

Post-1970s economic globalization consists of several elements. First, global trade generally increased, accelerating after the mid-1980s (Irwin 2005) until 2007. Second, new transportation and communication technologies enabled corporations in the global North to send their factories, and thus jobs, to states in the global South. U.S. and U.K. corporations conducted this "offshoring" the most, but Canadian, East Asian, and Western European corporations also did it. Asian, Latin American, and African states started manufacturing goods for wealthy capitalist countries, creating a complex network of global trade (Dicken 2015). These processes radically reduced prices on consumer goods in wealthy countries.

Global governance also changed. The IMF and the World Bank increasingly "advised" decolonized states to adopt a program reducing state power and expenditures and opening to the global economy. A new World Trade Organization replaced GATT in 1995, addressing new trade issues. The North Atlantic policy elite—global organizations, corporations, and governments—felt that the U.S.S.R.'s collapse vindicated capitalism; they provided programs and experts to former socialist states to become capitalist. By the mid-1990s, scholars and nonprofit international agencies increasingly criticized this program, as results in the former Soviet bloc were uneven (Hale and Turley 2013). Within a decade, the IMF and World Bank admitted, haltingly, that their programs had problems.

U.S. hegemony declined in the same period. While it still grew economically, its share of the global economy shrank as others, especially China, grew faster

(Maddison 2006). U.S. military involvement in Southwest Asian states also sparked resistance. In the 1990s, Europeans endured attacks from Islamic North African groups; the United States responded to the 2001 Al-Qaida attack by invading Afghanistan and Iraq. Both wars went well, then badly (Evans 2013).

Starting in the 1950s and accelerating in the 1990s, European states formed what seems to be a potentially federated state, the European Union. Considering the war-torn European past, the project proceeded surprisingly quickly. The E.U.'s most important function has been to create a large trade and labor migration zone, but recent events (see Chap. 11) increasingly highlight the political and military aspects of the project (Cameron 2006).

Some regions flourished in the same period. China's abandonment of state socialism in 1978 for market socialism helped the Chinese economy grow tremendously, as did its middle classes and military power. China's "Belt and Road" initiative, financing infrastructure for other developing states, competed with U.S. hegemony (Cai 2017). No other region experienced so much growth, though most have improved. After 1990, India dropped its mixed socialist–capitalist approach to development in favor of neoliberalism; it has been growing economically since. Sub-Saharan Africa faced difficulties in the 1980s and started improving after 1994 (Mkandawire 2014). Economic growth continued at 5 percent per year from 2000 to 2014, and a large urban African middle class now exists (Taylor 2015), though the region remains the poorest globally. The politics of petroleum have influenced Southwest Asia in the last few decades, with U.S. invasions, U.S.-supported Israeli dominance of Palestine, and the Syrian civil war all reducing regional stability, while many states there still experienced economic growth. Many neoliberal presidents gained power in Latin America in the 1990s, which opened the region to multinational corporations (MNCs), spurring economic growth but imposing "a bitter social cost" (Chasteen 2006:317).

Different definitions lead to different counts of "democracies," but the number of elected representative governments increased globally after 1990. States in Latin America, South Asia, Southeast Asia, and Africa ended military rule and returned to representative government (Kauffman 2010). The "Arab Spring" of 2011 in North Africa and Southwest Asia was short-lived, and that year, Syria devolved into a continuing civil war (Manar 2017).

Finally, the economic crisis of 2007–2009 had multiple effects. Even some of neoliberalism's strongest champions started to question it. Younger people in many countries became more open to socialism again, especially as economic inequality increased and global economic performance stagnated (World Bank 2016). Global migrations of people from developing countries to developed ones sparked backlashes, including increases in nationalist and racist ideologies (Solimano 2017). Lower rates of economic growth slowly returned.

In short, a global chronology of the century after 1914 seems to fall logically into three periods. The international system existing before 1914 gave way to a global war in four phases (1914–1945): war, recovery, depression, and more war. U.S. global hegemony and the breakup of European empires reorganized the world's social system (1945–1991). Economic intensification and Soviet collapse created a global capitalist economy (1991–2015).

Explaining European Ascendance: Debates

The account of social transformation, global conquest, and decolonization helps us understand why world historians worry about "Eurocentric" narratives. Europeans seem to have been the most innovative people in the world since 1800, and perhaps earlier. Yet, this is a tiny portion of world history; as described below and in the next two chapters, others are starting to surpass Europe's experience. Thus, we need neither denigrate nor celebrate Europe's global transformation; we need to *learn* from it because of its world-historical consequences.

Why did Europe transform itself and conquer the world? The question is contentious. Political economists tried to explain European economies in the nineteenth century. Their debate expanded after World War II, accelerating in the twenty-first century, in three phases. In the 1950s, growth theory in economics and modernization theory in sociology influenced international agencies' efforts at helping countries outside of "the West" (North Atlantic societies) "develop" economically. The failure of this effort and the rise of academic Marxism provided a second position starting in the 1970s: theories of dependent relations or of a capitalist world-system. Since 1990, the debate has both included and gone beyond these positions. In a debate named either after McNeill's world history *The Rise of the West* (1963) or after Pomeranz's Sino-European comparison in *The Great Divergence* (2000), economists, sociologists, geographers, political scientists, historians, and anthropologists (but few archaeologists) debate the question.

Modernization Theories

Theories of "modern" economic growth and social change originated in U.S. economics and sociology departments after World War II. Industry-led, intensive economic growth (Jones 1988) had transformed Europe and its offshoots since 1800, so the rest of the world could and should follow. As discussed, scholars made three assumptions: there were two types of societies, "traditional" (agrarian) and "modern" (industrial-capitalist); interstate trade is less important than internal ("endogenous") industrializing economic growth; and science and education support that growth (Berger 2000). Many economists followed and modified Solow's economic growth model (Berger 2000; Solow 2005), while Rostow (1960) posited five stages of economic growth to become richer. Gerschenkron (1962:30) argued that different kinds of "economic backwardness" and initial combinations of resources led to different paths of economic growth. Sociologists (Levy 1966; Eisenstadt 1973; Inkeles and Smith 1974) argued that other changes supported and resulted from industrialization: different attitudes to time, a scientific outlook, wage-based employment, participation in public life, and literacy and numeracy. U.S.-led international agencies implemented this approach around the world, but with disappointing results.

Neo-Marxist Approaches: Dependency and World-System Theories

This failure, and the rise of academic Marxism, led others to reject modernization theory. First, Latin American scholars developed "dependency" theory: Western societies grew rich by making others dependent, thus impoverishing them. Other scholars argued that the same dynamic operated between Europe and Africa (Rodney 1972; Limqueco and McFarlane 1983). Second, Wallerstein (1974, 2004) argued that underdevelopment occurred not merely between states, but within a global system. He argued that Europe created an Atlantic capitalist world-system in the 1500s of Europe and parts of the Americas. In the next four centuries, Europe globalized this capitalist world-economy. Capitalist production for sale to global markets involved trade between "core" industrialized states extracting surplus from "peripheral" raw-materials-producing areas, plus "semi-peripheral" states sinking into the periphery or industrializing into the core. Sociologists, anthropologists, and political scientists developed this theory; economists and some historians rejected it.

The "Rise of the West" or "The Great Divergence" Debates

Since 1990, the end of the Cold War and the global spread of capitalism, theoretical ferment in growth economics (Aghion and Durlauf 2005, 2014), further world-system research (Babones and Chase-Dunn 2012), and the development of world history (Stearns 2011) took the debate well beyond modernization and world-system theories. Names for the debate indicate scholarly emphases. In what they call the "Great Divergence," economists analyze "global differences in economic production levels" (Court 2020:633) that widened between Europe and the rest of the world after 1800; their program is therefore comparative. In what they call the "Rise of the West" debate, historians examine how northwestern Europe and its offshoots rose "to world preeminence" (Daly 2014:1), thus focusing on European global dominance. In practice, this is the same debate, asking "Why Europe?" but with different emphases. The discussion is so large and interdisciplinary that a comprehensive account is impossible (Daly 2014:10; Court 2020:634), but it shows that economists' theories of economic growth and sociologists' world-system analyses constitute a minority of answers to the question.

This debate revolves around at least three questions. First, was the change primarily cultural, involving ideas and institutions, or was it mostly material, relying on physical resources like land, forests, coal, and human labor? Second, did Europe start transforming before 1500, or between 1500 and 1800, or after 1800? Finally, were the changes primarily internal to Europe, or did European imperialism mostly make Europe rich? The answers to these three questions tend to cluster together for each author, often following disciplinary biases. For instance, cultural–institutional advocates usually argue changes were mostly internal, tending to start before or around 1500. Advocates of imperialism and trade as explanations are primarily materialist and see true transformation only after 1800. Still, there are exceptions.

A few examples illustrate answers to these three questions, and the exceptions. Economic historians always favored internal and longer-term explanations for European economic growth. For instance, O'Brien (2013) argues that external European trade was a small percentage of internal trade, so imperialism was a minor factor. Others agree but argue that the expansion of the European merchant class made it politically influential, especially in Britain (Acemoglu et al. 2005). North (1990) has long argued that institutions protecting property rights allowed capitalism to develop. Many are shifting to this institutional–cultural approach. Mokyr (2009) argues, complementarily, that British institutions—scientific ones and those limiting royal power—helped Europe transform. McCloskey (2011) argues that a shift in values toward dignifying bourgeois activities was crucial. Cipolla (1976) argues that Europe escaped the destructive Mongol invasions and engaged in continuous military innovation. McNeill (1992) and Rogers (1995) also make military arguments. Williamson (2011) argues that European changes—open borders, reduced costs of shipping, British global hegemony, and industrialization in Europe—created an ever-growing demand for industrial raw materials.

In materialist, non-world-system approaches, scholars argue that Europe looted the rest of the world. Frank (1967; 1998) argues for long Asian dominance, with a European interregnum. Pomeranz (2000) more subtly argues that coal in Britain and the riches of the Americas were most important. Inikori (2002) argues, *pace* O'Brien (2013), that external European trade was a large percentage of British output; the British Industrial Revolution came from profits of the slave trade, American colonial plantations, and colonial markets for British goods. Crosby (2004) emphasizes the Columbian exchange of plants, animals, and diseases in Europe's rise. Still, Wallerstein's students continue to analyze the capitalist world-system (Mielants 2008).

The "California school" of scholars steps outside the clustering of cultural-internal-early and material-external-late explanations. Goldstone (2009) suggests that six factors, mostly cultural (e.g., science, technology, the Reformation, and business institutions), led to Europe's transformation but argues that fundamental change occurred late. Weede (2014) focuses especially on how religion and values shape modern societies beyond economic growth. Pomeranz (2000) acknowledges a mix of material and cultural factors.

Some take a long-term approach. McNeill (1963) and Hodgson (1993) argue that human accumulation of knowledge supported European expansion. Diamond (1998) sees east–west orientation of Afro-Eurasia enabling biological and cultural diffusion, the north–south orientation of the Americas preventing this diffusion, and the Atlantic littoral of western Afro-Eurasia most likely being used to contact the Americas. Only Sanderson (1995), Lenski (2005), Christian (2011), and Chase-Dunn and Lerro (2014) compare European transformation with the Agricultural and Urban Revolutions.

Understanding the third social transformation is difficult. The Agricultural and Urban Revolutions happened independently in several places in the world. This makes comparison possible, supporting theory development; since globalizing industrial capitalism started in only one place, comparison to develop theories is difficult. Since in 1500, China was the most likely candidate for transformation aside from Europe, scholars keep comparing the two (e.g., Frank 1999; Pomeranz 2000; Goldstone 2009). The sheer size and extreme interdisciplinarity of the debate (Daly 2014; Court 2020) also hinders knowing all possible factors and deciding between them. Since the issue of "how Europe developed" relates to many issues today—inequalities and exploitation, economic development, and racism—the debate is often bitterly politicized.

Nevertheless, we must try to answer the question "Why Europe?" exactly because of these contemporary concerns. I think some scholars regard this transformation as a purely historical matter. I disagree. Global transformation is ongoing, so answers to this question have practical applicability now, especially for economic development. After all, the first two Revolutions took centuries or millennia to play out. Seeing the "Industrial Revolution" not as an English or European phenomenon but as a global transformation beginning centuries ago and still incomplete, continuing for the next few centuries, strikes me as a necessary reframing. For good reasons, historians hate asking historical questions for present purposes; we must try anyway, avoiding anachronism and teleology along the way, by proceeding slowly and consciously through the complexity and sheer volume of answers to this question.

Social Patterns: The Nascent Global Society as the Context for States, 1910s–2010s

The quick chronology above shows many trends that I argue add up to the beginnings of a global society. Seemingly nobody created it consciously, but states and businesses have created its institutional supports. Billions of others have lived in and been subject to this project.

Global Politics

The European governance system begun after 1500 became a more international system after 1850 and became a full-fledged global political system after 1945. States and other actors have responded to increasing collective problems by developing a complex structure of global political institutions. There is no global *government*, but there is global *governance*, a set of global political institutions to provide public goods that solve collective action problems. The most obvious collective problem is warfare, but others include economic development, environmental issues, trade, migration, gender inequalities, and technical standards.

Governance is complex. It consists of "collective efforts to identify, understand, or address worldwide problems that go beyond the capacities of individual states to solve," so it includes a "range of formal and informal values, rules, norms, practices, and organizations" (Weiss 2013:61). Thus, global governance consists of (a) actions to provide global public goods, i.e., solve collective problems, and (b) values, rules, norms, practices, and organizations to shape those actions. It comprises three components: the international state system; a cyclic pattern of global hegemony; and international organizations.

While social evolution led to diverse social forms before 1500, European imperialism and the two waves of independence (1775–1825 and 1945–1991) made "the state" the dominant social form globally. Those seeking independence used European-imposed colonial boundaries for new state boundaries and adapted colonial bureaucracies. While it seems naïve to mention it, leaders of independence movements seem never to have considered leaving small-scale, kin-ordered peoples alone. Only a few political scientists (e.g., Scott 2009) even recognize that kin-ordered peoples often resist living in states. Thus, most in the international system assume that "state equals society"—only states are legitimate forms of large-scale social organization.

Consequently, European diplomacy and law globalized during decolonization. Decolonizing states exchanged ambassadors, so now most states have relations with most others, creating a global network. Furthermore, the Havana Convention (1928) and the Vienna Convention on Diplomatic Relations (1961) formally codified previously informal norms of diplomacy (Siracusa 2010) as part of the larger globalization of international law. This included treaties, customs, principles, and "soft law," often from treaty preambles (Reuter 2011). Issues in international law include war, economic issues, the treatment of individuals, and others. Environmental law provides good examples. Multilateral environmental agreements (MEAs) started in the 1940s, the first on whaling (1946); others followed, such as those on wetlands (1971), wildlife (1973), migratory species (1979), and others (Simon 2007). Since then, conferences such as that of the World Commission on Environment and Development (1987) and the "Rio Summit" (1992) produced MEAs, some of which compel state obedience, while others are advisory (Simmons 2008). The United Nations has an Environment Program requiring other international organizations to monitor their environmental impacts (Simmons 2008). International environmental governance is often regional, not global. Finally, many international nongovernmental organizations (INGOs), such as Greenpeace, the Worldwatch Institute, and the World Wildlife Fund, work on environmental issues (Simon 2007). The story is similar for other issues.

There is no global government to compel obedience to international law, so scholars dispute its importance. Some argue that "most states observe most legal rules most of the time" (Reus-Smit 2005:350); other scholars agree, arguing that international law constrain state actions partly because noncompliance reduces credibility with other states. Others disagree (Krasner 2009), saying that global institutions do not really matter; powerful states make the rules, enforce them on less powerful states, and break them at will. Still, the trend toward democratization may solve the problem of warfare because despite objections, most political scientists argue that while democracies fight wars, they do not fight each other (Ray 2011; Kurian 2011).

Still, there is no global government to stop stronger states from imposing themselves on weaker ones; the state system is in the condition of "anarchy" (lack of rule). However, a second pattern, hegemony, seems to organize the global state system; again, it involves disproportionate influence, not total control. Some argue that hegemonic states provide public goods for the state system: Enforcing international rules, providing a main currency for international transactions, and leading on international problems (Kindleberger 1973; Modelski and Thompson 1996). Others say that dominance mostly benefits the hegemonic state (Wallerstein 1980; Arrighi 2010). As discussed, theorists suggest different periods of hegemonic power since the 1500s. However, all agree that Britain after 1815 and the United States after 1945 were globally dominant; many say that U.S. global power peaked between 1945 and 1970 (Wallerstein 1980; Modelski and Thompson 1996) and has declined since 1990 (Arrighi 2010).

International organizations collectively constitute the third component of global governance; since most issue regulations, they are effectively making global law. Existing for over 140 years, they continue to increase in number and influence. There are three types: intergovernmental organizations (IGOs), international nongovernmental organizations (INGOs) or "civil society" organizations, and multinational corporations (MNCs). Since these last are primarily economic, I discuss them later.

As the name implies, only states belong to IGOs. The first intergovernmental "organization" was the informal Concert of Europe; the first formal set of IGOs were the public international unions (PIUs) of the late nineteenth century (Murphy 1994). World War I led to the formation of IGOs with grander aspirations. The Paris Peace Conference formed a League of Nations to prevent further wars. Many European states joined it; the United States did not (Pirie 2021). The Conference, and later the League, also formed several other international organizations on air travel, harmonizing international law, health, refugees, and labor (Pirie 2021).

In the 1930s, the League proved powerless to stop another war; in 1945, the Allies replaced the League with the United Nations (UN). It does much more than the League ever did; for concrete problems, the UN somewhat operates as a global government. It incorporated both PIUs and League units as UN agencies; these proliferated after 1945, covering international justice, health, labor, air travel, finance, investment, shipping, mail, food, intellectual property, children, and weather. Still, scholars debate overall UN power, its record in enforcing international agreements, and its potential for further war prevention (Badie et al. 2011; Pirie 2021).

There is less debate about the power of three economic agencies: the IMF, the World Bank, and the World Trade Organization (WTO) (which superseded GATT). The Bank for International Settlements (BIS) is also powerful but narrower in scope (Pirie 2021). The IMF was to help international monetary flows and aid states with short-term financial problems. It ended the system of fixed exchange rates between national currencies in 1971. Since the 1980s, it has mostly helped developing states repay international loans. The World Bank handles long-term development projects that build physical infrastructure (hydroelectric dams, roads, and health clinics), improve education and healthcare, or build local financial systems. GATT regulated, and now the WTO regulates, international trade. GATT only covered trade in merchandise (manufactured goods); the WTO regulates merchandise, services, and intellectual property. The BIS promotes banking and insurance standards.

The establishment of state socialism in 1917 influenced how liberalism evolved in North Atlantic states, which influenced IMF, World Bank, and WTO policies; a digression is necessary to understand the issue fully. Nineteenth-century "classical" liberals acknowledged that individual actions always have social implications and that individuals rely on society to create wealth (Neill 1953). After the U.S.S.R. implemented a program of egalitarian industrialization, supposedly offering a realistic alternative to capitalism, liberal theorists also acknowledged that poverty reduced individual liberty and suggested government action to improve citizen well-being, a stance called "welfare liberalism" (Neill 1953). The resulting programs were called "social democracy" in Continental Europe and "liberalism" in the United States. Both involved expansion of the social insurance programs begun in Western Europe in the 1880s (Schmitt et al. 2015). Taxation of the rich funded them for the benefit of the poor. North Atlantic states also started regulating many areas of economic life in mid-century; these included industrial and labor policies and consumer and environmental protection (Vietor 2000; Dormois 2004).

In response to this growth of state power, in the 1950s and 1960s, some Anglo-American theorists revived and revised classical liberalism, calling it "neoliberalism" (e.g., Friedman 1951). Neoliberals accepted only a minimum of state economic intervention: to control the money supply, protect property rights, regulate business minimally, and "relieve acute misery" (Friedman 1951:90), implicitly accepting social insurance programs. Beyond this, however, they argued that state economic power should be reduced to allow private individuals to seek their own interests, ideally in markets: states should pare back regulations, privatize many of their functions, reduce social welfare programs, reduce or eliminate controls on financial flows, and reduce or end tariffs (Centeno and Cohen 2012).

This story helps us understand the global influence of the IMF, World Bank, and WTO. In the 1970s, owners of mostly U.S. multinational corporations constructed an international policy network of organizations—the International Chamber of Commerce, the World Economic Forum, and the Mont Pelerin Society, among others—to lobby governments to implement the neoliberal program (Carroll and Sapinski 2016). When neoliberal leaders won office in the United Kingdom and the United States in 1979 and 1980, respectively, they implemented their program domestically and internationally. These states controlled the IMF, World Bank, and WTO and had them implement neoliberal principles globally through "structural adjustment programs." Many developing states, whether capitalist, socialist, or mixed, had borrowed vast sums from banks in the global North in the 1970s. When they had trouble paying back loans in the recession of the early 1980s, the IMF helped these states renegotiate the loans, providing longer payment terms and lower interest rates; in return, developing states had to privatize state-owned enterprises, lower tariffs, reduce states food subsidies and other welfare programs, encourage exports, allow floating exchange rates, and reduce capital controls (i.e., open the state to foreign direct investment [FDI]). Structural Adjustment Programs became common after the mid-1980s (Mullings 2010).

Implementation of neoliberalism was one half of post-1970s intensified globalization, and neoliberal policies had varying effects across the globe. Trade accelerated in the mid-1980s (Irwin 2005), as did FDI by multinational corporations (Dicken 2015). In fact, "countries that participated in IMF-supported programs were able to attract more FDI [MNC investment] than countries that did not"—up to four times as much (Al-Sadiq 2015:5). Reduced state expenditures enabled repayment of the loans, shifting the cost to populations: unemployment and food prices increased, sometimes dramatically; welfare programs shrank; and exports increased and global prices decreased for many commodities, worsening states' trade situations. Living standards in many countries dropped to levels seen in the 1950s, and riots against these policies often occurred (Walton and Seddon 1994; Cohn 2016). Offshoring industries and the jobs in them that moved from industrial capitalist states to developing states created "rust belts" in the United States, the United Kingdom, and others, but the developing states increased raw materials extraction and upgraded manufacturing capabilities (Centeno and Cohen 2012; Hamilton and Gereffi 2009).

The collapse of state socialism (1989–1991) allowed the full globalization of neoliberalism, with mixed effects. After 1995, the new WTO held many rounds of multilateral negotiations that lowered tariffs globally (Mullings 2010). The IMF and World Bank provided neoliberal "transition" programs for post-socialist states in Eastern Europe and Central Asia, staffed by economists as consultants, who advocated a speedy end to state economic control. However, some states' economic situations improved, while others' worsened (Hale and Turley 2013). The failures forced the IMF and World Bank to listen, only partially, to the growing number of critics of neoliberalism (e.g., George 1988; Walton and Seddon 1994). The global financial crisis of 2007–2009 raised further doubts about neoliberalism among scholars, who have not yet developed an alternative (Lo 2012; Kyung–Sup et al. 2012).

Thus, the IMF, World Bank, and WTO have been the most influential IGOs. Still, many other IGOs exist globally and regionally, with their own areas of influence. Despite its considerable problems, it is remarkable that, after a millennium of internecine warfare, European states formed an unusually strong IGO, the European Union (EU), in only a few decades. Other regional associations of states exist, such as the North American Free Trade Agreement (NAFTA), the Mercado Común del Sur (Common Market of the South, or MERCOSUR) in South America, and elsewhere; none are as thoroughly integrated as the EU is, which itself is still not yet a fully integrated state.

Many perceive the system of global governance as so powerful that it is unstoppable. In contrast, people form and operate INGOs on the idea that individuals and small groups *can* influence the international system. Each INGO usually focuses on one issue, such as labor rights, human rights, education, healthcare, or local infrastructure. Examples include the International Red Cross, Doctors Without Borders, and Oxfam. INGOs lobby states, IGOs, and MNCs; build schools, clinics, roads, and waterways; and gather information and publish reports. Most operate in developing countries and are often explicitly political or religious. There are now almost 30,000 INGOs, but only a few thousand are global; while most operate in a few states on one continent, collectively regional INGOs influence the global system (Götz 2011; Zartner 2011). INGOs have formed an umbrella organization, the World Association of Non-Governmental Organizations (WANGO).

Some international organizations straddle the definitions of IGO and INGO. Many industry organizations set standards and conventions for international trade in one object, such as the International Diamond Manufacturers Association (Pirie 2021). Others manage public goods, such as the internet. The Internet Engineering Task Force (IETF), the Internet Society, and the World Wide Web Consortium (W3C), and all three coordinate with other international organizations. There are also international sports organizations (Pirie 2021).

Most of these organizations have practical purposes, but some have moral goals. For instance, establishing international standards for airplane, computer, and automobile parts, or making English the official language of air travel, improves practical coordination and safety. However, some international law has a moral agenda. The UN *Universal Declaration of Human Rights*—a new version of the Enlightenment-derived *Declaration of the Rights of Man and of the Citizen* (1789)—adopted partly in response to the horrors of World War II (Friesen 2015), has spurred work by states and INGOs to guarantee human rights around the world. Such rights include "life, liberty, and security of person," rights against slavery or torture, equal protection of the law, freedom of movement, and many others (United Nations 1948). Of course, these rights are "virtually unenforceable" (Pirie 2021:438), and many criticize them for imposing European values. Still, social movements globally rely on them for goals and strategies, as rights discourse "is a language that people in positions of power can hear" (Pirie 2021:441; Friesen 2015).

Global governance consists of a complex set of relations between the system of states and supporting structures (such as diplomacy and international law), with a dominating but currently declining hegemonic state, plus international organizations with tremendous influence. Together, these three elements comprise the global political system.

The Global Economy

Much global governance aims to regulate the global economy. While it is an abstraction constructed from multiple descriptions and data, the global economy has real effects. It consists of all commodity chains for all products, and all services supporting and controlling them. Over the twentieth century, despite downturns and a catastrophe, the global economy grew many times; trade grew faster than production;

and investment grew faster than trade. Most commodity chains currently end in Western Europe, North America, and East Asia. I will first trace the history and then analyze its various aspects, organizing it with the commodity chain concept.

Rates of global economic growth fluctuated in the last century. The global equivalent to gross domestic product (GDP) is gross global product (GGP): the value of all goods and services produced globally. GGP grew 1.82 percent per year from 1913 to 1950. During *les trentes glorieuses* (1945–1975), global economic growth averaged 4.90 percent per year, slowing to 3.05 percent per year from 1973 to 2001 (Maddison 2006:640). From 2001 to 2021, which includes the recessions of 2007–2009 and 2020, the global economy averaged growth of 2.85 percent per year (World Bank Open Data).

Importantly, trade grew much faster than production did. From 1913 to 1950, trade increased only around 40 percent, while world output almost doubled (Maddison 2006:361, 606). Capital invested abroad by European states fell by 20 percent from 1914 to 1938 (Maddison 2006:101). After World War II, global trade grew 5.7 times, doubled again by 1990, and grew 68 percent in the next eight years (Maddison 2006:101). Since 1985, trade has continued to grow faster than output, and investment has grown faster than trade.

Why did the global economy grow so much, and why is it based so much in trade and foreign investment? A deeper analysis of the commodity chain concept can help us understand. As documented throughout this book, such chains have existed for thousands of years; however, they have proliferated in the last century. While different disciplines have different names—commodity chains, supply chains, production networks, and commodity circuits—all denote system-based processes of raw material extraction, manufacturing, geographic distribution, and consumption (Bair 2009; Dicken 2015). Scholars are increasingly calling them "value" chains, as each step adds value. Relations between each stage constitute a market, so any value chain is a series of markets. Figure 8.2 presents a vastly oversimplified view of this process. (Dicken 2015:426–428, 452, and 467 offers examples of complex chains.)

The units along each chain, firms or households, relate in multiple ways. Corporations operating in multiple states are called "multinational" or "transnational" corporations (either MNCs or TNCs[3]). They have vast resources and therefore great power. Despite being technically private, many MNCs have annual revenues larger than some states, so they influence the international system. Ostensibly, MNCs do not provide public goods, and often create collective action problems rather than solve them. For instance, MNC investment in a developing state is often so large that host states informally allow MNCs to violate their labor and environmental laws. However, since MNCs bring jobs and money, they create positive economic "externalities," which constitute quasi-public goods in specific locales (Jenkins 2001; Hamilton et al. 2011). Most MNCs are based in the global North, often gaining support of their home state in international disputes. MNCs

[3] Some researchers distinguish between multinational corporations (MNCs) and transnational corporations (TNCs), while others do not. I will use the terms interchangeably.

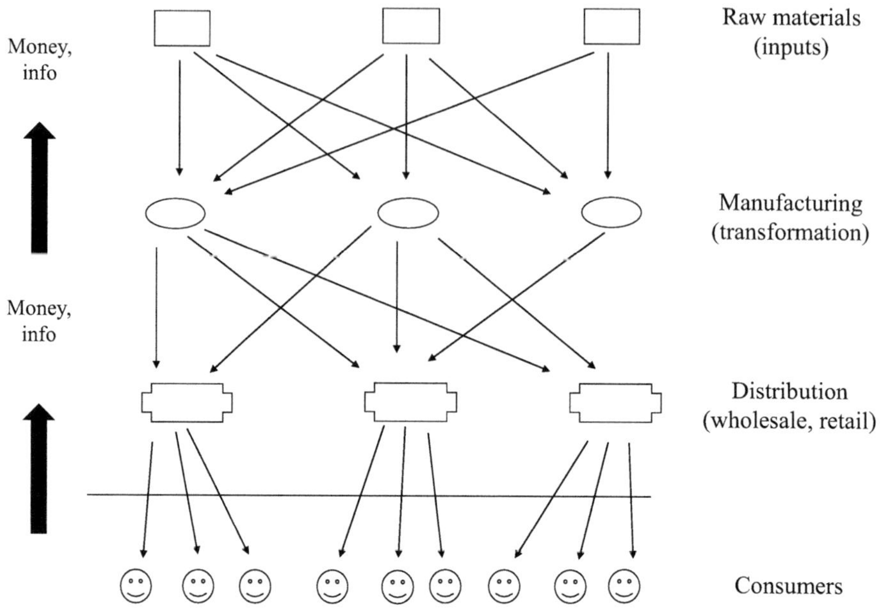

Fig. 8.2 Value Chain as a Series of Markets (simplified)

based in developing countries are increasing in size and power (Casanova 2009) but do not yet challenge global Northern MNCs.

Scholars posit two basic relationships in how corporations construct and manage global value chains. In the classic "producer-driven" model, a manufacturer sells a product to retailers; it has the power and structures the chain. "Outsourcing" production increasingly leads to the second, "buyer-driven" model. The corporation buys parts and services from small overseas firms, often piecemeal from many small companies. The buyer corporation retains product design, marketing, retailing, and overall control over the stages. Scholars argue that developed countries are increasingly moving to buyer-driven chains with a "lead firm" controlling small, overseas manufacturers (Gereffi et al. 2005; Hamilton et al. 2011), while marketing scholars still assume that the manufacturer, the "channel captain," controls retailers (Truell et al. 2009). In reality, there is a long but incomplete global shift toward retailer power (Sternquist 2007). Other scholars have suggested even more types of inter-corporate links, including markets (illustrated in Fig. 8.2); vertical integration, in which one company owns a working company at every commercial stage; and three types between them (Gereffi et al. 2005).

Value chains exist within support structures, which have elaborated lately (Dicken 2015). The first level involves technologies and energy inputs for transportation, which logistics, insurance, accounting, marketing, product design, and legal advice support. As mentioned, neoliberal policies constituted half of the acceleration of economic globalization after the 1970s; the other half consisted of the technological revolution called "containerization," which made intensified globalization physically possible. Into the 1950s, ships sat in port for weeks while dockhands walked or craned small bundles of goods into ships' holds, a process reversed at unloading. Containerization allowed firms to ship container boxes by rail or truck to port where huge machines load or unload them, often taking only two *days*, and then rail or truck takes them to retailers who open the container. This containerization created "outsourcing": cheap shipping gave corporations access to cheap labor in foreign countries, radically reducing product costs even after shipping (Levinson 2006). Most goods travel in containers and rely on their physical supports (revamped ports, container-carrying ships, and container-fitted trucks and railcars). Containerization only works with improved communications, which facilitate ordering, design discussions, shipment tracking, payments, and international regulation compliance. Satellite technology and undersea cables reduced phone-call costs steadily over the twentieth century (Dicken 2015). The internet further reduced costs, making possible remote video conferencing.

These cost reductions help explain the growth in global trade and investment since the 1970s. For instance, cotton grown in west Texas is shipped to China, made into blue jeans, and shipped back to U.S. retailers (Rivoli 2005). Products such as automobiles or electronics have much more complex chains: raw materials come some from some states and manufactured parts from others, all of which are then assembled in other states and, finally, sold in yet others. Supporting industries are not product-specific; while one insurance firm may specialize in one type of product, as entire industries, accounting, insurance, product design, and other functions support the value chains that make up the understructure of the global economy, making two "layers" of that economy (Dicken 2015).

A third layer makes both value chains and services possible. The global economy operates on state-issued currencies, especially the U.S. dollar, and all firms need money to operate, often borrowing it. Thus, financial institutions (e.g., banks, stock markets) heavily influence, even control, value chains and supporting industries. Also, management exists within rules issued by states and international organizations, including product standards; labor, environmental, and consumer protection regulations; and tariffs.

The needs and wants of final customer-users, or consumers—mostly consumers in rich, developed states—drive value chains (Dicken 2015:55). Thus, most of what global value chains produce and distribute ends up in East Asia, North America, and Western Europe. In 1999,

> The world's richest countries make up only a fifth of the global population, but account for 45% of all meat consumption, 57% of total energy use, 84% of paper use and 87% of vehicle ownership (Robins 1999:9).

The situation has not fundamentally changed since. For instance, in 2011, the United States, with only 4 percent of the world's population, used 17 percent of the world's energy. China used a little more (21 percent) than that, but it had four times the population. Europeans consume nearly 32 percent of the world's coffee; North Americans, over 19 percent; and the Japanese, 5.5 percent (International Trade Centre 2014).

To sum up, then, a value chain is a series of markets to produce and distribute a physical good, and all value chains depend on transportation and communication markets; logistics or "supply chain management" markets; insurance, accounting, and other services markets; financial markets; and regulation and control. All these markets rely on labor markets, and laborers must buy goods in consumer markets. Thus, I think the global economy consists of all value chains, all support processes, and all control structures together, all of which are fueled by labor markets, which support consumer markets. The global economy is, therefore, a vast, complex network of interconnected markets.

The global economy is so complex and abstract that it needs simplification; we can understand it either as categories of economic activity or as a global network of spatially interconnected flows of goods, services, and money. In the first, we can translate value chains, supporting services, and control structures (Dicken 2015:55) into an abstract diagram of three macroeconomic "sectors" making up the economy (Fig. 8.3). The primary sector includes all industries extracting raw materials: agriculture, mining, fishing, timber, and ranching. The secondary sector consists of all industries transforming raw materials into finished goods (e.g., food, furniture, clothing, and computers). The tertiary sector includes everything else: product distribution and other services. Services vary, but economists disagree on their categories. Some lump them together in one, others suggest two or three types, and yet others suggest four categories: product distribution, business support, social services, and personal services (Schettkat and Yocarini 2003).

The other simplification is to trace geographic interconnections. Most trade is between bordering states, yet there are significant flows between regions, and most world trade occurs within the global North. Figure 8.4 offers an abstract version of these interconnections.

Nearly all capitalist organizations, from the smallest shop to the largest MNC, are organized hierarchically. Organizational owners reap profits (earned or unearned—see below), while most people labor for wages as subordinates in these

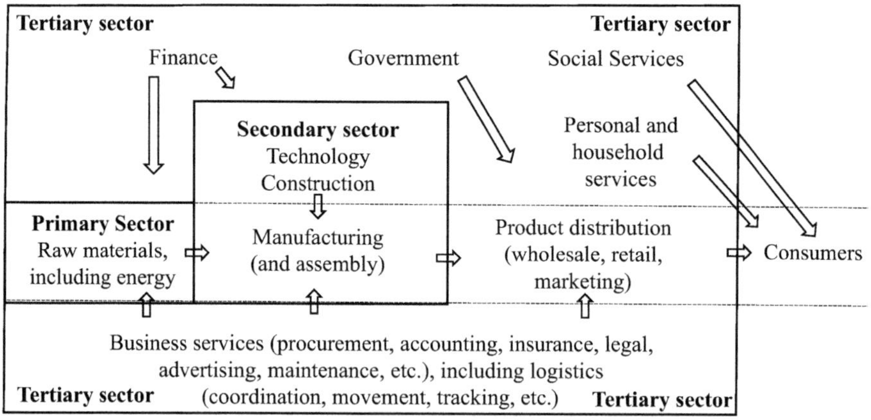

Fig. 8.3 The Three Sectors of the Global Economy

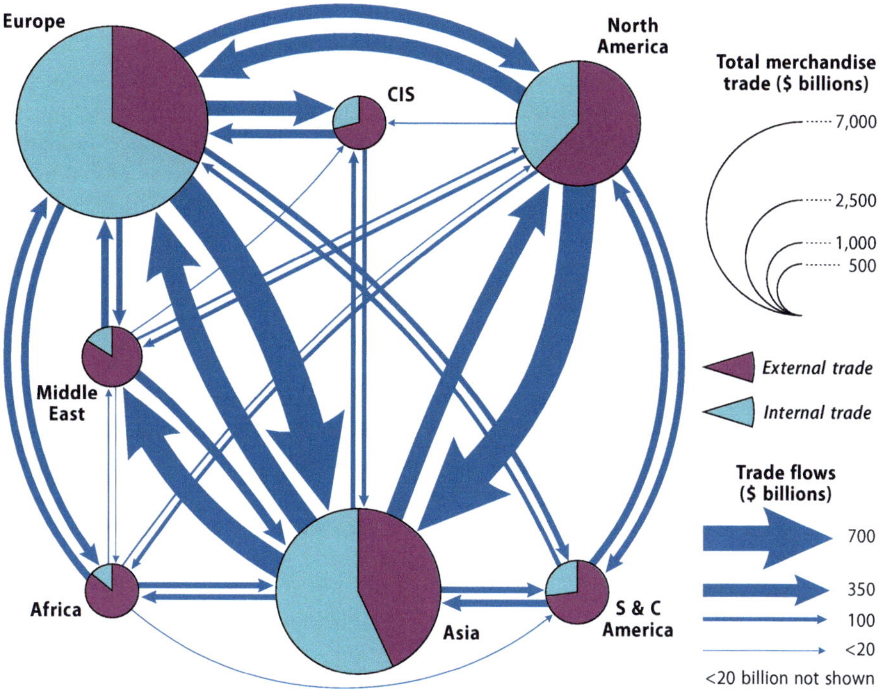

Fig. 8.4 The Global Economy as Flows between Regions. Source Dicken (2015:19)

organizations. Thus, in labor markets workers sell their labor to firms. With their earnings, they buy goods in consumer markets. Nearly all workers work for one firm at a time, but they buy a wide range of consumer goods. Thus, while the global economy consists of a flow of products, it also consists of flows of money—from companies to workers and back again. (See Chaps. 9–10 for more on this.)

Likewise, the point of capitalism is acquiring money. Money is wealth, but capital is wealth-*producing* wealth, which appears in many types: fixed capital, circulating capital, human capital, and so on. It is usually measured in money, so financiers (money managers) are central to the global economy. Due to regulatory (neoliberal) and technological (containerization and satellites/internet) changes, flows of money across international borders accelerated dramatically since the 1970s. Banks still manage much money, but international financial markets exist in many cities (e.g., London, New York, Singapore) in which one can buy and sell stocks, bonds, other securities, different national currencies, and short-term loans (Andrews 2001; Palan 2001).

There are two overall types of investments. "Portfolio investment" is investment as loan. Lenders do not manage the borrowing business; they just want their money back, with interest, at the stated time. In contrast to this hands-off attitude, FDI involves directly managing businesses in foreign states. FDI is what MNCs do: they build factories, hire and train labor, adhere to regulations, and coordinate production and distribution across many states. Consequently, measures of FDI are effectively measures of corporate investment across international borders.

The mix of the two types has shifted. Before 1914, most foreign investment was portfolio investment, such as British and French loans to South America. All investment shrank in the mid-twentieth century; since the 1980s, FDI has dominated. Between World War II and 1985, global FDI grew at the same rate as global trade; since then, FDI has grown more quickly. This "suggests that the primary mechanism of interconnectedness within the global economy has shifted from trade to FDI" since the "number of TNCs has grown exponentially over the past three decades. In 2009, there were around 82,000 parent company TNCs controlling 810,000 affiliates" (Dicken 2015:20). Thus, MNCs are the main drivers of globalization.

There is also a global division of labor. Different regions produce different products and specialize in different activities. This global division has shifted in the last few decades. "Outsourcing" meant moving industrial production, and therefore jobs, from developed to developing countries. The expensive employees in the global North ended up unemployed when their jobs went to the global South. In the United States, outsourcing created a devastated and deindustrialized "rust belt" across the Midwest, and it has affected other developed countries—less so but still painfully (e.g., the United Kingdom). However, outsourcing benefited some developing countries in "upgrading," the other side of outsourcing (Hamilton and Gereffi 2009). Production of raw materials is least profitable, so poor countries usually produce them. Processing raw materials is more profitable; designing, distributing, and marketing products adds even more value and is most profitable. Upgrading usually happens slowly and involves an entire region, often an entire state. Some argue that value-chain upgrading is now a path for economic development (Hamilton et al. 2011). Mostly economies oriented toward exports to existing developed countries, such as South Korea, Taiwan, and Singapore, have followed this path, but some African and other Asian countries are beginning to try it (Gereffi 2005).

Overall, then, the global capitalist economy has grown, despite severe reversals, over the twentieth century and into the twenty-first. All global value chains, the services that support them, and the flows of money and its control make up this economy. All these goods and services flow between capitalist organizations and their workers who are also consumers. The value of economic activity is highest in the global North, but this is changing.

Global Environmental Issues

Industrialized economic growth, growing global trade, and population growth have intensified global environmental problems. Transportation pollutes at every stage of any value chain and postconsumption disposal creates more problems. Still, raw materials extraction has the greatest environmental impact. Old extraction problems—salinization, waterlogging, deforestation, and desertification—have intensified, and new ones have appeared, such as accelerating species extinction. Industrialization also increases atmospheric carbon dioxide.

Extraction of raw materials from the earth—the starting point of all value chains—is the most environmentally destructive thing that humans do. In fact, "90 per cent of biodiversity loss and water stress are caused by resource extraction and processing. These same activities contribute to about half of global greenhouse gas emissions" (UNEP 2020:4). From 1969 to 2019, extraction of raw materials tripled. The situation has not changed in the last five years; resource extraction and use are still the main drivers of "the triple planetary crisis (climate change, biodiversity loss and pollution)" (UNEP 2024:3). Extraction of nonmetal minerals (sand, gravel, and clay) comprised 48 percent of what was extracted in 2020; biomass (crops, grazed materials, timber, and wild-caught fish) made up 26 percent of extractions; 16 percent of extractions consisted of fossil fuels; and 10 percent were metal ores (UNEP 2024:27–28). Interestingly, these increases are not due to expansion in the richest or the poorest states but from economic expansion in newly industrializing states, though rich states still extract disproportionately (UNEP 2020, 2024). It is worth examining some categories in more detail.

While biomass extraction is reducing as a share of overall resource extraction, it is still growing absolutely (UNEP 2024). Scholars define "collapse" of an animal population as a yield 10 percent of the original yield. Before 1900, fishing stocks collapsed around Europe, and the problem has spread. The global fish catch rose six times from 1950 to 2000; one-third of open sea fisheries have collapsed, mostly in the North Atlantic, and fishing fleets moved to the South Atlantic and Pacific Oceans (Ponting 2007). Early European imperialism destroyed stocks of wild fur-bearing animals in Europe and North America. Now 80 percent of fur, and one-third of fish consumed, are farmed. Increasing petroleum use ended whale population declines, but whales remain threatened. "The same pattern has been repeated in industry after industry" (Ponting 2007:168). Also, there has been "an average 69% [sic] decline in the relative abundance of monitored wildlife populations around the world between 1970 and 2018" (WWF 2022:4).

Value chains produce environmental impacts at every stage, mostly through energy use for transportation and manufacturing. Steam power continues to generate electricity, but most transportation relies on internal combustion engines, which use petroleum distillates (Smil 2017). The "annual extraction of fossil fuel carbon" increased "roughly 20-fold between 1900 and 2015" to about 9.7 gigatons (Smil 2017:269). Coal remains the most important source of energy by weight, but its share is falling in favor of petroleum, which produces less carbon dioxide and more energy by weight. Still, total atmospheric carbon dioxide keeps increasing (Smil 2017). Natural gas, nuclear power, and hydro-generation generate tremendous amounts of electricity with less pollution.

Increasing mechanized agriculture improves biomass extraction, but it worsens the global environment. Before 1950, land under cultivation increased 2.5 times in the United States, four times in Russia, eight times in Canada, and 27 times in Australia. Expansion halted mid-century, but it continues in Latin America, Africa, and Asia. This form of farming relies on massive amounts of energy, fertilizer, irrigation, and deforestation (Ponting 2007; Smil 2017), which contributed to severe soil erosion in the United States, Russia, China, and Australia over the twentieth century. Reforestation has begun in Europe and North America, but deforestation practices simply moved. In the last half-century, almost half the world's tropical forests have been cut down. This has caused silting of rivers and silt drops in riverbeds along the way, causing beds to rise, which leads to flooding. Pastoral nomads have been pushed to marginal lands, leading to overgrazing that contributes to soil erosion and desertification (Ponting 2007; Simmons 2008).

Increasing water use also degrades the environment. Large dams for irrigation and hydroelectricity bury ecosystems and force human outmigration. Irrigation still generates salinization and waterlogging, while drilling wells lowers water tables. In addition, "industrial effluent and the run-off of fertilisers, pesticides and herbicides from agricultural land" reduces water purity (Ponting 2007:345). Some estimate that 20 billion tons of sewage, agricultural waste, and industrial runoff discharges into the world's oceans annually (Palmer and Bose 2007).

Finally, industrialization is warming the atmosphere. The science behind this conclusion is firmly established (Brooke 2014:546–552; DiMento and Doughman 2014). Most "greenhouse" gases, which trap sunlight in the atmosphere, come from burning fossil fuels. The heavy reliance on coal in early industrialization meant "sulfate aerosols" entered the atmosphere, reflecting sunlight and producing a cooling effect, especially in the Northern Hemisphere. The combination of the shift from coal to petroleum and natural gas (1925–1950) and world war (1914–1945) seems to have reduced carbon emissions for several decades (Brooke 2014:405, 527). However, economic growth during *les trentes glorieuses* (1945–1975) greatly increased petroleum and natural gas burning; industrialization in developing states increased it more, and atmospheric sulfate aerosols declined. Consequently, the global climate warmed again and continues to do so (Brooke 2014:546–558). Predicted consequences include melting glaciers and ice caps, leading to rising seas and shrinking coastlines and intensifying hurricanes in some areas and desertification in others. Solutions focus mostly on increasing wind and solar power, the electrification of transportation, and other changes to reduce fossil-fuel energy production (Brooke 2014; Smil 2017).

Climate patterns ignore state boundaries, affecting everyone, so climate change is a classic collective action problem. However, many problems hinder global collective action. The global North historically produced and today produces most greenhouse gases, though this is changing. Convincing states in the global South that climate change is more pressing than economic development is understandably difficult. The distribution of climate effects also varies globally. U.S. support for action is internally divided, while Europe is more united (Sweet 2016).

Other environmental issues, such as mining pollution, also exist, but the overall point is clear. The global economy consumes the earth's resources. The production of food and other goods and energy generation for industry and transportation have depleted plant and animal stocks, is generating tremendous pollution in air and water, and is changing the planet's climate.

Global Economic Inequality

While growth in environmental extraction comes mostly from developing states, developed ones still use half of the world's raw materials, despite comprising a small percentage of the global population. The economic inequality underlying this environmental inequality is another consequence of the global Industrial Revolution. Indeed, the "Great Divergence" is about regional measures of income and wealth diverging since 1800; the need to redress this imbalance drives the push for economic development. Here I trace patterns of global inequality through the twentieth and early twenty-first centuries, then analyze debates about its consequences.

Quantitative evidence for economic inequality today is much better than that for the agrarian era. There are two basic approaches. One uses GDP per capita (GDP/cap, the value of a state's production per person per year) (Maddison 2006) adjusted for purchasing power variations across states. The other uses either aggregated global or national incomes or wealth. For income, one subtracts depreciation from, and adds net foreign income to, GDP (or GGP); then one divides the population into centiles (1 percent of the population) or deciles (10 percent) (Piketty 2014; Alvaredo et al. 2018; Chancel et al. 2022).

Scholars relying on the latter approach estimate that aggregate global inequality has increased in the last century. The top 10 percent of the global population took 50 percent of global income in 1820, 60 percent in 1920, and 55 percent in 2020. The middle 40 percent earned just over 35 percent of aggregate global income in 1820, over 30 percent in 1920, and almost 40 percent in 2020. The bottom 50 percent has lost ground relative to the two other groups, earning almost 15 percent of aggregate global income in 1820, around 8 percent in 1920, and the same in 2020 (Chancel et al. 2022:14). Aggregated global wealth inequalities are even more lopsided. The top 10 percent owns 76 percent of global wealth, the middle 40 percent owns 22 percent of global wealth, and the bottom 50 percent own only 2 percent—yes, *half* of humanity owns only 2 percent of its wealth (Chancel et al. 2022:10).

However, these are all shares of total global income and wealth; those totals have grown in the same period. Disaggregated (i.e., geographic) estimates by GDP/cap demonstrate this total global growth, but they also show how global inequality has grown since 1910. In 1913, the ratio of richest to poorest states was nearly 6:1. By 1950, the ratio was 10.4:1; by 1973, it was nearly 12:1; and by 2001, it was 18:1 (Table 8.1).

The deciles approach supports this picture. The rich countries are in North America, Western Europe, East Asia, Oceania, and the Persian Gulf. Most of Latin America, most of Eastern Europe, and much of Asia are in what scholars often call "middle income" countries, while South Asia, most of Africa, and parts of Latin America are relatively poor. This approach also shows how global inequality changed: "From 1900 to 1980, 70–80 percent of the global production of goods and services was concentrated in Europe and America" (Piketty 2014:59). However, by 2000, the Euro-American region produced only 50 percent of global output, so "we have embarked on a period of convergence" geographically (Piketty 2014:61) that continues.

Table 8.1 Per capita GDP (constant $), 1913–2001: regional comparisons

Region	1913	1950	1973	2001
European offshoots (U.S., Australia, etc.)	5233	9268	16,179	26,943
Western Europe	3458	4579	11,416	19,256
Japan	1387	1921	11,434	20,683
Eastern Europe	1695	2111	4988	6027
Latin America	1481	2506	4504	5811
China	552	439	839	3583
India	673	619	853	1957
Africa	637	894	1410	1489

Source: Maddison (2006:642)

Economic inequality within states also varies across the world. Again, scholars use two measures. One is the Gini coefficient, a decimal between 0 and 1. The higher the decimal (the closer to 1), the greater the inequality in wealth or income. The other is the decile/centile method applied nationally rather than globally (Piketty 2014; Alvaredo et al. 2018; Chancel et al. 2022). Scholars using the latter method argue that in the mid-twentieth century there was a "post-war egalitarian regime." In richer states, the share of income of the top 1 and 10 percents fell in the 1930s and 1940s. From the 1940s or 1950s, the top 1 percent's share of national income also fell in India, Argentina, South Africa, Indonesia, Colombia, and China (Piketty 2014; Scheidel 2017). Violent leveling was strong in socialist states: millions died after communist parties took over in the Soviet Union in the 1920s and in China in the 1950s and 1960s (Scheidel 2017). North Africa and Southwest Asia, South Africa (and perhaps much of sub-Saharan Africa), and some Latin American states remained high-inequality regions throughout the twentieth century. Inequality increased from the 1980s onward around the world (Alvaredo et al. 2018). Top decile and centile incomes, and top centile wealth, has grown in some regions since 1980 and in almost all regions since 1990 (Scheidel 2017). Two exceptions exist: in some European states, changes in incomes of the bottom 50 percent and increases in the top 1 percent have been minimal, and in some Latin American states, inequality lowered after 2000 (Scheidel 2017).

Gini coefficients for incomes since 1980 show a similar picture. Inequality has increased in all developed states since then (Scheidel 2017). Still, Canada, Australia, most of Western Europe, and many Asian countries have lower inequality than the United States, Russia, and China. The worst inequality is in relatively poor sub-Saharan Africa and in Latin America. Since much income of the wealthy is tax-sheltered offshore (Piketty 2014; Alvaredo et al. 2018), inequality may be higher in some places than figures show.

Combining the measures of GDP/cap and Gini coefficients for any state is even more illuminating. For instance, the Central African Republic has an extremely low GDP/cap and has a very high Gini coefficient, so the poorest live in desperate poverty. Norway has a very high GDP/cap, but it has a low Gini coefficient, so even its poor probably live reasonably well.

Some scholars distinguish between two types of poverty, absolute and relative (Beaudoin 2007). In absolute poverty (just above starvation), people's physical survival is threatened, and alternatives are often unknown. In relative poverty, people are poor compared to others around them and "lack the resources necessary to survive as full participants in the society in which they live" (Beaudoin 2007:110), a situation generating envy, shame, and resentment. Relative poverty in a poor country is worse: some people are doing better than others are, *and* absolute poverty threatens the latter's survival.

The concepts of absolute and relative poverty have only become useful in the last two centuries. As we have seen, under agrarian inequality, there was relative poverty, but most people lived in what we would call absolute poverty. Every growing season in agrarian states was a gamble on the weather, so "hunger remained a palpable fear throughout the premodern world" (Beaudoin 2007:16–17). One thousand years ago, estimated annual GDP/cap around the world was equivalent to $400 to $450 (Maddison 2006:264).[4] Until 1800 or so, more than one percent of any population died annually from hunger, malnutrition, and disease (Beaudoin 2007; Deaton 2013:73–81). Even in 1820, position in one's local social hierarchy was more important for one's well-being than geographic location (Milanovic 2011). Now, being born in a rich country is more important than being highborn (Milanovic 2011), though this is changing (Chancel et al. 2022).

Absolute poverty is arguably the most important human problem today. Almost one-sixth of the world's population is malnourished. From the 1970s to 2005, malnourishment fell in China, Southeast Asia, Latin America, and the Caribbean, but it worsened in sub-Saharan Africa. As of 2005, 8,760,000 people died per year from malnutrition and hunger: 24,000 people daily, 1000 people per hour, almost seventeen people a minute, or just under one person every three seconds, mostly children under age five. As of 2014, 7.7 million died of hunger per year: 21,000 every day or one every four seconds (Payne and Nassar 2006:89; http://www.poverty.com). This is a tremendous improvement in world history. By today's standards, however, saying that 7.7 million dead a year from hunger is "good" sounds monstrous—*and it is*, given global wealth and incomes. We can end absolute poverty right now, but do not (Fig. 8.5).

Ideas of absolute and relative poverty may explain why groups of scholars talk past each other on this issue. Most economists and some political scientists focus on global absolute poverty. To them, fewer people starving to death is *obviously* good, and they argue that recent decades of capitalist economic growth have accomplished this. Many sociologists and anthropologists, and some political scientists and economists, focus on global relative poverty. To them, a world in which some are rich and some are poor is *obviously* bad—incredibly unjust—and they attribute this to the intensification of the global capitalist system. Proposed solutions to global inequalities also differ. Most economists argue that more capitalist economic growth

[4]In 1990, U.S. dollars (adjusted for purchasing power parity). Clark (2010) has criticized Maddison's methodology and resulting estimates, especially on the basis of what he says are too-low estimates of subsistence before 1500.

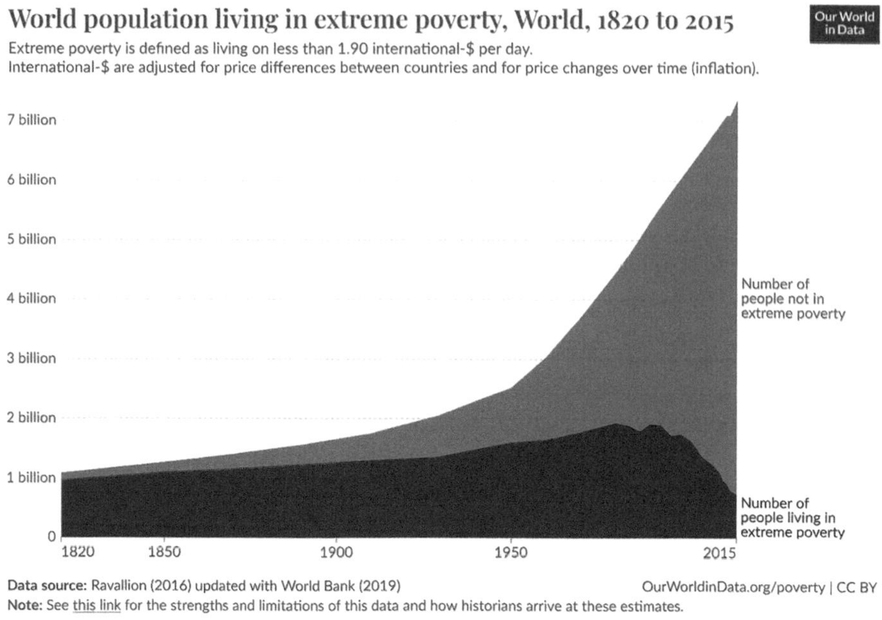

Fig. 8.5 Global Poverty and Number of Extreme Poor. Source Hasell et al. (2022), Chart 83

will end absolute poverty. Many economic historians argue that European institutional changes helped impel the Great Divergence; in partial parallel, other economists argue that neoliberal low tariffs aid economic growth. They acknowledge that this helps already high-income countries and that "low-income countries have not benefited much from trade liberalisation recently" (Winters and Masters 2013:1069). World-system theorists are unsurprised and go further: the system is fundamentally exploitive (e.g., Wallerstein 2004; McMichael 2012).

Some say a "transnational capitalist class based on the transnational corporation" is largely behind post-1970s globalization (Sklair 2001:5). It has supposedly been a U.S. project for the benefit of MNC owners and managers and has spread to European MNCs. This emerging class is composed of four groups: MNC executives; globalizing bureaucrats and politicians; globalizing technocrats; and merchants and media (2002:17). There is qualified support for Sklair's argument (Carroll et al. 2010): interlocks of corporate boards of directors are thinning nationally and growing transnationally. This transnational capitalist class is beginning to spread beyond North Atlantic states (Carroll et al. 2010:224–228).

We might call transnational capitalists "aggrandizers" since their actions have had both prosocial and antisocial effects. Just as some aggrandizing leaders in early chiefdoms and states created public goods (external security, dispute adjudication, and infrastructure), the decline of absolute poverty is a positive effect of neoliberal globalization; the great increase in global economic inequality is a negative effect. In general, over the early part of the twentieth century, inequality reduced in most places in the world but remained unchanged between regions; since 1980, inequality within states and regions has mostly increased, but global inequality has reduced.

Global Demography

In addition to globalizing global inequality, nineteenth-century population trends continued and intensified. First, more people than ever are on the move globally. Second, we are either approaching or have already reached a majority-urban planet. Finally, the second major human demographic transition, begun in Europe in the 1700s, has spread globally. Mortality has fallen dramatically worldwide, and average global lifespan has more than doubled. While fertility has also fallen, the lag between this and the mortality reduction increased global population several hundred percent in this century. Now the global total fertility rate (TFR) has fallen to just above replacement; in a majority of states and among a large minority of people, TFR is below replacement. People around the world experience these patterns—migration, urbanization, and demographic transition—extremely unevenly.

Global migration increased over the twentieth century, both in migration from the global South to the global North and migration between states in the global South. Global South-to-North migration changed over the twentieth century. After World War I, global migration decreased but it increased again after World War II. North American and European states are now migration destinations, as are Persian Gulf and East Asian states. Some migration is "circular"—people go back and forth between rural and urban areas—but most rural-to-urban migration still involves permanently moving. Examples include Mexicans and Central Americans to the United States, South Asians and Caribbean peoples to Britain, Indonesians to the Netherlands, and North Africans to France. This movement constitutes almost half of global migration, exacerbating the problem of "brain drain"—the loss of highly skilled workers from developing countries to developed ones. As just over half of global migration, South–South movement tends to be more circular. Sources are mostly high-population states of South and East Asia; destinations tend to be Southwest Asian oil-rich states. Nearly all migrants send money as "remittances" back home, making up a major source of income for source states (Gould 2015; Czaika and de Haas 2015).

Some patterns are new. The percentage of global population that migrates fell from 3.1 percent in 1960 to 2.7 percent in 2000 (Czaika and de Haas 2015). The absolute number of migrants has grown, but global population grew faster. Also, migrants are moving farther. Immigrant "populations come from an increasingly diverse array of origin countries" such as Western and Northern Africa, the Caribbean and Central America, ex-Soviet republics, western Asia, Eastern Europe, and parts of Oceania, but "they have tended to concentrate in an increasingly small number of destination countries" (Czaika and de Haas 2015:299). In contrast to the past, women make up roughly half of all migrants globally.

The resulting urbanization is a second major demographic change, but defining it is difficult. As discussed, before 10,000 BCE there were only a few complex hunter-gatherer villages. Only after regions of ongoing villages developed can we call anything "rural" or "urban." Cities in complex societies made the distinction more obvious; in agrarian states, cities became large enough that villages might have seemed rural. This problem of definition continues today: "there is no agreement on how cities are defined" (Satterthwaite 2015:283). Consequently, some argue that humanity is more than half-urban (Manning and Trimmer 2013; Gould 2015), while others argue that we have not reached this point (Satterthwaite 2015).

Yet cities clearly exist, so demographers chart global urbanization patterns. Over the twentieth century, rural-to-urban migration increased massively worldwide. In developing countries, "the largest migration flows … are to large cities, especially capital cities" (Gould 2015:150). This started before World War II, accelerating after decolonization. The largest cities in 1920 were much larger by 1990, rising from an average size of three million to fifteen million; furthermore, the world's largest cities are increasingly in the global South (Manning and Trimmer 2013:175–176). Cities offer many advantages: urban incomes can be ten times those of rural areas, even in squatter settlements (Gould 2015:150; also, see Chap. 11). Since farming globally is only partly commercialized (see Chap. 11), urbanization often incorporates people into the commercial economy, reducing subsistence production. Circular migration is increasing in sub-Saharan Africa but difficult in Latin America (Gould 2015).

Cities are important in the global ecumeme. First, several urban areas have more than ten million occupants (Chen and Fitts 2013), often with a central city surrounded by many smaller ones. In fact, "there are forty mega-regions that account for about 66 per cent of the world's economic output and 85 per cent of global innovation" (Chen and Fitts 2013:773). Second, as all global value chains rely on cities somewhat, some scholars characterize the global economy as a network of cities (Taylor and Derudder 2016), obviously continuing this pattern from agrarian world-systems. This city-based global economy is a web of relations not only between businesses (mostly MNCs); it also divides labor between politically oriented and economically oriented cities. This is a useful alternative to the concept of a system of interacting states; however, both views are incomplete since nearly all raw materials extraction occurs in rural areas, and cities have controlled large territories since the Urban Revolution.

Urban areas are now so massive because the global population grew five times in the last century. In the first demographic transition during the Agricultural Revolution, mortality increased, but fertility increased faster, so total population increased. In this second demographic transition, mortality *de*creased; after a while, fertility did as well, with total population increasing in the interim. The sequential mortality and fertility declines that started in Europe in the eighteenth century continued there and appeared in North America and in Japan in the nineteenth, spreading worldwide in the twentieth. Demographers usually divide changes in the last three centuries into four abstract stages. The first is the age-old situation of high mortality and high fertility. In the second stage, mortality started to fall. Since fertility remained high, total population increased rapidly. In the third stage, fertility started falling and population increases slowed. Demographers speculate that in stage four, mortality and fertility may reach a rough, now-low equilibrium, and population may stabilize. The process has, so far, resulted in vastly increased populations and dramatically longer average lifespans (Livi-Bacci 2017).

Again, different areas experience these changes at different times (Maddison 2006:183). East Asian, North American, and Western European states entered stage four late in the twentieth century, while other regions started stage two in the same century, initiating a truly global transition. As implied in Fig. 8.6 (and see Livi-Bacci 2017:129), there is a relationship between GDP/cap and lifespan. As GDP/cap increases, commercialized agriculture ends famines, states can fund public health measures such as vaccination and sanitation, and medical care improves (Livi-Bacci 2017:127), and these increase average lifespan. Additional increases in GDP/cap, and thus healthcare spending, increase lifespan somewhat more. However, the structure of healthcare systems influences rates of healthcare expenditure in states (Hickel 2021) so that the relationship is only loose.

In the last 60 years, these changes have reduced mortality and fertility outside of the global North. (The following comes from Livi-Bacci 2017:Chap. 5.) Even in 1950, the average life expectancy in decolonizing/developing states was somewhere

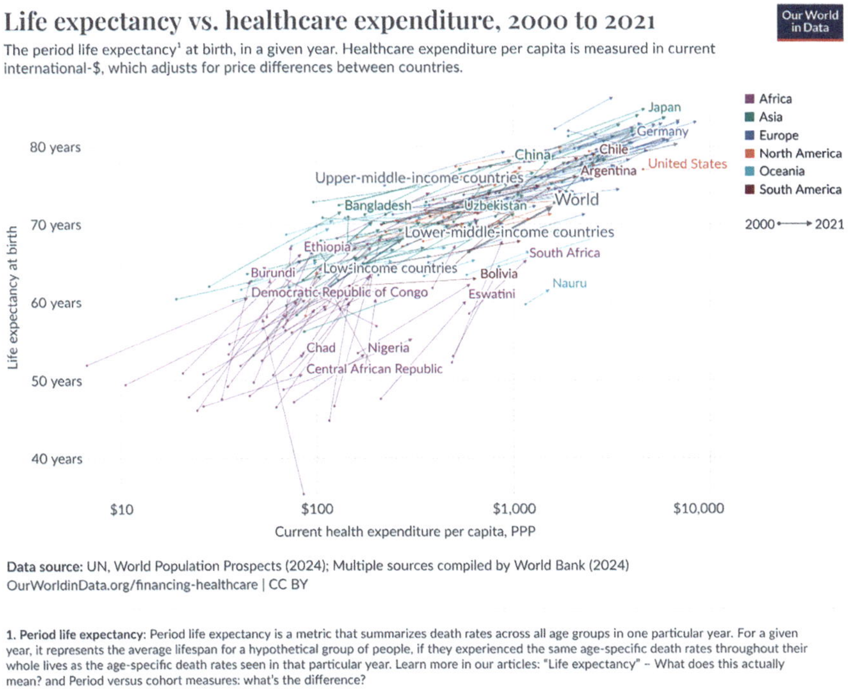

Fig. 8.6 Global Relationship between Life Expectancy and Healthcare Expenditure. Source: Ortiz-Ospina and Roser (2022)

around 40 years. Now global life expectancy approaches that of developed countries' expectancies but varies greatly: the average Chinese and Latin American person can expect to live around 75 years; an Indian 67 years; and an African not quite 60 years (Livi-Bacci 2017:170). This quick mortality transition resulted mainly from reducing infant and child mortality through immunization, improved sanitation, childbirth assistance, increasing female literacy, and reduced malnutrition, especially through actions of IGOs such as the United Nations International Children's Emergency Fund (UNICEF) and the World Health Organization (WHO). The global TFR was around 5 children per woman in 1950–55, but 2.5 in 2010–15. Africa's TFR is still the highest in the world, at 4.3, but this is down from 6. East Asia (including China), Europe, and North America are all below replacement TFR (2.1), while Latin America and the rest of Asia are between 2 and 3.

As a result of this global demographic transition, the global population grew roughly five times in the twentieth century. In 1900, it was about 1.6 billion; in 1950, 2.5 billion; and in 2000, 6.1 billion. As fertility decreased within a generation or two of the mortality decline, global population growth has slowed. Current estimated global population is just beyond 8 billion. World population will continue increasing, but estimates vary. Some estimates project global population between 9.6 and 12.3 billion by 2100 (Gerland et al. 2014); others estimate that it will reach 9.7 billion by 2050. Nearly all estimates project that most of the increase will occur in Africa (Livi-Bacci 2017:25; see also Brooke 2014:Chapter 13 and Gould 2015:21 *ff.*).

Rapidly dropping global fertility rates suggest that what stage four will look like remains unclear (Harper 2018:56, 63). As Fig. 8.7 shows, TFR has dropped below replacement in 110 states in the world; in the rest, it is projected to fall below replacement by 2075 (GBD 2024). The "high income" states of North America, Western Europe, and East Asia collectively fell below replacement in the 1970s, the former Soviet Union right around its end, and East and Southeast Asia and Oceania right after; South Asia and Latin America have only recently reached this milestone; Southwest Asia/North Africa is forecast to reach it around 2040; and sub-Saharan Africa should—if trends hold—reach it by 2075. However, previous predictions based on previous trends did not hold, so these predictions may also not pan out (Fig. 8.7).

As discussed above, this dramatic global transition does not mean that population globally will start falling soon. Instead, population "momentum" means that the global population will continue to grow, then peak, and finally start falling. This peak-and-decline process has already begun in a few states in the world, both developed and developing, with potentially disruptive political, economic, and gender implications. (See Global Gender Patterns below, and the politics, demography, and gender sections of Chaps. 9–10.)

In the last century, humanity has entered a new demographic reality. Some patterns, such as migration, are variations on previous ones. However, urbanization is creating a new situation globally, an average lifespan of 70 years is positively revolutionary, the planet's human population is massive compared to the entire past, and global fertility is falling, so our demographic future is uncertain.

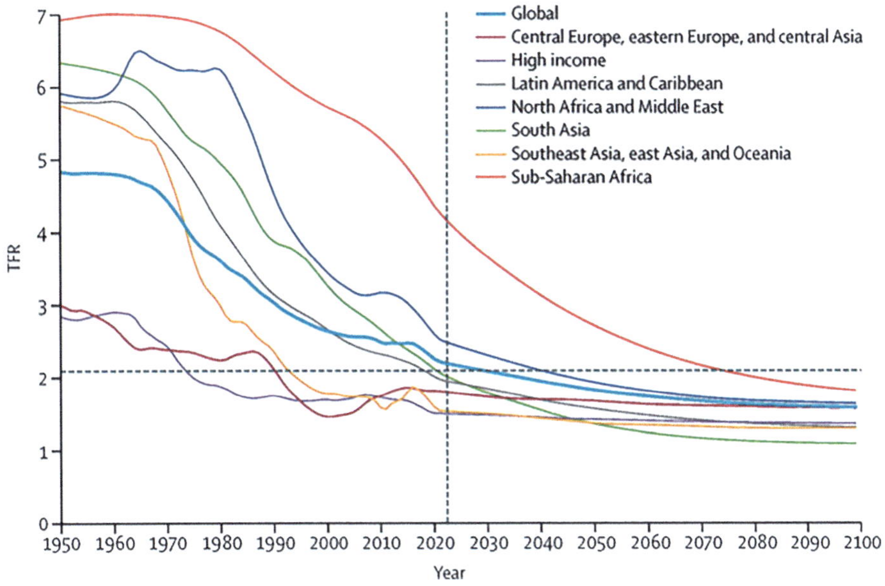

Fig. 8.7 Total Fertility Rates, 1950–2010. Source: GBD (2024). Published under Creative Commons CC-BY license

Global Ideas

The global flow of ideas also changed in the last century. The Cold War was partly an ideological struggle, and IGOs have implemented neoliberal policies since the 1970s. However, the global system has also diffused the English language, mass media, educational curricula, and scientific institutions across the globe into the existing religious environment. Tracing these patterns helps us understand cultural patterns, contradictions, and changes in the global ecumene.

One worldwide cultural project since World War II has been the diffusion of mass education, an Enlightenment idea. From 1914 to 1945, it spread incompletely. Western European, North American, and East Asian/Oceanian states all had public education systems, as did the Soviet Union. China upgraded its existing mass education system before 1914, with indifferent success. In Asia and Africa, colonial powers tried to impose mass education, "but it was a comparatively low priority" (Brint 2017:72; also, Barro and Lee 2015). Only 17 percent of sub-Saharan African children, 23.7 percent of children in Asia and the Pacific, and 25 percent of children in Southwest Asia and North Africa were enrolled in primary education by 1940. However, half of Latin American children and over 90 percent of children in developed countries were enrolled that year (Barro and Lee 2015:16). Most children of white colonial administrators attended universities; fewer than 1 percent of colonial peoples did (Brint 2017:72).

The situation changed greatly after World War II. African and Asian nationalist leaders were usually from educated colonial elites. States in Latin America, Africa, and Asia greatly expanded educational systems (Barro and Lee 2015). By 2010, 89.6 percent of children in sub-Saharan Africa attended primary school; everywhere else, at least 98 percent did. However, secondary education now lags behind primary education globally. Only 42.8 percent of sub-Saharan Africans are in secondary education, while percentages are 68.4 percent in North Africa and Southwest Asia, 72.4 percent in Oceania, and 82.8 percent in Latin America (Barro and Lee 2015:16). Actual attendance is probably lower.

There was a surprisingly consistent curriculum across the globe by the mid-1980s (Meyer et al. 1992). Schools devoted half the time to languages and math, with the rest to science, social studies, and the arts (Brint 2017:79). All scholars agree that after 1990, international agencies, especially the World Bank, pushed the global expansion of education; they debate whether states adopted educational systems and the global curriculum mostly for their own purposes or mostly because of this international effort (Barro and Lee 2015; Brint 2017).

Global linguistic change also supports globalization. Trade and diplomacy across language barriers incurs translation costs and risks miscommunication; one global language reduces these problems. Although a second language for most of its speakers, English has become humanity's first global language. While two consecutive centuries of Anglo-American hegemony formed the foundation for this situation (Northrup 2013), the change had complex roots. Over the twentieth century, five forces globalized English (Northrup 2013). First, most former British colonies in Africa chose English as their official language for practical reasons, so these states' elites, and many in Asia, strove to learn English (Northrup 2013). Second, international diplomacy increasingly relied on English over the twentieth century. Third, U.S. postwar hegemony influenced scientific practice: "By the early 1990s, 31 percent of all scientific articles appeared in journals based in the United States" (Northrup 2013:117), so scientists worldwide learned English. Fourth, British hegemony spread English along Indian Ocean coasts in the eighteenth and nineteenth centuries, and U.S. MNCs expanded the use of English in business even more. Fourth, novel writers worldwide increasingly write in English. Finally, popular music in English has diffused worldwide (Northrup 2013). Around 1990, four new influences further globalized English. One was the creation of the internet. Another was that citizens in many post-Soviet states chose English over Russian as a second language. Third, the Chinese state started a massive English teaching program after 1978. Popular demand there and in India outstripped state policy. Finally, most students who travel want to study English (Northrup 2013).

Institutions developed in the European Scientific Revolution have also globalized. The number of scientists and journals has "since the seventeenth century … [been] doubling in size about every 15 years" (McClellan and Dorn 1999:359, 361). Many fields plateaued after the 1960s; new fields still follow this growth pattern. While scientific INGOs continued to increase in number over the twentieth century (Drori et al. 2003:84–85), developed states' national science associations are often larger than international ones, and developing world scientists often rely more on international associations than their national ones (Drori et al. 2003:96).

Entrepreneurs and governments have also globalized mass media in the twentieth century, and technological change was a major factor. Radio stations in the United States started "broadcasting"—sending out signals in all directions—in the 1920s and 1930s and coalesced into national and international networks. States used radio for propaganda, often owning one or more stations. Film became an art form; after World War II, television transformed entertainment in homes. Recorded movies and television shows, cable networks for television, and the internet all transformed the experience of television, radio, and film. Media globalization proceeded slowly, then accelerated. The United States held a world conference to organize the post-World War I global communication system; it failed, and states developed their own media policies and traditions (Chapman 2005; Winseck and Pike 2007). During the Cold War, the United States and U.S.S.R. each exported their own media policies. Soviet-allied states owned all media, while U.S. allies followed a mixed state-private pattern or had mostly private media (Gorman and McLean 2009). Still, the United States was much more influential globally than the U.S.S.R. (Tunstall 1977). Even in 2001, the United States accounted "for 75 per cent of the value of world trade in television programs, as well as between two-thirds and three-quarters of the trade in television fiction by volume" (Gorman and McLean 2009:155). Film trade patterns were similar.

The Cold War's end transformed media globally. U.S. media influence continues, but it is declining (Winseck 2011). In 2009, ranked by annual revenue, U.S. internet, telecom, and ICT companies accounted for roughly half of all global media giants. The United States predominates in internet firms, while media, telecom, and ICT companies are divided roughly evenly between U.S. firms and European and East Asian firms. U.S. global cultural dominance peaked just after World War II (Tunstall 2008; Gorman and McLean 2009). Since then, "[m]ost people around the world" favor their own entertainment traditions, prefer that their media reflect their cultural patterns, and report on their own politics, sports, and weather (Tunstall 2008). The United States still dominates global popular culture exports, but regional giants such as India, China, Brazil, Mexico, France, Germany, and Nigeria export more media than they import; their neighbors import more.

Scholars debate patterns of control over global media systems. Some argue that U.S. and U.K. media corporations push neoliberal policies by packaging them within local cultural assumptions (Artz and Kamalipour 2003); this implicitly supports the transnational capitalist class argument mentioned above. Others point out the continuing erosion of U.S. political and media hegemony and expansion of other states' media (Winseck 2011). The vast increase in the sheer number of websites, TV stations, social media platforms, streaming services, and hardware companies raise questions about monolithic capitalist power: "concentration and fragmentation emerge as two sides of the same coin" (Winseck 2011:69).

All these cultural expansions occurred within preexisting agrarian-era religions. The Americas, Australia, and New Zealand are mostly Christian, with indigenous religions in some areas. Africa is a mix of indigenous, Christian, and Islamic influences. Europe and much of Siberia are Christian, the rest being indigenous. Islam covers much of western Asia and maritime Southeast Asia; India is mostly Hindu; China is Daoist, Confucian, and Buddhist; and Japan is Shinto and Buddhist. The details are more complicated. Buddhism has three major practices. North Atlantic Christianity is both Catholic and Protestant. Orthodox Christianity dominates Eastern Europe and the Caucasus, with significant numbers in Central Asia, Siberia, and East Africa. Shi'a Islam dominates Iran and areas around it, while Sunni Muslims numerically dominate the rest of the Islamic world.

Two trends complicate the issue. First, Enlightenment secularism—rejecting or downplaying religion—is expanding globally. Fewer than 25 percent of Western Europeans say that religion is important, and religious practice is declining generationally in the United States (Pew Research 2010), although the number of those who believe in God is increasing (Hout et al. 2013). China is relatively secular, and those stating "no religion" on surveys are the second-largest group in Turkey, in some Central Asian states, in three North African states, and in about half of former Spanish America. In contrast, sub-Saharan Africa is becoming more religious, not less (Bullard 2016). Others argue that there is a resurgence of religious belief in many developing states (Haynes 2007; Deneulin and Bano 2009). Second, what some scholars call *fundamentalism* is also increasing. This is a return to the "basics" of a faith (Boer 2005), often an effort to resist secularism (Strozier et al. 2010:5). Some argue that Muslim fundamentalism combines resistance to secularization with resistance to Western domination (Khosrokhavar 2010), especially U.S. oil-driven global policies (Anderson and Sloan 2009). Other scholars reject using the word "fundamentalism" to compare widely disparate movements, focusing instead on specifics of local religions (e.g., Wood and Watt 2014).

In addition to the previously discussed ideology of neoliberalism—which the global system turned into policy—education, mass media, and science have globalized, though scientific institutions lag the others in reach. Billions of people globally are still quite religious, some are increasingly secular, and some strive to return to the "fundamentals" of their beliefs.

Global Gender Patterns

The challenge of women's activism to male supremacy is also influencing the nascent global ecumene. These movements and changes in fertility are making women's lives very different from what they have previously been. Here, I trace global gender patterns, first with a narrative about political movements and then with aggregate statistics about gender inequality across all states. Political, economic, health, and educational inequalities are all changing. Overall, the gender inequality that originated in polygroups and states is beginning to reverse.

Women have increased their political power globally since 1914. Just as agrarian states were, the global political system of the 1920s was nearly exclusively male; today men still predominantly staff IGOs and states, but much less so. Early in the twentieth century, local women's movements[5] developed around the world; late in the century, these movement linked transnationally. They were suffrage-oriented in Europe and the Americas, integral to African decolonization, and expansionist in Asia (Friedman 2017; Poulomi 2017; Tripp 2017). Their first success was the global expansion of women's suffrage in four waves: in North America and Europe in the 1910s and 1920s; throughout the U.S.S.R. in the 1920s and 1930s; in Latin America between the 1920s and 1950s; and in Asia and Africa after decolonization from the 1940s through the 1960s (Towns 2010:Chapter 5). In every colony and independent state, "widespread suffrage activism" took place before actual legislation was passed. Some movements revived and relied on local traditions of female power before European colonialism.

Women's activism has also reshaped norms and staffing of international organizations. Women's INGOs, or "WINGOs," started forming in the 1880s. The number of these increased up to the Great Depression, then plateaued. Activists worked to include the goal of "reaffirm[ing] faith in … the equal rights of men and women" in the Preamble of the UN Charter (United Nations 1945). The General Assembly also recommended "granting to women the same political rights as men" to fulfill its charter (United Nations 1946, cited in Paxton and Hughes 2017:204). Women's movements repeatedly used this language to influence many individual states (Paxton and Hughes 2017:203–204). The UN has regularly held World Conferences on Women. WINGO formation increased again in the 1940s, and the rate of formation accelerated in the 1970s. Early on, most WINGOs were in the global North, but recently, most form in the global South.

Women have also increasingly staffed top state positions. The first female leader was elected in 1960, more were elected in the 1970s, and the trend accelerated in the 1980s—yet today these still constitute a minority of states' leaders (Paxton and Hughes 2017). They also increasingly lead state bureaucracies. The global average of these positions in 2015 was 18 percent female; the global North had the highest representation, and North Africa/Southwest Asia the lowest (Paxton and Hughes 2017:99). Few women take ministerial positions of defense or finance, either choosing or being shunted into family, health, or education services, though this is changing. Since 1990, many IGOs and INGOs have worked to increase female representation in states' legislatures, with some success (Towns 2010; Paxton and Hughes 2017). Many states have passed female representation quotas of 30, 40, or 50 percent for their national legislatures (Towns 2010). Actual representation often differs from these ideals, and regional patterns vary (Paxton and Hughes 2017). Overall, then, these efforts mean that the international system is enculturating developing states (Towns 2010).

Measuring other types of inequality between women and men globally is sometimes easier, but historical data is rarer. The World Economic Forum (WEF) has measured a "global gender gap" for the last decade, comparing men and women on

[5] Because people in developing states often reject "feminism" as global North import, Basu (2017:5) distinguishes between feminism and women's movements, which challenge gender inequalities with local cultural assumptions.

four dimensions: political empowerment, economic participation and opportunity, educational attainment, and health and survival.[6] The overall aggregate of these four scores, calculated together, leads to a global gender gap of 0.68, or 68 percent. That is, women around the world are doing roughly two-thirds as well as men are. Women fare best in Europe, North America, and Central Asia and worst in South and Southwest Asia and in North Africa (WEF 2016:18).

The four subindices are instructive. The WEF's Political Empowerment subindex includes ratios of female to male (1) seats in legislatures, (2) ministerial positions, and (3) heads of state (WEF 2016:6–7). The low score on this subindex, 0.23, is unsurprising, given long-term gender patterns in state structures; it is the lowest on the four subindices. Globally, the Economic Participation and Opportunity ratio is 0.58. This consists of labor force ratios, gender wage ratios for similar work, estimated earnings ratios, manager ratios, and professional ratios. Women globally are overrepresented in health, welfare, and education, while men are overrepresented in engineering, manufacturing, and construction. The WEF suggests that cultural factors still track men and women occupationally, especially as women in some fields experience discrimination (WEF 2016:30–31; World Bank 2012:22).

Estimation of gender disparities in wealth or incomes rather than in occupations produces a consistent picture. The oft-quoted statistic that women hold 1 percent of global wealth is false (Cohen 2013). Instead, women hold 30 percent of the world's wealth in stocks, bonds, cash, and other equities. In other words, men own over twice as much wealth as women do globally. The female share of wealth is slightly higher in developed states and a little lower in developing states (Beardsley et al. 2016).[7] Likewise, women earn about 35 percent of incomes globally, improving by 4 percent in the last 30 years. Women's incomes vary much more by state than women's wealth does; the latter is consistently lowest in most of Asia and in North Africa and highest in Russia and Central Asia, followed closely by North America and Western Europe (Chancel et al. 2022:17–18).

Women's global health situation is near parity with that of men. The health index score, 0.96, measures the sex ratio at birth and life expectancy. Other data suggests a more nuanced picture. Women live longer than men do: global life expectancy for women rose from 48 years in 1960 to 69 years in 2008, and for men, from 46 to 65 years in the same period (World Bank 2012:62). In 34 states, there are no health gaps between women and men. In other states, such as China and India, birth ratios are skewed. Also, mortality for females is slightly higher in developing states at all ages than it is for males, even accounting for underlying mortality conditions. Both overall lowered mortality and better medical care during childbirth contribute to less childbirth mortality for women (World Bank 2012:14–15, 62).

Women's education has also vastly improved. The WEF educational index score, 0.95, measures literacy and school enrollment rates. "For younger cohorts, the

[6] The Forum calculates the gap on each measure as the ratio of the female score to the male score. A score of 1:1 indicates no gap. The WEF report disaggregates dimensional subindices only by individual states, not by region.

[7] This excludes residential property, which is often held jointly.

gender gap in primary education enrollments has practically disappeared" (World Bank 2012:56), as has the literacy gender gap, though the gap remains wide among older people in some places (World Bank 2012:60). Women outnumber men in secondary and tertiary education in most developing states and nearly all developed ones (World Bank 2012:61; WEF 2016:12). Furthermore, girls have lower dropout rates and lower grade-repetition rates than boys.

Over the twentieth century, gendered inequalities in health, education, and politics—and perhaps the economy—have diminished, though quite unevenly. The UN Development Programme report (2020) on global gender norms notes a "backlash" against reductions in gender inequality from a survey of 31 countries in which "91 percent of men and 86 percent of women show at least one clear bias against gender equality" (UNDP 2020:8). Still, a century ago, women were rare as public officials anywhere, and few had the vote. Now, around 23 percent of state leaders globally are female, and nearly all women can vote—and gendered economic, health, and educational inequalities are decreasing.

After several millennia of increasing gender inequality in states, these changes in the now-global state system are astonishingly swift, but they are in line with the political, economic, demographic, and cultural changes of the last century.

Conclusion: A Global Society?

I have argued that the changes resulting from global conquest and internal transformation that Europeans unintentionally started in the 1490s accelerated after 1914, and even more so after 1945, and that this is tending to build a global society. Economic globalization seems to have intensified wealth-taking from the global South to the global North, but now states within the global South are transforming economically, which is beginning to reverse the trend. Through decolonization, the European state system became a global state system. A few people live lavish, powerful lives, many live in bare survival, but most experience some prosperity. Education, science, and mass media are globalizing, and secularism is intensifying in the global North and beginning to challenge religion outside it. Women's movements have begun to reverse the five-millennia trend of female subordination in states, aided in some ways by the globalization of the demographic transition and the consequent fall in fertility. Humans intensified their impact on their physical environment, but they have also begun to act collectively to reduce this impact.

Explaining the second globalization is difficult; the debate is politicized, and the stakes are high. Whether European conquests were central or peripheral in the second globalization, Europeans were clearly aggrandizing. And "development" can improve people's lives—and it can be destructive. The debate is massive and utterly unwieldy, but understanding the past helps us face our future. As we have seen, every group has always existed within a larger social context; now there is only one context for all groups, the global social system. Whether or not this system currently amounts to a global society, it will do so eventually if trends continue. Instead of cycles of imperial rise and fall, Europeans have developed the global social system through successive waves of global imperialism, each more powerful and far-reaching than the last. We can think of post-1970s "globalization" as the latest intensification. If we survive the next few decades (see Chap. 11), such processes are likely to intensify yet again.

The global ecumene operates only somewhat similarly to the way civilizations did in the past. It is still based in cities (Katzenstein 2007:36–37), and it tends to assimilate all people that it contacts. The international governance system, including the global hegemon, the United States, the UN and its agencies; the IMF, World Bank, and WTO; INGOs; and international law all together make up one influence. The global economy—including MNCs, containerization, and industrialization spreading physical and cultural products and practices around the world—is another. The flow of ideas through mass and social media and through migration are also slowly breaking down what coherence regional cultural areas used to have. Yet clear cultural differences remain (Inglehart 2018), and many scholars can still speak justifiably of civilizations (e.g., Rossi 2020). It is simply that, if present trends continue, at some point the global ecumene will be more influential than these cultural areas are in the lives of people and states. Those trends may not continue, but as I have emphasized, many trends operate on centuries-long cycles. Those forecasting from merely decades-long trends may draw false conclusions.

To further substantiate the idea that we are slowly tending toward a global society, I will now examine the different types of societies that the second globalization created.

References

Acemoglu, Daron, Simon Johnson, and James Robinson. 2005. The Rise of Europe: Atlantic Trade, Institutional Change, and Economic Growth. *American Economic Review* 95 (3): 546–579.

Aghion, Philippe, and Steven N. Durlauf, eds. 2005. *Handbook of Economic Growth*. Vols. IA and IB, *Handbooks in Economics 22*. Amsterdam and Boston: Elsevier/North-Holland.

Aghion, Philippe, and Steven N. Durlauf, eds. 2014. *Handbook of Economic Growth*. Vols. IIA and IIB, *Handbooks in Economics 22*. Amsterdam and Boston: Elsevier/North-Holland.

Al-Sadiq, Ali J. 2015. The Impact of IMF-Supported Programs on FDI in Low-Income Countries. *International Monetary Fund*, Working Paper WP/15/157. International Monetary Fund.

Alvaredo, Facundo, Lucas Chancel, Thomas Piketty, Emmanuel Saez, and Gabriel Zucman. 2018. *World Inequality Report 2018*. Paris: World Inequality Lab.

Anderson, Sean, and Stephen Sloan. 2009. Introduction. In *Historical Dictionary of Terrorism*, 3rd ed., li–lxxvi. Lanham, MD: Scarecrow Press.

Andrews, David M. 2001. Capital Mobility. In *Routledge Encyclopedia of International Political Economy*, ed. R. J. B. Jones, 126–127. London and New York: Routledge.

Arrighi, Giovanni. 2010. *The Long Twentieth Century: Money, Power, and the Origins of our Times*. New York: Verso.

Artz, Lee, and Yahya Kamalipour. 2003. *The Globalization of Corporate Media Hegemony*. Albany: State University of New York Press.

Babones, Salvatore, and Christopher Chase-Dunn. 2012. *Routledge Handbook of World-Systems Analysis*. New York: Routledge.

Badie, Bertrand, Dirk Berg-Schlosser, and Leonard Morlino. 2011. United Nations. In *International Encyclopedia of Political Science*. Thousand Oaks, CA: SAGE. SageReference: https://sk.sagepub.com/ency/edvol/intlpoliticalscience/toc.

Bair, Jennifer, ed. 2009. *Frontiers of Commodity Chain Research*. Stanford, CA: Stanford University Press.

Baratta, Joseph Preston. 2004. *The Politics of World Federation*. Westport, CT: Praeger.

Barro, Robert J., and Jong-Wha Lee. 2015. *Education Matters: Global Schooling Gains from the 19th to the 21st Century*. New York: Oxford University Press.

Basu, Amrita. 2017. Introduction. In *Women's Movements in the Global Era: The Power of Local Feminisms*, ed. A. Basu, 1–36. Boulder, CO: Westview Press.

Beardsley, Brent, Bruce Holley, Mariam Jaafar, Daniel Kessler, Federico Muxi, Mattias Naumann, Tjun Tang, et al. 2016. *Global Wealth 2016: Navigating the New Client Landscape*. Boston, MA: Boston Consulting Group.

Beaudoin, Steven M. 2007. *Poverty in World History*. London: Routledge.

Berger, Johannes. 2000. Modernization Theory and Economic Growth. In *Paradigms of Social Change: Modernization, Development, Transformation, Evolution*, ed. W. Schelke, W. Krauth, M. Kohli, and G. Elwert, 31–48. New York: St. Martin's Press.

Boer, Roland. 2005. Fundamentalism. In *New Keywords: A Revised Vocabulary of Culture and Society*, ed. T. Bennett, L. Grossberg, M. Morris, and R. Williams, 134–137. Cambridge, MA: Blackwell Publishing.

Brint, Steven. 2017. *Schools and Societies*. 3rd ed. Palo Alto, CA: Stanford University Press.

Brooke, John L. 2014. *Climate Change and the Course of Global History: A Rough Journey*. New York: Cambridge University Press.

Bullard, Gabe. 2016. The World's Newest Major Religion: No Religion. *National Geographic*. Retrieved February 17, 2018. https://news.nationalgeographic.com/2016/04/160422-atheism-agnostic-secular-nones-rising-religion/.

Cai, Peter. 2017. *Understanding China's Belt and Road Initiative*. Sydney: Lowy Institute for International Policy.

Cameron, David R. 2006. European Union. In *Europe Since 1914: Encyclopedia of the Age of War and Reconstruction*, ed. J. Merriman and J. Winter, 1022–1035. New York: Charles Scribner's Sons.

Carneiro, Robert L. 1970. A Theory of the Origin of the State. *Science* 169 (3947): 733–738.

Carroll, William K., and J. P. Sapinski. 2016. Neoliberalism and the Transnational Capitalist Class. In *The Handbook of Neoliberalism*, ed. S. Springer, K. Birch, and J. MacLeavy, 39–49. New York and London: Routledge.

Carroll, William K., with Colin Carson, Meindert Fennema, Eelke Heemskerk and J. P. Sapinski. 2010. *The Making of a Transnational Capitalist Class: Corporate Power in the Twenty-first Century*. London and New York: Zed Books.

Casanova, Lourdes. 2009. *Global Latinas: Latin America's Emerging Multinationals*. Houndmills, UK: Palgrave Macmillan.

Centeno, Miguel A., and Joseph N. Cohen. 2012. The Arc of Neoliberalism. *Annual Review of Sociology* 38:317–340.

Chancel, Lucas, Thomas Piketty, Emmanuel Saez, Gabriel Zucman, et al. 2022. *World Inequality Report 2022*. World Inequality Lab. wir2022.wid.world.

Chapman, Jane. 2005. *Comparative Media History: An Introduction: 1789 to the Present*. Cambridge, UK: Polity.

Chase-Dunn, Christopher, and Thomas Hall. 1997. *Rise and Demise: Comparing World Systems*. Boulder, CO: Westview Press.

Chase-Dunn, Christopher, and Bruce Lerro. 2014. *Social Change: Globalization from the Stone Age to the Present*. Boulder, CO: Paradigm.

Chasteen, John Charles. 2006. *Born in Blood and Fire: A Concise History of Latin America*. 2nd ed. New York: W.W. Norton.

Chen, Xiangming, and Henry Fitts. 2013. Contemporary Metropolitan Cities. In *The Oxford Handbook of Cities in World History*, ed. P. Clark, 770–790. New York: Oxford University Press.
Christian, David. 2011. *Maps of Time: An Introduction to Big History*. 2nd ed. Berkeley: University of California Press.
Cipolla, Carlo. 1976. *Before the Industrial Revolution: European Society and Economy, 1000-1700*. New York: W. W. Norton.
Clark, Gregory. 2010. Review of *Contours of the World Economy 1-2030 AD: Essays in Macro-Economic History* by Angus Maddison. H-Soz-u-Kult, H-Net Reviews.
Cohen, Philip. 2013. 'Women Own 1% of World Property': A Feminist Myth That Won't Die. *The Atlantic*. Retrieved February 7, 2018. https://www.theatlantic.com/sexes/archive/2013/03/women-own-1-of-world-property-a-feminist-myth-that-wont-die/273840/
Cohn, Theodore H. 2016. *Global Political Economy: Theory and Practice*. 7th ed. New York and London: Routledge.
Court, Victor. 2020. A Reassessment of the Great Divergence Debate: Towards a Reconciliation of Apparently Distinct Determinants. *European Review of Economic History* 24:633–674.
Crosby, Alfred. 2004. *Ecological Imperialism: The Biological Expansion of Europe, 900-1900*. 2nd ed. New York: Cambridge.
Czaika, Mathias, and Hein de Haas. 2015. The Globalization of Migration: Has the World Become More Migratory? *International Migration Review* 48 (2): 283–323.
Daly, Jonathan. 2014. *Historians Debate the Rise of the West*. London and New York: Routledge.
Deaton, Angus. 2013. *The Great Escape: Health, Wealth, and the Origins of Inequality*. Princeton, NJ and Oxford: Princeton University Press.
Deneulin, Séverine, and Masooda Bano. 2009. *Religion in Development: Rewriting the Secular Script*. London & New York: Zed Books.
Diamond, Jared. 1998. *Guns, Germs, and Steel: A Short History of Everybody for the Last 13,000 Years*. London: Vintage.
Dicken, Peter. 2015. *Global Shift: Mapping the Changing Contours of the World Economy*. 7th ed. New York: Guilford Press.
DiMento, Joseph F. C., and Pamela Doughman, eds. 2014. *Climate Change: What It Means for Us, Our Children, and Our Grandchildren*. Cambridge, MA: The MIT Press.
Dormois, Jean-Pierre. 2004. *The French Economy in the Twentieth Century*. New York: Cambridge University Press.
Drori, Gili S., John W. Meyer, Francisco O. Ramirez, and Evan Schofer. 2003. *Science in the Modern World Polity: Institutionalization and Globalization*. Stanford, CA: Stanford University Press.
Earle, Tim. 2016. Pathways to Power: Corporate and Network Strategies, Staple and Wealth Finance, and Primary and Secondary States. In *Alternative Pathways to Complexity: A Collection of Essays on Architecture, Economics, Power, and Cross-Cultural Analysis in Honor of Richard E. Blanton*, ed. L. Fargher and V. Y. Heredia Espinosa, 291–308. Boulder: University Press of Colorado.
Eisenstadt, S. N. 1973. *Tradition, Change, and Modernity*. New York: John Wiley & Sons.
Eisenstadt, S. N. 2004. The Civilizational Dimension of Modernity: Modernity as a Distinct Civilization. In *Rethinking Civilizational Analysis*, ed. S. Arjomand and E. A. Tiryakian, 48–66. Thousand Oaks: SAGE.
Evans, Kim Masters. 2013. *National Security*. Farmington Hills, MI: Gale.
Frank, Andre Gunder. 1967. *Capitalism and Underdevelopment in Latin America: Historical Studies of Chile and Brazil*. New York: Monthly Review Press.
Frank, Andre Gunder. 1998. *ReORIENT: Global Economy in the Asian Age*. Berkeley: University of California Press.
Friedman, Milton. 1951. Neo-Liberalism and Its Prospects. *Farmand* (Feb):89–93.
Friedman, Elizabeth Jay. 2017. Seeking Rights from the Left: Gender and Sexuality in Latin America. In *Women's Movements in the Global Era: The Power of Local Feminisms*, ed. A. Basu, 265–298. Boulder, CO: Westview Press.
Friesen, Bruce K. 2015. *Moral Systems and the Evolution of Human Rights*. New York: Springer.

GBD (Global Burden of Disease Study 2021). 2024. Global Fertility in 204 Countries and Territories, 1950–2021, with Forecasts to 2100: a Comprehensive Demographic Analysis for the Global Burden of Disease Study 2021. *The Lancet* 403 (10440): 2057–2099.
Gelvin, James. 2011. *The Modern Middle East: A History*. 3rd ed. New York: Oxford University Press.
George, Susan. 1988. *A Fate Worse Than Debt*. New York: Grove Press.
Gereffi, Gary. 2005. The Global Economy: Organization, Governance, and Development. In *Handbook of Economic Sociology*, ed. N. J. Smelser and R. Swedberg, 2nd ed., 160–182. Princeton, NJ: Princeton University Press.
Gereffi, Gary, John Humphrey, and Timothy Sturgeon. 2005. The Governance of Global Value Chains. *Review of International Political Economy* 12 (1): 78–104.
Gerland, Patrick, Adrian E. Raftery, Hana Ševčíková, Nan Li, Gu Danan, Thomas Spoorenberg, Leontine Alkema, et al. 2014. World Population Stabilization Unlikely This Century. *Science* 346 (6206): 234–237.
Gerschenkron, Alexander. 1962. *Economic Backwardness in Historical Perspective*. Cambridge, MA: Belknap.
Goldstone, Jack A. 2009. *Why Europe? The Rise of the West in World History, 1500–1850*. New York: McGraw-Hill.
Gorman, Lyn, and David McLean. 2009. *Media and Society into the 21st Century: A Historical Introduction*. 2nd ed. Malden, MA: WileyBlackwell.
Götz, Norbert. 2011. Nongovernmental Organizations (NGOs). In *Encyclopedia of Political Science*, ed. G. T. Kurian . Washington, DC: CQ Press.editor-in-chief
Gould, W. T. S. 2015. *Population and Development*. New York: Routledge.
Hale, Paul, and Gerard Turley, eds. 2013. *Handbook of the Economics and Political Economy of Transition*. New York: Routledge.
Hamilton, Gary, and Gary Gereffi. 2009. Global Commodity Chains, Market Makers, and the Rise of Demand-Responsive Economies. In *Frontiers of Commodity Chain Research*, ed. J. Bair, 136–162. Stanford, CA: Stanford University Press.
Hamilton, Gary, Misha Petrovic, and Benjamin Senauer, eds. 2011. *The Market Makers: How Retailers Are Changing the Global Economy*. Oxford: Oxford University Press.
Hannerz, Ulf. 1992. *Cultural Complexity: Studies in the Social Organization of Meaning*. New York: Columbia University Press.
Harper, Sarah. 2018. *Demography: A Very Short Introduction*. New York: Oxford University Press.
Hasell, Joe, Max Roser, Esteban Ortiz-Ospina, and Pablo Arriagada. 2022. Poverty. Retrieved January 2, 2023. https://ourworldindata.org/poverty
Haynes, Jeffrey. 2007. *Religion and Development: Conflict or Cooperation?* New York: Palgrave Macmillan.
Hickel, Jason. 2021. *Less Is More: How Degrowth Will Save the World*. New York: Penguin.
Hodgson, Marshall. 1993. *Rethinking World History: Essays on Europe, Islam, and World History*. Edited, with an introduction and conclusion, by Edmund, Burke, III. Cambridge, UK: Cambridge University Press.
Holcombe, Charles. 2017. *A History of East Asia: From the Origins of Civilization to the Twenty-First Century*. New York: Cambridge University Press.
Hout, Michael, Claude S. Fischer, and Mark A. Chaves. 2013. *More Americans Have No Religious Preference*. Berkeley, CA: Institute for the Study of Societal Issues.
Iliffe, John. 2017. *Africans: The History of a Continent*. 3rd ed. New York: Cambridge University Press.
Inglehart, Ronald F. 2018. *Cultural Evolution: People's Motivations are Changing, and Reshaping the World*. New York: Cambridge University Press.
Inikori, Joseph. 2002. *Africa and the Industrial Revolution in England: A Study in International Trade and Economic Development*. New York: Cambridge University Press.
Inkeles, Alex, and David H. Smith. 1974. *Becoming Modern: Individual Change in Six Developing Countries*. Cambridge, MA: Harvard University Press.

International Trade Centre. 2014. *The Coffee Guide*. Retrieved October 25, 2014. http://www.thecoffeeguide.org/

Irwin, Douglas. 2005. Trade and Globalization. In *Globalization: What's New?* ed. M. M. Weinstein, 19–35. New York: Columbia University Press.

Jansen, Jan, and Jürgen Osterhammel. 2017. *Decolonization: A Short History*. Translated by J. Riemer. Princeton, NJ: Princeton University Press.

Jenkins, Rhys. 2001. Multinational Corporations (MNCs). In *Routledge Encyclopedia of International Political Economy*, ed. R. J. B. Jones, 1057–1062. London and New York: Routledge.

Jones, Eric L. 1988. *Growth Recurring: Economic Change in World History*. Oxford: Clarendon Press.

Judge, Edward H., and John W. Langdon. 2018. *A Hard and Bitter Peace: A Global History of the Cold War*. 3rd ed. New York: Rowman & Littlefield.

Katzenstein, Peter J. 2007. A World of Plural and Pluralist Civilizations: Multiple Actors, Traditions, and Practices. In *Civilizations in World Politics: Plural and Pluralist Perspectives*, ed. P. J. Katzenstein, 1–40. London and New York: Routledge.

Kauffman, Craig M. 2010. Democratization. In *Encyclopedia of Political Theory*. Thousand Oaks, CA: SAGE.

Khosrokhavar, Farhad. 2010. The Psychology of the Global Jihadists. In *The Fundamentalist Mindset: Psychological Perspectives on Religion, Violence, and History*, ed. C. B. Strozier, D. M. Terman, J. W. Jones, and K. A. Boyd, 139–155. New York: Oxford University Press.

Kindleberger, Charles P. 1973. *The World in Depression, 1929-1939*. Berkeley: University of California Press.

Kivelson, Valerie A., and Ronald Grigor Suny. 2017. *Russia's Empires*. New York: Oxford University Press.

Krasner, Stephen D. 2009. *Power, the State, and Sovereignty: Essays on International Relations*. Milton Park, Abingdon, Oxon: Routledge.

Kurian, George Thomas. 2011. Democratic Peace. In *The Encyclopedia of Political Science*, ed. G. T. Kurian, 401. Washington, DC: CQ Press.

Kyung-Sup, Chang, Ben Fine, and Linda Weiss, eds. 2012. *Developmental Politics in Transition: The Neoliberal Era and Beyond*. New York: Palgrave Macmillan.

Lechner, Frank J. 2009. *Globalization: The Making of World Society*. Malden, MA: Wiley-Blackwell.

Lenski, Gerhard. 2005. *Ecological-Evolutionary Theory: Principles and Applications*. Boulder CO: Paradigm.

Levinson, Marc. 2006. *The Box: How the Shipping Container Made the World Smaller and the World Economy Bigger*. Princeton, NJ: Princeton University Press.

Levy, Marion. 1966. *Modernization and the Structure of Societies: A Setting for International Affairs*. Princeton, NJ: Princeton University Press.

Limqueco, Peter, and Bruce McFarlane, eds. 1983. *Neo-Marxist Theories of Development*. London: Croon Helm.

Livi-Bacci, Massimo. 2017. *A Concise History of World Population*. 6th ed. Malden, MA: John Wiley and Sons.

Lo, Dic. 2012. *Alternatives to Neoliberal Globalization: Studies in the Political Economy of Institutions and Late Development*. New York: Palgrave Macmillan.

Maddison, Angus. 2006. The World Economy, 1–2001 AD. In *The World Economy: Volume 1: A Millennial Perspective and Volume 2: Historical Statistics*. Paris: OECD Publishing.

Manar, Elizabeth. 2017. Syrian Civil War. In *Worldmark Modern Conflict and Diplomacy: Vol. II: Japanese Invasion of China to Yugoslav Wars*, 499–505. Farmington Hills, MI: Gale.

Manning, Patrick, with Tiffany Trimmer. 2013. *Migration in World History*, 2nd ed. London and New York: Routledge.

Mayhew, Susan. 2023. Ecumene. In *Oxford Dictionary of Geography*, 6th ed. New York: Oxford University Press. Oxford Reference: https://www.oxfordreference.com/.

McCauley, Martin. 2017. *The Cold War, 1949-2016*. New York: Routledge.

McClellan, James E., III, and Harold Dorn. 1999. *Science and Technology in World History: An Introduction*. Baltimore and London: The Johns Hopkins University Press.
McCloskey, Deirdre. 2011. *Bourgeois Dignity: Why Economics Can't Explain the Modern World*. Chicago: University of Chicago Press.
McMichael, Philip. 2012. *Development and Social Change: A Global Perspective*. 5th ed. Thousand Oaks, CA: SAGE Publications.
McNeill, William H. 1963. *The Rise of the West: A History of the Human Community*. Chicago & London: University of Chicago Press.
McNeill, William H. 1992. *The Pursuit of Power: Technology, Armed Force, and Society since A.D. 1000*. Chicago: University of Chicago Press.
Meyer, John W., David H. Kamens, Aaron Benavot, and with YunKyung Cha and SukYing Wong. 1992. *School Knowledge for the Masses: World Models and National Primary Curricular Categories in the Twentieth Century*. Washington, DC: The Falmer Press.
Meyer, John, John Boli, George M. Thomas, and Francisco O. Ramirez. 1997. World Society and the Nation-State. *American Journal of Sociology* 103 (1): 144–181.
Mielants, Eric M. 2008. *The Origins of Capitalism and the 'Rise of the West.'*. Philadelphia: Temple University Press.
Milanovic, Branko. 2011. *The Haves and the Have-Nots: A Brief and Idiosyncratic History of Global Inequality*. New York: Basic Books.
Mkandawire, Thandika. 2014. Can Africa Turn from Recovery to Development? *Current History* 113 (763): 171–177.
Modelski, George, and William R. Thompson. 1996. *Leading Sectors and World Powers: The Coevolution of Global Politics and Economics*. Columbia SC: University of South Carolina Press.
Mokyr, Joel. 2009. *The Enlightened Economy: An Economic History of Britain, 1700–1850*. New Haven, CT: Yale University Press.
Mullings, Beverly. 2010. Neoliberalism. In *Encyclopedia of Geography*, ed. B. Warf, 2011–2015. Thousand Oaks, CA: SAGE.
Murphy, Craig. 1994. *International Organization and Industrial Change: Global Governance Since 1850*. New York: Oxford University Press.
Neal, Larry, and Rondo Cameron. 2016. *A Concise Economic History of the World: From Paleolithic Times to the Present*. 5th ed. New York: Oxford University Press.
Neill, Thomas Patrick. 1953. *The Rise and Decline of Liberalism*. Milwaukee: Bruce Publishing Company.
North, Douglass C. 1990. *Institutions, Institutional Change and Economic Performance*. New York: Cambridge University Press.
Northrup, David. 2013. *How English Became the Global Language*. New York: Palgrave Macmillan.
O'Brien, Patrick. 2013. Fiscal, Financial and Monetary Foundations for the Formation of Nation States in the West Compared to Imperial States in the East c.1415–c.1839. *Journal of Chinese Economic and Business Studies* 11 (3): 161–168.
Ortiz-Ospina, Esteban, and Hans Roser. 2022. *Global Health*. Retrieved January 2, 2023. https://ourworldindata.org/health-meta
Palan, Ronen. 2001. Offshore Finance. In *Routledge Encyclopedia of International Political Economy*, ed. R. J. B. Jones, 1156–1158. London and New York: Routledge.
Paxton, Pamela, and Melanie M. Hughes. 2017. *Women, Politics, and Power: A Global Perspective*. Los Angeles: SAGE.
Payne, Richard J., and Jamal R. Nassar. 2006. *Politics and Culture in the Developing World: The Impact of Globalization*. 2nd ed. New York: Pearson/Longman.
Pew Research. 2010. Religion Among the Millennials. *Religion and Public Life Project*. Retrieved April 7, 2014. http://www.pewforum.org/2010/02/17/religion-among-the-millennials/
Piketty, Thomas. 2014. *Capital in the Twenty-First Century*. Cambridge, MA: Belknap Press.

Pirie, Fernanda. 2021. *The Rule of Laws: A 4,000-Year Quest to Order the World.* New York: Basic Books.
Pomeranz, Kenneth. 2000. *The Great Divergence: China, Europe, and the Making of the Modern World Economy*. Princeton, NJ: Princeton University Press.
Ponting, Clive. 2007. *A New Green History of the World: The Environment and the Collapse of Great Civilizations*. New York: Penguin.
Poulomi, Pal. 2017. The Indian Women's Movement Today: The Challenges of Addressing Genderbased Violence. In *Women's Movements in the Global Era: The Power of Local Feminisms*, ed. A. Basu, 129–154. Boulder CO: Westview Press.
Ray, James Lee. 2011. Democratic Peace. In *International Encyclopedia of Political Science*, ed. B. Badie, D. Berg-Schlosser, and L. Morlino, 621–622. Thousand Oaks, CA: SAGE.
Reus-Smit, Christian. 2005. International Law. In *The Globalization of World Politics: An Introduction to International Relations*, ed. J. Baylis, S. Smith, and with the assistance of P. Owens, 3rd ed., 349–368. New York: Oxford University Press.
Reuter, Tina Kempin. 2011. International Law. In *in 21st Century Political Science: A Reference Handbook*, ed. J. T. Ishiyama and M. Breuning, 431–439. Thousand Oaks, CA: SAGE.
Rivoli, Pietra. 2005. *The Travels of a T-Shirt in the Global Economy: An Economist Examines the Markets, Power, and Politics of World Trade*. New Jersey: John Wiley & Sons.
Robins, Nick. 1999. Making Sustainability Bite: Transforming Global Consumption Patterns. *The Journal of Sustainable Product Design* (July): 7–16.
Rodney, Walter. 1972. *How Europe Underdeveloped Africa*. London: Bogle-L'Ouverture Publications.
Rogers, Clifford J. 1995. *The Military Revolution Debate: Readings on the Military Transformation of Early Modern Europe*. Boulder, CO: Westview Press.
Rossi, Ino, ed. 2020. *Challenges of Globalization and Prospects for an Inter-Civilizational World Order*. Cham: Springer Nature Switzerland.
Rostow, W. W. 1960. *The Stages of Economic Growth: A Non-Communist Manifesto*. Cambridge, UK: Cambridge University Press.
Sanderson, Stephen K. 1995. *Social Transformations: A General Theory of Historical Development*. Cambridge, MA: Blackwell.
Satterthwaite, David. 2015. Urbanization in Low- and Middle-Income Nations in Africa, Asia and Latin America. In *The Companion to Development Studies*, ed. V. Desai and R. B. Potter, 3rd ed., 279–285. New York: Routledge.
Scheidel, Walter. 2017. *The Great Leveler: Violence and the History of Inequality from the Stone Age to the Twenty-First Century*. Princeton, NJ: Princeton University Press.
Schettkat, Ronald and Lara Yocarini. 2003. *The Shift to Services: A Review of the Literature* (IZA Discussion Paper No. 964). Bonn, Germany: Institute for the Study of Labor.
Schmitt, Carina, Hanna Lierse, Herbert Obinger, and Laura Seelkopf. 2015. The Global Emergence of Social Protection: Explaining Social Security Legislation, 1820–2013. *Politics and Society* 43 (4): 503–524.
Scott, James C. 2009. *The Art of Not Being Governed: An Anarchist History of Upland Southeast Asia*. New Haven, CT and London: Yale University Press.
Shipway, Martin. 2008. *Decolonization and Its Impact: A Comparative Approach to the End of the Colonial Empires*. Malden, MA: Blackwell.
Simmons, I. G. 2008. *Global Environmental History*. Chicago and London: University of Chicago Press.
Simon, Gregory. 2007. Globalization. In *Encyclopedia of Environment and Society*, ed. P. Robbins, 785–789. Thousand Oaks, CA: SAGE.
Siracusa, Joseph M. 2010. *Diplomacy: A Very Short Introduction*. New York: Oxford University Press.
Sklair, Leslie. 2001. *The Transnational Capitalist Class*. Malden, MA: Blackwell.
Smil, Vaclav. 2017. *Energy and Civilization: A History*. Cambridge, MA: The MIT Press.
Solimano, Andrés. 2017. *Global Capitalism in Disarray: Inequality, Debt, and Austerity*. New York: Oxford University Press.

Solow, Robert M. 2005. Reflections on Growth Theory. In *Handbook of Economic Growth, Vols. IA and IB. Handbooks in Economics 22*, ed. P. Aghion and S. N. Durlauf, 3–10. Amsterdam and Boston: Elsevier/North-Holland.

Stearns, Peter N. 1969. *Modern Europe, 1789–1914*. Glenview, IL: Scott, Foresman.

Sternquist, Brenda. 2007. *International Retailing*. 2nd ed. New York: Fairchild.

Strozier, Charles B., David M. Terman, James W. Jones, and with Katharine A. Boyd, eds. 2010. *The Fundamentalist Mindset: Psychological Perspectives on Religion, Violence, and History*. New York: Oxford University Press.

Sweet, William. 2016. *Climate Diplomacy from Rio to Paris: The Effort to Contain Global Warming*. New Haven, CT: Yale University Press.

Sztompka, Piotr. 1994. *The Sociology of Social Change*. Cambridge, MA: Blackwell.

Taylor, Peter, and Ben Derudder. 2016. *World City Network: A Global Urban Analysis*. New York: Routledge.

Taylor, Scott. 2015. The Warming Business Climate in Africa. *Current History* 114 (772): 163–169.

Thomas, George M. 1987. *Institutional Structure: Constituting State, Society, and the Individual*. Newbury Park, CA: Sage.

Towns, Ann E. 2010. *Women and States: Norms and Hierarchies in International Society*. New York: Cambridge University Press.

Trautmann, Thomas R. 2016. *India: A Brief History of a Civilization*. New York: Oxford University Press.

Tripp, Aili Mari. 2017. Women's Movements in Africa. In *Women's Movements in the Global Era: The Power of Local Feminisms*, ed. A. Basu, 37–64. Boulder, CO: Westview Press.

Truell, A. D., L. E. Pelton, and D. Strutton. 2009. Channels of Distribution. In *Encyclopedia of Business and Finance*, ed. B. S. Kaliski, 2nd ed., 108–112. New York: Thomson.

Tunstall, Jeremy. 1977. *The Media Are American: Anglo-American Media in the World*. London and New York: Columbia University Press.

Tunstall, Jeremy. 2008. *The Media Were American: U.S. Mass Media in Decline*. New York: Oxford University Press.

United Nations. 1945. *Charter*. Retrieved February 3, 2018. http://www.un.org/en/charter-united-nations/index.html.

United Nations. 1946. Political Rights of Women. *General Assembly Resolution 56 (1)*:

United Nations. 1948. *Universal Declaration of Human Rights*. Retrieved April 3, 2024. https://www.un.org/en/about-us/universal-declaration-of-human-rights.

United Nations Development Program (UNDP). 2020. *Tackling Social Norms: A Game Changer for Gender Inequalities*. New York: UNDP.

United Nations Environmental Program (UNEP). 2024. *Global Resource Trends: Bend the Trend: Pathways to a Liveable Planet as Resource Use Spikes*. New York: UNEP.

Vietor, R. H. K. 2000. Government Regulation of Business. In *Cambridge Economic History of the United States*, ed. S. L. Engerman and R. E. Gallman, vol. III, 969–1012. New York: Cambridge University Press.

Wallerstein, Immanuel. 1974. The Rise and Future Demise of the Capitalist World-Economy: Concepts for Comparative Analysis. *Comparative Studies in Society and History* 16 (4): 387–415.

Wallerstein, Immanuel. 1980. *The Modern World-System II: Mercantilism and the Consolidation of the European World-Economy, 1600–1750*. Berkeley: University of California Press.

Wallerstein, Immanuel. 2004. *World-Systems Analysis: An Introduction*. Durham, NC: Duke University Press.

Walton, John K., and David Seddon. 1994. *Free Markets and Food Riots: The Politics of Global Adjustment*. Oxford: Blackwell.

Weede, Erich. 2014. Economic Development and Growth. In *Concise Encyclopedia of Comparative Sociology*, ed. Masamichi Sasaki, Jack Goldstone, Ekkart Zimmermann, and Stephen K. Sanderson. Boston: Brill.

Weiss, Thomas. 2013. *Global Governance: What? Why? Whither?* Malden, MA: Polity.

Williamson, Jeffrey G. 2011. *Trade and Poverty: When the Third World Fell Behind.* Cambridge, MA: MIT Press.

Winseck, Dwayne R. 2011. The Political Economies of Media and the Transformation of the Global Media Industries. In *The Political Economies of Media: The Transformation of the Global Media Industries*, ed. D. R. Winseck and D. Y. Jin, 3–48. New York: Bloomsbury Academic.

Winseck, Dwayne R., and Robert M. Pike. 2007. *Communication and Empire: Media, Markets, and Globalization, 1860–1930.* Durham, NC and London: Duke University Press.

Winters, L. Alan, and Andrew Masters. 2013. Openness and Growth: Still an Open Question? *Journal of International Development* 25 (8): 1061–1070.

Wood, Simon A., and David Harrington Watt. 2014. *Fundamentalism: Perspectives on a Contested History.* Columbia: University of South Carolina Press.

World Bank. 2012. *World Development Report 2012: Gender Equality and Development.* Washington, DC: International Bank for Reconstruction and Development/The World Bank.

World Bank. 2016. *World Development Indicators 2016.* Washington, DC: International Bank for Reconstruction and Development/The World Bank.

World Economic Forum. 2016. *The Global Gender Gap Report 2017.* Geneva, Switzerland: World Economic Forum.

WWF. 2022. *Living Planet Report 2022: Building a Nature-Positive Society.* In *Gland*, ed. R. E. A. Almond, M. Grooten, D. J. Bignoli, and T. Petersen. World Wildlife Fund: Switzerland.

Zartner, Dana. 2011. International Organizations. In *Encyclopedia of Political Science*, ed. G. T. Kurian, 819–823. Washington, DC: CQ Press.

Chapter 9
Social Patterns of Developed States: Industrial Capitalist Democracies, 1910s–2010s

Abstract In the last century, industrial capitalist democracies became clearly different from previous societies. They are increasingly democratic, aspiring to universal suffrage, with representative legislatures, complex bureaucracies, and elaborate court systems. Through most of the twentieth century, they relied on manufacturing, but late in the century, employment shifted to services. Their consumption practices have a major global environmental impact. Economic inequality has increased in all such states, though variably, and state welfare systems to reduce these inequalities also vary. These societies have low mortality; most now have an average total fertility rate below replacement, creating an uncertain demographic future. Their citizens are well educated and increasingly secular, and their populations enjoy new artistic and entertainment experiences. Feminist social movements have greatly expanded opportunities for women, who are still far from full parity with men.

Keywords Modern societies · Global North · Industrialization · Capitalism · Democracy · Secularization · Welfare state · Demographic transition · Feminism

Introduction

In the previous chapter, I argued that a global society is beginning to form. This system is centered (for now) on Western Europe, North America, and East Asia/Oceania, which scholars call collectively "the global North." As discussed, scholars often call these states "modern." What exactly is modernity? Stated differently, what kind of societies are these?

Here, I argue that the global Industrial Revolution has produced a new type of society. We saw its basic outlines develop by 1850, but this new way of life achieved its fullest form only after 1950. I examine that new type, including the basic similarities between these societies and the cultural variations between them. They are wealthier than any previous societies, as they are industrialized, commercialized, and foster high consumption. Their governments operate (imperfectly) through democratic political institutions and tax the massive economic surplus to provide traditional and new public goods. They generate great economic inequality, yet they

T. C. Burr, *Societies Past and Present*, Emerging Globalities and Civilizational Perspectives, https://doi.org/10.1007/978-3-031-91532-1_9

mitigate it through state redistribution. Most are ethnoracially diverse, and all are urbanized; they have completed three stages of the demographic transition, but they face an uncertain demographic future. Their idea-lives are increasingly secular, educated, and scientific and are influenced by mass and social media. They are transitioning from the "separate spheres" family system to a more experimental sexual and gender system. Finally, their high consumption leads these societies to have a greater environmental impact than any other type of society.

These states also embody historical continuities or recent reversals. Massively increased production of physical goods has yet again increased human resource use, pollution, and waste. The slow agrarian trend toward increasingly literate populations intensified; now nearly every population is literate. Proliferating technologies allow the mixing of old religious ideas and moralities with scientific results and political ideologies by offering ways to express them in a wide range of artistic productions in mass and social media. The reversals are equally important. Despite tremendous imperfections, democratization is slowly reversing the long centralization of power in imperial monarchies. The most significant reversal in long-term trends, the demographic transition, is now leading to aging populations, a new and potentially disruptive phenomenon in world history. Women's movements are slowly and successfully challenging long-standing gender inequalities.

I again compare these societies with those in the past through seven aspects of human groups. Many points expand on patterns discussed in Chap. 8.

Democratic States

As discussed, many social scientists equate *states* with *societies*. States usually attempt to make them identical, and the Cold War and neoliberalism both shaped such attempts. State economic regulation and public goods provision increased greatly over the twentieth century, thus growing governments. Democracies are far more responsive to their populations than agrarian states ever were; all promise universal suffrage to elect governing officials, and most adults can in fact vote—though the institutional structures of governance vary by state.

The nineteenth-century provision of new public goods increased through the twentieth century. States continued to provide public goods such as external security, order and justice, and infrastructure. They started or increased social insurance programs, mass education, and stable banking systems, while they slowly reduced the provision of religion. Social insurance gradually included programs for sickness, injury, and unemployment; insurance for family issues; and old-age pensions. Eventually, "public welfare provision became a core task" of all states, not just developed ones (Schmitt et al. 2015:506). Educational systems increased in size as more children stayed in public education longer. After World War II, states greatly increased the provision of university education (Tanzi and Schuknecht 2000).

Nineteenth-century liberal reductions in economic regulation reversed in the twentieth century, but the regulatory increase sparked a reaction in some states.

Overall, capitalist states expanded greatly on agrarian states' regulation of markets and organizational forms to increase trust and thus willingness to transact (McMillan 2002:Chap. 5). State-owned central banks came to control each economy's money supply, supplementing or replacing the metal-based currencies of agrarian states (Bordo and Siklos 2017:2; Piketty 2014), and states also regulated national systems of commercial banks, enforced contracts, and extended the ability to form limited liability corporations (Neal and Cameron 2016). From the 1930s, states regulated their financial industries and labor markets more comprehensively. US government labor, consumer, and environmental regulation increased again in the 1960s (Vietor 2000), while France increased state regulation from the 1950s into the 1980s. The French and British states even owned some businesses (Dormois 2004; Hancké 2003). Furthermore, many states followed "Keynesian" policies in which the state intervened in the economy to achieve full employment, higher incomes, and increased purchasing power (Krieger 2012). As mentioned, neoliberalism was the reaction against this increase in state power. In response to economic difficulties in the 1970s, states started to implement neoliberalism after 1980. Reduction of social welfare programs proceeded more successfully in Anglo-American states (the UK and its "offshoots") than in Continental Europe or East Asia. Both France and Britain also sold state-owned businesses in the 1980s. Nevertheless, industrial capitalist states still regulate much of daily life and business activity (Vietor 2000; Hancké 2003; Dormois 2004).

Additional public goods required increasing the size of government. Developed states increased the percentages of gross domestic product (GDP) spent on government programs and services over the twentieth century (Tanzi and Schuknecht 2000) by increasing the number and size of their bureaucracies. For instance, the number of top executives of departments in Australia grew from seven to more than forty (Kirchner 2011:7). "Since 1965, six new federal cabinet agencies have been established … [and] many new sub-cabinet agencies have been created" in the USA (DiJulio 2014:14). Finally, states increased the number of people they employ, largely between World War II and 1980; since then, some have reduced state employment, while others have increased it slightly (Tanzi and Schuknecht 2000).

Modern states have grown so much because they have been able to. First, economic growth vastly increased the economic surplus. Second, the commercialization of the economy allows states to collect taxes in money rather than in grain. Finally, literate and numerate populations make possible exact calculation of those taxes (Ardant 1975). These have enabled industrial capitalist states to accomplish what leaders of agrarian states could not even imagine.

Populations now have a much greater voice in government operations than any time since small polygroups. As discussed, modern democracies are descendants of medieval noble assemblies, modified with ancient Greek and Roman models and Enlightenment theorizing about popular sovereignty. Social scientists define "democracy" in many ways. Some emphasize etymology: The Greek words *demos* (people) and *kratos* (rule) mean that in a democracy, the people rule. Others argue that democracy is a method of choosing political leadership through votes. Yet others list four characteristics: the rule of law, institutions of civil society, free elections, and accountability of state officials (Ishiyama et al. 2011).

Parliaments and presidencies continue to structure leadership in developed states. The USA and South Korea have powerful presidencies and heads of departments separate from their legislatures; Germany, Ireland, and others have only ceremonial presidencies, while others such as the UK, Spain, the Netherlands, Sweden, and more have ceremonial monarchies. Except for France, all the latter have parliaments: executives serve simultaneously in legislatures and as heads of departments. France has both a powerful president and a strong prime minister. Political scientists have found that parliamentary systems are more representative, spend more on public goods, and are more conducive to cooperation within the government. In presidential systems, presidents appoint their own senior executives and balance power between executives and legislatures (Robbins 2011).

Most parliamentary systems rely on proportional representation (PR) electoral systems, while most presidential systems use single-member district (SMD, or "winner take all" or "first past the post") systems. Exceptions, such as the UK's parliament being elected by a SMD system, are rare. PR systems, with each district sending multiple representatives, usually produce multiple political parties that better represent multiple interests, while SMD systems usually lead to two dominant parties. PR systems usually have higher electoral turnout than SMD systems do (Downs 2011; Robbins 2011) but require party coalitions to form legislative majorities.

Developed states also vary in their recruitment patterns, social status of bureaucrats, politicization versus administrative neutrality, and career tracks. Developed states have four different traditions of public administration: Anglo-American, Germanic, Napoleonic, and Scandinavian (Painter and Peters 2010). There is variation even with traditions. For instance, much of the US populace holds anti-statist attitudes, leading to the "submerged state" (Mettler 2011) in which the national government often operates through private actors. The US healthcare system is a mix of employer-offered programs, state subsidies for the poor, and state-offered markets. In contrast, the state in the UK runs the National Health Service (NHS), a government agency. Other states have private medical care paid for by the state ("single-payer" systems) or other systems.

Civil society and the public sphere also vary. While US anti-statist attitudes lead to submerged-state programs, US citizens are known for volunteering time and money (Anheier 2004), while the French populace distrusts the state, nevertheless expecting it to step in to solve social problems; the French political elite has long distrusted civil society organizations and worked to reduce their number (Glasius et al. 2004). In the public sphere, factors such as the number of media companies, the professionalization of journalism, the partisanship of media (or not), and other factors all shape how the media, the state, and civil society interact in any given society, producing greatly different systems (Hallin and Mancini 2004; Schudson 2012).

Still, critics question how "democracy" really works. They argue that it does not necessarily reveal the will of most voters, who are uninformed and emotional, not rational. Short-term economic conditions influence voters even when officials have little control over those conditions. Some argue that "'groupiness' ... is fundamental to thinking about the beliefs, preferences, and political behavior of democratic citizens" (Achen and Bartels 2016:231). Thus, people vote their group memberships and identities, which "are emotional attachments that transcend thinking" (2016:228). If so (and I agree), the ethnoracial composition of electorates discussed below is a major factor in constituting "the state."

As mentioned, the increase in size and complexity of societies over the millennia led to the "political economy" (Johnson and Earle 2000), the separation of governance structures from the general populace. Thus, recent popular participation in governing states of many millions of people is something new in world history. The growth of the state to increase regulation and provide new public goods has been hugely expensive, yet states have accomplished it. How?

Developed Economies

The economies on which these state structures rely and which they regulate are also new. Sometime after 1950, industrialization was far enough along that these societies could be called "developed." In this section, I will review the primary, secondary, and tertiary sectors of these economies (as diagrammed in Fig. 8.3), analyze consumption practices, and discuss macroeconomic issues, or the entire economies of states.

The Primary Sector

Industries that extract resources from the land or sea include agriculture, mining, timber, and fishing. These industries are now heavily industrialized, using massive machines such as cranes, tractors, combines, and large trucks, and the associated infrastructure. Consequently, they employ only a tiny percentage of the population in developed states.

Industry relies heavily on metals and petroleum, so mining is strategically important. It is also environmentally destructive, risky, nonrenewable, and obviously nonmovable, so states always get involved (Halland et al. 2015). In the 1970s, the USSR, the USA, South Africa, and Canada together produced over half the value of global mining output (Mikesell and Whitney 1987:6). Coal was central to European and North American industry in the nineteenth century, while petroleum drilling started in North America and Russia. Petroleum output expanded so much in the twentieth century that it replaced coal as the most important (Gregory 2001; Smil 2017).

Agriculture is more flexible but just as important. Another agricultural revolution in the twentieth century, "motomechanization," which replaced animals with complex, energy-intensive machines, led to intensive agriculture in developed countries. Bigger farms constantly outcompeted smaller farmers, most of whom sold out, so average farm size increased across the developed world. A massive increase in fertilizer use and a supporting dense network of research universities also increased output. Transportation improvements and commercialization helped a division of labor develop, so that each farm produced only a few products, either food crops, dairy, animals (ranching), or industrial crops. These changes together made possible productivity improvements of 100 times over previous systems (Mazoyer and Roudart 2006:Chap. 10).

The Secondary Sector

All Western European, North American, and East Asian states continued industrializing in the twentieth century. For a while, manufacturing was an economic mainstay. Technical process innovations, such as the assembly line for automobiles, greatly improved industrial productivity. Industries also developed new products, especially those for households such as "refrigerators, cars, and radios" (Chirot 2012:92), plus films with sound, television, and household appliances (e.g., washers and dryers) (Amatori and Colli 2020). In the 1920s, North American and Japanese economies grew quickly, while European economies recovered from World War I more slowly. The 1930s Depression reversed this boom, and many states intervened in their economies. World War II saw even faster economic growth: "Overall, almost everywhere (with the predictable exceptions of Germany and Japan) per capita GDP … recorded in 1950 was greater than in 1938, and the U.S. registered the greatest increase" (Amatori and Colli 2020:201). In *les trentes glorieuses* (1945–1975), East Asian, North American, and Western European states became "developed," or industrialized, commercialized, and consumerist.

After the 1970s, deindustrialization set in. Neoliberal corporate movement of industrial facilities and jobs from developed to developing economies through containerization and improved communications (Pomeranz and Topik 2018) aided economic upgrading in developing states. However, it turned prosperous areas in North America and Western Europe into "Rust Belts" with lowered wages and fewer jobs. Manufacturing has generally retained the same share of GDP, about 25% (Maddison 2006), across most developed states, but manufacturing employment kept falling for some decades.

The Tertiary Sector

As employment in services correspondingly grew, some argued that such societies are becoming "postindustrial" (Bell 1976). Originally, industrialization and commercializing markets reciprocally influenced each other: Factories needed raw materials, they produced more goods to sell, and merchants distributed those goods widely. Furthermore, raw materials producers, manufacturers, and merchants borrowed capital; employed wage workers, who spent their money in markets for household goods; and relied on supporting services such as transportation, accounting, and law. By the year 2000, developed economies were nearly fully commercialized, a new situation in world history. However, all developed economies are primarily service-oriented, and economists debate this shift. One explanation is the Baumol "cost-disease" hypothesis: Over time, productivity rises in industry (through improved technology) but not in services, so more employees are needed for increased output; this makes them more costly (Schettkat and Yocarini 2003). In this view, outsourcing did not cause the shift.

In any case, moving to a service economy means that formal product distribution now constitutes only a small share of all services. Over time, new kinds of retail stores appeared. In addition to the department stores and chain stores of the 1800s, grocers developed "supermarkets" in the 1930s; large "malls" put retail shops indoors. In the late 1900s, "big box" retail chains such as Walmart or Carrefour selling nearly everything a household needed also appeared, supplied by hundreds of global value chains (Hamilton et al. 2011). Lately, internet sales have competed with physical stores and malls. Originally meaning "bringing to market," marketing expanded and elaborated over the twentieth century. It has become a crucial business service, devoted mostly to persuading consumers to buy. It continued to include distribution (wholesale and retail) but also developed sophisticated market research and many promotional techniques, with varied national traditions (Berghoff et al. 2012).

Finance also became central. Previously, businesses financed expansions from retained profits; now they expand mostly by borrowing, so financiers are increasingly influential (Dicken 2015). Concurrently, all developed states left the gold standard and managed the money supply through central banks, thus heavily influencing private banking and finance. Anglo-American economies intensified regulation in the 1930s, deregulated after 1980, and reregulated somewhat after 2008. European and East Asian financial systems followed only somewhat similar paths. All made possible business and consumer financing, which supported economic stability and continuing economic growth (Davies 2016).

Business Forms

Formal, massive organizations structured all three economic sectors. By the late twentieth century, nearly all businesses organized as limited liability corporations, taking different forms. First in the USA, then in other states, companies could own other companies. Massive holding companies became common over the twentieth century (Prechel 2000). European forms differed from US versions. For instance, larger US firms often finance themselves through stock markets, consequently with shorter time horizons; German firms tend to use banks more, which often place a representative on the board of directors for a longer-term approach (Hall and Soskice

2001). Japanese business groups often have dozens of firms working together, while South Korean corporate groups are patrimonially organized (Orrù et al. 1997).

Workers' organizations also changed. Nineteenth-century unionizing efforts bore fruit in the twentieth century. European unionization was established by 1914 and expanded greatly afterward. The number of US trade unions grew from the 1930s to the 1970s, then declining precipitously from neoliberal resistance and outsourcing. East Asian unionization appeared after World War II in Japan, and only lately in South Korea. It is more company-based than industry-based (Jensen 2006).

Consumerism

Modern consumption is different from consumption throughout history. Vastly increased production, sale in markets, and higher wages made possible vastly increased consumption. While consumerism—purchasing and using objects beyond subsistence needs—had always existed in states, it has taken over developed economies in the twentieth century (Stearns 2006). Subsistence consumption—consuming what one has produced oneself—has nearly disappeared, replaced by wages and market purchases. Only a few still hunt animals, grow vegetables in gardens, and make their own clothes. Even people who make their food "from scratch" buy the ingredients in stores (Stearns 2006). In the last 200 years, food became an ever-decreasing percentage of family budgets, a trend so consistent that economists call it Engel's Law (Anker 2011). People increasingly spend money to buy products that they do not need but want: fashionable clothes, much furniture, special appliances, and leisure products and services.

The range of consumer goods greatly increased over the century. After 1920, women started using cosmetics, and movie watching became increasingly common. As radio stations developed, radios spread through households. Cigarette smoking increased among both sexes. After 1945, consumer practices changed even more. Working-class people fully participated in consumerism, as Europe started to match US automobility. Televisions and their supports (cable networks, VHS tapes, and DVDs) became ubiquitous, as did music technologies (vinyl records, cassette tapes, and CDs); internet streaming began to replace them all after 2010. Thus, artistic and entertainment choices have proliferated, disrupting media strategies (Napoli 2011), though increasing economic inequality (see below) lowered access to some (Stearns 2006).

The Macroeconomy

Most social scientists take states for granted, so they treat activity within all these markets within any state's borders as "the economy" of that state, always calculated in state monies. Following this assumption, the two (extremely abstract) diagrams

(Figs. 9.1 and 9.2) below provide overviews of capitalist economies, tracking the flow of *money*, not of goods, services, or labor.[1] The two main flows in the macroeconomy (the two bolded arrows in the central X) are wages and salaries from businesses to households, and money for products and services from households to businesses. The other flows consist of loans to and repayment from businesses, households, and government at all levels, and taxes to and expenditures from government.

Two types of services are worth noting. First, the increase in government spending discussed above is part of the trend of tertiarization. (While economists usually put the government outside their diagrams, states are also producers and consumers of goods and services.) It has only slowed, not stopped, in the neoliberal era. Second, as mentioned, finance constitutes the core of the capitalist economy, either for business expansion or major consumer purchases (e.g., housing and automobiles). By expanding on consumption to Fig. 8.3 and tracking money, not goods, we get Fig. 9.1, which also separates finance into its various functions.

The "balance of payments" measures whether the many monetary flows add up to a net income or a net outflow (Cohn 2016:128). Figure 9.2 splits up the traditional economists' categories, such as the current account (merchandise trade, services,

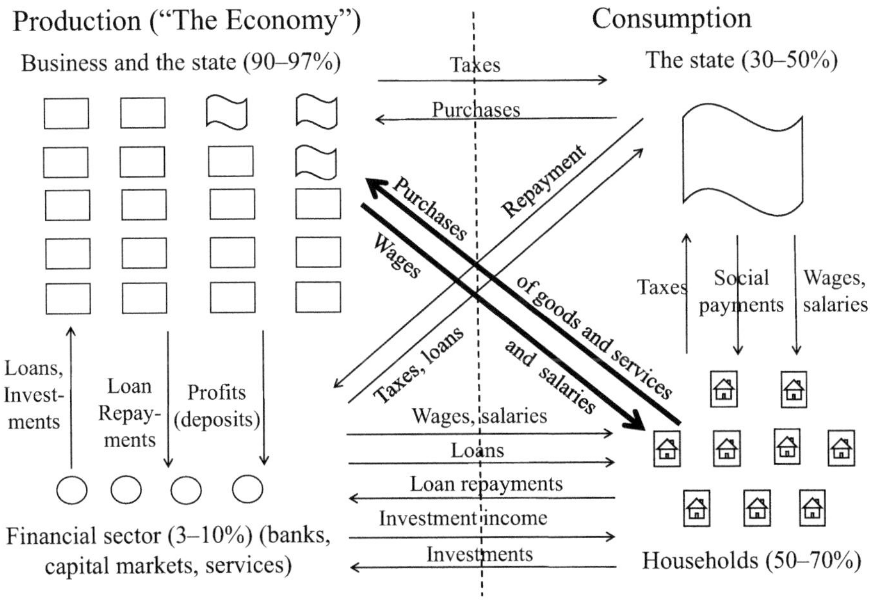

Fig. 9.1 Monetary flows in a market economy

[1] Figure 9.1 combines my adaptation of Dicken's (2015) value chain model (Figure 8.3) with a free interpretation of "circular flow" models common in macroeconomics, based on two of the three common methods for calculating GDP. Figure 9.2 separates out foreign from domestic flows, which circular flow models usually combine.

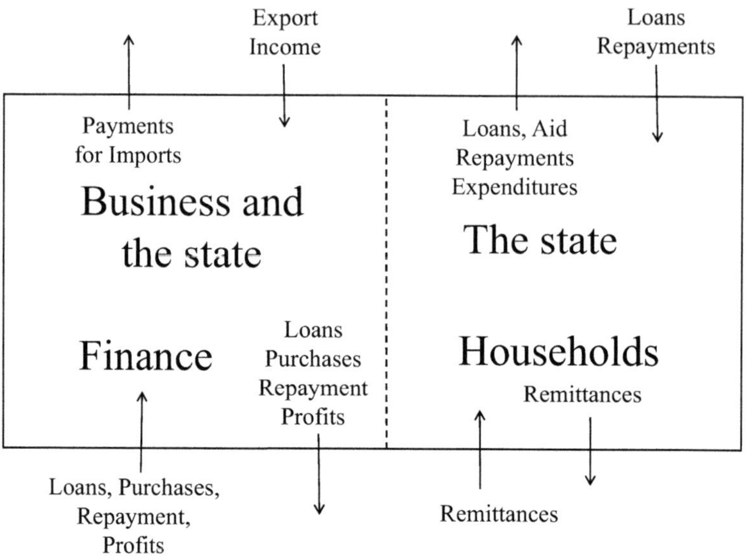

Fig. 9.2 Market economy embedded in the global economy

interest and dividend payments, and remittances), the financial account (portfolio and foreign direct investment), the capital account (of miscellaneous financial transactions), and the reserve account.

These reconfigured categories show that the balance of payments is complex, and economists debate the importance of credits and deficits between the categories. For instance, does income count the way investment does, since investments need repayment? Furthermore, state integration with the global economy varies. For instance, in 2017, foreign trade generated only 26% of US GDP and 31% of Japanese GDP, while many European states generate well over half their GDP from trade (World Bank Open Data 2018). The balance of payments is correspondingly more important for these states.

States cannot fully track all monetary movements. The informal economy consists of all "economic activities and income earned that circumvent government regulation, taxation or observation" (Medina and Schneider 2018:5). Some scholars only include in the informal economy legal economic activity hidden from the state to avoid taxes or regulation (Schneider et al. 2010); others include illegal activity (Medina and Schneider 2018:6, note 4). Concealment obviously makes estimating the informal economy difficult, but generally the wealthier the society, the lower the share of informal activity in the economy (Medina and Schneider 2018).

The economic growth rates of developed states over the twentieth century varied greatly (Table 9.1).

Since growth rates are compound, these economies were many times wealthier by 2001. Post-1973 growth rates seem slow only compared to the previous period, but advocates of neoliberalism seized on this "stagnation" to argue that strong states

Table 9.1 Economic growth rates (percent per year), developed states, 1870–2001

Period	Western Europe	European offshoots	Japan
1870–1913	2.11	3.92	2.44
1913–1950	1.19	2.83	2.21
1950–1973	4.79	4.03	9.29
1973–2001	2.21	2.95	2.71

Source: Maddison (2006:640)

Table 9.2 Structure of employment, 1820–1998 (percentage of total employment)

		United Kingdom	Netherlands	United States
1820	Agriculture	37	43	70
	Industry	33	26	15
	Services	30	31	15
1890	Agriculture	16	36	38
	Industry	43	32	24
	Services	41	32	38
1998	Agriculture	2	3	3
	Industry	26	22	23
	Services	72	75	74

Source: Maddison (2006:97)

were either worn out or never really worked. Economic growth improved after 2001; economic shrinkage occurred for a couple of years because of the 2007 financial collapse; and growth resumed until 2015 (World Bank Open Data 2018). The 2007–2009 Great Recession seemed to delegitimate neoliberal ideas, but there is no accepted alternative yet (Springer et al. 2016).

A sectoral analysis by employment of three states also helps us understand what it is to be "developed." All developed states have roughly these shares (Table 9.2).

The trend is clear. The primary sector[2] employs fewer people over time; only 2–3% of the population feeds the rest, most of whom are in services. The shares of GDP of agriculture and manufacturing in nearly all developed states have also shrunk: agriculture, from 5–6% around 1970 to around 2–3%, and manufacturing, from 30–35% in the 1990s to around 25% lately (Sepp et al. 2017).

Ongoing economic growth in the industrial capitalist economies of Western Europe, North America, and East Asia over the twentieth century confirmed these states as, collectively, the center of the global economy and a model for all other states.

[2] Scholars often include mining in industry. However, it is actually primary and is a very small part of it (Sepp et al. 2017:6), so I am using agriculture as an imperfect proxy for the primary sector.

Developed States and the Environment

Developed states are also a major source of environmental problems. One by-product of "development," full industrial capitalism, is high resource use and high waste output. Developed states use most of the resources, and produce most of the pollution, on the planet today. Raw materials extraction, transportation, and disposal exist on global and local scales.

Resource extraction expanded greatly with industrialization. Local mining sites cause tremendous problems, including land instability, pollution of water tables from metals, and atmospheric pollution. Firms sometimes decommission empty mines improperly (Simmons 2008). Petroleum production is a form of mining, one that still provides most energy for these societies (Simmons 2008:Chap. 5) and often causes local pollution problems. Industrial capitalism has also changed agriculture. After 1900, ammonia-based fertilizers replaced guano (Penna 2015). Agriculture in developed states requires large amounts of artificial fertilizer, pesticides, and most of all, energy. Tractors plow the land, combines harvest crops, and trucks ship them to market (Simmons 2008).

Environmental problems also appear in postconsumer disposal. Developed states produce massive quantities of waste—much more than poor states—and the USA produces more waste per capita than any other society. Most is brought to landfills (Moore 2007), which are increasingly distant from cities generating garbage; this travel generates more pollution (Palmer 2011). US cities have increasingly instituted curbside recycling, but compliance varies by household and by region (Moore 2007). Developed states in Europe tend to recycle much more than less-developed ones (European Environment Agency 2018: Fig. 2). Likewise, secondhand shops, flea markets, and websites for used goods have increased in number and sales volume in developed states so much that they are integral parts of their economies (Le Zotte 2017).

Environmental action started in the late nineteenth century as a "conservation" movement. Its first success was the formation of the US National Park System, preserving lands from resource extraction. The Anti-Pollution League formed in US cities in the 1920s and 1930s. Deadly smog outbreaks in the 1950s in the USA and the UK led to the first international air pollution conference in 1955 (Moni 2007). Books (e.g., Carson 1962), Congressional hearings, and rivers and lakes actually *catching fire* in the late 1960s led to an environmental movement. The US government established the Environmental Protection Agency (1970); now all developed states have environmental agencies. Many environmental international nongovernmental organizations (INGOs) were founded before the 1960s, and their number greatly increased afterward (Moni 2007; Laberge 2007).

Industrialized extraction of resources, including mining, logging, and agriculture, created major environmental problems in developed states. Carbon emissions and postconsumer disposal of waste created others. Secondhand sales, environmental movements, and recycling help mitigate these problems, but developed states still cause great stress on the environment.

Economic Inequality in Developed States

In agrarian states, elites derived their wealth mostly from landownership; most people engaged in subsistence production and consumption on that land or their own; and a few worked for wages. The nineteenth-century shift to elites owning businesses continued in the twentieth century; as nearly everyone works for wages, measuring economic inequality is much easier. Still, economic inequality in developed states varies by time and place.

As discussed, social scientists measure economic inequality within states with the national income method or by calculating Gini coefficients. Unfortunately, Gini data is historically shallow; for instance, World Bank data is available from 1991 for the USA and Germany, and only for the rest of North America, Western Europe, and East Asia consistently from 2003. Gini coefficients of developed states changed little in that period. The US measure, 0.415 in 2019, rose only a few points in the last 30 years; German economic inequality also remained stable. Since 2003, Gini coefficients for Norway, Sweden, and Denmark have been just under 0.30 and around 0.35 for Italy, the UK, France, Japan, and Canada.

National income data tells a more interesting story. Charting national income over the twentieth century shows a U-shaped curve: high inequality early in the century, lowered inequality mid-century, and growing inequality since 1980 (Piketty 2014). In the nineteenth century, Europe was more unequal; since the 1980s, the USA has been more so. A more detailed analysis demonstrates national variations. The top French centile (1%) had over 20% of national income in 1910, dropping to under 10% by 1945 (Piketty 2014:272–273). The top decile (10%) had 40–45% of national income from 1910 to the mid-1930s, falling to under 30% by the mid-1940s. The top US centile's income share was around 14–20% from 1910 to 1940, dropping to 11% in the 1940s and to about 8% by 1980; the top decile had around 45% of income from 1910 through 1940, dropping to 30–35% (Piketty 2014:291). Wealth ownership followed these income patterns. In 1914, the top centile owned 45%, 55%, and 66% of total national wealth in North America, France, and UK, respectively. By 1980, the corresponding figures were around 22%, 18%, and 19%, respectively (Piketty 2014:316; Alvaredo et al. 2018:16). Since 1980, inequality has increased in all developed states but at different rates. From 1980 to 2010, the top US centile's income share rose from just below 10% to more than 20% (Piketty 2014:292); the top centiles' income share in Canada, the UK, and Australia also rose, but less. The top US decile takes almost 50% of national income (Piketty 2014:291). In contrast, the shares of total national income of the top centile of France, Germany, Sweden, and Japan have risen only slightly since 1990 (Piketty 2014:317).

Quantitatively, wealth inequality also changed in twentieth-century developed states. From 1910 to 2010, agricultural land's share of total US national capital dropped to about 1–2%, similar to sectoral shares of agricultural labor. Percentages are slightly higher in Continental Europe, but agriculture's share of capital is still well under 10% in all developed states (Piketty 2014:116–117, 141, and 149). Now, the wealthy mostly own businesses.

The middle classes in all developed states went from owning about 5% of national wealth around 1910 to owning 25% of USA and 35% of European national wealth today. Middle-class prosperity is a new situation in world history (Piketty 2014), but it is fragile; the US middle class has begun to lose position. A long-term perspective suggests a different interpretation. In *les trentes glorieuses* (1945–1973), economic growth rates were unusually high historically (Maddison 2006:640; Piketty 2014:96–98). The period since the 1970s has seemed unjust by comparison, but instead we have simply returned to a historically long-standing situation: lower growth and very high inequality (Piketty 2014). Yet another interpretation exists. Economists usually argue that economic forces within firms set the distribution of incomes, and over much of the twentieth century, inequality reflected productivity (Acemoglu 2012). Other social scientists argue that political forces shape economic inequality. Trade unions became more powerful in Europe and North America in the middle of the twentieth century, but after 1980, "neoliberal administrations took strong anti-union positions" in the USA and UK (Centeno and Cohen 2012), and in the last 40 years, wages have been separated from productive skills (Acemoglu 2012). Union membership as a percentage of the workforce has remained high in Continental Europe, but it has fallen in Anglo-American countries, so many scholars argue that anti-union neoliberal policies have helped increase economic inequality (Mullings 2010; Centeno and Cohen 2012; Domhoff 2013) (Table 9.3).

Social insurance policies also vary by state. Table 9.3 shows the Gini coefficient for each country for 2013 twice: once before state insurance programs which tax parts of the population and give the proceeds to other parts of the population, and once after. Before social insurance programs operate, economic inequality is high throughout developed states, and there seems to be no pattern. However, after governments redistribute income, all Anglo-American states—the UK and its global offshoots—are in the top half of the list, and the USA has the highest Gini coefficient.

Thus, differences in social insurance programs show that inequality is at least somewhat political and cultural, but economists seem loath to admit cultural influences on economic phenomena. For instance, one economist unusually sympathetic to other disciplines argues that the difference between states' inequality profiles "has little to do a priori with eternal cultural differences: it seems to be explained mainly by differences in demographic structure and population growth" (Piketty 2014:428). Yet he also argues that US inequality results partly from boards of directors being willing to pay CEOs ever higher salaries (332), and there is a higher willingness in many European states to tax capital (372–376). From what does this "willingness," in either case, come from if not from widespread assumptions?

Also, one need not believe that culture is "eternal" to recognize that people transmit their assumptions through many generations. Anglo-American states have long post-Enlightenment records of suspicion of state power, the USA most of all. Additionally, white reluctance in the USA to share welfare state benefits with racial minorities (Bommes and Geddes 2000; Alesina et al. 2001) and the "catching-up"

Table 9.3 Gini coefficients, selected developed states, 2013: before and after taxes and transfers

State	Before	State	After
Ireland	0.57	United States	0.39
Greece	0.57	Israel	0.36
Portugal	0.56	United Kingdom	0.36
Spain	0.53	Australia	0.34
United Kingdom	0.53	Greece	0.34
Germany	0.51	Portugal	0.34
United States	0.51	Japan	0.33
Austria	0.50	New Zealand	0.33
France	0.50	Canada	0.32
Belgium	0.49	Ireland	0.31
Finland	0.49	South Korea	0.30
Japan	0.49	Spain	0.30
Australia	0.48	France	0.29
Luxembourg	0.48	Germany	0.29
Israel	0.46	Austria	0.28
New Zealand	0.46	Luxembourg	0.28
Canada	0.44	Netherlands	0.28
Netherlands	0.44	Sweden	0.28
Sweden	0.44	Belgium	0.27
Norway	0.41	Finland	0.26
Iceland	0.39	Norway	0.25
South Korea	0.34	Iceland	0.24

Source: *Our World in Data* (https://ourworldindata.org/grapher/inequality-of-incomes-before-and-after-taxes-and-transfers-scatter?tab=table), retrieved December 21, 2022

of US capital in comparison to Europe helps explain a smaller US welfare state, while European states' greater ethnoracial homogeneity and near-constant war making since the late 1400s help explain larger European welfare states (Bommes and Geddes 2000; Alesina et al. 2001). Since the 1970s, immigration from outside Europe has made European welfare programs politically contentious (Svallfors 2012).

Still, states have reasons to reduce economic inequality. Some argue that a range of outcomes in mental and physical health, drug abuse, education, imprisonment, obesity, and trust are better for the entire population, including the wealthy, in more equal societies than in more unequal societies (Wilkinson and Pickett 2009). Others argue that inequality slows economic growth (Berg and Ostry 2011; Stiglitz 2012). Some argue that in the USA, economic elites and business interests heavily influence political processes, while average citizens have "little or no independent influence" on them (Domhoff 2013; Gilens and Page 2014).

Despite these inequalities, life in developed states is dramatically better than life for nearly everyone in the past. While the relative poverty outlined above has anti-social effects, absolute poverty is minimal; malnutrition and food insecurity affect only a small percentage of these populations.[3] Diseases of long life (e.g., heart disease, cancer) are much more prevalent.

Economic inequality in developed states was high at the beginning of the twentieth century, reduced mid-century, and increased after 1980. Political efforts seem to have had an impact on both movements. Differences between states seem partly influenced by long-standing cultural assumptions and ethnoracial composition. Absolute poverty is nearly gone.

Demography in Developed States

Demographic patterns have changed in three dramatic ways within the last century. First, global migration patterns combined with economic inequality to create new ethnoracial stresses. Second, developed states are primarily urban societies, something completely new in world history. Finally, these states have "completed" the demographic transition but are moving to a new phase that nobody yet fully understands.

Ethnic and racial histories vary greatly among developed states. The groups first analyzed in this book as bands, villages, and polygroups, then re-termed "kin-ordered peoples," are often known today as "Indigenous peoples." The United Nations (UN) Commission on Human Rights defines an Indigenous people as a group with (1) first settlement of a territory; (2) voluntary perpetuation of cultural distinctiveness; (3) self-identification and recognition by others as a group; and (4) historical experience of subjugation, dispossession, or discrimination. Thousands of groups globally, totaling more or less 400 million people, fit this definition (Stavenhagen 2013). They generally remain more oriented to local interaction and are more gender-equal, more kin-oriented, and more spiritually oriented about land than people in industrial capitalist states are.

It is primarily the "offshoots" of the UK that have Indigenous populations, though the Ainu were village-based hunter-gatherers in Japan. Hunter-gatherers in clans existed throughout Australia. When Europeans arrived, what are now New Zealand and Hawaii were populated by villagers and polygroups, respectively, while the full range of kin-ordered societies, from simple foraging bands to complex chiefdoms, existed in what is now Canada and the USA. The majority populations of these states geographically displaced and socioeconomically marginalized their Indigenous populations, often trying to assimilate them forcibly. In Japan, the USA, Australia, and Canada, the Indigenous population is a small proportion of the current national population, but New Zealand estimates the Maori population at 17.4% of national population (stats.govt.nz).

---

[3] This website (https://www.worldlifeexpectancy.com/cause-of-death/malnutrition/by-country/) seems legitimate, claiming to rely on World Health Organization data for 2020, but I am still cautious. Unfortunately, Birn et al. (2017) devote only a single page to malnutrition and do not have comparative data.

Immigration into developed states has produced ethnoracial mosaics. The forced immigration of the slave trade brought Africans to North America. One Canadian province, Quebec, remains primarily French, the rest English. Nineteenth-century immigration brought multiple waves of Europeans and Asians to the USA and Canada; this almost completely ended in 1924 in the USA and in the Great Depression in Australia and Canada (Segal et al. 2010). Before World War I, most migration in Europe was between neighboring countries. After massive casualties led France and Germany to encourage immigration (Segal et al. 2010), immigration dropped during the Great Depression. After 1960, global migration increased again, bringing people from developing states to developed ones. Latin American migrants end up in North America, while African and Asian migrants usually go to Europe (Segal et al. 2010). Exceptions include South Korea and Japan, which are restrictive and ethnically homogeneous. Israel is almost 96% Jewish,[4] and Portugal, Italy, Poland, Greece, and the Netherlands have at least 90% of their citizens from one cultural group (Fearon 2003).[5] The UK, France, Germany, and the Nordic countries have more mixed populations.

Such ethnoracial diversity creates two problems in developed states, fitting theories of groupish politics (Achen and Bartels 2016). First, host populations resent any required adjustments because they see it as *their* country—ignoring the Indigenous populations' claims and, in Anglo-American offshoots, their own immigration histories; as in agrarian states (Turchin 2003), cultural assimilation requires decades. Second, underlying ethnoracial resentment leads to resentment of welfare state spending (Bommes and Geddes 2000; Alesina et al. 2001). These difficulties can create a vicious cycle: discrimination leads to segregation and lack of assimilation; this in turn leads to minority underemployment, poverty, and continuance on social insurance programs, which continues resentment, and so on. Populations often neither know nor care how state policies brought immigrants into host states in the first place.

A comparison is instructive. In 1924, the USA nearly stopped the flow of immigrants from Europe for the next 40 years, during which European groups partly assimilated. In 1965, immigration law loosened. Since then, migration increased dramatically, mostly from the Americas. The legal status of unauthorized immigrants (more overstayed visas than unauthorized entry) is a contentious political issue. The massive flow of people has altered the US demographic balance with over 18% of the population identifying as Latino. High immigration from East and South Asia changed the ethnoracial composition of the western USA (Segal 2010; US Census 2016). Prejudice and discrimination against people of color in housing and employment are well-documented. Underfunding of education and unequal treatment in the criminal justice system also continue (Lewis et al. 2004). Unequal, harsh treatment of African Americans is long-standing, but an equally long-standing

[4] This depends on Fearon's (2003) definition; Palestinians would strenuously disagree.

[5] Given continued migration since 2003, some of these societies are even more diverse than they were then.

tradition involves the descendants of immigrants discriminating against later waves of immigrants (Takaki 2008).

The French experience has been different. (This account is from Price 2005 and Price 2014.) Until the 1920s, immigration into France was a small flow of people, mostly from neighboring states. After high mortality in World War I, France loosened immigration laws, and immigration increased from countries farther away, such as Poland. After World War II, the state encouraged more immigration, this time from North African colonies, which became independent in the 1960s. Since then, French immigration policy has cycled through encouragement and discouragement based on business cycles, creating waves of immigration. First men came, then whole families; now France has a permanent population of North Africans. People from former French colonies in sub-Saharan Africa and Southeast Asia have also immigrated. The French political elite has long regarded counting immigrants as detrimental to cultural unity, so French demographers are forbidden to count them. The demographers call this "statistical invisibility," forcing them to estimate immigrant populations. These seem to be more than 10% of the French population, half from outside Europe. Interestingly, while sub-Saharan Africans experience some discrimination, surveys of white French people consistently show *much* greater animosity toward North Africans than toward sub-Saharan Africans. Islam may be a factor, but the bitterly fought Algerian war for independence still shapes French politics. Thus, each state has its own historically unique pattern, but the trend is clear: Since the 1960s, immigration of millions of people to developed states has generated major ethnoracial tensions.

Ethnoracial politics have overlapped with regional patterns of urbanization and suburbanization. The European urban pattern of wealthier residents in the city center and poorer ones outside continues today, but now suburbs usually contain ethnoracial minorities. The suburbs of Paris, Lyon, and other French cities have Arab French and African French populations. Offshoring leading to deindustrialization intensified their poverty. North American suburbanization also continued in the twentieth century, accelerating after World War II partly as a response to African American migration from the South to Northern cities, generating "white flight" to the suburbs, and the much stronger automobile culture in North America that created cities much larger (geographically) than in Europe. Offshoring and deindustrialization often led African Americans into ghettoes (Monkkonen 1988; Abbott 2013; Lees and Lees 2013).

The demographic transition has created a new situation in world history. The dramatic decrease in mortality and the consequent increase in average lifespan led to new age patterns. The old profile of many children, some young adults, fewer middle-aged adults, and very few elderly people is, in these states, a thing of the past. Now, mortality only seriously affects populations in these states after age 65; many people survive past their 70s. Thus, in many developed states there are as many middle-aged adults as there are young adults, and as many young adults as there are children. However, two trends modify this overall "modern" age pattern. First, there are generational "booms" and "busts" in births, so some older generations (age brackets) are larger than later ones. The second trend reveals a new and much more important problem with the demographic transition. As discussed (Chap. 8), fertility rates have fallen below the replacement rate of 2.11 lifetime births per woman in all developed states and many developing ones, so mortality is actually slightly higher than fertility. Long-term projections are that populations in these societies will decrease (GBD 2024). Indeed, Japan's population has already begun to shrink (Japan Times 2016). Thus, each generation is smaller than the last one, so in the overall population there are fewer children than there are young adults, and fewer young adults than middle-aged ones.

As population shrinkage eventually affects all developed states, it will create two political problems. First, how will developed states fund their popular and entrenched retirement programs? Increasingly, younger generations numerically smaller than older ones will have to pay for older generations' retirements. Ultimately, this will be "unsustainable" (Livi-Bacci 2017:145). Second, some scholars have argued that growing populations are one foundation for economic growth. As the working population decreases, economic growth will also slow—falling numbers reduce demand and reduce the number of workers who might develop innovations (Jones and Romer 2010).

This returns us to immigration. Only two ways exist to keep populations growing: Native-born women have more children, or countries let in more immigrants. Developed states are answering the first problem by instituting paid parental leave; only the USA has no paid national parental leave insurance program (Livingston 2019). However, these are having middling success at best. The second option, increasing immigration, increases population immediately and in the longer term since immigrant women tend to have higher birth rates than native-born women. However, this revives the explosive problem of ethnoracial resentment.

Developed states are demographically new in world history. Global South-North migration has changed their ethnoracial profiles in the last decades. Ethnoracial resentment among populations now influences economic inequality. Ethnoracial patterns are often spatially structured in cities that just keep growing. Drastically lower mortality and the resulting longer lifespan have become normal, but even lower fertility will create new political problems.

Ideas in Developed States

The idea-life of developed states is confusingly complex and varied. It includes group cultures influenced by racism and nationalism; the development of mass media, including the public sphere, entertainment, and social media; the continuing development of educational and scientific systems; technological proliferation; and the continuation of religion.

After four decades of neoliberal policies globally, most strongly in the Anglo-American world, socialism is experiencing a resurgence, especially in the USA (Ehrenfreund 2016). Some attribute this to a great increase in inequality in developed states in general and the global economic crisis of 2007–2009 in particular

(Solimano 2017). Socialist parties in Europe continue to be an electoral force, but they have had mixed experiences lately.

The course of nationalism has been different. Nationalist successes after World War I in Europe, and in Africa and Asia after World War II, were based on the desire to implement the equation, "cultural nation = political state." Some scholars distinguish between ethnic and civic nationalism, the former being based on preexisting cultural affinities, the latter on loyalty to a state, offering assimilation based on birth or on ethnic criteria (Calhoun 2005). Scholars also debate whether nationalism is a product of recent historical processes or rests on ancient, "primordial" group memberships. In practice, both ethnic and civic forms of nationalism influence all states. Every state now promotes nationalist beliefs and emotions through national history, public rituals, mass media, military service, and education (Green 1990; Calhoun 2005). Currently, the desire for "self-determination" (i.e., political independence) still fuels, actively or latently, European and North American separatist movements, such as those for Basques, Catholic Northern Irish, Scots, Catalonians, Quebecois, and Puerto Ricans (Kramer 2011; Norman 2006).

Interestingly, nationalism depends partly on the separation of ethnic groups, which requires the state to have clear boundaries. The gradually weakening zones of influence beyond agrarian state boundaries have given way to "modern political boundaries [that] are usually discrete, sharply and abruptly demarcating the end of effective jurisdiction" (Feuer 2016:51–52). States try to control passage through these borders strictly, with varying success; legal imports and exports, smuggling, and travel by persons make most borders somewhat porous (Feuer 2016). However, state control over these flows allows them to further their nationalist projects since the "sovereignty" within the international state system assumes that states can take laws and policies, educational systems, language, ceremonies, media reach, and flows of people up to that border, thus enculturating populations in a host of ways (Calhoun 2005).

Attendance in one of those enculturating institutions, mass education, grew after 1914. By 1910, 85.8% of children in developed states attended primary school. By 2010, 99.5% of children attended. In the same year, 93.4% of teens enrolled in secondary education (or "high school"), and 72.6% were in higher education (Barro and Lee 2015:16). Thus, all developed states now have very high literacy rates. Nevertheless, school systems vary. For instance, US public educational systems mostly prepare students in a similar general fashion for 12 years and then put them on different "tracks." German students, however, are tracked into different types of schools much earlier (Brint 2017).

Education also supports states' efforts to harness science and technology. After 1914, the links in developed states between education, science, technology, the state, and business intensified. Businesses increasingly applied scientific findings to technologies. World War II helped institutionalize the already existing relationship between science and the state in the USA and Germany. Afterward, all industrial capitalist states funded science, especially during the Cold War. Still, scientific research also aided the development and production of many new consumer products, thus supporting businesses. Eventually, states developed elaborate research and development (R&D) institutions: laboratories, professionalization, state funding, and encouragement of basic research for applied ends (Ede and Cormack 2012).

Economic growth is partly self-sustaining from science for two reasons. First, science attempts to learn *new* things about the world. Second, businesses and military establishments constantly search scientific findings for new ideas to turn into technologies since innovation reduces production and distribution costs and generates new products (Lipsey et al. 2005). Scholars call this constant application of scientific findings to technologies the "research pipeline." It is one support for modern economic growth (Trefil 2012) and has three parts: basic research, applied research, and R&D. Basic researchers examine "truths about nature [or society] without regard to possible future utility" (Trefil 2012:135). Applied research turns findings and concepts into prototypes. R&D turns prototypes into useful, sellable products or processes, such as televisions, household appliances, and public transit; the internet; modern medicines and contraceptives; and energy generation.

Innovators used the research pipeline to transform mass media over the twentieth century, with dramatic effects. Movies became popular by the mid-1910s, increasingly popular with sound by 1930. Radio and then television brought the immediacy of movies into people's homes. The internet has again transformed communication; literacy, mass media, and social media provide more information and entertainment to people in developed states than ever in world history, and they are all fueled by large capitalist businesses (Cullen 2002; Croteau et al. 2012). The relationship between media companies and the state varies by nation, such as heavier regulation in France and lighter regulation in the USA.

Agrarian religions are declining unevenly across the developed world. Scholars debate secularization in developed states contentiously because "the connection between faith in God and identifying with an organized faith … is far from simple" (Hout et al. 2013:4–5). Over the last century, religiosity measured by belief statements and church attendance has decreased from high to low in Europe but has increased in the USA. Fully half of US citizens say religion is important, while only 22% of Spanish, 21% of Germans, 17% of British, and 13% of French people agree (Pew Research 2011). However, younger US generations show lower religiosity on these measures (Fig. 9.3): Compared with their parents at the same age, US people who became adults after the year 2000 are less likely to be affiliated with a formal religion, attend religious services less often, and say that religion is important at lower rates (Pew Research 2010). Interestingly, the number of US citizens who believe in God or a higher power has increased in the last few years (Hout et al. 2013).

The vastly increased power of states means that law is a much more important form of morality, perhaps rivaling religion, in terms of how people behave. As previously discussed, much law is practical rather than moral (Pirie 2021). Still, many factors influence individual and group day-to-day moral behavior, including existing ethnoracial customs, religion and secularization, migration, and law—and all of these are influenced by group identities (Glenn 2010). In addition, as the economic security of people has increased, experimentation in value orientations has also increased (Inglehart 2018). Given these multiple influences, morality in developed states is often complex and contentious.

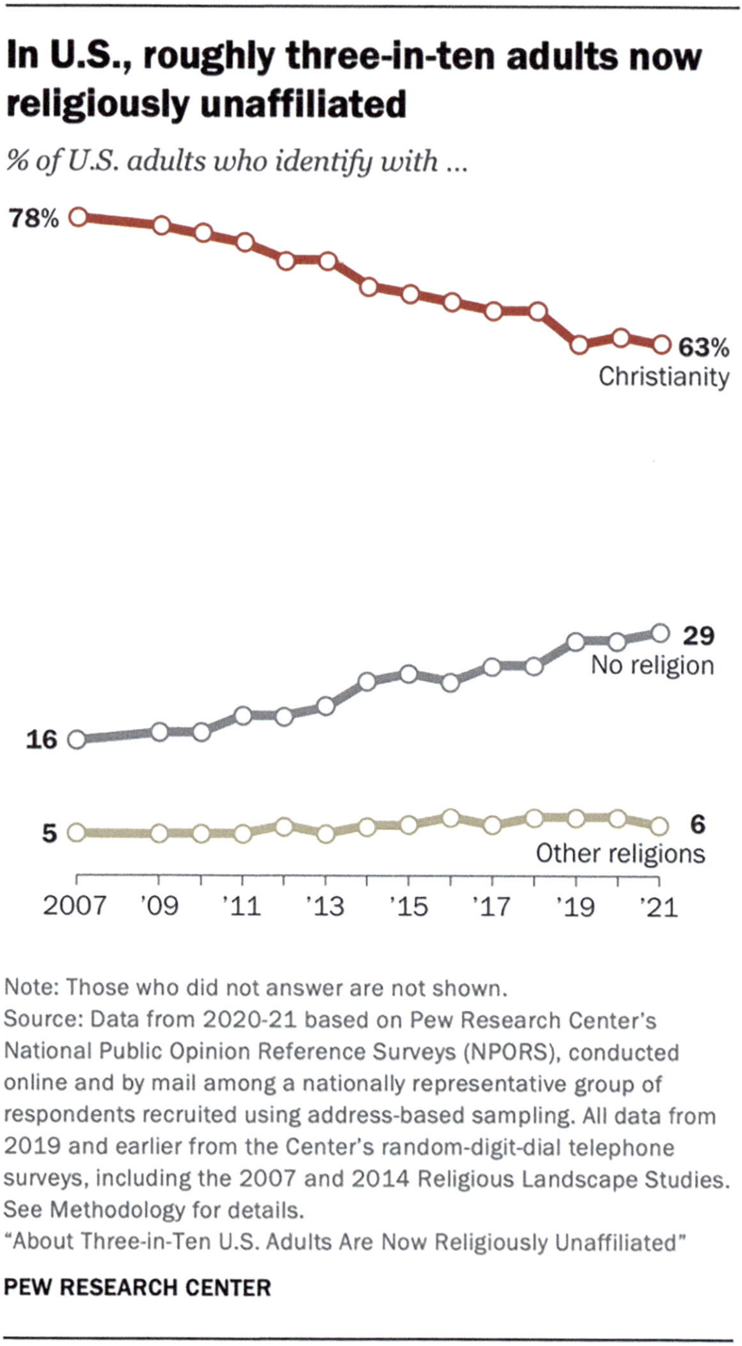

Fig. 9.3 Religious affiliations of adult Americans. Source: Smith (2021)

The complexity of idea-life of developed societies is much higher than it was in agrarian states. Forms of liberalism and socialism are only a few centuries old; mass media, the research pipeline, and mass education are over a century old. Mass and social media keep transforming, so they keep providing new experiences and complex influences.

Family, Gender, and the Sexual Revolution

Ideology and technology have also influenced gender relations, sex and marriage, and family structures. The consistent male domination of public life and institutions in nearly all states is reversing in developed states. Three changes have been important in increasing opportunities for women: the ideology of feminism, the industrialization of work, and technologies of contraception. These have also changed women's and men's roles in families.

In the early twentieth century, several changes in gender relations came together. After 1900, the ideology and practices of separate spheres became assumed in Europe and North America. Economic growth gradually increased incomes, so working-class women increasingly followed middle-class women in working only until marriage, then staying home, even without children (Anderson and Zinsser 2000; Nolan and Lenski 2011). However, middle-class women started attending college and usually worked until marriage. Eventually a new pattern developed among all classes, especially among educated, middle-class women: working before marriage; staying at home and raising a family; and returning to the workforce as children grew older, often in volunteer, part-time, and sex-segregated jobs (Nolan and Lenski 2011:289, chart, data for 1960). Most jobs were in the growing service sector of industrialized economies, especially in retail and office work (Anderson and Zinsser 2000:294). As female suffrage developed, other changes occurred. English law expanded the grounds for divorce. Scandinavian governments ended the family "headship" of the husband and the category of illegitimacy for children. US chaperonage of young people disappeared, and dating culture developed (Therborn 2004).

In the USA, World War II had an unexpected impact. Millions of American men joined the military, so women staffed factories and other "male" jobs, learning that they could do them and enjoying the autonomy and accomplishment of paid work. When men returned, employers mostly let women go, and there was a concerted effort in the late 1940s and 1950s to encourage women to be happy back in domestic roles with the consequent "baby boom." One historian argues that this sequence for US women of paid, "male" work, then enforced domesticity, was one of the roots of the 1960s feminist movement (Liftoff 2005).

Since the 1960s, developed states have experienced two major changes in gender relations: those in marriage and sexuality and those in political economy. Marriage rates and age at marriage varied over the twentieth century. In the middle third of the century, the highest proportion of the populace in history married; premarital births consequently declined; and age at marriage also dropped (Therborn 2004). This was the zenith of the "separate spheres" family. In the last third of the century, however, these patterns changed dramatically, and scholars debate why. Birth control has existed for centuries, used mostly (but not exclusively) within marriage. While prescription birth control pills partly changed fertility, unmarried women could not receive them until 1972. However, sexual norms started to change in the 1960s. One scholar argues that antibiotics in the 1950s led to a "sexual revolution" (Francis 2013). Sexual behavior outside marriage was dangerous before World War II, as syphilis killed thousands of people. The discovery of penicillin during the war meant that syphilis deaths fell to a low in 1957, and rates of premarital sexual intercourse started to rise. Additionally, an increase in unmarried cohabitation is a return to older patterns (Therborn 2004). It is often (but not always) a prelude to formal marriage, but it is much less normative in Southern Europe, so marriage rates and forms vary across Europe. In any case, now most people in the West, male and female, are sexually active before marriage, and the average age at marriage has risen by several years.

The sexual revolution coincided with a second "wave" of women's movements (Kent 2021). These movements, today called "feminist," varied by state. Religion influenced women's movements far more in Ireland, Italy, Spain, and other predominantly Catholic states than movements in largely Protestant and increasingly secular Scandinavia, which in turn had much stronger welfare programs, many of which already benefited women. Some (not all) French feminists emphasized fundamental differences between women and men, while US feminists emphasized similarities in primarily middle-class movements (Roth 2017a, b).

Second-wave feminism produced more legislation and court actions. Belgian and Dutch governments repealed laws requiring children to get their parents' permission to marry before age 25 (for men) and age 30 (for women). The German constitution declared women and men equal before the law, and the Constitutional Court enforced this provision. The Paris unrest of 1968 was partly resistance to patriarchal French laws, which the French state repealed. Finally, several US Supreme Court decisions supported gender equality, though the Equal Rights Amendment to the Constitution failed (Therborn 2004).

Feminists also worked to increase women's employment and incomes. Women have increasingly moved into paid employment and political positions in the last 50 years. In the USA, "the ratio of men to women in the labor force has … changed … from 70:30 in 1950 to nearly 50:50 in 2000" (Brownson et al. 2005:429). Women are increasingly employed in professional jobs and management, accomplishing this through legislation, lawsuits, and resisting male jobsite harassment. Still, women are managers and professionals at lower rates than men are, earning significantly less money despite graduating from graduate schools at higher rates than men in most professions. Only in STEM (science, technology, engineering, and math) professions do men still dominate graduate degrees (Nolan and Lenski 2011:273, 274, 291). Many jobs are still sex-segregated with "women's work" consistently paying less than jobs regarded as masculine. And women still occupy only a minority of state positions.

State policies and cultural traditions interact to shape women's work and household lives in varied ways (Lewis 2012). Scandinavian states offer generous paid leave after childbirth; the UK has shorter periods, while leave is unpaid in the USA. Most states offer fathers leave, yet only Swedish and Dutch men take them in large numbers. There is a strong tradition of part-time work for Northern European women, while Southern European women tend to work full-time or not at all. Northern European states tend to provide public day care, while Southern European families tend to provide it, and day care is mostly private in the USA. Leave policies and childcare programs are responses to total fertility rates below replacement (Lewis 2012).

Feminist activists also work toward eliminating the sexual double standard, reducing domestic violence, opening divorce laws, and aiding women's control of property, all of which are changing men's roles. As Fig. 9.4 shows, the average US father has doubled the amount of housework he does since the 1960s, while the average US mother has almost halved her housework duties. However, a gap remains, one even wider in Europe (Lewis 2012). As women are being educated

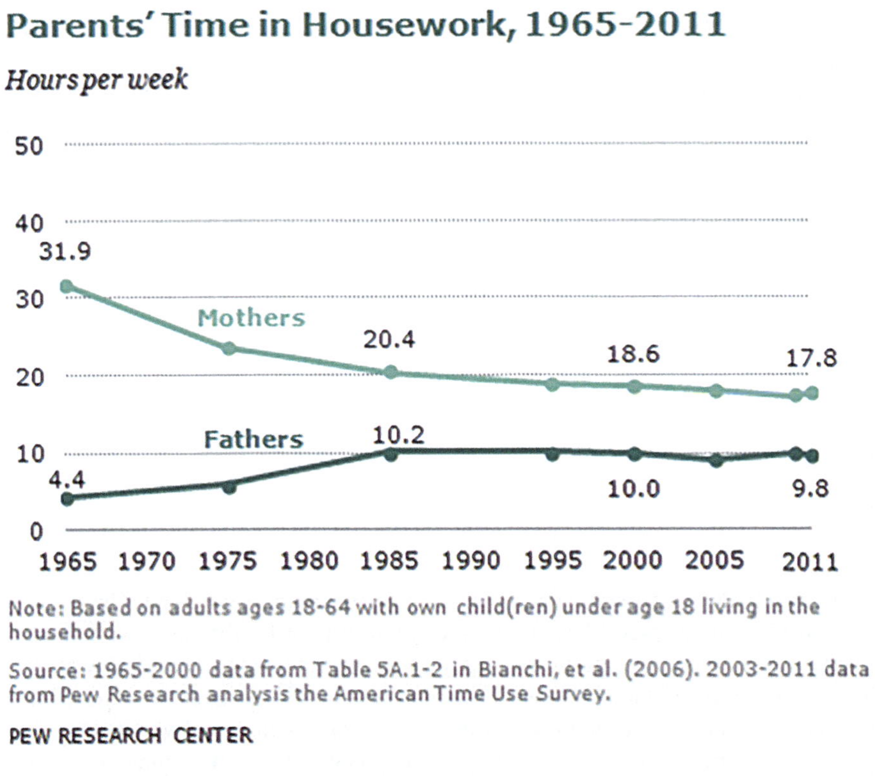

Fig. 9.4 US parents' time in housework, 1965–2011. Source: Pew Research (2013)

more than their proportion of the population, men are graduating less than their proportion in the population, but this has not yet translated into female economic dominance. Also, families have continued to lose functions. They are still economic units but are increasingly oriented toward consumption. For both sexes, production increasingly occurs outside the home. Mass education partly structures socialization from age six onward, while justice systems increasingly intervene in domestic quarrels. Scholars debate how influential parents are with teenagers, but as teens age, peers clearly increase in influence (Nolan and Lenski 2011).

Same-sex sexual behavior and gender diversity (or lesbian, gay, bisexual, and transgender [LGBT] people) also vary across developed states. A movement for research and rights originated in Germany in the 1920s, spreading to the rest of Europe and North America. It lost popularity in the 1930s, but revived in the 1940s and 1950s in North Atlantic states. LGBT social networks expanded—based in bars, clubs, and publications—also influenced by the growing feminist movement and changes in heterosexual behavior (Rayside 2020). Political movements based in multiple organizations became increasingly active in the 1970s and especially during the AIDS epidemic in the 1980s in Europe and North America (Herzog 2011; Rayside 2020), which marked the "turning point" in public acceptance (Rayside 2020:57). Since then, public opinion has gradually shifted in all North American and Western European states toward LGBT acceptance, though it is not complete (Rayside 2020).

Why did these changes occur? Is feminism possible only because of the decrease in fertility and women having fewer children to handle? Or has industrialization and the shift to a service economy made physical strength less necessary, making the intellectual equality between men and women obvious? Or have feminist movements been an independent cause of these changes? Did antibiotics, or contraception that women can control, change sexual norms? How important is sexuality in the state and the economy? Scholars will answer differently, but these changes will continue; how they will play out, and how long they will continue, remains to be seen.

Conclusion

I have argued that the third social transformation in human history resulted in developed states, a new type of society. While their economic foundation is industrial, commercialized capitalism, the transformation is much more involved. The tremendous increase in average lifespan and reduction in total fertility are revolutionary. The millennia-long increase in inequality between women and men is now reversing because of many factors. Mass literacy, access to a wide range of old religions, new entertainment opportunities and new(ish) idea-systems, and vast amounts of scientific information have created a completely new cultural environment in world history. Inequality now takes a monetary form rather than control of land, and the material standard of living is dramatically higher than ever. Political participation is greatly increased in developed states through elected representatives and officials. Since the subsistence economy is now an industrially commercialized economy, populations have unprecedented, if still minor, input into the political economy (Johnson and Earle 2000), which channels conflict (mostly) peacefully through political institutions. These states use up physical resources and transform them into pollution at a rapid rate, but they are beginning to change this.

The health and lifespan benefits of living in these societies have perhaps made them more functional than any previous society. Yet these societies take constant effort to hold together, especially through educational and legal systems; economic redistribution systems reduce misery and buy off resentments, but ethnoracial divisions still produce conflicts. The populations of these states comprise a small percentage of the global population, while most people live in the previous types of societies that were subsumed into the new global social system, a system that generally forces them into economic development, as we will now see.

References

Abbott, Carl. 2013. North America. In *The Oxford Handbook of Cities in World History*, ed. P. Clark, 504–521. New York: Oxford University Press.

Acemoglu, Daron. 2012. Five Books Interviews: Daron Acemoglu on Inequality. The Browser. Retrieved January 22, 2012. http://thebrowser.com/interviews/daron-acemoglu-on-inequality?page=full.

Achen, Christopher H., and Larry M. Bartels. 2016. *Democracy for Realists: Why Elections Do Not Produce Responsive Government*. Princeton, NJ and Oxford: Princeton University Press.

Alesina, Alberto, Edward Glaeser, and Bruce Sacerdote. 2001. Why Doesn't the United States Have a European Welfare State? National Bureau of Economic Research, Working Paper No. 8524.

Alvaredo, Facundo, Lucas Chancel, Thomas Piketty, Emmanuel Saez, and Gabriel Zucman. 2018. *World Inequality Report 2018*. Paris: World Inequality Lab.

Amatori, Franco, and Andrea Colli, eds. 2020. *The Global Economy: A Concise History*. Abingdon, Oxon: Routledge.

Anderson, Bonnie S., and Judith P. Zinsser. 2000. *A History of Their Own: Women in Europe from Prehistory to the Present*. Vol. II., revised ed. New York: Oxford University Press.

Anheier, Helmut. 2004. *Civil Society: Measurement, Evaluation, Policy*. London: CIVICUS.

Anker, Richard. 2011. *Engel's Law Around the World 150 Years Later*. Working Paper Series No. 247, Amherst: Political Economy Research Institute, University of Massachusetts.

Ardant, Gabriel. 1975. Financial Policy and Economic Infrastructure of Modern States and Nations. In *The Formation of National States in Western Europe*, ed. C. Tilly, 164–242. Princeton, NJ: Princeton University Press.

Barro, Robert J., and Jong-Wha Lee. 2015. *Education Matters: Global Schooling Gains from the 19th to the 21st Century*. New York: Oxford University Press.

Bell, Daniel. 1976. *The Coming of Post-Industrial Society: A Venture in Social Forecasting*. Vol. 40, 574. New York: Basic Books.

Berg, Andrew G., and Jonathan D. Ostry. 2011. *Inequality and Unsustainable Growth: Two Sides of the Same Coin? IMF Staff Discussion Note*. Washington, DC: International Monetary Fund.

Berghoff, Hartmut, Philip Scranton, and Uwe Spiekerman, eds. 2012. *The Rise of Marketing and Market Research*. New York: Palgrave Macmillan.

Birn, Anne-Emmanuelle, Yogan Pillay, and Timothy H. Holtz. 2017. *Textbook of Global Health*. 4th ed. New York: Oxford University Press.
Bommes, Michael, and Andrew Geddes. 2000. *Immigration and Welfare: Challenging the Borders of the Welfare State*. New York: Routledge.
Bordo, Michael D., and Pierre L. Siklos. 2017. *Central Banks: Evolution and Innovation in Historical Perspective*. Cambridge, MA: National Bureau of Economic Research.
Brint, Steven. 2017. *Schools and Societies*. 3rd ed. Stanford, CA: Stanford University Press.
Brownson, Ross C., Tegan K. Boehmer, and Douglas A. Luke. 2005. *Declining Rates of Physical Activity in the United States: What Are the Contributors? Annual Review of Public Health 26:421–443*. Palo Alto, CA: Annual Reviews, Inc.
Calhoun, Craig. 2005. Nationalism. In *Encyclopedia of Social Theory*, ed. G. Ritzer, 519–525. Thousand Oaks, CA: SAGE.
Carson, Rachel. 1962. *Silent Spring*. Boston: Houghton Mifflin.
Centeno, Miguel, and Joseph N. Cohen. 2012. The Arc of Neoliberalism. *Annual Review of Sociology* 38:317–340.
Chirot, Daniel. 2012. *How Societies Change*. 2nd ed. Thousand Oaks, CA: SAGE.
Cohn, Theodore H. 2016. *Global Political Economy*. New York: Routledge.
Croteau, David, William Hoynes, and Stefania Milan. 2012. *Media/Society: Industries, Images, and Audiences*. 4th ed. Thousand Oaks, CA: SAGE.
Cullen, Jim. 2002. *The Art of Democracy: A Concise History of U.S. Popular Culture*. New York: Monthly Review Press.
Davies, Glyn. 2016. *A History of Money*. 4th ed. Revised and updated by Duncan Connors. Cardiff: University of Wales Press.
Dicken, Peter. 2015. *Global Shift: Mapping the Changing Contours of the World Economy*. 7th ed. New York: The Guilford Press.
DiJulio, John J. 2014. *Bring Back the Bureaucrats! Why More Federal Workers Will Lead to Better (and Smaller!) Government*. West Conshohocken, PA: Templeton Press.
Domhoff, G. William. 2013. *Who Rules America? The Triumph of the Corporate Rich*. 7th ed. Boston: McGraw Hill Higher Education.
Dormois, Jean-Pierre. 2004. *The French Economy in the Twentieth Century*. Cambridge, UK: Cambridge University Press.
Downs, William M. 2011. Electoral Systems in Comparative Perspective. In *21st Century Political Science: A Reference Handbook*, ed. J. T. Ishiyama and M. Breuning, 159–167. Thousand Oaks, CA: SAGE.
Ede, Andrew, and Lesley B. Cormack. 2012. *A History of Science in Society: From Philosophy to Utility*. 2nd ed. Toronto: University of Toronto Press.
Ehrenfreund, Max. 2016. A Majority of Millennials Now Reject Capitalism, Poll Shows. W*ashington Post*. Retrieved June 21, 2018. https://www.washingtonpost.com/news/wonk/wp/2016/04/26/a-majority-of-millennials-now-reject-capitalism-poll-shows/?utm_term=.02a9aa5e889d.
European Environment Agency. 2018. *Waste Recycling*. Denmark: European Environment Agency. Retrieved December 26, 2018. https://www.eea.europa.eu/data-and-maps/indicators/waste-recycling-1/assessment.
Fearon, James D. 2003. Ethnic and Cultural Diversity by Country. *Journal of Economic Growth* 8 (2): 195–222.
Feuer, Bryan. 2016. *Boundaries, Borders and Frontiers in Archaeology: A Study of Spatial Relationships*. Jefferson, NC: McFarland & Co.
Francis, Andrew. 2013. The Wages of Sin: How the Discovery of Penicillin Reshaped Modern Sexuality. *Archives of Sexual Behavior* 42:5–13.
GBD (Global Burden of Disease Study 2021). 2024. Global Fertility in 204 Countries and Territories, 1950–2021, with Forecasts to 2100: A Comprehensive Demographic Analysis for the Global Burden of Disease Study 2021. *The Lancet* 403 (10440): 2057–2099.
Gilens, Martin, and Benjamin I. Page. 2014. Testing Theories of American Politics: Elites, Interest Groups, and Average Citizens. *Perspectives in Politics* 12 (3): 564–581.

Glasius, Marlies, David Lewis, and Hakan Seckinelgin. 2004. *Exploring Civil Society: Political and Cultural Contexts*. London: Routledge.

Glenn, H. Patrick. 2010. *Legal Traditions of the World: Sustainable Diversity in Law*. New York: Oxford University Press.

Green, Andy. 1990. *Education and State Formation: The Rise of Education Systems in England, France and the USA*. New York: St. Martin's Press.

Gregory, Cedric E. 2001. *A Concise History of Mining*. Revised ed. Lisse, The Netherlands: Swets & Zeitlinger B.V.

Hall, Peter, and David Soskice. 2001. *Varieties of Capitalism: The Institutional Foundations of Comparative Advantage*. New York: Oxford University Press.

Halland, Håvard, Martin Locanc, Arvind Nair, and with Sridar Padmanabhan Kannan. 2015. *The Extractive Industries Sector: Essentials for Economists, Public Finance Professionals, and Policymakers*. Washington, DC: World Bank.

Hallin, Daniel C., and Paolo Mancini. 2004. *Comparing Media Systems: Three Models of Media and Politics*. Cambridge, UK: Cambridge University Press.

Hamilton, Gary, Benjamin Senauer, and Misha Petrovic. 2011. *The Market Makers: How Retailers Are Changing the Global Economy*. Oxford: Oxford University Press.

Hancké, Bob. 2003. Revisiting the French Model: Coordination and Restructuring in French Industry in the 1980s and 1990s. In *Corporate Governance in a Changing Economic and Political Environment: Trajectories of Institutional Change*, ed. M. Federowicz and R. V. Aguilera, 195–219. New York: Palgrave Macmillan.

Herzog, Dagmar. 2011. *Sexuality in Europe: A Twentieth-Century History*. New York: Cambridge University Press.

Hout, Michael, Claude S. Fischer, and Mark A. Chaves. 2013. *More Americans Have No Religious Preference*. Berkeley, CA: Institute for the Study of Societal Issues.

Inglehart, Ronald. 2018. *Cultural Evolution: People's Motivations are Changing, and Reshaping the World*. New York: Cambridge University Press.

Ishiyama, John, Tatyana Kelman, and Anna Pechenina. 2011. Models of Democracy. In *21st Century Political Science: A Reference Handbook*, ed. J. T. Ishiyama and M. Breuning, 267–274. Thousand Oaks, CA: SAGE.

Japan Times. 2016. *Japan's Population Declines in 2015 for First Time Since 1920*. Retrieved December 23, 2022. https://www.japantimes.co.jp/news/2016/10/27/national/japans-population-declines-2015-first-time-since-1920/#.WBC3EtJ96M8.

Jensen, Carsten Strøby. 2006. Trade Unionism: Differences and Similarities—A Comparative View on Europe, USA and Asia. *Journal of Industrial Relations* 48 (1): 59–81.

Johnson, Allen W., and Timothy Earle. 2000. *The Evolution of Human Societies: From Foraging Group to Agrarian State*. 2nd ed. Stanford, CA: Stanford University Press.

Jones, Charles I., and Paul M. Romer. 2010. The New Kaldor Facts: Institutions, Population, and Human Capital. *American Economic Journal: Macroeconomics* 2 (1): 224–245.

Kent, Susan Kingsley. 2021. *Gender in World History*. New York: Oxford.

Kirchner, Stephen. 2011. *"Why Does Government Grow?" CIS Policy Monographs PM117*. Sydney: Center for Independent Studies.

Kramer, Lloyd. 2011. *Nationalism in Europe and America: Politics, Cultures, and Identities Since 1775*. Chapel Hill: University of North Carolina Press.

Krieger, Joel. 2012. *The Oxford Companion to Comparative Politics*. New York: Oxford University Press.

Laberge, Yves. 2007. Environmentalism. In *Encyclopedia of Environment and Society*, ed. P. Robbins. Thousand Oaks: SAGE.

Le Zotte, Jennifer. 2017. *From Goodwill to Grunge: A History of Secondhand Styles and Alternative Economies*. Chapel Hill: University of North Carolina Press.

Lees, Andrew, and Lynn Hollen Lees. 2013. Europe: 1800–2000. In *The Oxford Handbook of Cities in World History*, ed. P. Clark, 464–482. New York: Oxford University Press.

Lewis, Jane. 2012. Gender Equality and Work-Family Balance in a Cross-National Perspective. In *Gendered Lives: Gender Inequalities in Production and Reproduction*, ed. J. Scott et al., 206–224. Northampton, MA: Edward Elgar.

Lewis, Amanda E., Maria Krysan, Sharon M. Collins, Korie Edwards, and Geoff Ward. 2004. Institutional Patterns and Transformations: Race and Ethnicity in Housing, Education, Labor Markets, Religion, and Criminal Justice. In *The Changing Terrain of Race and Ethnicity*, 67–120. New York: Russell Sage.

Liftoff, Judy. 2005. Rosie the Riveter. In *Americans at War*, ed. J. Resch, vol. 3. Detroit: Macmillan Reference USA.

Lipsey, Richard G., Kenneth I. Carlaw, and Clifford T. Bekar. 2005. *Economic Transformations: General Purpose Technologies and Long-Term Economic Growth*. Oxford: Oxford University Press.

Livi-Bacci, Massimo. 2017. *A Concise History of World Population*. 6th ed. Malden, MA: Blackwell.

Livingston, Gretchen. 2019. Among 41 Countries, Only U.S. Lacks Paid Parental Leave. *Pew Research Center*. Retrieved December 23, 2022. https://www.pewresearch.org/fact-tank/2019/12/16/u-s-lacks-mandated-paid-parental-leave/.

Maddison, Angus. 2006. *The World Economy*. Paris: OECD Publishing.

Mazoyer, Marcel, and Laurence Roudart. 2006. *A History of World Agriculture: From the Neolithic Revolution to the Current Crisis*. Translated by J. H. Membrez. New York: Monthly Review Press.

McMillan, John. 2002. *Reinventing the Bazaar: A Natural History of Markets*. New York: W.W. Norton & Company.

Medina, Leandro, and Friedrich Schneider. 2018. Shadow Economies Around the World: What Did We Learn Over the Last 20 Years? In *International Monetary Fund Working Paper No. 2018/017*, vol. 18, 1. Washington, DC: International Monetary Fund.

Mettler, Suzanne. 2011. *The Submerged State: How Invisible Government Policies Undermine American Democracy*. Chicago: University of Chicago Press.

Mikesell, Raymond F., and John W. Whitney. 1987. *The World Mining Industry: Investment Strategy and Public Policy*. Boston: Allen & Unwin.

Moni, Monir Hassan. 2007. Environmental History. In *Encyclopedia of Environment and Society*, ed. P. Robbins, 872–876. Thousand Oaks, CA: SAGE.

Monkkonen, Eric. 1988. *America Becomes Urban: The Development of U.S. Cities and Towns, 1780–1980*. Berkeley: University of California Press.

Moore, Sarah. 2007. Garbage. In *Encyclopedia of Environment and Society*, ed. P. Robbins, 724–726. Thousand Oaks, CA: SAGE.

Mullings, Beverly. 2010. Neoliberalism. In *Encyclopedia of Geography*, ed. B. Warf, 2011–2015. Thousand Oaks, CA: SAGE.

Napoli, Philip. 2011. *Audience Evolution: New Technologies and the Transformation of Media Audiences*. New York: Columbia University Press.

Neal, Larry, and Rondo Cameron. 2016. *A Concise History of the World: From Paleolithic Times to the Present*. 5th ed. New York: Oxford University Press.

Nolan, Patrick, and Gerhard Lenski. 2011. *Human Societies: An Introduction to Sociology*. 11th ed. Boulder, CO: Paradigm.

Norman, Wayne. 2006. *Negotiating Nationalism: Nation-Building, Federalism, and Secession in the Multinational State*. Oxford: Oxford University Press.

Orrù, Marco, Nicole Woolsey Biggart, and Gary G. Hamilton. 1997. *The Economic Organization of East Asian Capitalism*. Thousand Oaks, CA: Sage.

Painter, Martin, and B. Guy Peters. 2010. *Tradition and Public Administration*. New York: Palgrave Macmillan.

Palmer, Brian. 2011. Go West, Garbage Can! Are We Running Out of Room for Our Garbage? *Slate*. Retrieved November 3, 2014. http://www.slate.com/articles/health_and_science/the_green_lantern/2011/02/go_west_garbage_can.html.

Penna, Anthony J. 2015. *The Human Footprint: A Global Environmental History*. 2nd ed. Malden, MA: Wiley Blackwell.

Pew Research. 2010. Religion Among the Millennials. *Religion and Public Life Project*. Retrieved April 7, 2014. http://www.pewforum.org/2010/02/17/religion-among-the-millennials/.

Pew Research. 2011. The American-Western European Values Gap. *Global Attitudes Project*. Retrieved April 7, 2014. http://www.pewglobal.org/2011/11/17/the-american-western-european-values-gap/

Pew Research. 2013. Americans' Time at Paid Work, Housework, Child Care, 1965 to 2011. *Modern Parenthood*. Retrieved January 2, 2023. https://www.pewresearch.org/social-trends/2013/03/14/chapter-5-americans-time-at-paid-work-housework-childcare-1965-to-2011/.

Piketty, Thomas. 2014. *Capital in the Twenty-First Century*. Translated by A. Goldhammer. Cambridge MA: Harvard University Press.

Pirie, Fernanda. 2021. *The Rule of Laws: A 4,000-Year Quest to Order the World*. New York: Basic Books.

Pomeranz, Kenneth, and Steven Topik. 2018. *The World That Trade Created: Society, Culture, and the World Economy, 1400 to the Present*. 4th ed. New York: Routledge.

Prechel, Harland. 2000. *Big Business and the State: Historical Transitions and Corporate Transformation, 1880s–1990s*. Albany: State University of New York Press.

Price, Roger. 2005. *A Concise History of France*. 2nd ed. Cambridge, UK: Cambridge University Press.

Price, Roger. 2014. *A Concise History of France*. 3rd ed. New York: Cambridge University Press.

Rayside, David. 2020. Early Advocacy for the Public Recognition of Sexual Diversity. In *The Oxford Handbook of Global LGBT and Sexual Diversity Politics*, ed. M. J. Bosia, S. M. McEvoy, and M. Rahman, 43–62. New York: Oxford University Press.

Robbins, Joseph W. 2011. Presidentialism Versus Parliamentarism. In *21st Century Political Science: A Reference Handbook*, ed. J. T. Ishiyama and M. Breuning, 177–185. Thousand Oaks, CA: SAGE.

Roth, Benita. 2017a. Women's and Feminist Movements in the United States: The Contradictory Effects of Class-based Success. In *Women's Movements in the Global Era: The Power of Local Feminisms*, ed. A. Basu, 241–264. Boulder, CO: Westview Press.

Roth, Silke. 2017b. Varieties of Women's Movements. In *Women's Movements in the Global Era: The Power of Local Feminisms*, ed. A. Basu, 185–212. Boulder, CO: Westview Press.

Schettkat, Ronald, and Lara Yocarini. 2003. "The Shift to Services: A Review of the Literature." IZA Discussion Paper 964. Bonn, Germany: Forschungsinstitut zur Zukunft der Arbeit GmbH (IZA).

Schmitt, Carina, Hanna Lierse, Herbert Obinger, and Laura Seelkopf. 2015. The Global Emergence of Social Protection: Explaining Social Security Legislation, 1820–2013. *Politics and Society* 43 (4): 503–524.

Schneider, Friedrich, Andreas Buehn, and Claudio E. Montenegro. 2010. *Shadow Economies all Over the World: New Estimates for 162 Countries from 1999 to 2007 (English)*. Policy Research, Working Paper No. WPS 5356. Washington, DC: World Bank. Retrieved June 13, 2018. http://documents.worldbank.org/curated/en/311991468037132740/Shadow-economies-all-over-the-world-new-estimates-for-162-countries-from-1999-to-2007.

Schudson, Michael. 2012. *The Sociology of News*. 2nd ed. New York: W.W. Norton.

Segal, Uma A. 2010. United States: The Changing Face of The United States of America. In *Immigration Worldwide: Policies, Practices, and Trends*, ed. U. A. Segal, D. Elliott, and N. S. Mayadas, 29–46. New York: Oxford University Press.

Segal, Uma A., Doreen Elliott, and Nazneen S. Mayadas. 2010. *Immigration Worldwide: Policies, Practices, and Trends*. New York: Oxford University Press.

Sepp, Jüri, Helje Kaldaru, and Uku Varblane. 2017. The Development and Typology of the Employment Structure in OECD Countries. *Ordnungspolitische Diskurse* No. 201701.

Simmons, I. G. 2008. *Global Environmental History*. Chicago and London: University of Chicago Press.

Smil, Vaclav. 2017. *Energy and Civilization: A History*. Cambridge, MA: The MIT Press.
Smith, Gregory A. 2021. About Three in Ten U.S. Adults Are Now Religiously Unaffiliated. *Pew Research Center*. Retrieved January 5, 2023. https://www.pewresearch.org/religion/2021/12/14/about-three-in-ten-u-s-adults-are-now-religiously-unaffiliated/.
Solimano, Andrés. 2017. *Global Capitalism in Disarray: Inequality, Debt, and Austerity*. New York: Oxford University Press.
Springer, Simon, Kean Birch, and Julie MacLeavy. 2016. *The Handbook of Neoliberalism*. New York and London: Routledge.
Stavenhagen, Rodolfo. 2013. The Emergence of Indigenous Peoples. In *Springer Briefs on Pioneers in Science and Practice, Texts and Protocols*, ed. H. G. Brausch, vol. 3. México, DF: Colegio de México.
Stearns, Peter. 2006. *Consumerism in World History*. 2nd ed. New York: Routledge.
Stiglitz, Joseph. 2012. *The Price of Inequality*. New York: W.W. Norton.
Svallfors, Stefan. 2012. *Contested Welfare States: Welfare Attitudes in Europe and Beyond*. Stanford, CA: Stanford University Press.
Takaki, Ronald. 2008. *A Different Mirror: A History of Multicultural America. 1st revised ed.* New York: Back Bay Books.
Tanzi, Vito, and Ludger Schuknecht. 2000. *Public Spending in the Twentieth Century: A Global Perspective*. New York: Cambridge University Press.
Therborn, Göran. 2004. *Between Sex and Power: Family in the World, 1900–2000*. London and New York: Routledge.
Trefil, James. 2012. *Science in World History*. London and New York: Routledge.
Turchin, Peter. 2003. *Historical Dynamics: Why States Rise and Fall*. Princeton, NJ and Oxford: Princeton University Press.
U.S. Census. 2016. 2016 American Community Survey 1-Year Estimates. *U.S. Census Fact Finder*. American Community Survey. Retrieved June 23, 2018. https://factfinder.census.gov/faces/tableservices/jsf/pages/productview.xhtml?pid=ACS_16_1YR_CP05&prodType=table.
Vietor, R. H. K. 2000. Government Regulation of Business. In *Cambridge Economic History of the United States*, ed. S. L. Engerman and R. E. Gallman, vol. III, 969–1012. New York: Cambridge University Press.
Wilkinson, Richard G., and Kate Pickett. 2009. *The Spirit Level: Why More Equal Societies Almost Always Do Better*. London: Allen Lane.
World Bank. 2018. GDP growth (annual %). *World Development Indicators*. Retrieved June 13, 2018. https://data.worldbank.org/indicator/NY.GDP.MKTP.KD.ZG?end=2016&locations=XD&name_desc=false&start=2001&view=chart.

Chapter 10
Social Patterns of Developing States: Globalizing the Industrial Revolution, 1910s–2010s

Abstract Economic "development" transforms societies. Sustained economic growth requires major, difficult social changes. Scholars and development advisors still disagree about the role of the state in development, although developing states tried many programs after decolonization. Agriculture must become vastly more productive, usually through commercialization, freeing most people to work in industry or services, almost always in cities. Developing states often face pressure to use up resources but face deforestation, pollution, and other problems. Economic growth reduces absolute poverty but increases economic inequality, sometimes making states politically unstable, as does ethnic fragmentation. All developing states reduced mortality over the twentieth century and most reduced fertility; however, demographic growth and urbanization put pressure on agriculture. Patriarchal patterns such as arranged marriages, girl marriage, lack of divorce, unequal property laws, and domestic violence have all been reduced, partly because of women's movements.

Keywords Economic growth · Economic development · Decolonization · Developing societies · Ethnic fragmentation · Absolute poverty · Relative poverty

Introduction

Here I pull together everything discussed so far. The Agricultural and Urban Revolutions never affected all societies in their eras; the global Industrial Revolution does, but it remains incomplete. Preindustrial social patterns still partly structure what were kin-ordered groups or agrarian states, but industrial economic growth, democracy, education, science, mass media, money-based classes, gender equalization, and demographic transitions are also changing these societies. This mix of ancient social structures and "modernity" influences all states in the world.

I argue that global and domestic forces push states to develop economically, a process that transforms societies but requires decades or centuries to play out. The global social system simultaneously fosters and impedes economic development; it may or may not be good for people and societies, but global and domestic forces are

T. C. Burr, *Societies Past and Present*, Emerging Globalities and Civilizational Perspectives, https://doi.org/10.1007/978-3-031-91532-1_10

so strong that changes occur anyway. The process is difficult to understand because it is so comprehensive and because each society deals with its own historical legacies. In a sense, development is a continuation of the old imperial tactic of assimilation, now on a global scale. European empires have been assimilating kin-ordered societies and agrarian states since the 1500s. This process intensified over the next centuries, and since the 1950s, states have continued it, though it is still quite incomplete.

Of course, the idea that some states are "developed" and others are "developing" provides us only a ludicrous binary. None of the various lists of states ranked by gross domestic product (GDP) per capita show any obvious point, any "magic moment," when a state moves from one category to another.[1] Yet all other terms, such as "emerging economies" or "lesser developed states," are worse, and the "developed/developing" categorization does point to two realities: some states are rich, some are poor, and some are in-between; and the international system subjects both the poor and in-between to similar pressures, with different outcomes rooted in historical legacies. The Rise of the West/Great Divergence debate addresses those legacies, and we will soon see how early development efforts were quite ahistorical. Only a few scholars ask the questions, posed in the discussion of that debate (Chap. 8), pointing to both history and contemporary realities: Were all societies before 1500 simply poor, and economic growth is possible for states with the "right" development strategies? Or did Europe underdevelop the rest of the world, and does that process of underdevelopment continue today? Either way, the state is the only form of society the global system seems to accept as legitimate, so that is where the analysis starts.

The Politics of Developing States

Imperialism and then decolonization imposed the European state form on all groups worldwide. Developing states are mostly organized in a way culturally alien to most peoples within them, and they do not have the resources of developed states. Through the twentieth century, academics, intergovernmental organizations (IGOs), and international non-governmental organizations (INGOs) debated the role of states in development, and different developing states tried different strategies. Scholars now agree that development requires a strong state, but many factors prevent this. I structure this story in three periods, following the chronology of Chap. 8; then I analyze the political consequences of these strategies.

[1] There is no obvious transition from developing to developed in the International Monetary Fund's World Economic Outlook Database, in the World Bank's dataset (https://data.worldbank.org/, search by "GDP, PPP (current international $)"), or in the "Real GDP per capita" table in the CIA's *World Factbook*.

Colonialism and Development, 1918–1945

By the 1920s, Latin America had been independent for a century, while Europeans still controlled most of Africa and much of Asia. Most colonies exported raw materials and imported finished goods. During the Great Depression, many Latin American states adopted what scholars call "import substitution industrialization" (ISI). The new strategy was to replace imports from developed countries with domestically manufactured goods, thus requiring development of domestic industries. States imposed high tariffs on imports to protect these new, "infant" industries and offered support for certain industries, such as providing capital to buy imported technology (Bruton 1998:909). Businesses in infant industries took time to produce better goods, so consumers dealt with substandard goods for a while. This strategy involved some failures along the way—industries that never took off (Chang and Grabel 2004; McKay 2016) (Fig. 10.1).

In Africa and Asia, European empires controlled economic policies, orienting them to serve Europe, which fueled desires for independence, but in Southwest Asia and North Africa, Egypt, Iran, and Saudi Arabia remained formally independent. Britain and France received a League of Nations "mandate" to control the Ottoman Levant. In sub-Saharan Africa, South Africa had gained partial independence just before World War I, while London and Paris maintained close control over colonies with majority native populations. Their policies favored European producers over African ones, providing few infrastructure improvements. This favoritism and neglect helped nationalist movements for independence, as did poor economic performance during the Great Depression. In South Asia, though the bureaucracy became increasingly indigenized, the British neglected infrastructure while retaining control, and the Depression also fostered sentiment for independence. Wealthy Indians funded the nationalist movement, and disgruntled peasants, running out of new land, joined in (Roy 2016).

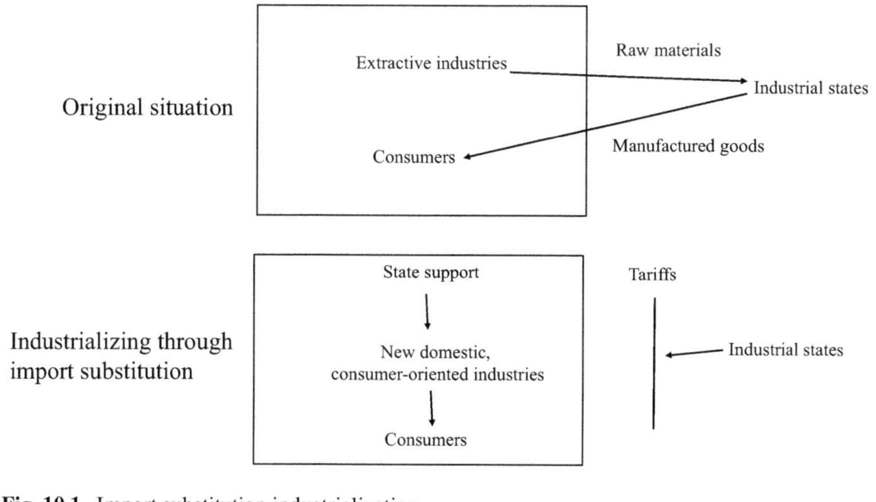

Fig. 10.1 Import substitution industrialization

Experiments in the Political Economy of Development, 1945–1990

In the decades after World War II, decolonizing states experimented with various models of political economy. These policies were not always by choice; Cold War superpowers, IGOs, INGOs, multinational corporations (MNCs), and scholar-consultants also influenced developing states' economic policies. From the 1950s to 1990, various developing states followed four basic approaches. (The following history relies on Chang and Grabel 2004; Hare and Turley 2013; Williams 2014; McKay 2016; and Kingsbury 2016.)

After World War II, Latin American states continued ISI as an industrial strategy, and others, such as India, followed. It was a capitalist strategy with socialist influence, partly rooted in dependency theory (Bruton 1998; McKay 2016), receiving a boost from two academic papers (Prebisch 1950; Singer 1950). This strategy would block industries in the global North from selling to their consumer markets, so Northern states criticized ISI. Also, many Anglo-American economists argue that ISI simply failed to foster development. However, other scholars argue that many developed countries, such as the United States, followed ISI principles well before the strategy had a name, and that ISI had a better record in Latin America than has been realized (Chang and Grabel 2004; see Bruton 1998 for a more mixed view).

The second program was socialism. During World War II, the Soviet Union pushed the German army back through Poland, Czechoslovakia, Bulgaria, Romania, Hungary, and part of Germany. The Soviets kept control over these countries, imposing state socialism on them. Soviet allies, including China, Vietnam, Cuba, South Yemen, and North Korea, implemented variations on Soviet socialism, as did nonaligned Yugoslavia. In state socialism, only the Communist party was allowed, so there was an "interweaving" of party and state (Kornai 1992:38). Only one state-owned enterprise produced any consumer product—a refrigerator, a radio, etc.—in an "organizational monopoly" (Kornai 1992:39). A state agency planned production levels for each enterprise in the famous "Five-Year Plans." The state also controlled local agricultural cooperatives, plus wholesale and retail. Very small-scale capitalist industry existed, along with large informal economy (Kornai 1992). Some countries (e.g., Tanzania and India) pursued quasi-socialist development with five-year plans, state ownership of some enterprises but private ownership of others, and other state controls. Socialist economic growth never kept up with capitalist growth; most states abandoned socialism after 1990. From 1978, Chinese and Vietnamese "market socialism" has allowed private firms to compete in markets with state-owned or state-directed enterprises, but the Communist Party still has control.

The third path was the "developmental state," a description of how some East Asian states developed. The state often invested in businesses. For instance, the Japanese had savings banks at post offices where nearly everyone saved their money. The state lent these savings to businesses, thus selecting industries to support. Only later did a large private Japanese banking sector develop. The state also supported businesses buying and importing factory machinery from North Atlantic states;

later, domestic industries started innovating technologically. Finally, the state set high tariff barriers and encouraged exports, encouraging businesses to enter value chains in manufacturing rather than raw materials production (Williams 2014).

The fourth developed strategy was neoliberalism. As discussed (Chap. 8), 1980s Structural Adjustment Programs (SAPs) helped many developing states renegotiate 1970s loans. In return, the International Monetary Fund (IMF) required "austerity" programs: privatization of state enterprises, cutting food subsidies, trade liberalization, an export orientation, and reduction of regulation. These neoliberal policies often replaced socialism and ISI in the global South in the 1980s and 1990s, sparking the most scholarly disagreement. Economists generally argue that neoliberal IMF and World Bank programs fostered economic growth (e.g., Shearmur 1992; Sumner 2010). Sociologists argue that states in the global North used these institutions to extract surplus from the periphery (e.g., Mullings 2010; McMichael 2012; Hickel 2015) as constant rescheduling of loans meant that many developing countries have paid many times more than they ever borrowed.

Consequences of "the Transition," 1990–2015

These four approaches amounted to a global process of experimentation. All raise the issue of state strength in relation to development. Economists since 1990 have increasingly appreciated the importance of states in capitalist economic growth, but this took time to develop.

Decolonization occurred during the Cold War when adherents of capitalism and socialism each thought their policies were correct. The post-1990 extension of neoliberal ideas to the post-socialist economies of Eastern Europe and Central Asia occurred in a climate of triumphalism, but the results raised uncomfortable questions about minimalist states. Eastern European countries, and newly decolonized countries in Central and Western Asia such as Kazakhstan, Georgia, and Armenia, abandoned state socialism and adopted capitalism; scholars and IGOs speak of *the* transition, always from socialism to capitalism. The World Bank and IMF were happy to advise them. The mostly Anglo-American economists who advised these new governments urged the usual program: privatization of state businesses, ending state subsidies and state planning, and reduction of regulations. Economists assumed that people naturally operate as capitalists, so removing state regulations would allow the economies to flourish.

In practice, the transition to capitalism was difficult. Eastern European states are similar to those such as Mexico, Brazil, and South Africa, as they are somewhat industrialized, have literate populations, and are "middle-income" countries, but their developmental paths varied. Some succeeded in the neoliberal transition to capitalism, but some failed completely. For instance, Poland performed very well, while Georgia has still not fully recovered from the transition. Some argue that this occurred because advisors applied a naïve "hyper-liberalism" (Nuti 2013:48). Also, neoliberal post-Soviet states have not grown economically as quickly as Chinese and Vietnamese market socialist economies have (Hare and Turley 2013).

These results led economists to appreciate institutions more. Sociologists have long argued for the importance of institutional arrangements such as physical infrastructure, education systems, good intellectual property law, and competent bureaucrats who are friendly to, yet skeptical of, business (Evans 1995). In contrast, economists had long taken for granted institutions of developed capitalist states such as private property rights, enforcement of business contracts, and specific regulations for specific product markets (Hare and Turley 2013:61–62); economists appreciated them more when they did not exist in Eastern European and Central Asian economies. They increasingly accept the need to socialize populations into capitalist institutions rather than such institutions working "naturally," a stance that is implicitly at odds with neoliberal assumptions. In particular, the rule of law—officials of the state enforcing laws equally on all, including on themselves—is crucial for economic growth (McMillan 2002:Chap. 5; Hare and Turley 2013) since it increases individual freedom. When it is absent, aggrandizers operating as corrupt officials, rebels, or organized crime leaders reduce trust between transaction partners, limiting freedom and ultimately reducing growth. Stated differently, while laws "support regimes of discipline, hierarchy, and centralized control," they also provide "a language that gives people a way to object to injustice and oppression and stand up to the powerful" (Pirie 2021:450, 451). It allows the population to fight the destructive capriciousness of leaders who hold themselves "above" the law, something that creates disorder, which is hardly helpful for any economic activity, much less modern economic growth.

While IGOs, INGOs, and even some MNCs apply this new appreciation of state institutions to capitalist development, other scholars recently apply an appreciation for strong states to questions of state stability. Civil wars as Cold War proxy wars usually made things worse for developing states. After the Cold War ended, most civil/proxy wars ended, often in negotiated settlements (The Economist 2013). Even so, not all developing states provide public goods such as public order, rule of law, physical infrastructure, dispute adjudication, and education that support economic growth. Political scientists developed a fourfold typology of states that measures public-goods provision: strong, weak, failed, and collapsed (Rotberg 2004).[2] Strong states provide all public goods, even if imperfectly. Some weak states provide some public goods: crime rises, corruption increases, state legitimacy falls, and infrastructure and schools weaken. Others deliver public goods fully, but only to a minority. Failed states provide few public goods: regimes prey on citizens, civil wars develop, and territorial secessions occur. Collapsed states might as well not exist; only warlords (effectively, modern, post-collapse chiefdoms) control local territories. Over a decade ago, the Fund for Peace, a U.S.-based INGO, developed the Failed States Index (renamed in 2014 the Fragile States Index), a listing and

[2] Diamond (2005) used this language of state "failure," yet his book shows no awareness of this literature.

analysis of all states globally (https://fragilestatesindex.org/). This applies a ten-item set of factors to analyze the "fragility" of each state, ranking them on the index of strong, weak, failed, or collapsed. The Fund provides this analysis to corporations, INGOs, and others.

One factor might be poverty. As discussed, three conditions—the economic surplus produced, taxability (usually in state-issued money), and literate and numerate populations (Ardant 1975)—help fund states. Many developing states only partly meet these conditions: producing a sufficient surplus is difficult; commercialization is often incomplete, inhibiting tax collection; and education is improving slowly. Yet, other problems affect developing states, not merely poverty. Rather than "failure," some scholars (e.g., Goldstone et al. 2010) analyze stability or instability (violent or nonviolent); they argue that regime type affects (in)stability.

Democracy? State Institutions and Forms

Developing states have borrowed, or experienced the imposition of, many political institutions of the global North, such as representative democracy. As partly discussed in Chap. 9, there have been three waves or clusters of democratization since the eighteenth century that outnumbered transitions away from democratic regimes (Huntington 1991:15): 1826–1926 (Western Europe, North America, and East Asia), 1943–1962 (southern Europe and some Latin American, Asian, and African states), and 1974 to the last few years (much of the rest of the world) (Huntington 1991; Márquez 2017). Organizations of global governance attempt to foster democracy (Grugel and Bishop 2014:Chap. 6), most obviously the World Bank's neoliberal-tinged "good governance" agenda enforced through loan requirements. Still, other international organizations offer election persuasion, advice, and monitoring. Even imperfect democracies took European and North American states over a century to achieve, so it may remain culturally shallow in developing states for decades. Also, some IGOs are not democratic; what standing do North Atlantic states, IGOs, and INGOs have to push a Western export called "democracy" on developing states? Despite these objections, nearly all states now at least claim to be democracies: "'democratic' ideas about legitimate authority, and the republican institutions with which they became associated, have become dominant in the world today" (Márquez 2017:29). Scholars have developed different ways to measure democratization: inclusiveness of suffrage, competitiveness of elections, and constraints on executives (Márquez 2017:Chap. 2). By these measures, the number of "true" democracies has continued to increase over time.

There are also regional differences in democratization. Soviet occupation affected Eastern European states, while the U.S. controlled Central American states much more than it did South American states (Grugel and Bishop 2014). Likewise, Latin American states sometimes alternate between two traditions, the strongman and democracy. Poverty, ethnic fragmentation, and neopatrimonialism make African democratization difficult. Few North African and Southwest Asian states are democratic, and China remains a communist dictatorship.

Scholars also analyze democratic stability through leadership forms: parliamentary and presidential forms, and hybrids. Many former British colonies are parliamentary, such as India and Pakistan. Most Latin American states (though few Caribbean states), some African states, and some Central Asian states have presidential systems. The number of parliamentary systems increased and then decreased globally from 1946 to 2002; the number of presidential systems fell and then rose; and the number of hybrid mixes remained constant until 1990, then increased (Cheibub 2007:42–43). Scholars ask which system is most stable. Many presidential developing states have collapsed into authoritarianism (Dahl 1998). Some think that parliamentary systems simply perform better (Gerring et al. 2009), but others argue that already unstable societies are more likely to adopt presidential systems (Cheibub 2007).

Systems of legislative representation also seem to affect democratic stability. Again, one-member "winner-take-all" districts for legislatures strongly tend to produce only two parties. Multiseat proportional representation systems usually produce three or more parties. Most Latin American states use proportional representation, while most African and Asian states use winner-take-all systems. Proportional representation generally increases voter turnout, helping to represent the range of beliefs and positions in a populace, which support legitimacy, while winner-take-all preserves the "one person, one vote" principle (Norris 1997).

Public administration (bureaucracy) is generally difficult in developing states. Some argue that bureaucratic structures in developing states are legacies of colonizers imposing European administrative patterns on widely varying societies so that they "are relatively incompatible with their indigenous societal contexts" (Haque 1996:316; Haque 2010). They assume individualism, ignoring different norms such as ascription, informality, a kinship orientation, and collective responsibility. Also, developing states' bureaucracies often end up being the most powerful elites rather than economic ones, and a politically neutral set of civil servants does not exist. Elite civil servants often have Western training, making them out of touch with most people. Finally, "race, caste, family, language and status" influence administrative recruitment far more than merit-based criteria do (Haque 1996:322). Caste is far more important in Southeast Asia, while patronage remains a major recruitment tradition in Latin America (Haque 1996:319). Still, expertise and education are beginning to matter more in some developing states in hiring and in popular opinion (Hyden et al. 2003).

Finally, "civil society" institutions—not part of the state, business, or the family—often try to influence democratic states in the public sphere. Some scholars focus on the nonprofit sector (Salamon and Sokolowski 1999), while others are more explicitly political. Mass and social media increasingly constitute the public sphere, and developing state governments often try to control both broadcasting and social media, with varying results. One measure of democracy is how much a citizenry can learn about and influence the state through the media, something particularly important for social movements, which have long existed in developing states.

Twentieth-century policy experimentation helped scholars and state leaders appreciate state stability and strength for fostering economic growth, but developing states' histories of colonization, decolonization, and restructuring put that stability and strength in question.

Developing Economies

Few question the underlying premise that developing states must seek economic growth; even socialist states attempted to industrialize. While experiences vary, all developing states face the same two problems. Industrialization happens mostly in cities where people rarely grow their own food, so developing states must massively increase their agricultural productivity. Consequently, states encourage commercial agriculture. Second, they must diversify their economies. Many rely heavily on raw materials exports when industrialization and services are far more enriching. Here, I address the primary sector (raw materials), secondary sector (industry), tertiary sector (services), consumption, and the entire economy.

The Primary Sector

Some developing states rely heavily on exporting primary products such as timber, minerals, or petroleum, and increasing agricultural productivity to support industrializing cities is difficult and complex. Here, I can address only some of the issues.

Mining has long been important in the global South; it remains so today. Southwest Asia exported little petroleum until 1940, but oil investment soared after World War II; by the 1970s, "petro-states" temporarily controlled the global oil market, but neoliberal policies lowered long-term prices (Ghanem and Baten 2016). Likewise, European powers invested in colonial mining early in the century, as did Latin American states. IGOs and Northern banks provided technical assistance and investment to developing states in the 1950s (Peterson 1987), so mining activity grew quickly, slowing in the 1970s (Mikesell and Whitney 1987). Multinational firms owned most mines in developing states through the 1950s. In the 1960s and 1970s, many Latin American and Africa states expropriated their mines, and by 1980, developing states together produced most minerals globally (Mikesell and Whitney 1987). When demand and prices fell, "[t]he 1980s and 1990s were generally grim years for mining" (Humphreys 2015:3). After 2001, Chinese economic development spurred demand for minerals, creating another global mining boom in Africa and Australia (Humphreys 2015). For most, "dependence on the extractive industries has increased … between 1996 and 2014" (Roe and Dodd 2018:41); "there are some thirty-five relatively low-income countries where extractive activity" and mineral exports constitute 30 percent of merchandise exports; in another 26 developing states, oil and gas make up 30 percent or more of merchandise exports (Roe and Dodd 2018).

Developing states also export timber commercially. During the nineteenth century, logging increased worldwide, and it increased more in the twentieth century and yet more after 1950 (Runyan and D'Odorico 2016). In tropical Africa, growth in "industrial-scale agriculture" required extensive logging (Runyan and D'Odorico 2016:20). Logging increased export revenues; as of 2005, wood was "the third most valuable primary commodity in world trade" (Williams 2006:467). Land management practices standard in developed states are not spreading to developing states. While the first three net exporters of wood products are Canada, Finland, and Sweden, nearly all the rest are developing states (Williams 2006).

Most logging makes land suitable for commercial agriculture, cattle ranching (commercial pastoralism), or subsistence production (Williams 2006; Runyan and D'Odorico 2016). Commercial plantations (*haciendas*) started in Spanish America in the 1500s, and an early 1800s "cash crop revolution" in Africa intensified during the colonial period (Roessler et al. 2022). Decolonizing states experimented with agricultural techniques; neoliberalism increased commercialization. Both scholars and states know that commercializing agriculture "is key for economic growth" (Carletto et al. 2017:106; World Bank 2006; Li et al. 2018), so IGOs, MNCs, and even some INGOs help subsistence farmers commercialize. Most farmers still grow their own food, in "staple" or "food crops," and are commercializing only slowly to grow "cash crops" (Rapsomanikis 2015; Todaro and Smith 2015; Carletto et al. 2017; Li et al. 2018).

Subsistence farmers' situations help us understand commercialization. First, small family farms constitute over 90 percent of all farms in the world and produce most of the world's food. Over 95 percent are smaller than five hectares (about 12 acres). Eighty percent of farms in South Asia and sub-Saharan Africa, 75 percent of farms in East Asia and the Pacific (excluding China), and 98 percent of Chinese farms are smaller than two hectares (FAO 2014; Rapsomanikis 2015). Latin American landownership was long highly concentrated; many states engaged in land reform to deconcentrate it mid-century, but neoliberal policies helped reconcentrate land ownership in agriculture, mining, and logging (Bravo 2011), a trend only beginning elsewhere.

Second, subsistence farmers face specific conditions. They work intensively, with higher yields than in developed states, though productivity per worker is lower. They usually grow several crops to reduce risk; most supplement farming with hunting, craft production, and part-time wage employment. In fact, most smallholders generate more household resources off the farm than on; the larger the farm, the more their income comes directly from farming. Most farmers are poorly connected to urban markets with inadequate roads and high transportation costs. Overall, then, most small farmers are poor (FAO 2014; Rapsomanikis 2015).

States can improve farming productivity. Research helps farmers innovate, though innovations must be specific to plants, soils, climate, cultural patterns, and local markets. Irrigation helps farmers raise yields, reduce risk of crop failure, and diversify crops; improved transportation and communication infrastructure increases commercialization; education and the consequent literacy and numeracy also improve yields; and all these changes raise household incomes. Larger farms also tend to mechanize. As farmers commercialize, they tend to specialize in fewer crops (Mazoyer and Roudart 2006; Waters 2007; FAO 2014; Rapsomanikis 2015).

Several things complicate these improvements. First, no primarily horticultural or pastoral society has ever fully industrialized; only agricultural or agropastoral states have (Nolan and Lenski 2011). Economists seem to assume agropastoralism, barely mentioning farming with hand tools, i.e., horticulture (e.g., FAO 2014; Rapsomanikis 2015), but millions in sub-Saharan Africa, Asia, and Latin America are still primarily horticulturalists or pastoralists. Moving to plow agriculture requires different crops, methods, and supports—a risky change (Mazoyer and Roudart 2006; Todaro and Smith 2015). Also, most subsistence farmers have non-capitalist values; abandoning subsistence production to commercialize usually involves some form of political, economic, or climatological coercion (Seavoy 2000; Waters 2007).

The global commercialization of agriculture partly created such coercion for subsistence farmers since the mid-nineteenth century (Mazoyer and Roudart 2006:Chap. 11; Todaro and Smith 2015). Globalizing agricultural markets put farmers in competition with each other globally, driving down crop prices. Competition with large-scale, mechanized farms disadvantages hand-tool horticulturists and single-plow farmers, in a typical sequence. First, horticulturalists often use fallow land more quickly than they did previously, leading to low harvests. Then they borrow money to survive. Another bad harvest prevents repayment of this debt, leading to debt slavery, the growth of illegal crops (e.g., opium), or migration to cities for wages. IGOs, states, and businesses aggravate these problems. Structural Adjustment Programs encourage open borders, putting farmers further into global competition. State clampdowns on illegal drug growing hurt partly commercialized farmers. Overvaluing national currencies makes crops too expensive on global markets. Finally, large landowners often buy distressed farms cheaply, a process most advanced in Latin America. All these tend to push farmers off their farms into urban wage labor.

The primary sector supports many developing states. While mining, timber, and agriculture produce goods for export earnings, commercialization of agriculture has many problems, and it is often inadequate to feed developing states' populations.

The Secondary Sector

Economic growth occurs primarily in the secondary and tertiary sectors. When subsistence farmers leave their land, they usually migrate to cities where industrialization mostly occurs. Industrialization since decolonization has often been difficult for developing states.

In the early twentieth century, Latin American industrialization continued. Many states implemented ISI before the 1930s. The Depression decreased demand for raw materials, so large states expanded ISI policies, focusing on light industry for mass-consumption items, while smaller states had domestic markets too small to sustain entire industries. Leaders supported quasi-socialist policies such as state ownership of "banks, public utilities, and key industries" such as petroleum (Chasteen 2006:230, 237–238), plus price regulation and labor protection. U.S. policy toward Latin America changed from the 1920s militarily interventions to a 1930s "Good Neighbor" policy, and during World War II, most Latin American states allied with the United States.

Despite political fragmentation after 1911, the Chinese economy grew to 1936. MNCs invested there, though domestic investment also grew (Ma 2016). Rail systems finally expanded, and the introduction of a new currency helped circulation. The Great Depression reduced export demand, the Japanese invaded Manchuria, and World War II devastated the Chinese economy. Industrialization and reindustrialization proceeded more slowly in European empires. Africans struggled against European producer collusion without success (Austin 2016), while local markets were more competitive. Agricultural commercialization continued gradually, but few colonial governments improved transportation or communication infrastructure. India started to reindustrialize (Roy 2016:255), while Southwest Asia and North Africa had little industry through World War II (Ghanem and Baten 2016).

After the war, developing states attempted industrialization on their own terms. Oil exports from Southwest Asian states increased (Ghanem and Baten 2016), helping economic growth in the 1950s and 1960s; in the 1970s, when revenues increased markedly, effects were mixed. Oil wealth sparked wars and led to resource-curse effects (see below). Indian economic growth was always highest in its secondary sector, but it was higher in the 1950s and the early 1960s, then slowed. Secondary sector growth increased again after 1986 (Roy 2016:268) (Table 10.1).

Overall, developing states' manufacturing growth rates increased greatly in the twentieth century. However, despite growth *rates*, in some places, the value of industrial output started at such a low level that many states are still extremely poor and not very industrialized (Table 10.2).

Industry in East Asia and the Pacific (excluding Japan, South Korea, Taiwan, etc.) has grown almost five times in the last 20 years. South Asian industry has grown in value over 350 percent, while it has more than doubled in non-oil-producing Middle East and North African states, not quite doubled in sub-Saharan Africa, and grown around only 40 percent in Latin America and the Caribbean.

All this raises the question: Why does industrialization occur quickly or slowly in different states? Neoliberals argue that trade liberalization fostered economic growth; world-system scholars strenuously disagree. As discussed, foreign direct investment (FDI), a measure of MNC activity, has been rising in the global South. Some scholars argue that FDI is beneficial: MNCs bring new knowledge into developing states, creating "spillovers" into other areas of the economy and forcing the upgrading of business infrastructure (Rugman and Doh 2008; Sturgeon et al. 2011). Early on, this helped states grow small parts production; later they moved to larger-scale manufacturing or assembly, or outsourced to other countries (Sturgeon et al. 2011). Other scholars criticize MNCs' impacts on developing states. In agriculture,

Table 10.1 Average industrial growth rates (percent, per annum)

Groups	1870–96	1896–1913	1913–20	1920–38	1938–50	1950–73	1973–90	1990–2007
European periphery	4.6	4.4	−6.3	4.7	3.1	8.8	3.1	2.8
Asia	3.0	4.4	4.3	4.3	−1.1	8.5	5.8	4.2
Asia[a]					−0.7	8.3	5.9	4.3
Lat. Am. and Carib.	4.1	4.1	2.5	2.7	5.3	5.7	2.7	2.3
MENA	1.0	1.5	−5.4	4.7	5.0	6.3	5.0	4.4
SS Africa			13.4	4.6	8.6	5.5	3.5	3.9

Source: Bénétrix et al. (2017:16)
[a] Excluding Japan after 1939

Table 10.2 Industrial value added (in millions of constant 2010 US$)

Region	1996	2016
Europe and Central Asia	500,945.06	1,098,331.43
East Asia/Pacific	1,110,127.48	5,356,624.32
Latin America/Caribbean	1,162,379.00	1,641,285.35
Middle East/North Africa	301,659.73	643,024.45
South Asia	229,834.79	833,937.23
Sub-Saharan Africa	250,775.23	452,295.60

Source: World Bank Open Data (all "non-high-income" states)

MNCs sell seeds, fertilizers, pesticides; dominate global markets; and threaten smallholders. They engage in environmentally destructive logging, overfishing, and mining. They offer low manufacturing wages, long hours, and poor working conditions (Madeley 2008). Trade-Related Investment Measures (TRIMs) compromise developing states' sovereignty by reducing the power of host governments to impose performance requirements on MNCs (McMichael 2012). The General Agreement on Trade in Services (GATS) regime compels governments "to provide unlimited market access to foreign service providers [i.e., MNCs], without regard for social and environmental impacts" (McMichael 2012:145). Perhaps both sides are right and again are talking past each other. Certainly, MNCs are environmentally destructive, ignore labor laws and exploit workers, and threaten small farmers; they also create commercial spillovers, improve business infrastructure, and increase GDP. The primary casualties are probably rural people and the urban poor, the great majority of populations (Madeley 2008). The primary beneficiaries are likely urban elites. And some heterodox economists argue for resisting MNCs and neoliberalism completely in favor of renewing ISI strategies (Chang and Grabel 2004).

The Tertiary Sector

Services are also growing in developing states. They now constitute the majority of GDP in all world regions, employing most workers everywhere except in South Asia and sub-Saharan Africa. Thus, tertiarization is a major component of economic development (Joshi 2008) since business and consumer services become more important as markets develop. Some scholars divide services into distribution (wholesale and retail); logistics (transport, storage, and communication); business services (finance, insurance, and real estate, or FIRE), and social and personal services (Joshi 2008). From 1970 to around 2000, the shares of GDP of FIRE and social/personal services grew in most states, while distribution's share shrank. Employment in all four sub-sectors also grew in this period. Developing states' services often cross international borders. International services trade generated at least 10 percent of GDP for most world regions in 2007 (Cattaneo et al. 2010:5). Some services require producers and consumers to be in the same physical space at the same time, raising issues of migration, border controls, and education.

Compared to industrialization, distribution is an understudied phenomenon (Dicken 2011), yet markets (distribution channels) are a prerequisite for economic growth. A few scholars recognize the obvious: commercialization must exist at least somewhat before industrialization because "[t]here can be no production for sale without an outlet and an accessible conduit to it" (Bauer 2000:8; see also Hamilton et al. 2011). However, market distribution is often greatly underdeveloped. For instance, transportation and communication infrastructure in developing economies is often deficient. Few small businesses own automobiles or can afford rail shipping, which also may be lacking (Rapsomanikis 2015). Some use bicycles, bicycle-rickshaws, and scooters, especially for small loads. While email and internet access is often inadequate due to low computer ownership and limited electricity, where electricity exists, cell phones are creating a revolution. Cell phone towers set up by private companies are far cheaper than state-utility landline networks. Even poor people use cell phones for their financial needs (Singh 2013).

Wholesale and retail practices are often problematic. Markets in West Africa, bazaars in North Africa and Southwest Asia, and East Asian local, regional, and national market systems have existed for over a thousand years. European colonialism brought local stores and marketing boards; decolonization brought more, and neoliberalism brought wholesale and retail FDI into developing states. For instance, in Vietnam, increases in incomes, urbanization, education, female employment, and consumer time pressures shape the consumer environment; lack of experience with customer service, lack of trained employees, upgrades in product standards, complex logistics, and competition from multinational retailers tend to worsen the business environment for traditional Vietnamese retailers (Maruyama and Trung 2012). State support for domestic retailers helps them, but that support varies. The situation seems similar in Botswana (Phambuka-Nsimbi et al. 2015); thus, it is probably difficult in many developing states.

Third, many subsistence producers, perhaps more than a billion globally, stay mostly out of markets. Money is merely supplementary in their lives, lowering general commercial activity. Even people reliant on wages often make little. Since spending is low, wages are low, so retail purchases are low, generally holding back economic growth.

Finally, financial institutions for small businesses and consumers are often undeveloped. European colonialism almost uniformly brought about the shift from local commodity monies (cowrie shells, cattle, and many other objects [Singh 2013]) to state-issued currencies. Scholars, IGOs, INGOs, and businesses all assume state currencies in commercialization, but since "[n]early half the people in the world aged fifteen and above do not have an account with a bank, credit union, cooperative, post office, or microfinance institution" (Singh 2013:42), economic growth is greatly hindered. Access would help people secure their money, smooth out cash flow (Knoop 2013), ease payments, and send money to relatives. Many in sub-Saharan Africa rely on centuries-old institutions such as rotating savings and credit associations or informal family networks. Most banks are foreign-owned, and over 80 percent of the population have no access to formal finance. Islamic financial institutions have existed in Southwest Asia and North Africa for over a millennium;

many states there set up banking systems after decolonization, plus post office savings accounts. In contrast, the Chinese Communist Party set up a state-owned banking system after the 1949 revolution, while India has a mix of public and private banks and has better developed stock markets than China. Finally, foreign banks play a large role in Latin America; the region has a long history of financial instability (Knoop 2013). Financial services are also an equity issue. Unequal access to banking and finance usually disadvantages women in developing states (Singh 2013). Well-constructed financial systems help the poor more than the rich but only in well-regulated institutional environments (Knoop 2013). State financial systems in developing states are improving only slowly.

Despite the centrality of industrialization, the tertiary sector is a crucial factor in developing states. Transport, communications, wholesale and retail institutions, and finance are often inadequate for business-to-business and consumer markets, but they are improving.

Business Forms

As discussed repeatedly, part of the shift to "capitalism" involves moving from landownership or partnership power to ownership of a business. Most businesses in developing states are called "small- and medium-sized enterprises" (SMEs). Many SMEs follow a typical growth strategy. Early on, they rely on personal networks; later, or as institutions and networks improve, they increasingly address customer needs and competitors (Farashahi and Muhammad 2014). While global North MNCs still dominate developing states, some SMEs have grown so much that they are now regional MNCs. Latin American and Indian MNCs are expanding by combining business models from the global North with local traditions, using deep knowledge of domestic markets (Casanova 2009; Kumar and Mhapatra 2009).

Businesses in developing states also borrow management practices from developed ones—effectively, borrowing a cultural system and adapting it to local situations. They focus on the concrete practices of running businesses (Murinde 2007; McGee 2009; Boubaker and Nguyen 2014). These include board composition, differing legal contexts, international standards, accounting and financial reporting, financial markets, and shareholder and stakeholder rights. Organizational forms, and relationships between corporations and states, vary tremendously. Studies analyzing both in developing countries address institutional patterns of specific states, orientations to export-led growth or to growth through domestic consumption, and the influences of global economic forces. Still, there are regional patterns: Latin American firms and economies concentrating on "the non-tradable and resource-extracting sectors" which leave manufacturing to MNCs; "strong protection against external financial volatilities" in China, India, and sometimes Brazil; and "patrimonial capitalism … particularly in Russia and the Arab region" with informal patron–client relations structuring most businesses (Schedelik et al. 2021:516).

Labor unions have long been part of developing states. Trade unions early on developed among dock workers, railway workers, and miners (Levine et al. 2022).

Unions increased in number and size with industrialization, especially in Brazil, South Africa, and the Philippines. Nevertheless, unionizing efforts respond to global economic forces, and "[t]heir success. .. is sporadic and very uneven across countries" (Levine et al. 2022). Organizing is easier in public sectors and more difficult in the informal economy (see below). Overall, the more developed a state is, the denser unions are (Levine et al. 2022).

Consumerism?

Industrialization increased consumerism, which is spreading globally (Stearns 2006). It often manifests differently in developing states than it does in developed ones.

As mentioned, consumerism started to spread beyond North Atlantic states by 1914. This trend intensified in the twenty-first century. The Soviet Union improved living standards, yet most people knew that Western living standards were much higher, and this played a delegitimating role in 1989. Before 1949, urban Chinese increasingly bought Western products, but the revolution reversed consumerism. It returned after economic reforms started in 1978 and spread throughout the country. Latin American imitation of Europe shifted to partial imitation of the United States, combined with traditional forms such as the Brazilian *Carnaval*, music, and foods. In cities, Africans started serious consumerism, with a traditionalist backlash. Southwest Asia and North Africa have been simultaneously religiously resistant and eagerly imitative (Stearns 2006).

The current situation in developing states varies. In the poorest countries, low purchasing power limits production and retail, which limits jobs, which of course limits purchasing power. Rural subsistence producers are also subsistence consumers, thus mostly staying out of the commercial economy. In cities, nearly everyone is integrated into it, but most consumption supports survival, not consumerism. In cities, poor and wealthy neighborhoods exist side-by-side; among Indian, Chinese, Mexican, Brazilian, Russian, and Nigerian elites, consumption is similar to that of the global North: middle classes afford good clothing, cars, houses, furniture, and leisure activities.

Most people in developing states cannot enjoy this lifestyle, and there are good reasons to question it. Still, increasing consumption is part of economic growth and thus development.

The Macroeconomy

Concepts applied to developed states can help us understand the economies of developing state. The diagram of the macroeconomy from Chap. 9 requires two modifications.

The first difference is that the informal economy in most developing states is much larger than it is in developed states. For instance, the U.S. informal sector was estimated as almost 9 percent of the economy in 2000, while the informal sectors (legal activities only) of Georgia, Bolivia, Panama, and Azerbaijan constituted over half the economy; indeed, shadow economy estimates in the year 2000 ranged from 13 percent of GDP in China to 35 percent in Algeria to almost 68 percent in Bolivia (Schneider et al. 2010). Informal employment is often a starting point for rural-to-urban migrants and the unemployed. Often it involves subcontracting with businesses in the formal economy. The dotted line through production in Fig. 10.2 indicates that informal economic activity varies greatly among developing states. As SMEs grow, become more capital-intensive, and more connected to banks, suppliers, and customers, they tend to formalize, obtaining licenses, following labor regulations, and paying taxes (Chant 2014). Also, compliance with applicable laws can be high in manufacturing and low in services in one country and the reverse in another (Chant 2014), depending on institutional environments.

Scholars, international agencies, and states used to criticize the informal economy. It kept taxes from the state, imposed dangerous working conditions on workers, and competed with the formal sector. Informal economic activity in developing countries grew greatly after 1990, often due to neoliberal economic policies and recessions, sparking a rethinking. Formal employment opportunities are often limited, and informal businesses increase a state's economic potential. Also, avoidance is sometimes a response to overregulation of the formal sector (Chant 2014).

The second difference is that subsistence production is still important to many developing economies. In Fig. 10.2, it appears in consumption: people consume what they produce (what they grow), so what they produce never reaches the commercial economy. The dotted line more represents divisions within households rather than between them: some consumption depends on household production, and some depends on earning wages. This is even more obscured in official estimates than informal production is. It hinders development because economic growth involves commercial economic activity, but it is difficult to change.

Developing states are deeply intertwined in the global economy. Figure 10.3 is unrevised from Chap. 9, but it is even more important in understanding developing states than for understanding developed ones. On the one hand, most developed states generate higher shares of their GDP from foreign trade than developing states do. On the other, many economists argue that developing states' economies usually rely too strongly on a few primary-sector exports (e.g., mining, timber), a situation economists call the "resource curse." They say that this distorts both economy and society. Super-profits ("rents") from mining, timber, or agriculture entrenches the owning elite, distorting innovation and political activity. Short-term excessive earnings lead to badly planned state spending and extreme vulnerability to global markets. Finally, "Dutch disease" involves currency appreciation from profits of overproduction of the resource, thus overpricing the economy's other exports (Halland et al. 2015:4–7[3]). Others think that mining had positive local economic

[3] This edited volume focuses on mining oil, gas, or minerals, yet these general problems extend to agricultural monocultures, even though the specifics are very different.

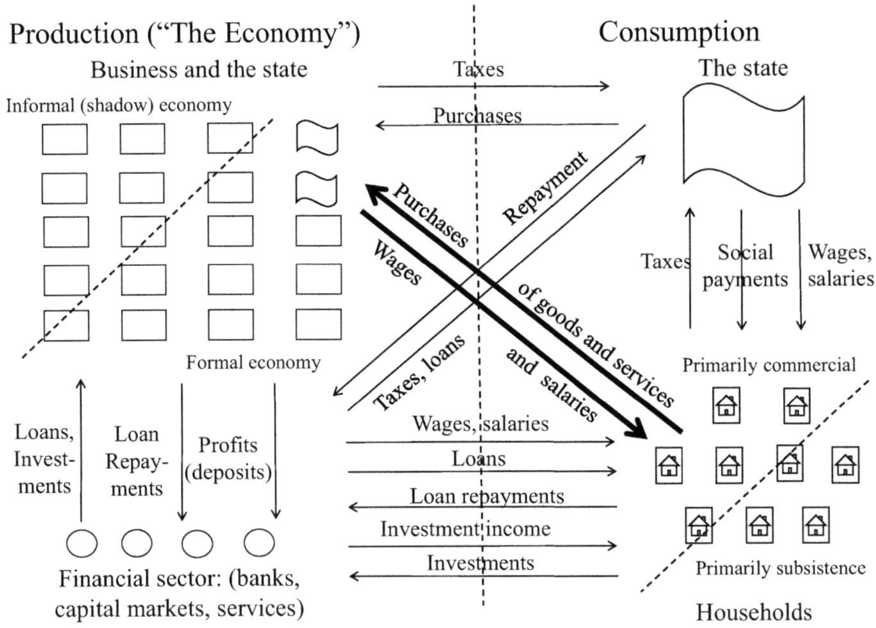

Fig. 10.2 The macroeconomy of developing states

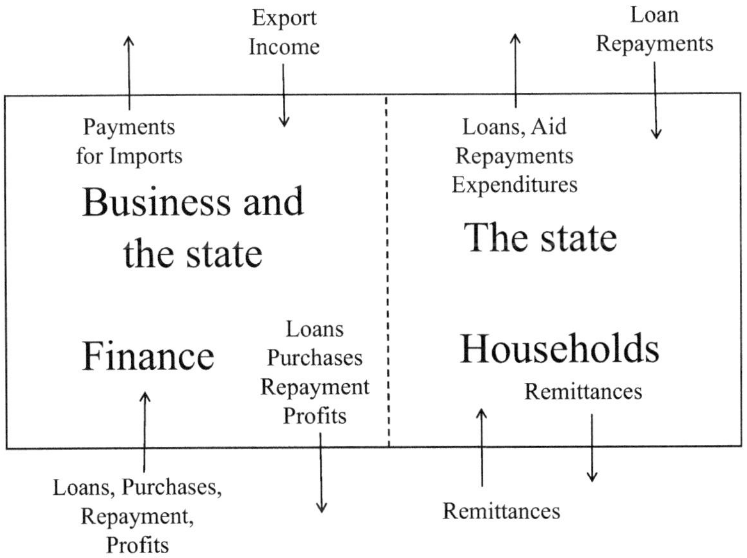

Fig. 10.3 The balance of payments for developing states

effects from reliance on local suppliers, expertise, repairs, and labor (Crowson 2009). Since many developing states have increased dependence on mining for export revenues (Roe and Dodd 2018), perhaps this analysis is incomplete.

Economists identify other macroeconomic implications of overreliance on the primary sector. When a developing state relies on export earnings from one physical resource, its global price influences nearly everyone's prosperity (Todaro and Smith 2015). Also, states in the global North benefit from many developing states together experiencing the resource curse since it lowers global commodity prices. Commercial agriculture faces a variant of this problem. Commercial farmers often grow only few crops for export. This again increases foreign earnings for the entire economy. However, when global market prices for a crop are high, farmers in many developing countries devote a greater percentage of the entire economy to it, flooding global markets and driving prices down. Then developing states' economies crash (Mazoyer and Roudart 2006). Also, much agricultural produce that could go to domestic cities goes instead to exports to bring in money that is then resent overseas to banks to pay back restructured loans. Many scholars see this as a neo-imperial trap set by the global North for developing states (McMichael 2012).

Overreliance on primary sector exports demonstrates the importance of diversifying an economy, and industrialization leads to economic diversity. Fully industrialized economies never specialize in one product. While textiles were the central industry of the British Industrial Revolution, it generated "spillovers," including cloth dyes that changed chemistry, increased commercialization, increasing demand for iron, increased and steam engine innovation, which affected transportation. Indeed, several Persian Gulf states have high GDP per capita (GDP/cap) measures, so they purportedly qualify as developed states, but they also rely extremely heavily on petroleum exports, thus being more like developing states. For instance, as of 2012, 45 percent of Qatar's GDP came from petroleum and natural gas (Ibrahim and Harrigan 2012).

Finally, three problems combined—rural-to-urban migration, low agricultural productivity, and cash crop production for export—strain developing states' economies. Cash crops could be sold in cities but are exported; rural-to-urban migration literally eats up improvements in agricultural productivity; and both may require food imports (ironically, to agriculturally oriented states), possibly increasing a state's dependence on the global economy.

The informal economy, subsistence production and consumption, and export concentration help us understand, and be more cautious of, statistics on rates of economic growth over the twentieth century. Table 10.3 shows that between the two world wars, economic growth was quite low in what later became developing states. During *les trentes glorieuses* (1945–1975), economic growth was high worldwide. It stagnated in the 1970s, though China kept growing quickly (Table 10.4).

Since 2001, East Asia and the Pacific, South Asia, and Southwest Asia and North Africa have continued high growth, while Latin America and sub-Saharan Africa have experienced slowing growth (World Bank Open Data). Tables 10.5 and 10.6 allow comparison of sectoral shares of GDP and show great regional differences.

Table 10.3 Economic growth rates by region, 1913–2001

Region	1913–1950	1950–1973	1973–2001
Eastern Europe	0.86	4.86	1.01
Latin America	3.42	5.38	2.89
China	−0.02	5.02	6.72
India	0.23	3.54	5.12
Other Asia (excluding Japan)	0.82	5.17	5.41
Africa	2.57	4.43	2.89

Source: Maddison (2006:640)

Table 10.4 GDP/cap (by region, non-high-income) (constant 2010 US$)

Region	1996	2016
Europe and Central Asia	4460.4	9027.3
East Asia/Pacific	1531.6	5798.8
Latin America/Caribbean	7029.1	9239.2
Middle East/North Africa	2852.1	4226.7
South Asia	659.8	1691.2
Sub-Saharan Africa	1160.4	1638.5

Source: World Bank Open Data

Indeed, many world regions seem to be moving toward being "developed."[4] Every region increased the share of services in the economy, and all decreased shares of agriculture and industry. East Asia and Latin America seem to be approaching developed country shares of GDP for agriculture, while South Asia and sub-Saharan Africa lag behind, but they do seem to be on the path toward this situation.

Data on employment shares of GDP in sectors (Table 10.6) tell a different story. Many regions seem to be reducing primary and even secondary employment, while South Asia and sub-Saharan Africa are doing this much more slowly; most sub-Saharan Africans are still mostly subsistence producers. Also, these are shares of a very small pie. South Asia has grown 2.5 times in 20 years, while sub-Saharan African economies have not even come near doubling (Table 10.7).

The preceding discussion tends to obscure the tremendous diversity between developing states. It derives from physical factors—such as different climates, soils and minerals, land profiles (mountains, plains), plants and animals, and shapes (coastlines/landlocked, rivers)—and from unique institutional histories—including combinations of foraging, horticultural, pastoral, agricultural, and industrial elements, politics, and gender and kinship structures.

There are variations inside established cultural regions. The post-socialist or "transition" economies of Eastern Europe are middle-income, somewhat industrialized economies with educated populations, but with little history of market operations and profit-seeking businesses. Some, such as Slovenia, are doing very well; others, such as Albania, are not. East Asia is seeing tremendous development, while

[4] Again, I use agriculture, industry, and services as imperfect proxies for primary, secondary, and tertiary sectors.

Table 10.5 Value added (constant 2010 US$) (millions) (non-high-income)

	Agriculture		Industry		Services, etc.	
Region	1996	2016	1996	2016	1996	2016
Eastern Europe and Central Asia	142,879.83	220,733.46	500,945.06	1,098,331.43	907,466.67	1,985,346.95
East Asia/Pacific	513,614.80	1,025,600.28	1,110,127.48	5,356,624.32	1,071,044.73	5,474,080.97
Latin America/Caribbean	173,972.45	284,010.15	1,162,379.00	1,641,285.35	1,720,179.05	3,105,352.33
Middle East/North Africa	77,790.39	139,284.84	301,659.73	643,024.45	318,364.53	760,118.93
South Asia	247,884.91	432,046.21	229,834.79	833,937.23	298,953.19	1,486,117.17
Sub-Saharan Africa	117,809.30	280,664.77	250,775.23	452,295.60	313,125.36	893,509.78

Source: World Bank Open Data

Table 10.6 Share of GDP by sector (non-high-income states in regions)

	Agriculture value added (% of GDP)		Industry value added (% of GDP)		Services, etc., value added (% of GDP)	
Region	1996	2016	1996	2016	1996	2016
Europe and Central Asia	12.6	6.2	35.8	32.1	51.6	61.7
East Asia/Pacific	9.7	5.1	45.8	39.2	35.4	51.4
Latin America/ Caribbean	6.4	5.6	31.0	26.9	62.7	67.4
Middle East/North Africa	14.4	10.2	40.4	34.2	45.2	55.7
South Asia	27.6	17.8	30.3	27.8	42.1	54.4
Sub-Saharan Africa	22.4	17.5	34.9	24.4	42.6	58.1

Source: World Bank Open Data

Table 10.7 Structure of employment, 1996–2016 (percentage of total employment, non-high-income states, modeled International Labor Organization [ILO] estimate)

	Agriculture		Industry		Services, etc.	
Region	1996	2016	1996	2016	1996	2016
Europe and Central Asia	25	15	29	27	46	58
East Asia/Pacific	50	24	26	25	25	51
Latin America/Caribbean	23	14	21	21	56	64
Middle East/North Africa	29	22	25	28	45	50
South Asia	61	44	15	23	24	33
Sub-Saharan Africa	67	57	9	11	24	31

Source: World Bank Open Data

parts of South Asia are rich, and others remain very poor. Latin American countries vary from very poor (Paraguay) to middle income (Brazil and Mexico). Sub-Saharan Africa is the poorest region on the planet, but it has recently experienced strong economic growth. North African countries are much richer. Even within these regions, however, there is variation: Morocco, Algeria, and Egypt are very different from each other.

Economic development is difficult. Commercialization is difficult in agriculture and may cause overreliance on the primary sector. There are several paths to industrialization, markets are often inadequate, and consumers often stay out of markets—or are forced into them. The global economy reduces developing states' autonomy through global commodity prices, exportation of business traditions, Structural Adjustment Programs, and encouragement of consumerism.

Development and the Environment

Reliance on primary products creates environmental problems for developing states different from those in developed ones. A 1960s "Green Revolution" improved agricultural yields, but it also caused problems. Deforestation contributes to climate change (Penna 2015:307), while mechanized commercial agriculture and transport increase carbon emissions. Developing states also face animal habitat destruction, soil erosion, and urban pollution.

Deforestation is a major issue. It increased in Asia and in Latin America in the 1920s, while tropical Africa was being reforested (Williams 2006). Globally to 1950, about 235 million hectares of forest was cut, mostly in sub-Saharan Africa and Latin America; from 1950 to 1980, 318 million hectares were cut (Williams 2006:373). This continued until recently: "Between 1990 and 2005, there was a net reduction in global forest area of 66.4 million" hectares (Runyon and D'Odorico 2016:10), mostly in South America and Africa, with a slight net gain in Asia.

Deforestation has varied purposes and effects. Over half of tree cutting in developing states goes to household fuelwood; the rest goes to the global timber industry (Williams 2006:463) for lumber, paper, furniture, and other products. In developed states, tree regrowth is standard; in developing states, most deforested land goes to either commercial agriculture or ranching, both by small farmers and by large corporations. From the 1920s into the 1950s, just over half of deforested land in what are now developing states went to commercial agriculture, the rest to grassland. From then to the 1980s, the great majority went to new cropland (Williams 2006:373; Runyon and D'Odorico 2016). This helps drive global climate change, both from reduced carbon sequestering and from intensified use of fossil fuels (Simmons 2008; Ponting 2007; Penna 2015; Headrick 2020). Deforestation also creates added rain runoff, reduced precipitation, and desertification (Runyon and D'Odorico 2016).

The "Green Revolution" had major effects. This massive research program developed and transferred new farming techniques to developing states, often funded by U.S. private foundations. Researchers developed high-yield cash crops, including new strains of rice, wheat, maize, and soy; encouraged the increased use of fertilizers and mechanization common in developed states; and provided new cultivation methods. The effort neglected "secondary" products such as millet or sorghum, many kinds of livestock, and labor-intensive mixed methods (Mazoyer and Roudart 2006:450; Brooke 2014). Through the 1960s and 1970s, it led to a corporatization of agriculture, requiring "huge amounts of water" that caused water shortages (Brooke 2014:533), plus massive increases in nitrogen fertilizer that release greenhouse gases and end up in rivers and oceans, creating algae blooms and dead zones (Brooke 2014:562–563). Mechanization and cash crop exports increased carbon emissions (Simmons 2008). The Green Revolution thus mostly benefited large farmers, and its benefits have leveled off (Ponting 2007).

Urbanization and economic growth also increased energy use in developing states. In 1920, people in the Unites States emitted 17 times "as much CO_2 as all other people combined." However, this ratio was 11:1 in 1950 and 5:1 by 2010, and it continues to fall (Brooke 2014:551; see also Penna 2015). Automobile ownership, air conditioning, and refrigeration is rising quickly across South, Southeast, and East Asia, the most populated region in the world. This requires increased use of oil, coal, and other energy sources, all of which accelerate climate change (Penna 2015:263–264, 298–299).

New consumer uses and industrialization increased air pollution rapidly in developing states. For instance, the "expansion of heavy industry in eastern Europe after 1945" led to air concentrations of heavy metals far above those safe for humans. Other problems included dead rivers, contaminated food, and unsafe water. The collapse of Soviet socialism reduced pollution levels "dramatically" (Ponting 2007:358 and 359, respectively). Industry polluted the region around Cubatão and São Paolo, Brazil, so badly that infant mortality increased; "there were no fish in the rivers, no birds in the area, nearly all the insects had died and all the trees were dead" (Ponting 2007:359). Urban sanitation is usually inadequate, leading to unsafe drinking water and polluted air, which increases mortality in and around informal settlements (Lees 2015).

As in the global North, environmental movements developed across the global South (Bello 2007). Grassroots movements in China, the Philippines, India, and others appeared in response to projects that promised to be environmentally destructive or already had such a track record, such as large hydroelectric dam projects. Other movements have responded to industrial pollution, deforestation, groundwater depletion, and climate change (Bello 2007). Movements by Indigenous peoples usually have a strong environmental component (Stavenhagen 2013).

Environmental problems in developing states include deforestation and agricultural commercialization, both contributing to global climate change. Urbanization and economic growth also produce local problems such as desertification and air and water pollution.

Inequalities in Developing States

Another difference between developing and developed states is economic inequality. I first provide historical background, then discuss absolute and relative poverty. Finally, I discuss how economic inequality relates to ethnic divisions. Lack of data prevents much historical analysis before 1950 or so (Piketty 2014; Alvaredo et al. 2018). Still, forcing kin-ordered peoples with subsistence agriculturalists, commercial farmers, urban merchants and wage workers into colonies and then into states automatically produced vast inequalities.

Economic inequality in developing states only sometimes followed that of developed states. From 1910 to 1950, the top centile (1 percent) took about 15–18 percent of all income in India and 22–25 percent in South Africa, the Dutch East Indies, and Argentina (Piketty 2014:327). From decolonization into the 1980s, the income share of the top centile fell only in these countries (the Dutch East Indies becoming Indonesia) (Piketty 2014:327). Economic inequality was also low in socialist states, but inequality "took other, non-monetary forms, such as privileged access to particular shops and vacation centers for the political elite" (Alvaredo et al. 2018:73). In some Latin American states, all of Africa, and Southwest Asia, inequality remained high through the twentieth century (Haller 2015:229; Alvaredo et al. 2018). Since the 1980s, income inequality has either remained high or increased in most developing states (Piketty 2014:327). Wealth inequality in Southwest Asia, North Africa, and parts of Latin America has lowered a bit; income distribution has become less unequal in China and India as both experience capitalist economic growth (Alvaredo et al. 2018:52–53). Today, the highest inequalities are in southern sub-Saharan Africa and Latin America, but they are also variable. For instance, Egypt and India are less unequal than the United States is, though they are not as rich either. Stated differently, relative poverty is bad in the United States, but almost nobody is starving to death. Many are starving in sub-Saharan Africa. And a few countries are so poor that they do not collect statistics.

Again, absolute poverty occurs nearly exclusively in developing states. The international poverty line is at $2.15 a day in U.S. dollars, adjusted for purchasing power (Hasell et al. 2022). Economists estimate subsistence production equivalents for this standard (Nafziger 2006:27–30). Over one billion people are below this poverty line. At this level, over 75 percent of a family's budget is spent on food. About half of the absolute poor are malnourished, and hundreds of millions are severely malnourished. This saps their immune systems, so diseases are common. Infant mortality is about 10 percent. Two-thirds lack access to clean water; more have problems handling solid human waste. Over half are illiterate and innumerate. Average life expectancy is 45 years (Beaudoin 2007; Todaro and Smith 2015). Over half of the world's absolute poor are in Africa, which relies more on horticultural subsistence production than any other continent. In fact, most of the absolute poor are rural; only a fifth live in cities.

Inequality is not monolithic. Every state has an economic elite, a middle class, and the poor. In one sense, the proportions of people in each category define "developed" or "developing" states. And in many developing states (e.g., Kyrgyzstan or Sri Lanka), starvation to death is less likely than it is in some developed states (worldlifeexpectancy.com).

Development scholars actively debate economic inequality. Economists assume or argue that subsistence producers should commercialize, arguing that market-based economic growth has alleviated more poverty than any other historical change (e.g., Seavoy 2000). The number of people globally living on $1 a day (in inflation-adjusted U.S. dollars) was dramatically reduced since 1970 by hundreds of millions; state-by-state analyses show dramatic reductions in poverty always follow periods of capitalist economic growth (Helpman 2004). As discussed, critics argue that inequality under capitalism is inherently political, especially in the developing world, since the global North makes the rules in which developing states operate. They (McMichael 2012; Hickel 2015) argue that surplus extraction occurs primarily on a global scale, though of course on national, regional, and local levels too. A few claim (Hickel 2015) that the percentage of people in absolute poverty has fallen, but absolute numbers have not.[5]

Scholars debate whether inequality supports or impedes economic growth. Empirical results are mixed. They previously argued that inequality creates

---

[5] This contradicts Fig. 8.5, which shows a fall in absolute numbers of those in absolute poverty.

incentives for entrepreneurial activity and innovation; redistributive welfare systems reduce entrepreneurial incentives; and high-income individuals save more, providing money for investment. Others argued that developing countries first became more unequal, then became less unequal with full economic development (Kuznets 1955). Recent studies show that inequality slows economic growth. Not everyone agrees; some argue that it is a long-run relationship (Herzer and Vollmer 2012); that highly unequal land ownership lowers economic growth (Helpman 2004); or that badly enforced property rights for the poor lower growth and well-enforced rights improve it (Davis and Hopkins 2011). Others argue that high inequality simply does not reduce economic growth (See Helpman 2004; Todaro and Smith 2015:Chap. 5.) Critics also object to capitalist economic growth. Industrialization causes tremendous ecological damage. Kin-ordered groups operate with strong social obligations, while capitalism thrives on individualistic competition, and commercialization mostly benefits global Northern MNCs, not developing states much (Hickel 2015).

Developing states do have social welfare programs. Socialist states helped egalitarianism: the second-lowest Gini coefficients globally are still in Eastern Europe. The lowest are in Scandinavia, with significant state redistribution programs. While nearly all states globally introduced social welfare programs (Schmitt et al. 2015), many poor states introduced only some, which often do not reach everyone, and they experience colonial legacies: former British colonies are less likely, and former French colonies are more likely, to have such programs (Schmitt et al. 2015). Some say redistribution has no net effect on economic growth (Lindert 2004).

While most measures of economic inequality use individual or household data, inequalities between groups are also important. In contrast to economists' money-measured "vertical" measures, some scholars discuss "horizontal" inequality in which ethnic groups have differential access to resources (Cederman et al. 2013). Previously, research on civil wars only addressed individual-level or state-level data. Instead, data "on interactions between the state and actors at intermediate levels of aggregation" show that ethnic groups are important in these conflicts (Cederman et al. 2013:3). In some developing states, usually in Latin America, race is the major issue. In Africa and Asia, however, the problem is often between two culturally different groups who have lived in the same land for centuries. Ethnicity is composed primarily of descent (real or fictive), language, and religion; it is somewhere between "primordial" and "in eternal flux"; and political activity often structures it (Haller 2015:32). In fact, "economic inequality is produced by a triad of three processes: class formation, social stratification and ethnic differentiation" (Haller 2015:43). Sub-Saharan Africa is the most ethnically heterogeneous region of the world; the slave trade, colonization, and decolonization often remade ethnic identities, especially through ethnic jockeying for position and state-heavy clientelism (or neopatrimonialism).

Scholars worry that ethnic fragmentation weakens developing states, especially when ethnic groups are shut out of state power or economic resources. The poorest excluded groups in each state are most likely to fight, especially large groups. The Fragile States Index tracks many indicators of potential and actual state failure (http://fsi.fundforpeace.org/indicators), and at least two, "Group Grievance" and "Uneven Economic Development," are ethnic. Other indicators, such as "Factionalized Elites," "Refugees and IDPs" (internally displaced persons), and "State Legitimacy" are indirectly related to ethnicity. Ethnic fragmentation usually causes conflict when combined with other problems, which disrupts economic development.

Most developing states are not currently experiencing civil wars; few situations break out into open conflict, and problems stay low-level or are managed. Democratization allows ethnic groups to channel their grievances into electoral efforts (Achen and Bartels 2016), and many developing states' political parties are based on ethnicity. Inter-ethnic coalitions and deals are a normal part of African politics, especially since post-1990 democratization (Arriola 2012). Group-based patronage, funded in part by resources from businesses, is also typical.

Thus, economic inequality in developing states is different from that of developed states. While the global number of absolute poor is shrinking over time, many millions still live in absolute poverty. If high inequality does depress economic growth, areas like sub-Saharan Africa and Latin America face difficult prospects. And horizontal inequality, based not on immigration but on ethnic groups being on their own lands, can be destructive.

Demography and Development

Ethnic fragmentation is part of larger demographic patterns. Migration accelerates urbanization, pressuring agriculture and creating environmental problems. Constantly growing populations strain family and state resources. Reducing mortality has been mostly, not totally, accomplished; reducing fertility is more difficult, but it may enable a demographic "dividend" that helps stimulate economic growth.

Ethnic diversity varies regionally, resulting from historical patterns. Long-standing agrarian patterns still influence Asia today. Most Chinese people usually speak either Mandarin or Cantonese, yet many minority populations exist in China. Repeated invasions over millennia created the ethnic diversity of South Asia, with hundreds of languages (Trautmann 2016). Spanish and Portuguese colonization produced contemporary Latin American ethnoracial hierarchies, generally composed of a European-ish elite, a Euro-Indian (*mestizo*) or Afro-European middle group, and Indigenous peoples or African descendants, all overlapping imperfectly with economic inequalities. Still, each state is unique. The African presence dominates Brazil, while indigenous influence is more important in Peru (Chasteen 2006). Nineteenth-century immigration from southern and eastern Europe went to North America and to Chile, Argentina, Uruguay, and southern Brazil. Central Asian states were Russian possessions, then Soviet republics, depositing a Russian-speaking minority in each.

European colonization had even more effects on today's African and Southwest Asian states. As discussed, during decolonization, all political leaders involved assumed states as societies, partly because of the potential results of renegotiating borders. Most leaders knew of the Latin American postindependence wars and the communal violence in the partition of British India, in which hundreds of thousands, perhaps millions, had died (Kulke and Rothermund 2016). Undoing colonial borders might create a multitude of civil wars, so the new Organization of African Unity (OAU), an IGO composed of new African states,[6] unanimously confirmed the old colonial borders. However, the clearer and more rigid borders of states today (Feuer 2016) meant that borders join many linguistic and ethnic groups together, and they split many other groups. The ethnic map of Nigeria below (Fig. 10.4) illustrates the joining problem; many of those groups are in adjacent states. Also, some groups controlled resources such as fertile land, oil fields, mines, or ports while others were poor. Most sub-Saharan states face the same problems of border splitting and state-joined groups, as do many Asian states. Avoiding civil wars at independence meant ongoing problems, such as identification with the central government. Religious patterns overlap with these linguistic ones, with Christian,

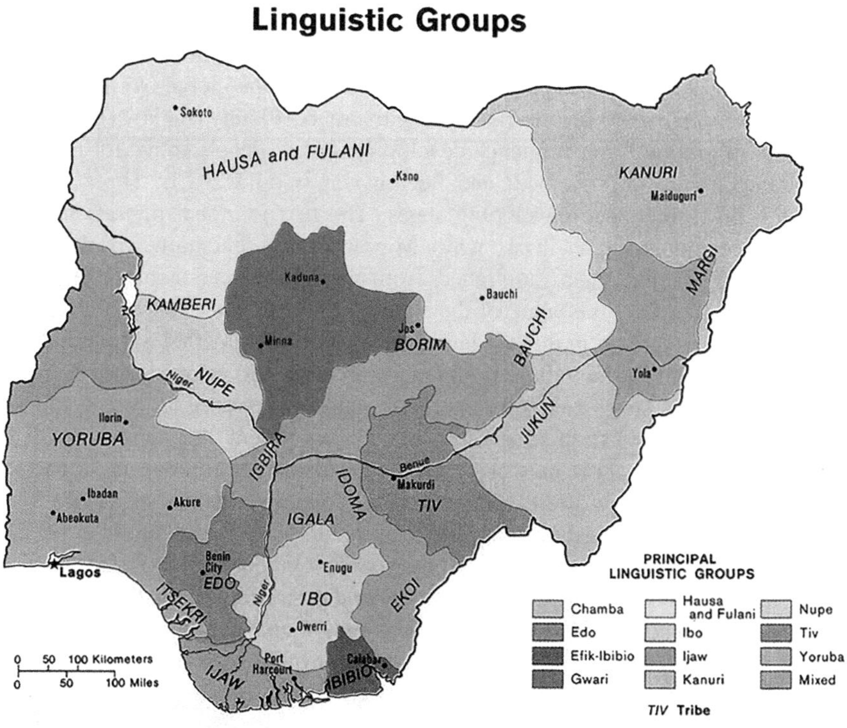

Fig. 10.4 Nigerian linguistic groups and international borders. Source: Central Intelligence Agency (1979)

---

[6]The African Union, another IGO with much the same membership, replaced the OAU in 2001.

Muslim, and traditional belief-systems all coexisting, often uncomfortably, throughout the region. Sub-Saharan Africa is the most ethnically diverse region in the world (Fearon 2003).

Indigenous peoples in Africa, Asia, and Latin America complicate demographic situations even more (Stavenhagen 2013). A few groups, mostly in Brazil and Papua New Guinea, remain completely uncontacted by the global system (Survival International 2018). Furthermore, the definition of "indigenous peoples" is murky in developing states. Despite Latin American *mestizaje*, some peoples have survived more or less "pure," racially and ethnically, into the present, however oppressed they have been. The definition of "Indigenous" excludes state-based or state-dominant peoples in Africa and Asia, and Asian governments sometimes say "tribal" rather than "Indigenous" (Stavenhagen 2013).

Protecting Indigenous peoples is difficult. Corporations and governments want Indigenous lands for mining, commercial agriculture, dams, roads, or tourism. Most Indigenous peoples feel a deep spiritual connection to their lands; this shapes their subsistence practices, beliefs, kinship and political structures, and collective identities, so political action to keep their land is vital to their survival. They organized in the 1980s with the support of the United Nations Commission on Human Rights and since then have formed international alliances to lobby governments for protection, carry out strikes and protests, and mobilize boycotts in the global North. Many are adapting to capitalism to survive but often with great difficulty (Iankova et al. 2016).

Global migration flows influence developing states. People move for two basic reasons, economic improvement and fleeing war (Gould 2015). Three kinds of migration flows influence developing states. The first is global migration, which influences developing states in two ways. Migrants often send money back home as remittances. For some states, remittances from abroad are a major source of income, sometimes higher than the value of the state's principal export (Gould 2015:169). Remittances have grown in the last few decades, but only in 2004 did international agencies recommend including remittances in national balances of payments. Second, "source" states and "destination" states experience different impacts. Destination states receive productive people who work and thus increase GDP, but as mentioned, these people may experience ethnoracial discrimination. Impacts on source states vary. They often lose their most productive people, especially those who migrate for more education—the aforementioned "brain drain"—but out-migration from high-density areas can ease unemployment (Gould 2015).

The second kind of migration flow, rural–rural migration, is mostly state-driven. Indonesia, Malaysia, China, Brazil, Ethiopia, and others have sponsored such projects. It rarely seems to have much impact on development. In one type, states develop underused land. These programs often try to relieve pressure on highly populated areas (Gould 2015). In the other, states restructure existing land use, which is often part of land reform in which private lands were redistributed.

Finally, the rural-to-urban flow is far more important than rural-rural migration, and it tends to originate in family decisions, not state policies. Unfortunately, the data are "notoriously difficult to obtain" (Leone 2009:383). Transportation systems make circular migration more common than ever; millions move between their home rural areas and urban areas seasonally or on short visits. Circular migration is more common in Africa; Latin American rural–urban migration tends to be permanent (Gould 2015). Of course, permanent migration increases city sizes (Leone 2009). In the last century, cities in developing states grew at rapid rates. In 1940, one-third of the population of Latin America lived in cities; by 1980, two-thirds did. By 2010, some, such as Argentina, were 90 percent urban (Gilbert 2013). China has rapidly urbanized. Official definitions of "urban" changed in 1985. In that year, 21.1 percent of the population was urban; by 2010, almost half was (Stapleton 2013). South Asia has urbanized more slowly. In 1950, India was approximately 17 percent urban, while Nepal and Bhutan were less than 3 percent. By 2007, Pakistan was 41.4 percent and India was 33.4 percent urban, while Nepal was 25.5 and Bhutan was 21.2 percent urban (Kidambi 2013). In 1960, Southwest Asia and North Africa were 35 percent urban; in 2016, they were 65 percent urban. Finally, sub-Saharan Africa was 15 percent urbanized in 1960 and, in 2016, was 38 percent urban (World Bank Open Data). In the past, rural-urban movement caused most urban population increase. Now urban births are the main driver of urban population growth. Consequently, urbanization has been much faster in the global South than it was in the global North (Gould 2015).

Since people's lives in developing states are usually better in cities than in the country (Elbow 1998; Gould 2015), rural–urban migration continues. Still, urbanization causes problems in all developing states. The most obvious one is informal or "squatter" settlements. These usually start on vacant land where people camp; construction of temporary buildings draws others, soon creating communities. They face collective problems such as finding clean water, sewage and garbage disposal, lack of electricity, and crime. They may provide themselves public goods illegally, such as tapping into electricity or cable supplies or creating a trash dump (Elbow 1998; Gould 2015). As they grow, eviction becomes harder, as it provides governments with negative publicity. Sometimes the state buys the land under squatter settlements and grants it to the squatters, which makes possible further investment in homes, leading to reasonably safe and healthy neighborhoods. The process is decades-long and may halt at any time by recalcitrant landowners or shifting government policies. Also, sometimes building codes are too expensive to follow, so upgrading does not happen (Elbow 1998; Gould 2015).

Urbanization creates both state-wide problems and solutions. Since city people rarely grow food but easily organize politically, developing states' leaders sometimes appease them by importing or subsidizing food. Mortality and fertility are lower in cities, so urbanization is a key component of the demographic transition (Gould 2015). Nearly all states have reduced their mortality rates through sanitation and immunization; oral rehydration of children; and improved nutrition and parental education (Livi-Bacci 2017:172–176). From 1950–1955 to 2010–2015, life expectancy increased around the world from 46.5 years to 70.6 years. In the 1950s, people in developed states could expect to live into their 60s, but in the decolonizing world, average life expectancies were in the 40s, with African expectancies the lowest, averaging 37.8 years. Now the average life expectancy in developing states is over 65 years and almost 60 years in Africa (Livi-Bacci 2017:170). Fertility still remains higher than mortality, so the global demographic transition is not yet complete. Global total fertility rate (TFR) was 5 children in 1950–1955. In 2010–2015, however, it was 2.5, a dramatic drop in 60 years. TFR in Africa was 6.6 in 1950–1955, 5.9 in India, 6.3 in Western Asia, and 5.9 in Latin America. In 2010–2015, it was 4.3 in Africa, 2.5 in India, 2.9 in Western Asia, and 2.2 in Latin America (Livi-Bacci 2017).

The demographic transition is having age-structure effects in some developing states similar to the effects it has in developed ones, but it has hardly changed the age-structures of others. As economic development proceeds, age-structures tend to shift slowly from the older, preindustrial pattern to the newer modern pattern. For instance, in Argentina, each successively older generation is only a little smaller than younger ones until after age 50, when mortality rates rise quickly. However, Zambia continues to experience the preindustrial pattern of many children, some young adults, fewer middle-aged adults, and very few elderly (U.S. Census 2024).

Because TFR is still above replacement in many developing states, their populations keep growing. In 1950, Africa had 229 million people, Latin America 169 million, and Asia 1.39 billion. In 2015, the equivalent numbers were 1.19 billion, 634 million, and 4.39 billion. Projections are that, by 2050, Africa will have almost 2.5 billion people, Latin America 784 million, and Asia 5.267 billion (Livi-Bacci 2017:229) because of what demographers call "population momentum." This means that while fertility rates fall, population still grows for some decades since people enter their reproductive years and the elderly tend to live longer. Population only stops growing a generation or so after TFR falls to replacement.

Population growth is much-debated in development. Some say it has benefits: poverty leads to large families for old-age security, developed states use far more resources, and people are too concentrated in cities or too sparse on farmland. Others reply that population growth slows economic development because the economy must grow as quickly as the population just to keep up, and then more so to achieve true growth. Agricultural productivity must likewise keep pace with population growth. Finally, population growth requires spending on education to be spread more thinly, lowering its quality (Nolan and Lenski 2011; Todaro and Smith 2015).

Recently, demographers are pondering a demographic "dividend," defined as "socioeconomic gains arising from fertility transitions" (Eloundou-Enyegue 2013). Children are expensive; having fewer frees up resources such as extended education, more food, and improved health for existing children, investment in self-employment for the parents, and spending on consumer goods—all of which support economic growth. Previous generations still experience higher mortality than subsequent, healthier ones, which also frees resources. Finally, falling fertility enables women to work more. Thus, economists increasingly argue that the demographic transition has been an under-recognized yet major component of modern economic growth. Its dividend has appeared in recent history, such as in East Asia and Ireland. In contrast, sub-Saharan Africa has had a substantial mortality decline but has only started a fertility decline, which has led to a dramatic population increase and economic stagnation. Research continues on the demographic dividend.

Developing states experience ethnic fragmentation, rural–urban migration and urbanization, and the demographic transition possibly leading to a "dividend" very differently, yet these historical demographic trends continue to influence all developing states in some way.

Ideas in Development

Ethnic fragmentation is clearly important politically. While education supports economic growth, it also aids state cohesion. The influences of science, mass media, and religion in developing states are both similar to and different from those in developed world states.

The "national question" was a major issue right after decolonization. Leaders of Asian and African new states faced group–splitting and group-collecting patterns from the European colonial borders, which colonizers had worsened by pitting groups against each other (Ezrow and Frantz 2013:56–58). Many African and Asian states are thus ethnically fragmented (Fearon 2003). International agencies proposed "nation-building" to foster civic nationalism—inculcate a common national identity—among ethnically varied peoples. These agencies missed other issues militating against national identities: subsistence systems were inadequate to fund states; small-scale peoples are deeply kin-oriented, rarely affiliated with large, abstract entities; and horizontal ethnic inequalities are always potential problems.

The concept of nationalism in developing states seems to have fared badly since decolonization,[7] in part because of regional specifics. For instance, Latin American states generally share a political style that includes Catholicism, the *hacienda* (commercialized plantation), authoritarianism, militarism, restrictions on civil liberties, and international economic dependence, though democratic traditions seem to be taking root (Wiarda and Mott 2003). Many Asian states contain "sub-state nationalist" movements (Bertrand and Laliberté 2010). These groups see themselves as nations aspiring to self-determination, sometimes to independence, and are territorially based. Asian states rarely accommodate such groups. Sub-Saharan African nationalist leaders faced so much cultural fragmentation that they relied on colonizers' languages (e.g., English and French) as foundations for national unity.

Mass education is a new version of agrarian cultural assimilation. The global diffusion of education after 1950 may seem imposed by the global system, but developing states had reasons to adopt it. The old reasons still applied: it increases cohesion (Turchin 2003), causing leaders fewer problems, and it inculcates national values and loyalty to the state (Green 1990). Economists argue that mass education builds skills and knowledge and spurs innovation, supporting economic growth (Maclure et al. 2009; Barro and Yi 2015) which increases tax revenues and

[7] A scan of indexes of recent political science books focusing on the trends in global regions shows very few entries for "nationalism." Instead, they mostly focus on democratization and state building.

strengthens leaders' power. Finally, female education reduces fertility, aiding the demographic dividend (Barro and Yi 2015).

As discussed, schools globally have taught broadly the same general curriculum since the 1980s. It consists of language, mathematics, natural science, social science, arts, and physical education. About one-third of the curriculum was and is devoted to language (reading and writing), and math takes up close to 20 percent. Regional variations in the curriculum exist; for instance, in the 1980s, North African and Southwest Asian states devoted about 10 percent of instructional time to religion, while Eastern European states omitted it (Benavot et al. 1991:51).

The details are more complex. Some states have recently started "mother tongue" education, i.e., in people's own language instead of official state languages. By international (i.e., North Atlantic) standards, work before age 16 is exploitation; in most developing states, people believe that children should work to contribute income to improve household well-being, especially girls (Colclough 2014)—but those working are not in school. Furthermore, terms of SAPs of the 1980s meant that after 1990, many developing countries scaled back educational expenditures; many charge families fees for attendance, limiting enrollments. Developing states report enrollment data, while household surveys reveal actual attendance, and this reveals that official enrollment tends to be about 5 percent higher than actual attendance (Roser and Ortiz-Ospina 2016). Still, for the first time ever in world history, over half of humanity is literate and numerate in some language (Brint 2017).

In developed states, scientific development supports innovation that drives economic growth and improves quality of life. In developing states, three problems inhibit turning enhanced literacy into support for scientific institutions. While primary school enrollments are increasing, secondary enrollments are minimal (Roser and Ortiz-Ospina 2016), so few people go to university. An even smaller proportion earn PhDs, so the pool of scientists is tiny. Also, those who qualify often go abroad for graduate training; many do not return, exacerbating brain drain, though providing remittances. Developing states can rarely afford universities and scientific facilities, but some have them anyway, which reduces funding for primary and secondary education.

The practice of science itself in developing states is also difficult. Europe and North America publish 38.6 percent and 28.4 percent of scientific papers, respectively; Asian publications constitute 24.3 percent, almost half of them originating in China. Latin American and African papers (both north and south of the Sahara) together constitute 8.7 percent of scientific papers globally. Difficulties for scientists in the Arab world (Hanafi and Arvanitis 2016) sometimes apply to Latin America, sub-Saharan Africa, and parts of Asia. Developing states often do not support institutions that provide incentives for entrepreneurship, an educated population, an innovation system (or the "research pipeline") of firms, research centers, and universities. These are rare across Africa and in Southwestern Asia, and only somewhat more prevalent in Latin America. Sometimes developing-state scholars are connected to each other more through scientific INGOs than through national associations (Drori et al. 2003).

Mass media in developing states are much more vibrant than science is. After World War II, U.S. media dominated globally (Tunstall 1977), so scholars and policymakers discussed U.S. "cultural imperialism." In recent decades, ownership and broadcast content of mass media has localized (Tunstall 2008). For instance, the United States provided corporate investment, technological transmission, and show formats to radio and television systems in Latin America. Now media systems owners in Mexico and Brazil are Mexican and Brazilian, and a major show format is the *telenovela*, which shares only some elements of the U.S. soap opera (Sinclair 2013). For-profit ownership dominates media systems in Latin America, while Cold-War, state-run systems in India and China spawned large for-profit systems side-by-side with state media (Sinclair 2013). One *Times of India* executive admitted copying U.S. journalistic practice but said that Indians had "indigenized it to suit the Indian social and political market." This seems successful; the *Times* and its subsidiaries are financially healthy (Thussu 2012:277). Some argue that privatization and deregulation have fostered a "vibrant and expanding media sphere" in India (Thussu 2012). The Indian film industry, "Bollywood" (Bombay-Hollywood), produces three times the films that Hollywood does and exports them around South Asia. Also, the Indian television industry grew from a few channels in the 1970s to 461 channels by 2009 (Thussu 2012).

This "indigenizing" trend also affected the public sphere. Arab states' radio broadcasting in the 1920s and television broadcasting in the 1950s helped media organizations get around low literacy rates. These were usually state-owned, following the British Broadcasting Corporation model rather than U.S. commercialism. Foreign news and satellite dishes brought competition in the 1990s, and states loosened their censorship regimes (Gunter and Dickinson 2013). As mentioned, Latin American broadcasting heavily borrowed U.S. formats and commercial traditions: private companies owned most newspapers and radio and television stations, and there were weak, publicly owned stations. Government control increased during mid-twentieth-century dictatorships. In the 1990s, international media corporations moved in and created large media conglomerates in Brazil and Mexico and national producers in smaller states, with governments ceding control. Since then, states have tried to make media more democratic by supporting public broadcasting and increasing commercial competition (Matos 2012).

In contrast to the secular discourse common in media in the global North, religious discourse is far more influential in the public sphere in developing states. As mentioned, scholars debate whether secularization is making headway in the developing world (Haynes 2007; Deneulin and Bano 2009; Bullard 2016). Both Christian and Muslim proselytization is succeeding in many states. Some argue a version of the common idea that religion is a comfort: the most vulnerable populations are the most religious, while the most economically secure either are or are becoming increasingly secular (Norris and Inglehart 2004).

Sources of moral behavior in developing states are diverse and complex. All have adopted some version of European law. Even some who had not been colonized "chose the French *Code Civil* as the basis for new legal systems," often as a means to "participate more readily in the new economic and commercial order," thus emphasizing practical law rather than morality. Latin American states had already adopted Spanish and Portuguese legal traditions. African and Asian states, even China, adopted "European forms" of law (all quotes, Pirie 2021:364). However, the content of laws in many states was very different from these European forms. In Southwest Asia, Shari'a is the law of the land in Saudi Arabia, Iran, Sudan, and Pakistan, and it predominates in some areas, such as Northern Nigeria, while only Shari'a family laws—but not other principles—are applicable to Muslims in India (Glenn 2010:227). Western and Islamic law are mixed in many areas of Southeast Asia. During colonization, Hindu sources of law actually became "officially recognized law" in British India (Glenn 2010:312), influencing Indian law today. Chinese legal scholars adapted Japanese and German models in the 1930s and Soviet socialist law in the 1950s and 1960s. Chinese law still remains heavily Confucian, being less rights-oriented and more obligations-oriented (Glenn 2010; Ruskola 2012). Law and morality in much of sub-Saharan Africa is diverse and derives from multiple sources: traditional law, Christianity and Islam, and European law, modified by traditions of military power over the state and the concentration of power in state leadership (Sacco 2012).

Among many, perhaps most, Indigenous peoples, belief and behavior "are tightly interwoven," so "there is no separation of law and morals" (Glenn 2010:73). They usually have a spiritual relationship to their physical environments, so one must treat the land, plants, and animals morally (Glenn 2010). Between people, "[i]njury to a member is an injury to the group," and restitution, not punishment, is often the goal (Glenn 2010:72). Some groups must obey state laws, while others may govern themselves semiautonomously. Still, some peoples have used state courts to reinforce their local traditions (Glenn 2010), just as they have used United Nations resolutions and international law to protect themselves, as mentioned (Stavenhagen 2013).

Global influences have exported education, science, mass media, and law to developing states, which hybridize them, but the assimilating power of educational and legal systems has, so far, been no match for ethnic fragmentation, lower scientific progress, and religious identities.

Gender, Kinship, and Development

Many factors are changing gender and kinship in developing states. Women's political movements developed over the twentieth century and have had important effects on marriage and family policies. As discussed, some were trying to change millennia-old traditions, while others were attempting to reclaim recently lost autonomy. In the last few decades IGOs, INGOs, and researchers have focused on gender inequalities, many of which continue.

Women's political movements developed across Africa, Asia, and Latin America right after World War I. They usually allied with independence movements and responded closely to local conditions. In Africa, women were "among the most vocal advocates of independence" (Tripp 2017:39). They rioted and protested tax policies in Nigeria, Tanganyika, and Bujumbura. They were also active in Mozambique, Guinea, Mali, and Cameroon. In Asia, the Indian independence movement encouraged women's mobilization (Poulomi 2017). Women were active in the Chinese May Fourth Movement from 1919. The reconstituted Chinese state (1949) implemented a new Marriage Law in 1950; this forbade arranged marriage, provided the freedom to divorce and remarry, and mandated equality in work (Zheng 2017). Early Latin American movements were suffrage-oriented, emphasizing women's different life-mission within Catholic beliefs, thus often gaining Church support (Friedman 2017).

The 1960s and 1970s feminist movements in the global North also sparked efforts in the international system to reduce gender inequality. Three overlapping groups—academic researchers studying gender and development; women activists in developing states; and IGOs and INGOs—have worked together since then. Boserup's research (1965) helped start the academic field "women in development"; those who followed her argued that excluding women's issues impeded economic development. Awareness that this approach did not address underlying gender inequalities led to the "gender and development" (GAD) approach, including men as part of analysis (Duflo 2012; Bradshaw 2013). There are multiple theoretical positions.[8]

International agencies and movement organizations (IGOs and women's INGOs, or WINGOs) include gender in all programs because academic research findings have practical implications. For instance, "societies that discriminate by gender tend to experience less rapid economic growth and poverty reduction than societies that treat males and females more equally" (Bradshaw 2013:4–5). Duflo (2012) argues that this is a reciprocal relationship: improving women's lives increases economic growth and reduces poverty, and economic growth and reducing poverty improve women's lives.

Education is possibly the most important improvement. Girls' education tends to result in later female marriage, lower fertility, and improved child health and nutrition, all of which improve a state's chances for economic development as well as improving personal and familial outcomes. In turn, lowering fertility increases the demographic dividend, reduces the strain on women's household resources, and increases women's bargaining power in forming marriages (Therborn 2004; Bradshaw 2013; Todaro and Smith 2015). Conversely, "educational discrimination against women hinders economic development" (Todaro and Smith 2015:397). Nevertheless, the emphasis is on primary education. Interestingly, women in universities perform better than men do (Bradshaw 2013), and more women than men are enrolled in universities around the world (World Bank 2012).

The World Bank is probably the IGO that consistently focuses attention on gender in development. The Bank certainly supports and acts upon the claims about girls' education lowering mortality and increasing economic growth (Duflo

[8] An enormous literature addresses gender and development. Duflo (2012) and Bradshaw (2013) are only a start.

2012:1064). However, its program is utilitarian and instrumentalist: the goal is economic growth "with equity concerns being somewhat secondary" (Bradshaw 2013:5). Critics argue that improving women's lives and social positions is worthwhile in itself (Duflo 2012:1053), which the Bank acknowledges, though it immediately says, "It is also smart economics" (World Bank 2012:xx). To be fair, the World Bank is in fact a bank, so it should address economic issues. While gender equality is only one issue among many for the Bank (Independent Evaluation Group 2010:ix–x), its emphasis on gender equality matters: the World Bank influences nearly all developing states and other IGOs. Even WINGOs skeptical of the Bank can use its stance on gender to increase their own legitimacy with states and IGOs.

In contrast to this international effort, women's movements in developing states mostly lost influence during the 1960s and 1970s wave of dictatorships across developing states. In Africa, women's movements were often depoliticized; ruling parties incorporated them and forced them to focus on practical women's issues (Moghadam 2017; Tripp 2017). After Mao Zedong's death, there was an anti-feminist reaction, implemented partly through privatization and market socialism (Zheng 2017). The "household responsibility" system pressured women to have sons; combined with the one-child policy, sex-selective abortions and infanticide increased (Zheng 2017). Women often mobilized against repressive Latin American right-wing regimes, though sometimes they supported them (Friedman 2017). The resurgent Indian women's movement in the 1970s was largely upper-caste, mostly working with courts and administrators.

Women's movements in developing states revived during democratization and neoliberal globalization. In response to "market socialism," women's movements returned in China. Activists influenced policy mostly behind the scenes; recently, activists are largely independent of the state, networking online to engage in disruptive, public activities (Zheng 2017). The Indian women's movement increasingly includes Dalit, poor, and minority women, focusing especially on fighting sexual violence (Poulomi 2017). Responding to increased Islamic fundamentalism, Moroccan and Tunisian women's movements allied with government officials and civil society groups to enact policy changes (Moghadam 2017:357); in Algeria, women acted independently of the state. In sub-Saharan Africa, movements mobilized against civil wars, acted as go-between negotiators, or worked at democratization (Tripp 2017). Latin American women's movements now work with Indigenous peoples or on racial issues (Friedman 2017).

In addition to activism, commercialization of subsistence producers to foster economic growth also has gendered effects. Globally, women work slightly more often on farms than men do (World Bank 2012); however, this varies by country. For instance, in Nepal and Albania, women work more on farms; in Ethiopia and Nicaragua, men do. However, in smallholder farms around the world, women always do a great deal of labor, and the tasks vary:

> In Sub Saharan Africa, women tend to engage in the production of maize, fruits and vegetables and they also rear small animals. In Asia, women are heavily involved in rice production, and in Latin America they tend to manage small plots on the farm, cultivating food crops and raising poultry and small livestock (Rapsomanikis 2015:17).

Women in Africa and Asia more often cultivate subsistence crops, especially in African horticultural groups, while men more often farm commercially (World Bank 2012). Also, women as farmers have less access to economic and social capital and market information than men do, and men more often control property and income rights, though this varies greatly by state and by global region (Duflo 2012; World Bank 2012).

Women face difficulties when they seek wage employment. As women's labor is more often subsistence-oriented than men's is, it is more difficult to track. Around the world, women earn less than men do even with similar educational qualifications (Duflo 2012), more so in developing states than developed ones. Women take part-time and informal jobs in childcare and household work more often than men do (World Bank 2012) since in every society, women do more household work than men. In the formal sector, laws in many developing states restrict women's hours and the industries in which they can work. Childbearing imposes a penalty on women's productivity, reducing their experience and lowering their wages around the world (World Bank 2012). Both employment segregation by gender—women work more in services, while men work more in industry—and gender discrimination reduce women's wages. These overall patterns obscure a tremendous diversity of women's experiences. In some places, women are industrial workers, especially in textiles. Elsewhere, women often work in services as domestic servants, retail salespeople, or office workers. In West Africa, women are usually traders in local markets, not men. Women are very restricted in Saudi Arabia: they are required to have a chaperone everywhere in public and only lately are allowed to drive. In contrast, women in Turkey are members of government, professors, and businesswomen.

Activism for legal change, girls' education, and commercialization are all having effects on marriage patterns and family structures, though only slowly (Therborn 2004). Latin American men have long resisted increasing women's rights, but many laws changed after 1980, and parental influence on marriages has decreased. In Africa, there has been a "drift away from parental arrangement of marriages to mix of individual choice and parental involvement" (Therborn 2004:117). Increasingly, African constitutions and court systems recognize statutory law over customary law when they conflict, affecting family law, which includes genital cutting and political office quotas (Tripp 2017). However, brideprice is still common, and polygyny and wife-beating are still somewhat accepted. Patriarchal norms remain strong in North Africa and Southwest Asia: although Turkish Kemalist revolutionaries passed legislation against arranged marriages in the 1920s, possibly half of marriages were arranged in the 1960s (Therborn 2004). More Egyptian women than not were in arranged marriages in the 1990s, and genital cutting was nearly universal. Divorce across the Maghreb remains easy for men and difficult for women, parental consent for marriage remained important, polygyny is still accepted, and husbands still control wives' public movements. The Indian Supreme Court increasingly favors bilateral over patrilineal kinship and supports "daughters' rights of inheritance" (Therborn 2004:96), but old practices continue in rural areas. Even in urban areas, arranged marriages flourish, though prospective couples have more refusal rights. Parentally arranged marriages occur less in China, mostly in rural areas only. In urban areas, neolocality increasingly replaces patrilocality. Finally, Soviet bloc states all passed gender-equal laws, though men mostly remained in charge of the state and organizations (Therborn 2004).

Gender inequality still occurs in families in developing states, but girls' and boys' treatment varies by economic situation. When families are in economic crisis, they treat boys better (Duflo 2012). Girls die among droughts and die of diarrhea more than boys do, and during droughts, women in some African states are accused of being witches and are harassed or killed. In South Asia, "families are far more likely to take an ill boy than an ill girl to a health center" (Todaro and Smith 2015:398). In contrast, when parents have adequate resources, they tend to treat children equally. For instance, households reduce consumption by the same amount whether a girl or boy is born; households that have assets to handle a drought do not show increased female mortality; and, even in India, vaccination rates between boys and girls are nearly identical. When fertility declines, fathers want daughters to have more opportunities (Duflo 2012).

Still, neither education nor poverty reduction leads to full gender equality. Sex-selective abortion continues even in affluent households based on kinship structures and laws (e.g., dowry in India; China's one-child policy, plus patriarchal norms). For instance, in China, Pakistan, and India, the sex ratio is 94:100; in Saudi Arabia, it is 80:100; all are below the statistical norm of 103 or 105 females to every 100 males.[9] These result partly from ultrasound and sex-selective abortion. Additionally, families tend to believe boys need more education than girls (Duflo 2012:1057).

As we have seen, same-sex sexual relations and diverse gender expressions have existed throughout the past around the world. However, the situation for LGBT people in developing states is extremely diverse today, influenced by conceptual and political models from the global North in the last few decades. The LGBT political movements of the 1970s and 1980s in global Northern states have spread throughout the world since then (Rayside 2020). Legal rights for LGBT people have expanded in South America but not in "much of Africa, the Caribbean, Russia, and the Middle East" or in South Asia (Adam 2020:37) despite long histories of such practices in those areas. Even where formal legal rights exist, LGBT people often face discrimination and violence, and where populations resist accepting them, political leaders often use anti-LGBT rhetoric and foster moral panics (Adam 2020).

Women's activism, the international system, commercialization, and state efforts are all shifting gender relations in developing states, but the situation is extremely varied. Still, the twentieth and twenty-first century global trend was toward gender equality.

[9] The *birth* ratio is skewed male at roughly 105:100 around the world (with these exceptions). Males start dying at faster rates than females right away, so including all ages means normal ratios skew slightly female-dominant.

Conclusion

Efforts at helping states develop in the last century are part of global Industrial Revolution, the third transformation in world history. This history of "developing states," an artificial concept itself, demonstrates how the global ecumene operates. The most obvious change has been the globalization of the state, creating strange and difficult ethnoracial combinations in the global South. Both the sham emphasis on democracy early in the twentieth century and its increasing reality later, plus the tremendous emphasis on economic growth—with its seemingly beneficial effects for health, longevity, and gender, and its dire environmental potentials and socially embittering increases in economic inequalities—show how important the global social system is to "development." The demographic transition and ending absolute poverty seem to have similar trajectories: great progress but not full accomplishment. Education spreads widely at the primary level, and mass media follow closely, but science is often difficult to develop. Still, all such changes are incomplete, consisting of the adoption of alien traditions and the adaptation of them locally, which leads to mixing, hybridization, and syncretism.

Furthermore, the story illustrates very long-term trends in world history. The state is now so universal that many social scientists take it for granted, but it is the consequence of a long trend: the growth and elaboration of human groups. "Development" means other long continuities are reversing, such as mortality and fertility or the long-lasting inequality between women and men in states. Economic inequalities continue to increase, another very long-term trend. Finally, tremendous economic growth in East Asia already shows that Europe's worldwide dominance will be, over the long term, ephemeral. This brings us to the issue of humanity's future.

References

Achen, Christopher H., and Larry M. Bartels. 2016. *Democracy for Realists: Why Elections Do Not Produce Responsive Government*. Princeton, NJ and Oxford: Princeton University Press.

Adam, Barry. 2020. Political Economy, Sexuality, and Intimacy. In *The Oxford Handbook of Global LGBT and Sexual Diversity Politics*, ed. M. J. Bosia, S. M. McEvoy, and M. Rahman, 31–42. New York: Oxford University Press.

Alvaredo, Facundo, Lucas Chancel, Thomas Piketty, Emmanuel Saez, and Gabriel Zucman. 2018. *World Inequality Report 2018*. Paris: World Inequality Lab.

Ardant, Gabriel. 1975. Financial Policy and Economic Infrastructure of Modern States and Nations. In *The Formation of National States in Western Europe*, ed. C. Tilly, 164–242. Princeton, NJ: Princeton University Press.

Arriola, Leonardo R. 2012. *Multiethnic Coalitions in Africa: Business Financing of Opposition Election Campaigns*. New York: Cambridge University Press.

Austin, Gareth. 2016. Sub-Saharan Africa. In *A History of the Global Economy: From 1500 to the Present*, ed. J. Baten, 316–352. Cambridge, UK: Cambridge University Press.

Barro, Robert, and Chong-hwa Yi. 2015. *Education Matters: Global Schooling Gains from the 19th to the 21st Century*. New York: Oxford University Press.

Bauer, Peter. 2000. *From Subsistence to Exchange: And Other Essays*. Princeton, NJ: Princeton University Press.

Beaudoin, Steven M. 2007. *Poverty in World History*. London: Routledge.

Bello, Walden. 2007. *The Environmental Movement in the Global South*. Amsterdam, Netherlands: The Transnational Institute. Retrieved December 30, 2022. https://www.tni.org/en/article/the-environmental-movement-in-the-global-south.

Benavot, Aaron, Yun-Kyung Cha, David H. Kamens, John Meyer, and Suk-ying Wong. 1991. Knowledge for the Masses: World Models and National Curricula, 1920–1986. *American Sociological Review* 56 (1): 85–100.

Bénétrix, Agustín S., Kevin Hjortshøj O'Rourke, and Jeffrey G. Williamson. 2017. Measuring the Spread of Modern Manufacturing to the Poor Periphery. In *The Spread of Modern Industry to the Periphery Since 1871*, ed. K. H. O'Rourke and J. G. Williamson, 13–32. New York: Oxford University Press.

Bertrand, Jacques, and André Laliberté, eds. 2010. *Multination States in Asia: Accommodation or Resistance*. New York: Cambridge University Press.

Boserup, Ester. 1965. *The Conditions of Agricultural Growth: The Economics of Agrarian Change Under Population Pressure*. New York: Aldine Publishing.

Boubaker, Sabri, and Duc Khuong Nguyen, eds. 2014. *Corporate Governance in Emerging Markets: Theories, Practices and Cases*. New York: Springer.

Bradshaw, Sarah. 2013. *Women's Role in Economic Development: Overcoming the Constraints*. Paris: United Nations Sustainable Development Solutions Network.

Bravo, Elisa Wiener. 2011. *The Concentration of Land Ownership in Latin America: An Approach to Current Problems*. Rome: International Land Coalition.

Brint, Steven. 2017. *Schools and Societies*. 3rd ed. Palo Alto, CA: Stanford University Press.

Brooke, John L. 2014. *Climate Change and the Course of Global History: A Rough Journey*. New York: Cambridge University Press.

Bruton, Henry J. 1998. A Reconsideration of Import Substitution. *Journal of Economic Literature* 36 (2): 903–936.

Bullard, Gabe. 2016. The World's Newest Major Religion: No Religion. *National Geographic.*. Retrieved February 17, 2018. https://news.nationalgeographic.com/2016/04/160422-atheism-agnostic-secular-nones-rising-religion/.

Carletto, Calogero, Paul Corral, and Anita Guelfi. 2017. Agricultural Commercialization and Nutrition Revisited: Empirical Evidence from Three African Countries. *Food Policy* 67:106–118.

Casanova, Lourdes. 2009. *Global Latinas: Latin America's Emerging Multinationals*. Houndmills, UK: Palgrave Macmillan.

Cattaneo, Olivier, Michael Engman, Sebastián Sáez, and Robert M. Stern, eds. 2010. *International Trade in Services: New Trends and Opportunities for Developing Countries*. Washington, DC: The World Bank.

Cederman, Lars-Erik, Kristian Skrede Gleditsch, and Halvard Buhaug. 2013. *Inequality, Grievances, and Civil War*. New York: Cambridge University Press.

Central Intelligence Agency. 1979. "Ethnolinguistic Map of Nigeria." https://maps.lib.utexas.edu/maps/africa/nigeria_linguistic_1979.jpg. Retrieved June 21, 2023. From Perry-Castañeda Map Collection, University of Texas, Austin (https://maps.lib.utexas.edu/maps/nigeria.html).

Chang, Ha-Joon, and Ilene Grabel. 2004. *Reclaiming Development: An Alternative Economic Policy Manual*. London: Zed Books.

Chant, Sylvia. 2014. The Informal Economy in Cities of the South. In *The Companion to Development Studies*, ed. V. Desai and R. Potter, 200–206. New York: Routledge.

Chasteen, John Charles. 2006. *Born in Blood & Fire: A Concise History of Latin America*. 2nd ed. New York: W.W. Norton.

Cheibub, José Antonio. 2007. *Presidentialism, Parliamentarism, and Democracy*. New York: Cambridge University Press.

Colclough, Christopher. 2014. Female Participation in Education. In *The Companion to Developmental Studies*, ed. V. Desai and R. Potter, 475–480. London, UK: Hodder Education.

Crowson, Philip. 2009. The Resource Curse: A Modern Myth? In *Mining, Society, and a Sustainable World*, ed. J. P. Richards, 3–36. Heidelberg, Germany: Springer.
Dahl, Robert A. 1998. *On Democracy*. New Haven: Yale University Press.
Davis, Lewis, and Mark Hopkins. 2011. The Institutional Foundations of Inequality and Growth. *Journal of Development Studies* 47 (7): 977–997.
Deneulin, Séverine, and with Masooda Bano. 2009. *Religion in Development: Rewriting the Secular Script*. London & New York: Zed Books.
Dicken, Peter. 2011. *Global Shift: Mapping the Changing Contours of the World Economy*. 6th ed. New York: The Guilford Press.
Drori, Gili S., John W. Meyer, Francisco O. Ramirez, and Evan Schofer. 2003. *Science in the Modern World Polity: Institutionalization and Globalization*. Stanford, CA: Stanford University Press.
Duflo, Esther. 2012. Women Empowerment and Economic Development. *Journal of Economic Literature* 50 (4): 1051–1079.
Elbow, Gary. 1998. Coming Out of the Country. In *The New Third World*, ed. A. Gonzalez and J. Norwine, 2nd ed., 52–74. Boulder, CO: Westview Press.
Eloundou-Enyegue, Parfait M. 2013. Harnessing a Demographic Dividend: Challenges and Opportunities in High and Intermediate Fertility Countries. United Nations Department of Economic and Social Affairs, Expert Paper No. 2013/7. New York: United Nations.
Evans, Peter. 1995. *Embedded Autonomy: States and Industrial Transformation*. Princeton, NJ: Princeton University Press.
Ezrow, Natasha, and Erica Frantz. 2013. *Failed States and Institutional Decay: Understanding Instability and Poverty in the Developing World*. New York: Bloomsbury.
Farashahi, Mehdi, and Faiza Muhammad. 2014. Strategies of SMEs in Emerging Economies: A Dynamic Approach Based on Local/Global Concerns. In *Building Businesses in Emerging and Developing Countries: Challenges and Opportunities*, ed. E. V. Chrystosome and R. Molz, 175–189. London and New York: Routledge.
Fearon, James D. 2003. Ethnic and Cultural Diversity by Country. *Journal of Economic Growth* 8:195–222.
Feuer, Bryan. 2016. *Boundaries, Borders and Frontiers in Archaeology: A Study of Spatial Relationships*. Jefferson, NC: McFarland & Co.
Food and Agriculture Organization of the United Nations. 2014. *"The State of Food and Agriculture" in Brief*. Rome: Food and Agriculture Organization.
Friedman, Elizabeth Jay. 2017. Seeking Rights from the Left: Gender and Sexuality in Latin America. In *Women's Movements in the Global Era: The Power of Local Feminisms*, ed. A. Basu, 265–298. Boulder, CO: Westview Press.
Gerring, John, Strom C. Thacker, and Carola Moreno. 2009. Are Parliamentary Systems Better? *Comparative Political Studies* 42 (3): 327–359.
Ghanem, Rima, and Joerg Baten. 2016. Middle East, North Africa, and Central Asia. In *A History of the Global Economy: From 1500 to the Present*, ed. J. Baten, 208–239. New York: Cambridge University Press.
Gilbert, Alan. 2013. Latin America. In *The Oxford Handbook of Cities in World History*, ed. P. Clark, 483–503. New York: Oxford University Press.
Glenn, H. Patrick. 2010. *Legal Traditions of the World: Sustainable Diversity in Law*. New York: Oxford University Press.
Goldstone, Jack A., Robert H. Bates, David L. Epstein, Ted Robert Gurr, Michael B. Lustik, Monty G. Marshall, Jay Ulfelder, and Mark Woodward. 2010. A Global Model for Forecasting Political Instability. *American Journal of Political Science* 54 (1): 190–208.
Gould, W. T. S. 2015. *Population and Development*. 2nd ed. New York: Routledge.
Green, Andy. 1990. *Education and State Formation: The Rise of Education Systems in England, France, and the USA*. Blasingstoke, Hampshire: Macmillan.
Grugel, Jean, and Matthew Louis Bishop. 2014. *Democratization: A Critical Introduction*. 2nd ed. New York: Palgrave Macmillan.
Gunter, Barrie, and Roger Dickinson. 2013. The Changing Nature of News and the Arab World. In *News Media in the Arab World: A Study of 10 Arab and Muslim Countries*, ed. B. Gunter and R. Dickinson, 1–20. New York: Bloomsbury Academic.

Halland, Håvard, Martin Lokanc, Arvind Nair, and with Sridar Padmanabhan Kannan. 2015. *The Extractive Industries Sector: Essentials for Economists, Public Finance Professionals, and Policy Makers*. Washington, DC: World Bank.

Haller, Max, and in collaboration with Anja Eder. 2015. *Ethnic Stratification and Economic Inequality Around the World: The End of Exploitation and Exclusion?* Burlington, VT: Ashgate.

Hamilton, Gary G., Misha Petrovic, and Benjamin Senauer, eds. 2011. *The Market Makers: How Retailers are Reshaping the Global Economy*. New York: Oxford.

Hanafi, Sari, and Rigas Arvanitis. 2016. *Knowledge Production in the Arab World: The Impossible Promise*. New York: Routledge.

Hare, Paul, and Gerard Turley, eds. 2013. *Handbook of the Economics and Political Economy of Transition*. London and New York: Routledge.

Hasell, Joe, Max Roser, Esteban Ortiz-Ospina, and Pablo Arriagada. 2022. "Poverty." Retrieved November 19, 2022. https://ourworldindata.org/poverty.

Haynes, Jeffrey. 2007. *Religion and Development: Conflict or Cooperation?* New York: Palgrave Macmillan.

Haque, M. Shamsul. 1996. The Contextless Nature of Public Administration in Third World Countries. *International Review of Administrative Sciences* 62 (3): 315–329.

Haque, M. Shamsul. 2010. Rethinking Development Administration and Remembering Fred W. Riggs. *International Review of Administrative Sciences* 76 (4): 767–773.

Headrick, Daniel R. 2020. *Humans Versus Nature: A Global Environmental History*. New York: Oxford University Press.

Helpman, Elhanan. 2004. *The Mystery of Economic Growth*. Cambridge, MA: Harvard University Press.

Herzer, Dierk, and Sebastian Vollmer. 2012. Inequality and Growth: Evidence from Panel Cointegration. *Journal of Economic Inequality* 10:489–503.

Hickel, Jason. 2015. *The Divide: Global Inequality from Conquest to Free Markets*. New York: W.W. Norton.

Humphreys, David. 2015. *The Remaking of the Mining Industry*. London: Palgrave Macmillan.

Huntington, Samuel P. 1991. *The Third Wave: Democratization in the Late Twentieth Century*. Norman and London: University of Oklahoma Press.

Hyden, Goran, Julius Court, and Ken Mease. 2003. *The Bureaucracy and Governance in 16 Developing Countries*, World Governance Survey, Discussion Paper 7. London: Overseas Development Institute.

Iankova, Katia, Azizul Hassan, and Rachel L'Abbé. 2016. *Indigenous People and Economic Development: An International Perspective*. New York: Routledge.

Ibrahim, Ibrahim, and Frank Harrigan. 2012. *Qatar's Economy: Past, Present, and Future*. Doha, Qatar: General Secretariat for Development Planning. Retrieved May 23, 2024. https://www.psa.gov.qa/en/knowledge/Doc/Studies/Qatars_Economy_Past_Present_and_Future_2012_EN.pdf.

Independent Evaluation Group (IEG). 2010. *An Evaluation of World Bank Support 2002–08: Gender and Development*. Washington, DC: The World Bank.

Joshi, Seema. 2008. *Growth and Structure of Tertiary Sector in Developing Economies*. New Delhi: Academic Foundation.

Kidambi, Prashant. 2013. South Asia. In *The Oxford Handbook of Cities in World History*, ed. P. Clark, 561–580. New York: Oxford University Press.

Kingsbury, Damien. 2016. Introduction. In *International Development: Issues and Challenges*, ed. D. Kingsbury, J. McKay, J. Hunt, M. McGillivray, and M. Clarke, 12–20. New York: St. Martin's.

Knoop, Todd A. 2013. *Global Finance in Emerging Market Economies*. New York: Routledge.

Kornai, János. 1992. *The Socialist System: The Political Economy of Communism*. Princeton, NJ: Princeton University Press.

Kulke, Hermann, and Dietmar Rothermund. 2016. *A History of India*. 6th ed. New York: Routledge.

Kumar, Nirmalya, Pradipta K. Mhapatra, and Suj Chandrasekhar. 2009. *India's Global Powerhouses: How They Are Taking on the World*. Boston: Harvard Business Press.

Kuznets, Simon. 1955. Economic Growth and Income Inequality. *American Economic Review* 45 (1): 1–28.
Lees, Andrew. 2015. *The City: A World History*. New York: Oxford University Press.
Leone, Tiziana. 2009. Population Trends in Developing Countries. In *The Companion to Development Studies*, ed. V. Desai and R. B. Potter, 2nd ed., 378–385. New York: Routledge.
Levine, S. B., Diane P. Koenker, Ronaldo Munck, John Christopher Lovell, Wolfgang Streeck, and David Brody. 2022. Organized Labour. *Encyclopedia Britannica*. .
Li, Jing, Zelie Zhang, Xianfeng Jin, Jiaquan Chen, Shaojia Zhang, Zong He, Sheng Li, Zhiming He, Haipeng Zhang, and He Xiao. 2018. Exploring the Socioeconomic and Ecological Consequences of Cash Crop Cultivation for Policy Implications. *Land Use Policy* 76:46–57.
Lindert, Peter H. 2004. *Growing Public: Social Spending and Economic Growth Since the Eighteenth Century*. New York: Cambridge University Press.
Livi-Bacci, Massimo. 2017. *A Concise History of World Population*. 6th ed. Malden, MA: Blackwell.
Ma, Debin. 2016. China. In *A History of the Global Economy: From 1500 to the Present*, ed. J. Baten, 188–207. Cambridge, UK: Cambridge University Press.
Maclure, Richard, Refaat Sabbah, and Daniel Lavan. 2009. Education and Development: The Perennial Contradictions of Policy Discourse. In *Introduction to International Development: Approaches, Actors, and Issues*, ed. P. A. Haslam, J. Schafer, and P. Beaudet, 367–383. New York: Oxford University Press.
Maddison, Angus. 2006. *The World Economy, 1–2001 AD. The World Economy: Volume 1: A Millennial Perspective and Volume 2: Historical Statistics*. Paris: OECD Publishing.
Madeley, John. 2008. *Big Business, Poor Peoples: How Transnational Corporations Damage the World's Poor*. 2nd ed. London: Zed Books.
Márquez, Xavier. 2017. *Non-Democratic Politics: Authoritarianism, Dictatorship and Democratization*. London: Palgrave.
Maruyama, Masayoshi, and Le Viet Trung. 2012. Modern Retailers in Transition Economies: The Case of Vietnam. *Journal of Macromarketing* 32 (1): 31–51.
Matos, Carolina. 2012. *Media and Politics in Latin America: Globalization, Democracy and Identity*. New York and London: I.B. Tauris.
Mazoyer, Marcel, and Laurence Roudart. 2006. A History of World Agriculture: From the Neolithic Revolution to the Current Crisis. Translated by J. H. Membrez. New York: Monthly Review Press.
McGee, Robert W., ed. 2009. *Corporate Governance in Developing Economies: Country Studies of Africa, Asia and Latin America*. New York: Springer.
McKay, John. 2016. Reassessing Development Theory. In *International Development: Issues and Challenges*, ed. D. Kingsbury, J. McKay, J. Hunt, M. McGillivray, and M. Clarke, 50–77. New York: St. Martin's.
McMichael, Philip. 2012. *Development and Social Change: A Global Perspective*. 5th ed. Los Angeles: SAGE.
McMillan, John. 2002. *Reinventing the Bazaar: A Natural History of Markets*. New York: W.W. Norton & Company.
Mikesell, Raymond F., and John W. Whitney. 1987. *The World Mining Industry: Investment Strategy and Public Policy*. Boston: Allen & Unwin.
Moghadam, Valentine M. 2017. Feminist Movements in the Maghreb. In *Women's Movements in the Global Era: The Power of Local Feminisms*, ed. A. Basu, 333–360. Boulder CO: Westview Press.
Mullings, Beverly. 2010. Neoliberalism. In *Encyclopedia of Geography*, ed. B. Warf, 2011–2015. Thousand Oaks, CA: SAGE.
Murinde, Victor. 2007. Accounting, Banking and Corporate Financial Management in Emerging Economies. In *Research in Accounting in Emerging Economies*, vol. 7. Boston: Elsevier.
Nafziger, E. Wayne. 2006. *Economic Development*. 4th ed. New York: Cambridge University Press.

Nolan, Patrick, and Gerhard Lenski. 2011. *Human Societies: An Introduction to Macrosociology*. 11th ed. Boulder, CO: Paradigm.
Norris, Pippa, and Ronald Inglehart. 2004. *Sacred and Secular: Religion and Politics Worldwide*. New York: Cambridge University Press.
Norris, Pippa. 1997. Choosing Electoral Systems: Proportional, Majoritarian, and Mixed Systems. *International Political Science Review* 18 (3): 297–312.
Nuti, D. Mario. 2013. Did We Go About Transition in the Right Way? In *Handbook of the Economics and Political Economy of Transition*, ed. P. Hare and G. Turley, 46–58. New York: Routledge.
Penna, Anthony J. 2015. *The Human Footprint: A Global Environmental History*. 2nd ed. Malden, MA: Wiley Blackwell.
Peterson, Jack G. 1987. The Economic, Geopolitical, and Business Roots of a Global Minerals Industry: An Analysis of a National Economic and Industrial Debate. In *Mineral Resource Development: Geopolitics, Economics, and Policy*, ed. H. E. Johansen, O. P. Matthews, and G. Rudzitis, 9–22. Boulder, CO and London: Westview Press.
Phambuka-Nsimbi, Catherine, Rina Makgosa, Godfrey Themba, and Olukunle Iyanda. 2015. Modern Retailing and its Implications for Developing Countries: Insights from Retail Managers. *Business Management and Strategy* 6 (1): 1–24.
Piketty, Thomas. 2014. *Capital in the Twenty-First Century*. Translated by A. Goldhammer. Cambridge, MA: Harvard University Press.
Pirie, Fernanda. 2021. *The Rule of Laws: A 4,000-Year Quest to Order the World*. New York: Basic Books.
Ponting, Clive. 2007. *A New Green History of the World: The Environment and the Collapse of Great Civilizations*. New York: Penguin.
Poulomi, Pal. 2017. The Indian Women's Movement Today: The Challenges of Addressing Genderbased Violence. In *Women's Movements in the Global Era: The Power of Local Feminisms*, ed. A. Basu, 129–154. Boulder, CO: Westview Press.
Prebisch, Raul. 1950. *The Economic Development of Latin America and Its Principal Problems*. Lake Success, NY: Economic Commission for Latin America, United National Department of Economic Affairs.
Rapsomanikis, George. 2015. *The Economic Lives of Smallholder Farmers: An Analysis Based on Household Data from Nine Countries*. Rome: Food and Agriculture Organization of the United Nations.
Rayside, David. 2020. Early Advocacy for the Public Recognition of Sexual Diversity. In *The Oxford Handbook of Global LGBT and Sexual Diversity Politics*, ed. M. J. Bosia, S. M. McEvoy, and M. Rahman, 43–62. New York: Oxford University Press.
Roe, Alan, and Samantha Dodd. 2018. Dependence on Extractive Industries in Lower-Income Countries: The Statistical Tendencies. In *Extractive Industries: The Management of Resources as a Driver of Sustainable Development*, ed. T. Addison and A. Roe, 33–50. New York: Oxford University Press.
Roessler, Philip, Yannick I. Pengl, Robert Marty, Kyle Sorlie Titlow, and Nicolas van de Walle. 2022. The Cash Crop Revolution, Colonialism and Economic Reorganization in Africa. *World Development* 158:1–17.
Roser, Max, and Enrique Ortiz-Ospina. 2016. "Global Education." *Our World in Data*. Retrieved November 25, 2022. https://ourworldindata.org/global-education.
Rotberg, Robert I. 2004. The Failure and Collapse of Nation-States: Breakdown, Prevention, and Repair. In *When States Fail: Causes and Consequences*, ed. R. I. Rotberg, 1–50. Princeton, NJ: Princeton University Press.
Roy, Tirthankar. 2016. South Asia. In *A History of the Global Economy: From 1500 to the Present*, ed. J. Baten, 249–281. Cambridge, UK: Cambridge University Press.
Rugman, Alan M., and Jonathan P. Doh. 2008. *Multinationals and Development*. New Haven, CT: Yale University Press.
Runyan, Christiane, and Paolo D'Odorico. 2016. *Global Deforestation*. New York: Cambridge University Press.

Ruskola, Teemu. 2012. The East Asian Legal Tradition. In *The Cambridge Companion to Comparative Law*, ed. M. Bussani and U. Mattei, 257–277. New York: Cambridge University Press.

Sacco, Rodolfo. 2012. The Sub-Saharan Legal Tradition. In *The Cambridge Companion to Comparative Law*, ed. M. Bussani and U. Mattei, 313–343. New York: Cambridge University Press.

Salamon, Lester, and S. Wojciech Sokolowski, eds. 1999. *Global Civil Society: Dimensions of the Nonprofit Sector*. Baltimore, MD: Johns Hopkins Center for Civil Society Studies.

Schedelik, Michael, Andreas Nölke, Daniel Mertens, and Christian May. 2021. Comparative Capitalism, Growth Models and Emerging Markets: The Development of the Field. *New Political Economy* 26 (4): 514–526.

Schmitt, Carina, Hanna Lierse, Herbert Obinger, and Laura Seelkopf. 2015. The Global Emergence of Social Protection: Explaining Social Security Legislation 1820–2013. *Politics and Society* 43 (4): 503–524.

Schneider, Friedrich, Andreas Buehn, and Claudio E. Montenegro. 2010. "Shadow Economies All Over the World: New Estimates for 162 Countries from 1999 to 2007." Policy Research, Working Paper 5356. The World Bank.

Seavoy, Ronald E. 2000. *Subsistence and Economic Development*. Westport, CT: Praeger.

Shearmur, Jeremy. 1992. In Defense of Neoliberalism. *Journal of Democracy* 3 (3): 75–81.

Simmons, I. G. 2008. *Global Environmental History*. Chicago and London: University of Chicago Press.

Sinclair, John. 2013. 'Cultural Imperialism' Revisited: Broadcasting in Latin America, India, and China. In *The Handbook of Communication History*, ed. P. Simonson, J. Peck, R. T. Craig, and J. P. Jackson Jr., 412–423. London and New York: Routledge.

Singer, Hans. 1950. U.S. Foreign Investment in Underdeveloped Areas: The Distribution of Gains between Investing and Borrowing Countries. *American Economic Review, Papers and Proceedings* 40 (2): 473–485.

Singh, Supriya. 2013. *Globalization and Money: A Global South Perspective*. Lanham, MD: Rowman & Littlefield.

Stapleton, Kristin. 2013. China. In *The Oxford Handbook of Cities in World History*, ed. P. Clark, 522–541. New York: Oxford University Press.

Stavenhagen, Rodolfo. 2013. The Emergence of Indigenous Peoples. In *of Springer Briefs on Pioneers in Science and Practice, Texts and Protocols*, ed. H. G. Brausch, vol. 3. México, DF: Colegio de México.

Stearns, Peter. 2006. *Consumerism in World History: The Global Transformation of Desire*. 2nd ed. New York: Routledge.

Sturgeon, Timothy, John Humphrey, and Gary Gereffi. 2011. Making the Global Supply Base. In *The Market Makers: How Retailers are Reshaping the Global Economy*, ed. G. G. Hamilton, M. Petrovic, and B. Senauer, 231–254. New York: Oxford University Press.

Sumner, Scott. 2010. "The Unacknowledged Success of Neoliberalism." Library of Economics and Liberty. Retrieved March 10, 2018. http://www.econlib.org/library/Columns/y2010/Sumnerneoliberalism.html#affiliation.

Survival International. 2018. "Uncontacted tribes: Who they are." Retrieved May 20, 2024. https://www.survivalinternational.org/uncontactedtribes/who-they-are.

The Economist. 2013. How to Stop the Fighting, Sometimes. May 10, accessed June 4, 2025. https://www.economist.com/briefing/2013/11/10/how-to-stop-the-fighting-sometimes

Therborn, Göran. 2004. *Between Sex and Power: Family in the World, 1900–2000*. London and New York: Routledge.

Thussu, Daya Kishan. 2012. India and a New Cartography of Global Communication. In *The Handbook of Global Media Research*, ed. I. Volkmer, 276–288. Malden, MA: WileyBlackwell.

Todaro, Michael P., and Stephen C. Smith. 2015. *Economic Development*. 12th ed. Boston: Pearson.

Trautmann, Thomas R. 2016. *India: Brief History of a Civilization*. 2nd ed. New York: Oxford University Press.

Tripp, Aili Mari. 2017. Women's Movements in Africa. In *Women's Movements in the Global Era: The Power of Local Feminisms*, ed. A. Basu, 37–64. Boulder, CO: Westview Press.
Tunstall, Jeremy. 1977. *The Media Are American: Anglo-American Media in the World.* London and New York: Columbia University Press.
Tunstall, Jeremy. 2008. *The Media Were American: U.S. Mass Media in Decline.* New York: Oxford University Press.
Turchin, Peter. 2003. *Historical Dynamics: Why States Rise and Fall.* Princeton, NJ and Oxford: Princeton University Press.
U.S. Census. 2024. https://www.census.gov/data-tools/demo/idb/#/dashboard?COUNTRY_YEAR=2024&COUNTRY_YR_ANIM=2024&dashboard_page=country
Waters, Tony. 2007. *The Persistence of Subsistence Agriculture: Life Beneath the Level of the Marketplace.* New York: Lexington (Rowan and Littlefield).
Wiarda, Howard J., and Margaret Mac Leish Mott, eds. 2003. *Politics and Social Change in Latin America: Still a Distinct Tradition?* Westport, CT: Praeger.
Williams, Michael. 2006. *Deforesting the Earth: From Prehistory to Global Crisis: An Abridgement.* Chicago: The University of Chicago Press.
Williams, Michelle, ed. 2014. *The End of the Developmental State?* New York: Routledge.
World Bank. 2006. *World Bank Assistance to Agriculture in Sub-Saharan Africa: An IEG Review.* Washington, DC: The World Bank.
World Bank. 2012. *World Development Report 2012: Gender Equality and Development.* Washington, DC: The World Bank.
Zheng, Wang. 2017. Feminist Struggles in a Changing China. In *Women's Movements in the Global Era: The Power of Local Feminisms*, ed. A. Basu, 155–184. Boulder, CO: Westview Press.

Chapter 11
Conclusion: The Future, Social Evolution, and Theorizing Humans

Abstract The last 10–15 years have seen a reversal of promising trends in human welfare, such as democratization and absolute poverty reduction. Yet, within the long span of human history, this is a tiny period. We can extrapolate worrisome long-term trends, such as environmental destruction or warfare, into the future, yet the accuracy of these extrapolations as predictions is uncertain since so many in the international system are aware of their destructive potentials. Less worrisome but still risky trends, such as capitalist economic growth, a gradual shift away from Eurocentrism, and the reversal of high mortality and high fertility and of women's subordination also may have unpredictable effects. Three social transformations—the Agricultural, Urban, and Industrial Revolutions—produced three new basic kinds of societies, and three separate scholarly debates analyze their origins. Improving our understanding of all three together—theorizing about social evolution as a whole—requires addressing our assumptions about people. Hypersociality as a paradigm was useful in tracing social evolution; the obvious next step is to develop it as a theory. This will likely be a messy process, but we can hope that it will generate a better understanding of humans and their societies than we currently have.

Keywords The future · Long term · Social evolution · Paradigms · Social theory · Human nature

Introduction: Globalization Reversing?

In contrast to recent trends of globally reducing absolute poverty and increasing lifespan, the last 10 years or so seem to have been worse. Perhaps the huge downsides of globalization, long ignored by policymakers and economists, have now gained the upper hand. The continuing increase in global inequality and its local variations has reached a critical point. The last decade or so has produced a recession of democracy (Freedom House 2020), including the rise of authoritarian leaders in China, Russia, India, Brazil, the Philippines, Hungary, Turkey, and the United States. The number of wars in the world, especially internationalized civil wars, has

T. C. Burr, *Societies Past and Present*, Emerging Globalities and Civilizational Perspectives, https://doi.org/10.1007/978-3-031-91532-1_11

increased in the last 15 years; examples include Syria; Yemen; Ethiopia; Mali; Armenia and Azerbaijan; the Congo; Russia and Ukraine; and Israel and Palestine. These wars are bringing death and misery to millions (Institute for Economics and Peace 2023). Since 2015, the number of people in absolute poverty has started to increase, reversing decades of reduction. The COVID-19 pandemic was devastating to the poor globally, it increased economic inequality around the world (FAO et al. 2021), and it killed millions. Several world leaders have started discussing nuclear war again, something the international state system has ignored since 1990. Some predict that climate change is simply too late to reverse.

It is easy to argue that globalization has "failed," but again, this depends upon a definition of globalization as the effort only of the last few decades. A longer-term view suggests an alternative. Hegemony in the first three centuries of European global expansion is debated (Wallerstein 1980; Arrighi [1994] 2010; Modelski and Thompson 1996), but the Napoleonic Wars clearly halted and somewhat reversed eighteenth-century European global expansion. After 1815, European expansion restarted and accelerated under British hegemony, even more so after 1870. The closure of the global ecumene by 1910 resulted in another reversal or slowing of globalization processes, most obviously in the Great Depression. Post-1945 U.S. hegemony resulted in yet more intensified globalization—first political (decolonization), then economic (neoliberal expansion). That we are today living through another negative result of that period should not, at this point, be surprising. This history suggests that after a rocky period of some decades, globalization processes may accelerate yet again, perhaps after 2050 or so.

Will we get through those next few decades, or will we face a much larger disaster? As the globalization of the last five centuries is new (and brief) within the over 50-millennia perspective of Big Human History, perhaps globalization itself is subject to macrocycles. Instead of another hegemonic cycle, perhaps instead we face a global collapse of a few centuries—an utter disaster of billions of deaths—and then a new expansion of a millennium or two leading to re-globalization, maybe one more successful. Or maybe we will get through this phase with global society generally intact, ready to grow again. We really do not know (and we, ourselves, will never know). Temporal cosmopolitanism (Robbins 2016) calls us to consider the present within that long-term past and, as I have just implied, this suggests seeing the present as part of the flow of time into the long-term future.

This consideration of the future underlines my main argument: very long-term global history helps us understand the present. I separated this claim into three arguments. First, since I believe in making our assumptions about humans clear, I conceptualized humans as groupish individualists. Then I told two stories about human groups. The main story traced how three geographically and demographically uneven transformations of groups into societies created our contemporary global situation. I therefore traced social evolution: the expansion and elaboration of societies over more than ten millennia leading to one global social system. Along the way, I reviewed theories of each of these changes. I also argued that another story was threaded through these transformations: several very long trends in human societies have continued throughout the three big changes; recent reversals of a few of them are part of the third social revolution.

Overall, my purpose has been to ask three theoretical questions: Who are humans? What are human societies and how do they change over time? And over that time, what stays the same in societies, intensifies, or reverses? The bulk of this book was descriptive, but I wrote it using the concepts in Chap. 2, so description had a theoretical purpose, as did the reviews of debates along the way. Here I review and comment on the descriptive stories, the debates, and the conceptualization in reverse order to address these theoretical questions. I start by extrapolating long-term trends from the past into the future in several directions: some ominous, some potentially promising, yet also risky. Then I discuss theorization. I review several long-term trends again, two of them reversing, and this time with a different purpose. I ponder how human societies evolved, discussing other long-term trends along the way. I end by suggesting plans for turning my version of the hypersociality paradigm into a theory of societies.

Forecasting the Future from the Long-Term Past

Few people think in terms of millennia; most understandably think only of the future in terms of their lives or their children's. Yet, we make our lives within deep, ongoing trends, and to ignore them is to misunderstand our present situation. There are many books (e.g., Gidley 2017), journals, associations, and institutes on futures studies. Rather than dive into that—a huge project itself—I will confine myself to sketching out a few scenarios of the future from long-term patterns identified above. I will also identify what issues might make extrapolation useless.

The transformation I call the second globalization or the global Industrial Revolution is still incomplete, perhaps even in its early phases, and the type of society it has produced is quite new in world-historical terms. Since so many changes—the transformation of science and technology, the mostly violent unification of humanity, the intellectual stimulation of science, education, and mass media, the massive wealth that industrial capitalism generates, popular participation in governance ("democracy"), demographic transition and long average lifetimes, and the beginnings of gender (re-)equalization—are either new or are the return of old patterns in new guises, we have only a few principles to guide us. Thus, the risks are high; one could say that a bundle of human-generated environmental problems is one risk, and the combination of nuclear weapons and hegemonic decline is another. If we face down these dangers in the next few decades, we might succeed. That might lead to other, yet-to-be-discovered risks, or humanity might face a reasonably bright future. I do not know.

Let us start with the worst possibilities, then sound a more hopeful tone. The first two long-term trends to discuss are environmental destruction and war. Scholars still debate whether humans caused extinctions 10–15 millennia ago. Deforestation for buildings and firewood started with villages and intensified over time. Scholars debate whether agrarian states undermined their own power through desertification, waterlogging, and pollution or managed resources reasonably well over long periods (Chew 2001; Ponting 2007; Brooke 2014). Still, industrial capitalist states are much more environmentally destructive than previous societies. Humans gradually moved away from the environmental stewardship orientation of animism, and industrialization made it far easier to extract resources and modify the land. Capitalism separates landowners from the consequences of their land policies even more than agrarian landownership did: Multinational corporations (MNCs) treat their land thousands of miles away merely as profit-extraction opportunities.

While salinization, waterlogging, and deforestation are long-standing problems now intensified, global warming and resource overuse have resulted directly from the Industrial Revolution. The science is so solid that we know global warming will have certain effects; rising seas and more intense weather are already happening, while scientists forecast more intense storms in some areas, the disruption of agricultural systems from desertification, or the potential shutdown of the North Atlantic global "pump" which sends heat around the globe through the oceans, which would induce a frozen Europe and other problems. Likewise, recent research confirms studies done in the 1970s predicting that we continue a path toward resource overuse and consequent social collapse in the mid-twenty-first century (Turner 2014; Herrington 2022). This is only two decades or so in the future.

I think it doubtful that either global warming or resource overuse threatens human extinction, but the worst case is still bad. Perhaps a global collapse will reverse social evolution (with, again, a lot of death) and return humans to fewer, smaller, kin-oriented groups who are nevertheless literate, technologically sophisticated, and oriented toward scavenging and improvising from the detritus of global society. Such a collapse may or may not occur because of the newness of industrialization. Climate change and resource use are classic collective action problems, and while the international state system is taking collective action to address them, that system is so new, and humanity is so fractured vertically and horizontally, that whether the speed of this collective action will be sufficient to avoid disaster is an open question.

Another long-term trend is warfare. Seemingly starting in the early Holocene, Carneiro's (1970, 2017) theory of population-driven circumscription must have applied at least sometimes, making warfare a conflict over land, the source of survival in the high-mortality circumstances of village life. In states and empires, war became an aggrandizing tactic. Empires mostly expanded through warfare and conquest; millions of people died in agrarian wars. From 1500 to 1900 CE, Europeans engaged in global conquest and fought global wars, killing millions more. What does this tell us about global politics over the next century? Are scholars (e.g., Arrighi [1994] 2010; Modelski and Thompson 1996) correct that single-state dominance, or systemic hegemony, runs in long cycles? Do unusually large global wars decide global hegemony? Regional wars do enough damage on their own; the intersection of nuclear weapons with hegemonic change might be much worse. Deaths of several billion people are possible. Global value chains would not merely be "disrupted"; the global economy would collapse. People in cities and suburbs fed by food systems would starve, while semi-commercialized subsistence farmers could shift back to full subsistence production. Since the 1980s, climate scientists have also talked about a "nuclear winter" in which hundreds of post-nuclear war urban firestorms generate massive atmospheric soot which blocks sunlight. The worst-case estimates posit that the soot cools the climate rapidly back to temperatures last seen in the Last Glacial Maximum, making agriculture almost impossible (Robock et al. 2007). Most survivors of nuclear war would starve to death.

Other possibilities exist. If population growth combined with a high mortality rate and geographic circumscription does lead to wars over resources (Carneiro 1970, 2017), then industrialization and the demographic transition helps us hold out the hope that warfare can be eliminated. Does the empirical generalization that democracies do not fight each other (Ray 2011; Kurian 2011) reflect the fact that most democracies are wealthy? In turn, does that mean that ending war requires global economic development? The disastrous environmental consequences of ten billion people consuming the way North Americans do now would dwarf the problems that global warming poses. Declining populations ease the problems of circumscription, but introduce other problems (see below).

Environmental and political-military disasters are the direst possibilities. Their likely two-phase sequences underline the importance of human social systems. In both scenarios, the disaster itself will be only the first phase. If enough people die, the second phase will start: the social systems feeding whole societies will collapse, so the survivors of the initial disaster will face starvation. Subsistence producers will not, though their lives will worsen.

Again, these are global collective problems requiring collective action. Some scholars think we are tending toward world government (e.g., Carneiro 2004). Global government more powerful than a single state's hegemony may be unrealistic, but if the problems are pressing enough, it might be possible. However, given the long history of states, perhaps this will result in the repression that libertarians fear rather than the peaceful world that world federalists seek.

Other trends are not as dire, and some provide hope. The massive expansion of trade from before the Last Glacial Maximum to today raises several questions about capitalism. I think those who use the phrase "late capitalism" are being historically myopic and too optimistic. The industrial capitalist-democratic period is only two centuries old, while states and empires took at least 5000 years to develop their fullest forms. Capitalism has not fully spread globally; there is plenty of room for it to grow. Critical scholars might be ambivalent about rise of global South MNCs, as their continuing growth portends a deeper entrenchment of capitalism. However, societies might evolve enough to find an alternative, or a different form of capitalism. Just as many, perhaps most, early states engaged in human sacrifice and evolved beyond it, capitalism might improve over centuries. Starting with Bismarck, European leaders learned to buy off workers enough to gain stability. It is currently immigration into Europe that puts strain on those systems, not merely capitalist "contradictions." Other changes may also improve it.

Whether this is a hopeful possibility depends on the reader; my own stance is ambivalence, because of the lifespan improvements capitalism has financed versus the environmental disaster it portends. Focusing solely upon the social justice (relative poverty) implications of global capitalism ignores the more than doubling of mean lifespan in the last three centuries. Because of this, I wrote Chap. 9 and especially Chap. 10 in a vein that seemingly supports capitalist economic growth with only a few scattered hesitations, but those chapters underplayed my ambivalence. Hickel (2021) argues that the world economy needs "degrowth," a strategy to restructure capitalist economies into another mode of organization in order to reduce economic inequality as the endless elite drive for accumulation intensifies our environmental problems. The strategy includes principles such as ending planned obsolescence, cutting advertising, moving from ownership to usership, and scaling down ecologically destructive industries (Hickel 2021:Chap. 5). He is like others (e.g., Piketty 2014) who make excellent suggestions about reducing inequality but offer no suggested practical political path for achieving them against the power of economic elites. Perhaps instead, capitalism will evolve to such a policy regime anyway. (I am not betting on it.)

In addition to vertical inequality, I think scholars must address horizontal inequalities more. States and empires were always multicultural, so what we now call horizontal inequality always existed within them. In the last 500 years, racism, an ideology about bodily physical features, became another foundation for groups. Migration in that half-millennium has also added to the welter of different ethnoracial groups in different states. As argued (Chap. 10), many social scientists underappreciate or ignore ethnicity (cultural membership) as a driver of state instability. If ethnocentrism is "built in" to humans—and I think many scholars actively reject this idea—it helps us understand why so many states engaged in assimilation (incorporating *them* into *us*), and it explains a lot about our current global situation. Ethnicity is never, on its own, enough to explain conflict, which is perhaps why some agrarian states did not require assimilation, but it is always a crucial background factor, ready for activation. Acknowledging the normality of ethnocentrism, understanding it (developing and testing theories), and learning to manage it seems more useful than ignoring it.

Along with global vertical inequality, there seems to be global horizontal inequality. Europeans created and have dominated the global system, but as Bang (2021:12) points out, this is fading: "The world was made Eurocentric for a brief while. .. but before this, human society moved to a different rhythm and. .. the world is [now] becoming steadily more modern and again increasingly less centered on Europe." Recall McNeill's (1990, 1998) empirical generalization that continental regions in Afro-Eurasia have tended to be the most innovative and richest for several centuries each—in sequence, the Mediterranean, India, Southwest Asia, China, and Europe. Although McNeill refused to predict the future from the implications of this pattern, I think it suggests that China's rapid economic growth seems to be shifting the power center of the human world from the North Atlantic region to East Asia. This would be a conjuncture of hegemonic cycles (Arrighi [1994] 2010; Modelski and Thompson 1996) with McNeill's idea of half-millennium continental shifts.

Debates about this, of course, already exist. As Kennedy (1989) and Modelski and Thompson (1996) argued, great-power politics is no longer European but inherently global. The world is moving speedily away from a Eurocentric system because economic development is spreading globally. Some argue that the United States is declining, and China will be its hegemonic successor (e.g., Fishman 2006), or at least the two states are on a course for war (Allison 2017). Others agree that the United States is declining, especially due to financialization, but are unsure what will supervene. Arrighi [1994] 2010) says that East Asia, especially China and its Southeast Asian diaspora, is likely to take over globally but with a different approach—though he is not sure. Yet others argue that the United States has "delusions of decline" (Brown 2013) and that it will remain dominant for a few more decades; at least one scholar argues that the Chinese state's two-millennia takeover of society, continuing today, may prevent its ability to take advantage of U.S. decline (Huang 2023). And its falling population creates real unpredictability. Still, China's economic development has altered the global balance of power.

Whether we are in the middle of a global hegemonic shift or not, if we somehow avoid a global collapse, East Asia will likely develop economically and eventually dominate the world for the next few hundred years, so this also means that cultural Eurocentrism will fade. As societies around the world develop, they will have the wealth to develop other, more creative ways to live, perhaps superseding European industrial capitalist democracy—or perhaps not. If they do, I do not know what these new ways will look like. I do believe that they will be different, and I think they could be better than what we have now.

Two long-term trends, one demographic and another cultural-political, are distinguished not by their intensification but by their reversal. One is the post-1700 mortality transition, and its sequel, the fertility transition—together, "the demographic transition." The average death rate for most of human history was dramatically higher than it is now, and I repeatedly suggested that most people experienced much more grief and fear than they do today. The demographic transition constitutes one of the most profound changes in humanity's existence, with several possible consequences. First, we have a vastly increased global population on every continent; combined with industrial capitalism, the population of the global North contributes to half of our massive environmental problems. As the other seven billion people start to use resources the way people in rich states do, things will get worse, as mentioned above. Still, lower mortality has made possible lower fertility, which makes population shrinkage a possibility in most places in the next few decades, and the next century in Africa. This takes us into unknown territory. It may ease environmental pressures, but it may also depress innovation and therefore economic growth. As more and more of the aged retire, fewer working-age people paying for more retirees and sick elderly people will lead to strain on social insurance systems. Potential solutions are higher taxes on the wealthy; improved health for the elderly; and innovations we obviously do not know about yet, and which may or may not occur.

What is the longer-term future of below-replacement birthrates? If present trends continue, global population will peak and start to drop. The theoretical endpoint of women *en masse* refusing to give birth is a humanity dwindling away to nothing. Few leaders, female or male, will likely want that; probably many populations may not either. Furthermore, many governments that want women to maintain a birthrate of at least replacement (total fertility rate [TFR] ≈ 2.1) are starting to provide parental leave and other economic incentives. However, this does not seem to be working.

The reversal of the long demographic trend in mortality and fertility obviously intersects with the reversal of another long-term cultural–political trend: male dominance and female subordination. The falling birthrate is having a major effect on women's lives; women's movements have started only within falling fertility systems, so it is difficult to know if such movements would have succeeded in high-fertility systems. (I doubt it, but I could be wrong.) Reliable birth control has helped fertility fall, but so has women's political action, guided by ideas, to accomplish and implement changes we have seen in law and practice. Women's lives may or may not have improved in European empires, but since 1950, their lives have definitely improved. Women are generally becoming better educated than men are, women are increasingly moving into positions of political power, and even intergovernmental organizations (IGOs) such as the World Bank agree that development requires improvement in women's lives and is pushing a gender-equal agenda. The results of women's movements may take several centuries to play out. After all, it took thousands of years for male supremacy to develop and thousands more to become institutionalized around the world. Feminists alive now will obviously find "several centuries" extremely frustrating; looking back 7000 or 8000 years, a few hundred years is extremely fast.

Like many projections above, this outcome is not guaranteed. Selective abortion in India and China has led to a scarcity of women in those states that does not seem to have increased women's bargaining power. As the United Nations Development Program (UNDP 2020) report shows, attitudes around the world still support at least some male dominance. However, this survey takes a very short-term approach, at least from the "term" considered in this book. TFR below replacement may be a powerful bargaining chip for women globally, but men may simply not be able to cede power to women in order keep the birthrate up. Again, I do not know. However, as women increasingly enter and rule political systems, they themselves will see the issue; they may come up with better solutions than men have. Thus, the path locally, regionally, nationally, and internationally for a gender-equal world is not automatic; if it ever arrives, it is likely to involve multiple paths.

I have only sketched possibilities. Readers might find this frustrating, but as I have said (too often?) about people in the past, we too are making it up as we go along. I do not think any of the problems that these trends create are insoluble, though I understand the pessimism of those who do. These are all collective problems amenable to collective solutions. Humans have learned, and set up institutions, to act collectively on a global scale. Will we continue to do so enough to solve these problems?

Theorizing from Long-Term Trends

While speculation *may* help prepare us for the future, theorization—attempting to explain the world in order to understand it—is something we *can* do now. Some long-term continuities offer a fertile field for retheorization of contemporary issues.

Here, I discuss three trends that need more theorization: population growth, gender inequality, and economic inequality.

In researching this book, I was struck by how few historical scholars in all disciplines appreciate demographic issues. Life before 1700 CE was physically and emotionally harsh. People saw a great deal of death from childhood onward, so I have argued that they generally lived with a tremendous amount of grief and fear. Perhaps familiarity somewhat inured them to the deaths of others, or grief and fear were easily turned to rage, supporting violence. The modern demographic transition has completely changed the material and emotional environment for nearly everyone. Nearly all children survive, most people survive most of their adulthoods, and it is primarily the old who die—those who may feel ready to go.

It seems that there is a tendency for human groups to increase in size over time, and population growth both made possible and generally led to larger and more complex groups. Yet, with the modern demographic transition, we are seeing the beginnings of global depopulation, even while the global population continues to rise. Does the forecast decrease in global population mean that human groups do not, in fact, have a long-term tendency to increase their size? This suggests that any "natural" tendency to increase may have been rooted in high mortality. And cultural cumulation has led to technologies and social organization that allows for the control of fertility, raising the question of whether women are in fact "natural" mothers. Or is the nascent global depopulation tendency rooted in massive economic inequality, tremendous anxiety about the global future, and confusingly disrupted gender norms?

As discussed, over the last seven or eight millennia, half of the population in most societies has increasingly subjugated the other half. Foraging women seem to have been effectively equal with foraging men in influence and autonomy, yet they had a high birthrate. The original transition from roughly gender-equal societies to male-dominant ones developed after the Agricultural Revolution started. Many of the authors I relied on to discuss female subordination tried hard to show how women found creative ways to retain personal agency; however, the basic gender inequality in states forced that creativity. Why did it happen? One argument is that it originated in the long, gradual increase in warfare (Bonvillain 2001); others have a more economic explanation (e.g., Lerner 1986). I am likely missing recent discussions of this topic, but we must make the question of the origins, and the potential fading, of male dominance one of the central questions in world history. After all, we cannot honestly claim to study either history or society adequately when we downplay both the behavior and the well-being of half of humanity. Also, a better understanding of sex and gender will have theoretical and practical effects, given women's increasing power and the uncertain demographic future we face.

Scholars are also retheorizing economic inequality, which is reversing globally (mostly because of Chinese economic development). I think an expanded application of the term "aggrandizing" might help. Database searches in psychology, sociology, and economics revealed nothing on developing "aggrandizers" as a construct; the term only seems to appear in anthropology. Archaeologists who use this term seem to drop it once they study established polygroups and states. We should not;

instead, we should understand inequality as a verb, so to speak. We would do well to study aggrandizing as an active attempt to increase inequality. We could start with the ideas of natural, situational, and cultural aggrandizing: individuals occasionally born wanting to take over; zero-sum competition between two expanding polities; and elite political–ideological movements to gain power. In large-scale societies, "natural" aggrandizers are born at every social level—yet as mentioned (Chap. 5), elite situations also generated aggrandizers. Henrich's (2015) distinction between dominance and prestige could help our understanding, since successful aggrandizing requires social skills.

The new global inequality—"new" being two to five centuries within the context of seven millennia of local and regional inequalities—raises other questions, one about inequalities within societies and another about that global inequality itself. First, can elites learn to restrain themselves for their own long-term viability? Politically induced low inequality in some European states suggests that it is possible, while cultural homogeneity seems to help, and cultural diversity seems to hurt. Still, in the long run, Scheidel's (2017) argument does not offer much hope; increasing inequality has always reversed itself violently. Scheidel is agnostic about whether inequality causes its own backlash and reversal. Culture can make situations appear fair, but at some point, people know that the system is rigged far beyond anything called "fair," so I suggest excessive inequality may bring about its own destruction—elites do not learn the long-term lesson and always overdo the taking.

Scholars increasingly appreciate how cyclic economic inequality is. Scheidel (2017) documents this for thousands of years in multiple places, a tendency implicit in the record of imperial rise-and-fall, while Piketty (2014) theorizes how it varied so far under capitalism. Given any reduction in inequality, how do elites "go back to. .. reconquering their ruling position" (Milanovic 2023:291)? It is in the "rise" of inequality that applying the concept of aggrandizing might help. While Piketty argues for some automaticity in his famous formula $r > g$ (elite wealth always grows faster than mass wealth does) he also acknowledges political factors. After all, neoliberalism did not "just happen"; a movement implemented it. I suggest that elites in both agrarian and industrial capitalist states work actively to ensure that $r > g$ occurs at some point in cycles of expansion and collapse. On the other side of this cycle, Goldstone ([1991] 2016) criticizes "selfish elites" today, especially in the United States. But under what conditions do elites become selfish?

Perhaps Marxists have derailed this issue a bit because they seem to assume that elite selfishness is a constant. Since the previous chapters show that inequality is a variable, elite selfishness may be also. For instance, Mizruchi (2013) argues that elites can take a longer-term, "enlightened" self-interest stance or a shorter-term, more selfish stance. He argues that the U.S. economic elite was unified and somewhat prosocial after World War II, tolerating high tax rates, increased government regulation, and strong labor unions. After *les trentes glorieuses* (1945–1975) ended, global competition increased, so the U.S. corporate elite increased its domestic extraction activity, reducing the power of state regulation and of trade unions; however, this fragmented the corporate elite, undermining its power. Is Mizruchi's argument the germ of another component in explanations of agrarian rise and fall?

Perhaps agrarian elites acted cohesively and somewhat prosocially during expansion (e.g., setting up public goods after conquest); when gains from conquest diminished, they turned on each other and on their populations, increasing extraction. Such an explanation might also apply at both larger and smaller scales, in varying conditions, in the capitalist era.

Three long-term trends—population growth, and gender and economic inequalities—need fresh analysis, especially since two, perhaps all three, are reversing. I could not untangle other long-term trends from the issue of social evolution since they suffered no reversal but simply continued, so I will discuss them as part of the three social revolutions in human history.

Explaining Social Evolution: Theorizing Social Transformation

I defined social evolution neutrally, as a long-term increase in the size and institutional complexity of any group. Three massive and slow social transformations occurred in the human past. Explaining social evolution is difficult, and most scholars who study it focus on just one social transformation. I will review those three changes separately, then together. I did not offer an overall definition of "society"—what becomes larger and more complex—partly because societies kept changing in world history. Sztompka (1994) defines society as "a multi-level, intermeshed set of processes" (1994:41 and 56), and processual archaeologists agree (e.g., Earle 2016), so I treated societies as both entities and processes. Thus, I suggested that a *society* is more complex than a *group*; any society over time is formed of multiple, interpenetrating groups.

But when did one shift to another? For instance, was the kin-structured interband group a "society"? Perhaps. Certainly there were three intermeshed, constantly changing levels: families, bands, and the interband, kin-defined community. Still, these seem to have had, and have now, a maximum size of 500 people (Read 2012). Also, the first globalization was not a social transformation. It merely populated the planet with that kind of group. Toward the end of that process, the Agricultural Revolution—settlement in villages, domestication, ceramics, warfare, elaborating kinship, and so on—started, though complex hunter-gatherers had existed off and on for millennia. This was the first real change in what societies were, and it raises two questions.

First, scholars in the debate about the origins of war (e.g., LeBlanc 2014; Gat 2017) bundle "hunter-gatherer" groups in one big category, which neither of the Kellys (Raymond 2000 and Robert 2013) do; this means LeBlanc and Gat (and others) do not consider seriously the full implications of Kelly's (2000) argument. What constitutes a "group"? Is his definition of war—intergroup violence founded on "social substitutability"—accurate? If so, then Kelly (2000) is right that the most simply organized hunter-gatherer groups were not cohesive enough for collective responsibility and thus social substitutability, and therefore did not engage in warfare. Consequently, such competition did not drive human biological evolution, so true warfare (intergroup violence) appeared only after the Agricultural Revolution. The debate seems to illustrate well the warnings about caution while making ethnographic inferences (Wobst 1978; Kelly 2013; Lemke 2018).

On the other hand, paleoanthropologists increasingly argue that many kinds of humans existed at the same time. Evidence of *sapiens-neandertalensis* interaction—and *sapiens* with other subspecies and perhaps species—is increasingly available. Conflict between them might have constituted true intergroup violence. Perhaps they looked different enough from each other that either social substitutability operated, or perhaps Kelly (2000) is incorrect and some other cognitive mechanism worked to make two subspecies fight with each other in collective violence. If this is the case, war did indeed drive human biological evolution well before the Agricultural Revolution. This takes warfare from being a long-term issue to a *very* long-term one, raising the possibility that war is "built-in" to humans (LeBlanc 2014). Obviously more and more careful research is necessary.

Second, no society domesticated plants, or even did much "management" of them, before the Holocene. Yet long before it, humans domesticated wolves into dogs, perhaps in multiple places. Does this have any bearing on our understanding of animal domestication once the Holocene started? I cannot imagine that paleoanthropologists have ignored this link, but I saw no indication that they have addressed it. Perhaps it is simply that the surviving evidence is far too insufficient to address parallels between wolf-to-dog and herd animal domestication. (Or perhaps I missed a body of research.)

The transition to social complexity seems more difficult to explain. The resulting social structures started to resemble "societies" as we know them, with different kinds of groups within, divisions of labor, and vertical inequalities. However, the sheer diversity of social forms—so that some scholars focus on "chiefdoms" (e.g., Flannery and Marcus 2012; Carneiro 2017), while others emphasize alternative forms (McIntosh 1999; Grinin and Korotayev 2017; Graeber and Wengrow 2021)—tend to imply several pathways to complexity, as indeed some argue (Peterson and Drennan 2012). Because there are so many instances of new complexity, understanding is difficult. In this Urban Revolution, populations concentrated in cities almost everywhere there was social complexity—but not everywhere (McIntosh 1999). The role of belief-systems, which in large-scale societies seemed to have kept getting ever-further from their animist roots millennia before, is not clear.

The transition to agrarian states seems less difficult to understand, as processes of complexity seem to have simply continued on ever-larger scales. Why these societies always came apart seems more confusing, while reconstitution mostly through conquest seems easier to explain. However, a question Sanderson (1995) asked is still apt: why did social evolution "slow down" after the development of states? For the next 5000 years, states simply became bigger, as did state systems and world-systems—usually with terrible, centuries-long reversals—and occupied more and more areas of the planet's land surface. Efforts to understand this period continue, such as one that attempts to compile a comprehensive quantitative database ("Seshat") about complex societies (Turchin et al. 2015).

It seems impossible to analyze fully how the global Industrial Revolution, or second globalization, happened. As detailed earlier, the Rise of the West/Great Divergence debate sprawls massively throughout the social sciences (excluding psychology and archaeology), overwhelming anyone who tries to handle its conceptual and informational requirements. And it only happened once. Why? As Turchin et al. (2015) point out, everyone in this debate simply accepts the existence of societies around 1500 CE without asking about the theoretical implications of their pre-existence. Maybe the millennia-long expansion of states, empires, civilizations, and world-systems was required for Europe's—or another region's—transformation, as Sanderson (1995) implied. Perhaps commercialization and industrialization were so difficult that there needed to be some critical mass of resources—sheer size of populations and accumulated knowledge, techniques, and sufficient variety in mental forms ("culture")—to stimulate this transformation. Europe's marginal position in the Afro-Eurasian trade system no longer seems like a new idea (given Abu-Lughod 1989; Frank 1998; and so on), but it still seems greatly underappreciated in that debate.

Finally, in many areas, societies did *not* evolve. People lived broadly as their ancestors did for thousands of years, a fact that supports anti-teleological arguments most strongly. Social evolution was not pre-ordained. Instead, people did what made sense locally, in the moment. Indeed, devolution occurred in some places; frontier processes occurred to initiate social evolution in others. People generally did not get more organized until they had to, so this raises the question of which teleology to critique. If humans have a long-term tendency to increase in population, then that would be enough produce some sort of social evolution without any goal.

In any case, three separate theoretical debates analyze these three social revolutions. Archaeologists try to figure out why the first main two changes happened, and historians and social scientists join archaeologists in trying to understand the second change. In contrast, prehistorians largely ignore the third one, and psychologists are unfortunately absent from all three debates. The literature that theoretically analyzes all three revolutions together is remarkably thin and, as mentioned, occurs mostly in sociology (e.g., Sanderson 1995; Lenski 2005; Chase-Dunn and Lerro 2014; only Christian 2011 is a historian). I think we need more works that address all three transformations together. Peterson and Drennan (2012) argue that once three changes in the Agricultural Revolution—domestication, sedentism, and ceramics—all appeared together in any location, social complexity, the Urban Revolution, started there within two millennia. One revolution led to another; with no exogenous force stopping it, complexity ultimately took over much of each hemisphere. As the discussion above implies, that was the foundation for the third social transformation, the global Industrial Revolution. In other words, the three massive changes seem all of a piece, though the timescale is daunting to consider.

Theorizing all human history is more than ambitious; it is grandiose. It seems necessary anyway. The mere attempt would help us understand not only human history, but also our situation today, much better. Part of that is, again, understanding what a "society" is, which means that questions proliferate. Do states equal societies? Economists do not even ask this question; political scientists usually answer "yes"; sociologists do ask it but disagree; and I think most anthropologists would answer "no." We can answer this question several ways. Legally enforced state boundaries force people into interaction with others within the boundary, all subject to national laws. In that sense, states are societies. However, boundaries often cut across cultural groups, creating resentment. "Society" to many of the members of those groups is the cultural group, not the political state. Relations between groups within the boundary can range from friendly to hostile, and anything in between. In that sense, states are not societies. When (if) the answer is yes, and *states* are collective actors (however coercively so), are *societies* collective actors? For that matter, the global Industrial Revolution created a global social system. Is it a global "society"? It is certainly an intermeshed set of multi-level entities in multiple processes. Or do societies exist within it, and it is something new, beyond "society"? Is global society a collective actor? I would generally answer "no," but some actions taken by global governance structures may suggest "yes" occasionally.

Thus, understanding our current situation—what social entities we live in, how they operate—is a time-bound issue: the answer to "What is society?" has changed for ten or more millennia. An explanation of social evolution and the long-term trends intertwined with it, and those that reversed, cries out for understanding—for explanation—for *theory*.

Explaining Our Common Humanity: Future Theorization

To develop a theory that explains, or at least helps us explain, how all three social transformations happened and that also helps us explain how certain trends continued through those transformations is obviously a tall order, perhaps an outrageously arduous one. Maybe there is no possible theory that can handle the demands placed upon it. Yet, several scholars think it possible and have tried to do it (Sanderson 1995; Lenski 2005; Christian 2011; Chase-Dunn and Lerro 2014). However, they tended to start with social systems, then reverse engineer, so to speak, a model of a human being that would help the systems make sense, and then theorize about the system. I prefer to start with individuals in small groups and work outward. That, however, is daunting in and of itself, with vast numbers of books, whole subdisciplines, and even a journal (*Human Nature*) devoted to it. Indeed, the effort may take several theoretical steps: a model of individual human capacities and variations; a theorization of group behavior; and—what was mentioned but inadequately addressed in Chap. 2—a deeper theory of group change, especially evolution.

In that chapter, I laid out my assumptions about human beings. I did not and do not consider it "the truth" about humans; it is simply a set of assumptions I supported with empirical findings from many disciplines. Everyone makes assumptions about humans, though as multiple scholars pointed out for multiple disciplines (most notably history and sociology), many scholars are reluctant even to discuss the basics of "the human." I argued that making explicit or implicit assumptions is good for the social sciences, since all theories rest on sets of assumptions, or paradigms (Kuhn [1962] 2012). It may seem arrogant to say what it human. Perhaps this is why scholars in these disciplines are so hesitant, reluctant, or opposed to discussing it. Given the tremendous variation in understanding what is human, what is right and proper, and so on, who has the right to say how humans operate and what they need, both individually and collectively? On the other hand, those who think human rights are important must have assumptions about how humans operate, what they need, and what makes for flourishing individuals and groups. Indeed, any assertion that everyone has "human rights" contains within it some assumptions of "the human." Let us all start making explicit our assumptions about what *human* actually is.

As I argued in Chap. 1, I found the paradigm of assumptions about humans called "hypersociality" compelling yet incomplete, so I tried to develop it more in Chap. 2. Despite findings in social psychology in both psychology and sociology, and economic psychology in behavioral economics, few in the hypersocial paradigm seemed ready to borrow from or use findings in psychology in general. I thought we needed to use many findings in psychology as a foundation for theorizing. I conceptualized what individual humans have in common as a set of capabilities, tendencies, and possibilities, not a fixed set of behaviors. As these included being group-capable, I also laid out a set of assumptions about human groups. I applied this conceptualization to human history with the supports of social evolution and institutionalism.

Throughout the book, I discussed human emotions and their consequences in given social contexts. As promised, I did this only in an *ad hoc* way; these discussions were incomplete and merely suggestive. Still, I carried this theme across most chapters to demonstrate how they could influence behaviors within societies. Grief and fear in high-mortality environments seemed to have made life harder for everyone and helped drive dehumanization of others, which also derived from ethnically driven moral disgust; envy increased in agrarian and industrial consumerism; aggrandizers strove for "glory" (inflated self-concept) in conquest, causing the deaths of thousands; ethnocentric and racist disgust seem to have influenced modern democracies; and horizontal inequality, producing anger, now influences state stability.

I also argued that culture is not simply "information" or "norms"; it involves *ways* to store information—concepts—which shape both emotions and behaviors. Concepts not only make possible the cumulative and collective nature of human learning (Richerson and Boyd 2005; Henrich 2015) a keystone of social evolution; they also make cognition possible. People make honest attempts to make sense of, to understand, the physical and social worlds in order to operate in them. Beliefs are mostly implicit, rarely well-developed or perfectly logical. Concepts, and beliefs formed from them, help humans imagine the future in ways that animals seem unable to do; they are especially important in innovation. Groups bundle norms in collective scripts for dividing labor—institutions—to get things done.

Groupishness and emotion, especially, are not particularly theorized or often used in "big picture" political economy studies, and the role of culture as a causal factor is heavily debated. Many concepts from social psychology go underused. For instance, individual self-concept and its impact on individual self-esteem (emotional self-regard) could be deployed in explanation much more usefully, especially since individuals' self-concepts are always structured culturally.

From these points, I suggest here further developments of my version of the hypersociality paradigm. While I accepted rational-choice principles but argued that they were extremely and thus distortingly incomplete, I offered another "side" to humans: groupishness, emotion, and culture. However, thinking that individuals have two sides is entirely artificial, an artifact of my finding it most convenient to start with rational-choice thinking in the social sciences. But rather than keep talking about humans as two-sided entities, I want to develop an integrated model of individuals to provide a firm foundation for understanding human groups and therefore understanding societies. I see three areas of improvement of the conceptualization I offered in Chap. 2 to turn it into a theory. We can develop groupishness, culture, and emotion; we can introduce other elements to create a dynamic model, one that explains change; and normal academic criticism can help fill in the (possibly considerable) gaps.

The issue of groupishness is central. How valid is this as an understanding of people? I think the main issue to address is under what circumstances individuals are individualistic, and under what others are they groupish. How much of those conditions involve enculturation or socialization, and how many involve material circumstances? Inglehart (2018) has argued that people's values shift to greater self-expression and individuality as survival is increasingly assured, though this is a decades-long process. I think this means that scholars must rethink their attribution of political-economic motivations; attributing a "materialistic" orientation to social actors in a poor, high-mortality environment seems obvious, but in a wealthy, low-mortality one, perhaps individual and collective ego-needs, structured culturally, take precedence.

How can we understand how important culture is in explanation? Further study of culture can definitely improve our understanding of institutions and cognition, which would provide a firmer foundation for theorizing about people. Culture also seems important in understanding social change. To what degree do people's ideas change after material conditions change? To what degree can we imagine futures and impose our new ideas on the social world? And many theorists of complex societies (Chap. 5) discussed how would-be leaders used existing belief-systems to gain power. How important were they?

Yet, people are not pure thought-creatures. What feelings and physical needs motivated people to follow leaders? What motivates people in general? There are four ways we can develop our understanding of emotion to understand human societies: understanding individual and collective motivation; learning more about cognition and emotion (Damasio 1994; Turner and Machalek 2018); understanding how emotions promote prosociality; and analyzing how they lead to antisocial behaviors, both individual and collective. In particular, the new understanding of emotional trauma (e.g., van der Kolk 2014) offers real possibilities for increased knowledge. Social workers have a saying: "Hurt people hurt people." That is, traumatized people ease their pain by injuring others. This could be a key component of a theory of social behavior, both for intergroup relations and intergenerational patterns. Here is where culture is also important: How do small groups handle emotional traumas? Do they pass them on intergenerationally, through "dysfunctional" behaviors? Likewise, the rational-choice argument that humans are indifferent to others (e.g., Kirchgässner 2008) seems unrealistic. Under what circumstances do people care about others, and under which ones do they not, or actively dislike them? In other words, analyzing prosociality and antisociality go together. *Why* do people engage in dehumanization? Grief and rage from loss of others (Rosaldo 1989), fear of one's own death, and moral disgust seem to be important motivators in intergroup relations, yet few historical political economy scholars try to understand or theorize these psychological realities.

Development of these three factors—groupishness, culture, and emotion—merely extends the model I developed in Chap. 2. However, the historical section, the bulk of the book, showed that it is incomplete. That chapter offers a static model of human groups; despite discussing change, it does not include enough factors to account for the evolution of human groups. The definition of "social evolution"—the expansion and elaboration of human groups, and their increased interconnections— becomes crucial here.

The first additional factor would be population growth. The definition above, especially its "expansion" part, relies on the addition of people to a group. While Carneiro's (1970, 2017) critics and Goldstone ([1991] 2016) show that there is no simple one-to-one relationship between population growth and social evolution, in *some* way population growth was part of social evolution. If groups do not grow, why elaborate institutionally? Subdivisions are unnecessary; interconnections do not increase. In particular, demographic transitions are excellent sources for theorizing. The broad pattern of mortality and fertility seemed to be constant, yet it changed during or after the Agricultural Revolution and has changed much more drastically in the last three centuries. What is the role of population change in social evolution?

The second addition would be the physical environment. The advent of the Holocene seems crucial in "kicking off" social evolution, even though complex hunter-gatherers existed millennia before the Last Glacial Maximum. Humans domesticated dogs in the Pleistocene era, but was the much broader scale of domestication that ultimately occurred even possible in that climate? And Brooke's (2014) account shows that climate patterns and especially their changes seem to have shaped specific patterns of settlement and development of pastoral, horticultural, and especially agrarian peoples. This, too, was greatly underdeveloped in Chap. 2, so it needs more development.

A third modification would be to develop the implications of two sexes in humans. As I argued above, the overall structure of societies and recent demographic and political-cultural reversals make the issue of sex and gender central in understanding societies. This is a basic division in all societies, but that division is always particular in each society. The two obvious issues are physical: the strength and height differentials, and the differing roles in reproduction. I mentioned this basic difference in Chap. 2 but did little with it. I then followed, from a wide range of evidence in two chapters, the story of how social power between the sexes seemed roughly equal and then how it started to vary. Again, I think this receives less attention in social evolution studies and long-term world history than it should; exceptions (e.g., Graeber and Wengrow 2021) are rare. Given the demographic and political-cultural reversals outlined above, how sex and gender organize societies seems central. Thus, I would analyze them theoretically much more carefully.

Fourth, the narrative showed that societies experienced cycles of expansion and decline, or cohesion and disunity, at multiple scales. They also seem different after deep change—social transformations or revolutions (Childe [1936] 1996; Sanderson 1995)—occurred. Chiefly, "oscillation" was not nearly on the same scale as agrarian rise and fall, which was different from industrial capitalist business cycles within much longer hegemonic sequences. While macrosocial (macroeconomic, macropolitical) scholars in specific areas study cyclic experience (I did not even mention "K-waves" and other such topics—which I will develop in the future), world historians seem more reluctant to acknowledge such patterns. I think we need much more study of very long-term cycles in world history as well as in historical social science.

Finally, the collective academic process of review, criticism, suggestion, and development is especially appropriate for developing a paradigm about collective processes. Being extremely aware of how limited one human mind is, I am very open to this process.

Such a theory must also rest on a few principles. The first principle must be cross-cultural comparison. Any theory purporting to be about all human beings must avoid sampling only from people in WEIRD (Western, educated, industrialized, rich democratic) societies; instead, we must always compare people from around the world whenever possible. This would require chronological comparison in addition to geographic-cultural comparison. Indeed, as I wrote the history, it occurred to me that human capacities are revealed as human societies evolve. Previously untapped aspects of our "nature" appeared when more complex forms of social organization and consequent new situations developed. Even evolutionary psychologists increasingly understand the evolutionary environment of adaptedness (EEA) as extending into recent history (Volk and Atkinson 2013). Stated differently, social evolution revealed human capacities not visible among—really, not necessary to—foragers. Furthermore, the many forms of social organization in the modern world, such as clubs, communities, formal organizations, and societies reveal other human capacities (Turner and Machalek 2018).

The second principle is that this paradigm is rooted in the human body. As I traced it early in Chap. 2, hominin biological evolution seems to have led to gene-culture co-evolution and thus to the hyper-groupish *Homo sapiens sapiens*. Biological evolution made us cultural; now we cannot function without culture, as "feral children" research showed. It is easy to assume that Sapiens stopped evolving, but I think this would be mistaken. Clearly, biological evolution continued during the first globalization; differing skin colors and body shapes were the most obvious adaptations to new physical environments. Yet, physical evolution influencing behaviors probably continued in other ways. Ethnocentrism may have been "selected for"—groups that were more ethnocentric may have survived better than less groupish ones in episodes of intergroup violence, perhaps before, but certainly after, the Agricultural Revolution. This again raises the thorny issue of group selection in evolution studies. Tomasello's research on children (2019) seems to indicate that ethnocentrism is "built into" humans. How else could a strong group orientation be built into individuals without group selection?

Worries that social scientists will have little to say if biology "takes over" are wrong; biologists themselves say that humans need culture to survive, and they admit that they are ill-equipped to study it. Equally so, I think that the understandable fear that biological arguments will justify and reinforce various inequalities, especially gender inequality, overlooks the possibility that studying human beings as physical creatures may show how those inequalities distort our societies and reduce human well-being. Instead of following beliefs in "natural" inequalities that came from complex societies, both the original condition of humanity and the need for socialization suggest that we start with the assumption of basic equality between individuals—that is, a human is a human is a human; only after that assumption is firmly established should we add physical differences between the sexes and cultural differences between genders to develop this theory.

Third, rather than striving for the simplest model possible, studying normal variations (e.g., same-sex attraction, autism, etc.) should be part of this theorization. Although they seem abnormal within each generation, they keep showing up generation after generation, so we might be better off seeing them as normal variation. What are the statistically normal variations that appear consistently? How do they influence societies, and what do different groups do with these differences?

Again, such a set of sequential models seems grandiose. More importantly, it probably seems too complex, even messy. A psychologically realistic theory would have far too many factors to count as "parsimonious." I recognize what drives scholars to value parsimony. In addition to being an application of Occam's razor, researchers' resources are extremely limited; problems are pressing because people are suffering *now, today*; and the human world is subject to a complicated, confusing welter of influences. Thus, the root purpose of parsimonious theory is efficiency, or getting the most explanatory gain from applying the fewest resources (Sanderson 2001). Still, parsimony seems to be a fetish for some scholars at this point, so I am suspicious of it. Leaving out important factors in explanation can lead to false conclusions; in such a case, all resources have been wasted. Including more explanatory factors, not fewer, will be time-consuming and perhaps chaotic for a while; however, it may lead to greatly improved understanding of societies and eventually a simple (even parsimonious) explanation.

The ultimate purpose of developing this new paradigm should be human betterment: what is human well-being, and how is it achieved? This means that the development of the hypersocial paradigm should not simply be applied to understanding the past; we can use it to improve the present and future. If it really is true that humans are groupish, then the social environment in which individuals live is crucial to their well-being. Hrdy (2009), Junger (2016), and Graeber and Wengrow (2021) all argue, in various ways, that human well-being is found in community, and that modern societies lack such community. The fields of positive psychology (Lomas et al. 2021) and well-being economics (Dalziel et al. 2018) address this. The former increasingly focuses on groups and systems for individual well-being, and on different cultural contexts within a global scope, thus questioning psychology's focus on WEIRD societies; the latter posits a complex sense of nested spheres of activity from individual action to global systems that affect individual well-being. Indeed, there is an entire *Journal of Happiness Studies*. Sociologists, anthropologists, and historians seem less inclined than psychologists or economists to be either predictive or prescriptive. Still, some sociologists do discuss what a "good society" should look like. Three decades ago, Bellah (1992) argued that the United States was failing at providing a good society; Christakis (2019)[1] argues that we have biologically evolved proclivities toward good societies; and Kenworthy (2024) has recently laid out a vision of such a society.

There is much to do to develop the hypersocial paradigm in the social sciences. It will be, perhaps, much more congenial to those who reject rational-choice theory or mainstream evolutionary psychology, but who want to understand all facets of human experience.

Conclusion

The first part of this chapter outlined the potential for several extremely gloomy futures for humanity, then it traced some less somber, maybe hopeful, possibilities. I then suggested several recently reversed long-term patterns that could use rethinking, and made some suggestions about understanding human history, especially the expansion and elaboration of human societies over time in three big jumps, along with long-term trends that simply continue. Finally, I made some suggestions about developing a foundation for both continuity and change: understanding humans as groupish. Such a theoretical project will likely not have much of an impact on humanity's future for a while, and it will have to be a collective one. Still, I think it offers the potential to help us understand ourselves and maybe even build, eventually, a better future.

---

[1] Christakis uses the phrase "our common humanity," as I do, though I arrived at the phrase separately. In fact, I discovered his book extremely lately in writing this one, and being on deadline, I was unable to integrate it into Chap. 2. I will certainly use it in future theorizing.

References

Abu-Lughod, Janet. 1989. *Before European Hegemony: The World-System, A.D. 1250–1350*. New York: Oxford University Press.

Allison, Graham. 2017. *Destined for War: Can America and China Escape Thucydides's Trap?* New York: Houghton Mifflin.

Arrighi, Giovanni. (1994) 2010. *The Long Twentieth Century: Money, Power, and the Origins of our Times*. New York: Verso.

Bang, Peter Fibiger. 2021. Empire–A World History. In *The Oxford World History of Empire*, ed. P. F. Bang, C. A. Bayly, and W. Scheidel. New York: Oxford University Press.

Bellah, Robert N. 1992. *The Good Society*. New York: Vintage Books.

Bonvillain, Nancy M. 2001. *Women and Men: Cultural Constructs of Gender*. Upper Saddle River, NJ: Prentice Hall.

Brooke, John L. 2014. *Climate Change and the Course of Global History: A Rough Journey*. New York: Cambridge University Press.

Brown, Stuart Scott. 2013. *The Future of U.S. Global Power: Delusions of Decline*. Houndsmills, UK: Palgrave Macmillan.

Carneiro, Robert. 1970. A Theory of the Origin of the State. *Science* 169 (3947): 733–738.

Carneiro, Robert. 2004. The Political Unification of the World. *Cross Cultural Survey* 38 (2): 162–177.

Carneiro, Robert L. 2017. The Chiefdom in Evolutionary Perspective. In *Chiefdoms: Yesterday and Today*, ed. R. L. Carneiro, L. E. Grinin, and A. V. Korotayev, 15–60. Clinton Corners, NY: Eliot Werner.

Chase-Dunn, Christopher, and Bruce Lerro. 2014. *Social Change: Globalization from the Stone Age to the Present*. Boulder, CO: Paradigm.

Chew, Sing C. 2001. *World Ecological Degradation: Accumulation, Urbanization, and Deforestation, 3000 BC–A.D. 2000*. Walnut Creek, CA: AltaMira Press.

Childe, V. Gordon. (1936) 1966. *Man Makes Himself*. 4th ed. Bungay, Suffolk: Fontana.

Christakis, Nicholas A. 2019. *Blueprint: The Evolutionary Origins of a Good Society*. New York: Little, Brown.

Christian, David. 2011. *Maps of Time: An Introduction to Big History*. 2nd ed. Berkeley: University of California Press.

Dalziel, Paul, Caroline Saunders, and Joe Saunders. 2018. *Wellbeing Economics: The Capabilities Approach to Prosperity*. Cham, Switzerland: Palgrave Macmillan.

Damasio, Antonio. 1994. *Descartes' Error: Emotion, Reason, and the Human Brain*. New York: G. P. Putnam.

Earle, Tim. 2016. Pathways to Power: Corporate and Network Strategies, Staple and Wealth Finance, and Primary and Secondary States. In *Alternative Pathways to Complexity: A Collection of Essays on Architecture, Economics, Power, and Cross-Cultural Analysis in Honor of Richard E. Blanton*, ed. L. Fargher and V. Y. Heredia Espinosa, 291–308. Boulder: University Press of Colorado.

FAO, IFAD, UNICEF, WFP, and WHO. 2021. *The State of Food Security and Nutrition in the World 2021: Transforming Food Systems for Food Security, Improved Nutrition and Affordable Healthy Diets for All*. Rome: FAO.

Fishman, Ted. 2006. *China, Inc.: How the Rise of the Next Superpower Challenges America and the World*. New York: Scribner.

Flannery, Kent, and Joyce Marcus. 2012. *The Creation of Inequality: How Our Prehistoric Ancestors Set the Stage for Monarchy, Slavery, and Empire*. Cambridge, MA: Harvard University Press.

Frank, Andre Gunder. 1998. *ReORIENT: Global Economy in the Asian Age*. Berkeley: University of California Press.

Freedom House. 2020. Retrieved January 3, 2023, from https://freedomhouse.org/report/freedom-world/2020/leaderless-struggle-democracy

Gat, Azar. 2017. *The Causes of War and the Spread of Peace: But Will War Rebound?* New York: Oxford University Press.

Gidley, Jennifer M. 2017. *The Future: A Very Short Introduction*. New York: Oxford University Press.

Goldstone, Jack A. (1991) 2016. *Revolution and Rebellion in the Early Modern World*. Twenty-fifth anniversary edition. New York: Routledge.

Graeber, David, and David Wengrow. 2021. *The Dawn of Everything: A New History of Humanity*. New York: Farrar, Straus, and Giroux.

Grinin, Leonid E., and Andrey V. Korotayev. 2017. Chiefdoms and Their Analogues: Alternatives of Social Evolution at the Societal Level of Medium Complexity. In *Chiefdoms: Yesterday and Today*, ed. R. L. Carneiro, L. E. Grinin, and A. V. Korotayev, 63–108. Clinton Corners, NY: Eliot Werner.

Henrich, Joseph. 2015. *The Secret of Our Success: How Culture Is Driving Human Evolution, Domesticating Our Species, and Making Us Smarter*. Princeton, NJ and Oxford: Princeton University Press.

Herrington, Gaya. 2022. *The Limits to Growth Model: Still Prescient 50 Years Later*. Winterthur, Switzerland: Club of Rome.

Hickel, Jason. 2021. *Less Is More: How Degrowth Will Save the World*. New York: Penguin.

Hrdy, Sarah Blaffer. 2009. *Mothers and Others: The Evolutionary Origins of Mutual Understanding*. Cambridge, MA: Harvard University Press.

Huang, Yasheng. 2023. *The Rise and Fall of the EAST: How Exams, Autocracy, Stability, and Technology Brought China Success, and Why They Might Lead to Its Decline*. New Haven, CT: Yale University Press.

Inglehart, Ronald F. 2018. *Cultural Evolution: People's Motivations are Changing, and Reshaping the World*. New York: Cambridge University Press.

Institute for Economics and Peace. 2023. *Global Peace Index 2023: Measuring Peace in a Complex World*. Sydney. Retrieved May 27, 2024, from http://visionofhumanity.org/resources

Junger, Sebastian. 2016. *Tribe: On Homecoming and Belonging*. New York: Simon & Schuster.

Kelly, Raymond C. 2000. *Warless Societies and the Origin of War*. Ann Arbor: The University of Michigan Press.

Kelly, Robert L. 2013. *The Lifeways of Hunter-Gatherers: The Foraging Spectrum*. New York: Cambridge University Press.

Kennedy, Paul. 1989. *The Rise and Fall of the Great Powers: Economic Change and Military Conflict from 1500 to 2000*. London: Fontana Press.

Kenworthy, Lane. 2024. *The Good Society*. Retrieved December 26, 2024, from Lanekenworthy.net

Kirchgässner, Gebhard. 2008. *Homo Economicus: The Economic Model of Behaviour and Its Applications in Economics and Other Social Sciences*. New York: Springer.

Kuhn, Thomas S. (1962) 2012. *The Structure of Scientific Revolutions*. 4th ed. Chicago and London: The University of Chicago Press.

Kurian, George Thomas. 2011. Democratic Peace. In *The Encyclopedia of Political Science*, ed. G. T. Kurian, 401. Washington, DC: CQ Press.

LeBlanc, Steven A. 2014. Forager Warfare and Our Evolutionary Past. In *Violence and Warfare Among Hunter-Gatherers*, ed. M. W. Allen and T. L. Jones, 26–48. Walnut Creek, CA: Left Coast Press.

Lemke, Ashley K. 2018. Hunter-Gatherers and Archaeology. In *Foraging in the Past: Archaeological Studies of Hunter-Gatherer Diversity*, ed. A. K. Lemke, 3–18. Louisville CO: University Press of Colorado.

Lenski, Gerhard. 2005. *Ecological-Evolutionary Theory: Principles and Applications*. Boulder, CO: Paradigm.

Lerner, Gerda. 1986. *The Creation of Patriarchy*. New York: Oxford University Press.

Lomas, Tim, Lea Waters, Paige Williams, Lindsay G. Oades, and Margaret L. Kern. 2021. Third Wave Positive Psychology: Broadening Towards Complexity. *Journal of Positive Psychology* 16 (5): 600–674.

McIntosh, Susan Keech. 1999. Pathways to Complexity: An African Perspective. In *Beyond Chiefdoms: Pathways to Complexity in Africa*, ed. S. K. McIntosh, 1–30. New York: Cambridge University Press.
McNeill, William H. 1990. The *Rise of the West* After Twenty-Five Years. *Journal of World History* 1 (1): 1–21.
McNeill, William H. 1998. World History and the Rise and Fall of the West. *Journal of World History* 1 (1): 1–21.
Milanovic, Branko. 2023. *Visions of Inequality: From the French Revolution to the End of the Cold War*. Cambridge, MA: Harvard University Press.
Mizruchi, Mark S. 2013. *The Fracturing of the American Corporate Elite*. Cambridge, MA: Harvard University Press.
Modelski, George, and William R. Thompson. 1996. *Leading Sectors and World Powers: The Coevolution of Global Politics and Economics*. Columbia: University of South Carolina Press.
Peterson, Christian E., and Robert D. Drennan. 2012. Patterned Variation in Regional Trajectories of Community Growth. In *The Comparative Archaeology of Complex Societies*, ed. M. E. Smith, 88–137. New York: Cambridge University Press.
Piketty, Thomas. 2014. *Capital in the Twenty-First Century*. Cambridge, MA: Belknap Press.
Ponting, Clive. 2007. *A New Green History of the World: The Environment and the Collapse of Great Civilizations*. New York: Penguin.
Ray, James Lee. 2011. Democratic Peace. In *International Encyclopedia of Political Science*, ed. B. Badie, D. Berg-Schlosser, and L. Morlino, 621–622. Thousand Oaks, CA: SAGE.
Read, Dwight W. 2012. *How Culture Makes Us Human: Primate Social Evolution and the Formation of Human Societies*. Walnut Creek, CA: Left Coast Press.
Richerson, Peter J., and Robert Boyd. 2005. *Not by Genes Alone: How Culture Transformed Human Evolution*. Chicago and London: The University of Chicago Press.
Robbins, Bruce. 2016. Prolegomena to a Cosmopolitanism in Deep Time. *Interventions* 18 (2): 172–186.
Robock, Alan, Luke Oman, and Georgiy L. Stenchikov. 2007. Nuclear Winter Revisited with a Modern Climate Model and Current Nuclear Arsenals: Still Catastrophic Consequences. *Journal of Geophysical Research* 112 (D13107): 1–14. https://doi.org/10.1029/2006JD008235.
Rosaldo, Renato. 1989. *Culture and Truth: The Remaking of Social Analysis*. Boston: Beacon Press.
Sanderson, Stephen K. 1995. *Social Transformations: A General Theory of Historical Development*. Cambridge, MA: Blackwell.
Sanderson, Stephen K. 2001. *The Evolution of Human Sociality: A Darwinian Conflict Perspective*. Lanham, MD: Rowman and Littlefield.
Scheidel, Walter. 2017. *The Great Leveler: Violence and the History of Inequality from the Stone Age to the Twenty-First Century*. Princeton, NJ and Oxford: Princeton University Press.
Sztompka, Piotr. 1994. *The Sociology of Social Change*. Cambridge, MA: Blackwell.
Tomasello, Michael. 2019. *Becoming Human: A Theory of Ontogeny*. Cambridge, MA: Harvard University Press.
Turchin, Peter, Rob Brennan, Thomas E. Currie, Kevin C. Feeney, Pieter François, Daniel Hoyer, J. G. Manning, et al. 2015. Seshat: The Global History Databank. *Cliodynamics: The Journal of Quantitative History and Cultural Evolution* 6 (1): 78–107.
Turner, Graham. 2014. Is Global Collapse Imminent? An Updated Comparison of *The Limits to Growth* with Historical Data. MSSI Research Paper No. 4. University of Melbourne: Melbourne Sustainable Society Institute.
Turner, Jonathan H., and Richard Machalek. 2018. *The New Evolutionary Sociology: Recent and Revitalized Theoretical and Methodological Approaches*. New York and London: Routledge.
United Nations Development Program (UNDP). 2020. *Tackling Social Norms: A Game Changer for Gender Inequalities*. New York: UNDP.
van der Kolk, Bessel. 2014. *The Body Keeps the Score: Brain, Mind, and Body in the Healing of Trauma*. New York: Viking Press.

Volk, Anthony A., and Jeremy A. Atkinson. 2013. Infant and Child Death in the Human Environment of Evolutionary Adaptation. *Evolution and Human Behavior* 34 (3): 182–192.

Wallerstein, Immanuel. 1980. *Mercantilism and the Consolidation of the European World-Economy, 1600–1750*. New York: Academic Press.

Wobst, H. Martin. 1978. The Archaeo-Ethnology of Hunter-Gatherers or the Tyranny of the Ethnographic Record in Archaeology. *American Antiquity* 43 (2): 303–309.

GPSR Compliance

The European Union's (EU) General Product Safety Regulation (GPSR) is a set of rules that requires consumer products to be safe and our obligations to ensure this.

If you have any concerns about our products, you can contact us on ProductSafety@springernature.com

In case Publisher is established outside the EU, the EU authorized representative is:

Springer Nature Customer Service Center GmbH
Europaplatz 3
69115 Heidelberg, Germany

Batch number: 10755329

Printed by Printforce, the Netherlands